Springer Handbook of Auditory Research

For other titles published in this series, go to
www.springer.com/series/2506

David K. Ryugo • Richard R. Fay
Arthur N. Popper
Editors

Auditory and Vestibular Efferents

Springer

Editors
David K. Ryugo
Garvan Institute of Medical Research
Program in Neuroscience
384 Victoria St., Level 7
Darlinghurst, NSW 2010
Australia
d.ryugo@garvan.org.au

Richard R. Fay
Department of Psychology
Loyola University of Chicago
6525 N. Sheridan Rd.
Chicago, Illinois 60626
USA
rfay@luc.edu

Arthur N. Popper
University of Maryland
Department of Biology
College Park, Maryland
20742-4415
USA
apopper@umd.edu

ISSN 0947-2657
ISBN 978-1-4419-7069-5 e-ISBN 978-1-4419-7070-1
DOI 10.1007/978-1-4419-7070-1
Springer New York Dordrecht Heidelberg London

Library of Congress Control Number: 2010937633

Printed on acid-free paper

Springer is part of Springer Science+Business Media (www.springer.com)

This volume is dedicated to our families:
Kay and Masako Ryugo; Karen, Ben, Dan, and Nick Ryugo
Catherine Fay; Christian and Kara Fay, Amanda Fay,
and Nate Evan and Stella Fay
Helen Popper; Michelle, Roman, and Emma Levit; Melissa, Jeff,
Ethan, and Sophie Levinsohn

Series Preface

The Springer Handbook of Auditory Research presents a series of comprehensive and synthetic reviews of the fundamental topics in modern auditory research. The volumes are aimed at all individuals with interests in hearing research including advanced graduate students, post-doctoral researchers, and clinical investigators. The volumes are intended to introduce new investigators to important aspects of hearing science and to help established investigators to better understand the fundamental theories and data in fields of hearing that they may not normally follow closely.

Each volume presents a particular topic comprehensively, and each serves as a synthetic overview and guide to the literature. As such, the chapters present neither exhaustive data reviews nor original research that has not yet appeared in peer-reviewed journals. The volumes focus on topics that have developed a solid data and conceptual foundation rather than on those for which a literature is only beginning to develop. New research areas will be covered on a timely basis in the series as they begin to mature.

Each volume in the series consists of a few substantial chapters on a particular topic. In some cases, the topics will be ones of traditional interest for which there is a substantial body of data and theory, such as auditory neuroanatomy (Vol. 1) and neurophysiology (Vol. 2). Other volumes in the series deal with topics that have begun to mature more recently, such as development, plasticity, and computational models of neural processing. In many cases, the series editors are joined by a co-editor having special expertise in the topic of the volume.

Richard R. Fay, Chicago, IL
Arthur N. Popper, College Park, MC

Volume Preface

Research on sensory efferents has entered a renaissance period, particularly with respect to the auditory and vestibular systems. Since the discovery of auditory efferents by Grant Rasmussen, understanding of its significance to sensory processing has grown. To accentuate the whole efferent system associated with the ear, this volume covers a wide range of topics addressing the biology of auditory and vestibular efferents. Basic research reviews of the anatomy, electrophysiology, and pharmacology lead into discussions of cellular and molecular features of the inner ear. Chapters on the development and evolution of efferent systems illuminate key phylogenetic stages and ontogenetic mechanisms that have given rise to present-day efferent systems. The final chapters provide an overview of central efferent anatomy and neuronal responses and plasticity to efferent activation.

The first chapter by David Ryugo introduces the idea of sensory efferents and explores the concept with respect to biological mechanisms and behavior. The behavioral responses of organisms when confronted by sensory challenges are often best explained by invoking a functioning efferent system. When considering what a nervous system must do, one can design experiments to test hypotheses about what the nervous system actually does. This context sets the stage for the rest of the volume.

Chapter 2 by Chris Brown exploits the basic relationship between structure and function to establish an anatomical foundation for understanding olivocochlear neurons. This discussion is followed by a summary of the physiological response properties of the efferent neurons in Chapter 3 by John Guinan. The anatomical distinctions outlined in Chapter 2 are consistent with the different mechanisms utilized by the lateral and medial olivocochlear systems to alter cochlear function. The separate olivocochlear systems use different chemical mechanisms that are discussed in detail by Bill Sewell in Chapter 4. Sewell's introduction to cochlear efferent neurochemistry is followed by a consideration of the role of special nicotinic receptors and various ion channels by Eleonora Katz, Ana Belén Elgoyhen, and Paul A. Fuchs in Chapter 5 to explain how acetylcholine mediates fast inhibition.

Chapter 6 by Joseph C. Holt, Anna Lysakowski, and Jay M. Goldberg introduce the vestibular component of inner ear efferents. This update on the current knowledge of vestibular efferents emphasizes the complicated nature of the system and hints at new directions and questions.

Dwayne Simmons, Jeremy Duncan, Dominique Crapon de Caprona, and Bernd Fritzsch review development of the vestibulocochlear efferent system in Chapter 7 and reveal provocative findings that shape our understanding about mechanisms of inner ear development. This treatise is followed by Chapter 8 in which Christine Köppl addresses efferent system diversity in terms of evolutionary concepts.

Chapter 9 by Brett Schofield describes the descending auditory circuitry that forms long descending projections as well as short feedback loops within the central auditory system. Chapter 10 by Donald Robertson and Wilhelmina Mulders discusses the central effects of efferent activation on physiological response properties of auditory neurons, and Nobuo Suga, Weiqing Ji, Xiaofeng Ma, Jie Tang, Zhongju Xiao, and Jun Yan summarize in Chapter 11 how many forms of brain and behavioral plasticity depend on efferent systems.

As is often the case, chapters in a newer SHAR volume are complemented by, and complimentary to, chapters in earlier volumes. Although there have been few chapters in earlier volumes that were specifically on efferent systems, the issue was critical as parts of chapters in volumes such as *The Cochlea* (Vol. 8), *Integrative Functions in the Mammalian Auditory Pathway* (Vol. 15), and *The Vestibular System* (Vol. 19). In addition, the anatomy of the olivocochlear vestibular system was specifically discussed by W. Bruce Warr in Vol. 1 of this series, *The Mammalian Auditory Pathway: Neuroanatomy* and by Russell and Lukashkin in *Active Processes and Otoacoustic Emissions* (Vol. 30).

David Ryugo, Darlinghurst, NSW, Australia
Richard R. Fay, Chicago, IL
Arthur N. Popper, College Park, MD

Contents

Contributors

M. Christian Brown
Department of Otology and Laryngology, Harvard Medical School,
and
Eaton-Peabody Laboratory, Massachusetts Eye and Ear Infirmary, Boston,
MA 02114, USA
Chris_Brown@meei.harvard.edu

Dominique Crapon de Caprona
Department of Biology, University of Iowa, Iowa City, IA, 52242, USA
dominique-decaprona@uiowa.edu

Jeremy Duncan
Department of Biology, University of Iowa, Iowa City, IA, 52242, USA
jeremy-duncan@uiowa.edu

Ana Belén Elgoyhen
Instituto de Investigaciones en Ingeniería Genética y Biología Molecular
(Consejo Nacional de Investigaciones Científicas y Técnicas), 1428,
Buenos Aires, Argentina
elgoyhen@dna.uba.ar

Bernd Fritzsch
Department of Biology, University of Iowa, Iowa City, IA 52242, USA
bernd-fritzsch@uiowa.edu

Paul Albert Fuchs
Department of Otolaryngology, Head and Neck Surgery,
Johns Hopkins University School of Medicine,
Baltimore, MD, 21205, USA
pfuchs@jhmi.edu

Jay M. Goldberg
Department of Neurobiology, Pharmacology, and Physiology,
University of Chicago, Chicago, IL 60637, USA
jgoldber@bsd.uchicago.edu

John J. GuinanJr
Eaton Peabody Laboratory, Massachusetts Eye and Ear Infirmary, Boston,
MA 02114, USA
Department of Otology and Laryngology, Harvard Medical School, Boston,
MA, USA
Speech and Hearing Bioscience and Technology Program, Harvard-MIT Division
of Health Sciences and Technology, Boston, MA, USA
jjg@epl.meei.harvard.edu

Joseph C. Holt
Department of Otolaryngology/Neurobiology and Anatomy,
University of Rochester, Rochester, NY 14642, USA
joseph_holt@urmc.rochester.edu

Weiqing Ji
Department of Biology, Washington University, St. Louis, MO 63130, USA
ji@biology.wustl.edu

Eleonora Katz
Instituto de Investigaciones en Ingeniería Genética y Biología Molecular
(Consejo Nacional de Investigaciones Científicas y Técnicas),
1428, Buenos Aires, Argentina
ekatz@dna.uba.ar

Christine Köppl
Cochlear and Auditory Brainstem Physiology, IBU, Carl von Ossietzky
University Oldenburg, 26111, Oldenburg, Germany
christine.koeppl@uni-oldenburg.de

Anna Lysakowski
Department of Anatomy and Cell Biology, University of Illinois at Chicago,
Chicago, IL 60612, USA
alysakow@uic.edu

Xiaofeng Ma
Department of Biology, Washington University, St. Louis, MO 63130, USA
xfma@biology.wustl.edu

Wilhelmina H.A.M. Mulders
The Auditory Laboratory, Discipline of Physiology, M311 School of Biomedical Biomolecular and Chemical Sciences, The University of Western Australia, Crawley, WA 6009, Australia
hmulders@cyllene.uwa.edu.au

Donald Robertson
The Auditory Laboratory, Discipline of Physiology, M311 School of Biomedical Biomolecular and Chemical Sciences, The University of Western Australia, Crawley, WA 6009, Australia
drobed@cyllene.uwa.edu.au

David K. Ryugo
Garvan Institute of Medical Research, Program in Neuroscience, 384 Victoria St., Level 7, Darlinghurst, NSW, Australia
and
Center for Hearing and Balance, 720 Rutland Ave., Johns Hopkins University School of Medicine, Baltimore, MD 21205, USA
d.ryugo@garvan.org.au

Brett R. Schofield
Department of Anatomy and Neurobiology, Northeastern Ohio Universities Colleges of Medicine and Pharmacy, Rootstown, OH 44272, USA
bschofie@neoucom.edu

William F. Sewell
Department of Otology and Laryngology, Harvard Medical School, Boston, MA 02114, USA
and
Eaton-Peabody Laboratory, Massachusetts Eye and Ear Infirmary, Boston, MA 02114, USA
wfs@epl.meei.harvard.edu

Dwayne Simmons
Department of Physiological Science, University of California, Los Angeles, CA, 90095-1606, USA
dd.simmons@ucla.edu

Nobuo Suga
Department of Biology, Washington University, St. Louis, MO 63130, USA
suga@biology.wustl.edu

Jie Tang
Department of Biology, Washington University, St. Louis, MO 63130, USA
ietang@creighton.edu

Zhongju Xiao
Department of Physiology, Southern Medical University, Guangzhou 510515, P.R., China
Xiaozj@fimmu.com

Jun Yan
Department of Physiology and Biophysics, University of Calgary, N.W. Calgary, Alberta, Canada, T2N 4N1
juyan@ucalgary.ca

Chapter 1
Introduction to Efferent Systems

David K. Ryugo

1.1 Introduction and Overview

Organisms must learn what representations in the world are important – that is, which sights, smells, and sounds indicate safety, food, or danger. Knowledge of what is and is not important is acquired by information arising from the sensory organs, and this knowledge is then acted upon by the motor system, expressed by approach or avoidance behavior. A loud "Hey you!" will evoke a strikingly different motor and autonomic response compared to that of a sultry "Hello, handsome." Likewise, a patron can ignore the sounds inside a busy restaurant but not when his name is being called. Stimuli that have no immediate significance become relegated to "background noise" and can be disregarded. During our lifetimes, we learn about stimuli and stimulus context. The sound and sight of gunshots in the street are generally different from those experienced in a movie theater. Stimulus content and context are presumably processed in the cerebral hemispheres, where significance is established.

Historically, sensory information had been thought to access the cerebral hemispheres by ascending the neuraxis via successive links to reach the forebrain where a hypothetical "central processor" resided. Implicit in this conceptualization was the notion that subsequent to cortical processing, descending motor signals were generated to produce a response. The discovery that the central nervous system (CNS) initiated neuronal projections that terminated upon auditory receptors (Rasmussen 1946, 1953) and muscle spindles (Hagbarth and Kerr 1954) contributed to a revolution of thought, which introduced the idea that the brain could control or at least modulate signals arising *from* a sense organ (Granit 1955;

D.K. Ryugo (✉)
Garvan Institute of Medical Research, Program in Neuroscience, 384 Victoria St., Level 7, Darlinghurst, NSW, Australia
and
Center for Hearing and Balance, 720 Rutland Ave., Johns Hopkins University School of Medicine, Baltimore, MD 21205, USA
e-mail: d.ryugo@garvan.org.au

D.K. Ryugo et al. (eds.), *Auditory and Vestibular Efferents*,
Springer Handbook of Auditory Research 38, DOI 10.1007/978-1-4419-7070-1_1,

Galambos 1956). These new data challenged classical thinking that sensory pathways conveyed information about the external world without modification. Instead, there were indications that sensory information might be modified at multiple levels of the nervous system, and these data ignited the concept of central control of sensory input (Galambos 1956). In short, the process of "priority setting" could be initiated at potentially every synaptic station along the ascending sensory pathways by feedback circuits.

As an outgrowth of this new approach to sensory processing, Rasmussen (1953) introduced a more expansive use of the term "efferent" to describe the centrifugal pathway to the inner ear. Until this time, the term "efferent systems" applied to motoneurons that carried nerve impulses away from the CNS to act on glands, organs, or muscles. These efferent motoneurons represented the final common pathway to target structures in the periphery that initiated autonomic (e.g., change in pupillary tension, heart rate, or glandular secretion) and/or voluntary (e.g., skeletal muscle contractions) responses. Rasmussen extended the definition of the term "efferent" to include his sensory pathway that conducted impulses away from the CNS to a sensory organ in a manner analogous to motoneuron projections to skeletal muscle. Thus, the term "sensory efferent" was born.

This more inclusive approach to sensory processing begged the question: Why would a sensory system have descending circuits? One answer was that it needed to influence the information that was ascending (Granit 1955; Galambos 1956). In this case, modulation might consist of feedback enhancement or suppression of ascending information. For example, binaural hearing refers to the auditory processing involved in the comparison of sounds received by one ear to the sounds received by the other ear. The interaction between these sounds provides important spatial cues for determining the direction and the distance of sound sources. Interaural time and intensity differences plus head-related transfer functions are the dominant cues for identifying the location of a sound source within three-dimensional auditory space (Popper and Fay 2005). Essential to this process is the assumption that the binaural system is functionally and structurally symmetrical such that sensitivity, rate of response, numbers of neurons, and magnitude of response from each ear are equal for equal stimulation. Biological systems, however, only approximate symmetry. One role of descending systems could be to equilibrate the response of each ear with respect to the midline. Another might be to balance the sensitivity of each ear (Cullen and Minor 2002; Darrow et al. 2006). Experience would calibrate the responses of the organism when a sound occurred on the midline. Descending systems might augment a smaller signal or depress a stronger signal to balance the output of a structure pair for midline stimuli. This balancing act would establish a reference from which more lateralized sounds could be compared.

Environmental sounds may be described by their composition of different frequencies. Because sounds of different frequencies can vary systematically across other physical parameters when perturbed (e.g., low frequencies are less disrupted by large objects compared to high frequencies), animals can learn about their auditory environment by using such spectral information. There are spectral cues that are created by the interactions among sound, the head, and the pinnae that are used to resolve front–back locations, to determine sound elevation, and to localize sound using one ear alone. Head and pinnae position, movement of the sound source, and feedback from other sensory detectors (visual, vestibular, and proprioception) also play into perception of space (Oertel and Young 2004). Organisms learn to distinguish between near and distant sounds via cues that include sound clarity, overall sound level, the amount of reverberation relative to the original signal, and timbre. In light of individual differences and asymmetries in head shapes, external ear morphology, and cochleae, the relationship between cue values and sound location is presumed to be learned through experience. Moreover, as the organism's head and body grow with maturation, cue values associated with particular locations in space will change. Sound motion and/or organism motion must also be entered into the equation. The brain must constantly recalibrate its three-dimensional coordinate system to preserve auditory space, and descending feedback circuits could facilitate the constant adjustments.

We listen, detect, localize, identify, and then attend to those sounds we have deemed important. In addition, we can deemphasize elements of our auditory environment if they are routine or uninteresting. Each of these activities is performed across a number of auditory streams, both in parallel and in tandem. At any instant in time, sounds from multiple sources impinge upon our ears, and humans have the ability to separate sound streams from each other. Unique characteristics bind sounds to a common source, such as the intonation of a French horn during a symphonic performance, or a hawk's screech among a cacophony of bird songs. It is the extraction of these characteristics, coupled with learned significance of some sounds over others, that provides certain survival advantages for attending one signal while ignoring others. Mechanisms for these activities, however, remain poorly understood. Complex neural circuits extract learned information from memory in real time to focus and/or switch attention in hearing. For example, we follow a conversation in a noisy restaurant or "eavesdrop" on a different conversation or flip back and forth between the two.

The process of hearing is initiated within a remarkably complex sensory organ, the inner ear (also known as the cochlea). Sensory hair cells reside in the bony cochlea and function to convert environmental sound into neural signals used by the brain, to separate sounds into elemental bands of their constituent frequencies, and to compress the amplitudes of sounds so that it is possible to process the huge range of sound intensities that are encountered during a normal day. It is the feature of "gain control" where soft sounds are amplified and loud sounds are dampened that is so crucial to how we hear. In this case, the term "gain" resembles the volume on stereo amplifiers and iPods. It is the process of selective gain that facilitates the ability to discriminate one sound in the presence of competing sounds.

In the auditory system, central pathways are initiated from the cochlear nucleus and ascend the neuraxis through a series of parallel lines and serial synaptic stations (Fig. 1.1). The processing of sound by the brain starts from the analysis of very basic attributes (Kiang et al. 1965; Evans 1975), and it becomes progressively more complex as one ascends the progressive hierarchy of auditory stations (Tsuchitani 1977;

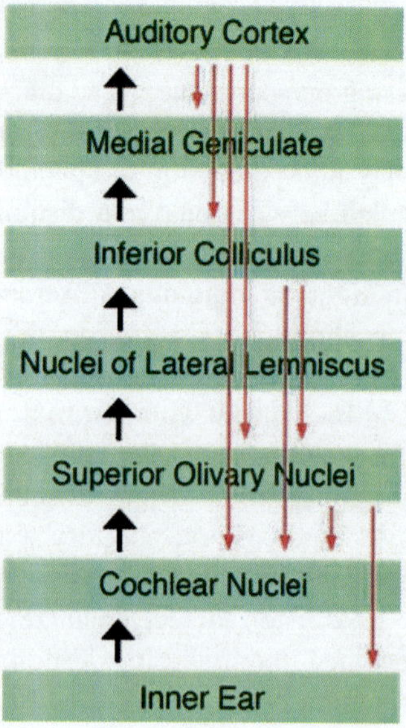

Fig. 1.1 Simplified box diagram of central auditory system. The main components of the system are labeled and begin with the cochlear nuclei. Subdivisions of nuclei are not included. *UP arrows* indicate sequence of progression of ascending auditory information. *Down arrows* indicate the start and termination of descending projections

Aitkin et al. 1984). This linear and sequential processing of auditory information as it ascends toward the cerebral hemispheres has been referred to as a feedforward mechanism (Mackay 1956). The construction of an auditory percept is based on such processing and depends upon prior knowledge and situational context. Feedforward mechanisms alone, however, do not seem adequate for achieving stimulus recognition that is simultaneously invariant and flexible for our rapidly changing environment.

Hearing is a dynamic process. Cognitive functions including attention, memory, and expectation modulate the nature of the sensory information reaching consciousness. The function of any cortical area must be influenced by ongoing perceptual requirements. In this context, there is no starting point for information flow. Streams of information from cortical centers interact with each other and in turn modulate the information ascending from the sensory periphery. Presumably, these descending and lateral feedback circuits enable the rapid discrimination of signals from noise, the sharpening of tuning curves, and the switching of attention. What remains remarkable is that this process is constantly engaged.

Knowledge about how the brain modulates auditory processing has grown significantly over the years. Although most research still tends to focus on the ascending central pathways, it is apparent that a parallel system of descending pathways exists and that it has an important role in hearing (Fig. 1.1). In fact, the descending corticothalamic projections greatly exceed the ascending thalamocortical projections (Jones 2002). Descending systems are found that involve higher control of visceral reflexes (Menétrey and Basbaum 1987; Card et al. 2006), gating of sensory information (Wiederhold and Kiang 1970; Dewson 1967; Sherman and Guillery 2002;

Waleszczyk et al. 2005; Wang and Wall 2005), regulation of postural reflexes (Matesz et al. 2002; Barmack 2003), and modulation of motor behavior (Canedo 1997). It is therefore timely to review what we know about auditory and vestibular efferents, how current research has contributed to our understanding of brain function, and to indicate avenues for future research.

1.2 Overview of the Volume

In the more than 60 years since Rasmussen reported his discovery of efferents, the significance of his work continues to grow. The ten chapters that follow in this volume cover a wide range of topics addressing the biology of auditory and vestibular efferents. Basic research summaries of the anatomy, electrophysiology, and pharmacology segue into cellular and molecular features of the inner ear. Chapters on the development and evolution of efferent systems illuminate key phylogenetic stages and ontogenetic mechanisms that have given rise to present-day efferent systems. The final chapters provide an overview of central efferent anatomy and neuronal responses and plasticity to efferent activation.

The fundamental relationship between structure and function represents a starting point for understanding any biological system. We use anatomical methods to study how systems are constructed, and to infer how interconnected structural components work together. Chapter 2 by Chris Brown exploits this strategy to discuss the anatomy of olivocochlear neurons. Two major groups of olivocochlear neurons have been identified (Warr and Guinan 1979). The lateral olivocochlear neurons reside in the lateral part of the superior olivary complex and send unmyelinated axons that terminate primarily on the peripheral processes of auditory nerve fibers under the inner hair cells. In contrast, medial olivocochlear neurons inhabit the medial part of the superior olivary complex and send myelinated axons that terminate on the cell bodies of outer hair cells (OHCs). The obvious structural differences in these two systems support the notion that they will subserve different functions (Brown 1987). There is also growing evidence suggesting that there are distinct subgroups that comprise the lateral and medial olivocochlear system. It seems that with increased sophistication and resolution of methods, more components will emerge to help us understand how this system of relatively few neurons is capable of serving the function of gain control in hearing.

The physiological response properties of the efferent neurons are described in Chap. 3 by John Guinan. Efferent spike-trains affect auditory nerve responses by modulating basilar membrane motion and hair cell status. In light of the anatomical differences outlined in Chap. 2, it is not surprising that the lateral and medial olivocochlear systems utilize different mechanisms to alter the operation of the cochlea. The lateral effects have been parsed into separate dopaminergic and cholinergic components. Indirect activation of these efferents can increase or decrease auditory nerve responses, and the separate groups of lateral efferent neurons presumably mediate these different effects. It has been proposed that one function of the lateral efferent system is to balance the outputs from the separate ears to enable sound localization based on interaural level differences (Guinan 1996; Darrow et al. 2006). The lateral efferents are unmyelinated and have slow conduction velocities, and so exert effects that have a long (minutes) time course.

In contrast to the lateral efferent system, the medial efferent system operates on a fast (on the order of 100 ms) or a medium (tens of seconds) time course. The fast effect dampens the cochlear amplifier by reducing receptor currents and by hyperpolarizing the OHCs (Guinan 1996). In turn, basilar membrane motion is depressed and the spike rate in the affected auditory nerve fibers is reduced (Cooper and Guinan 2006); both effects lower the sensitivity of the ear to sound. It remains to be determined whether these effects are completely separate or a result of cascading, sequential mechanisms.

The medium-speed effect is hypothesized to help protect the ear against overstimulation (Maison and Liberman 2000). An independent measure of medial efferent action is its effect on otoacoustic emissions. Otoacoustic emissions are sounds that originate in the cochlea, travel back through the middle ear and into the ear canal, where they can be detected. These emissions get most of their energy from the OHCs and medial efferent activation reduces the amplitude of the emissions. In addition to the fast inhibition, there is a slow inhibitory effect that lasts for tens of seconds that is attributable a decrease in OHC stiffness (Dallos 1997). It is this change in OHC stiffness that influences the overall sensitivity of the inner ear, and a less sensitive ear is less prone to damage by loud sounds.

The separate synaptic mechanisms utilized by the lateral and medial olivocochlear efferents within the inner ear represent an important component in describing cellular events that mediate efferent action. Understanding the cellular and molecular basis of neurotransmission will not only expand our knowledge of brain function but also reveal potential intervention strategies (e.g., pharmacological or transgenic treatment) to redress abnormalities in hearing. Neurotransmitter release occurs when an action potential invades the ending of an efferent terminal. Chemical mechanisms involving efferent synapses are discussed in detail by Bill Sewell in Chap. 4, where cochlear efferent neurochemistry is highlighted. The main chemical released by efferents is acetylcholine but other neuroactive substances are involved as well, including opioid peptides, calcitonin-gene related protein, dopamine, GABA, and serotonin (Schrott-Fischer et al. 2007). The medial efferent fibers terminate as large endings primarily against the OHCs and release acetylcholine when activated. Acetylcholine binds to nicotinic cholinergic receptors located on the postsynaptic OHCs and causes the OHCs to become permeable to calcium. The calcium that enters is thought to activate the release of more calcium. This calcium activates a class of calcium-activated potassium channels where potassium leaves the OHC and hyperpolarizes the cell (Fuchs and Murrow 1992). The lateral efferents are cholinergic too, and are associated with additional neuroactive substances, but little is known about the function of these other chemicals.

The role of special nicotinic receptors and various ion channels are discussed by Eleonora Katz, Ana Belén Elgoyhen, and Paul Fuchs in Chap. 5 to explain how acetylcholine mediates fast inhibition. At issue is how acetylcholine triggers inhibition of hair cells. The chapter explains cholinergic inhibition by revealing an unusual ionic mechanism that appears mediated by two separate families of ion channels. First, activation of a particular type of nicotinic receptor, the α9 nicotinic receptor, permits cationic calcium ions to enter the cell. The calcium in turn activates calcium-dependent potassium channels, and the egress of potassium swamps the entry of calcium so that the cell hyperpolarizes. This basic mechanism has been highly conserved across vertebrates, and the hyperpolarization is associated with diminished excitation of afferent fibers (Flock and Russell 1976).

The vestibular component of the inner ear, like the auditory receptor cells, is special among the sensory organs because it receives direct efferent innervation. The vestibular system is phylogenetically ancient, yet our basic understanding of the role of vestibular efferents is greatly lacking. Only recently have data been accumulating about the basic anatomy and physiology of this system. Joseph Holt, Anna Lysakowski, and Jay Goldberg update our current knowledge on this topic in Chap. 6, where the physiological consequences of efferent activation remain complicated. There are fast and slow mechanisms where excitatory effects are seen in the background discharges of vestibular afferent fibers. These effects are different for regular and irregular afferents, and such results are further complicated depending on whether the afferents arise from the central or striolar zones of the sensory epithelium (Goldberg and Fernández 1980; McCue and Guinan 1994; Marlinski et al. 2004). This field has many more questions than answers at present, but progress is being made.

One idea is that vestibular efferents respond to motor signals so that incoming information about the organisms own movements are suppressed in order to modulate afferent sensitivity during head movements (Highstein 1992; Brichta and Goldberg 2000). Another idea is that the vestibular efferents serve to balance the afferent output of the end organs from both sides of the head (Cullen and Minor 2002). A bilaterally balanced system is important because of the relative symmetry of the two end organs around a central axis. The output of one end organ is equal and opposite in sign to that of the other end organ as referenced to the central axis.

Dwayne Simmons, Jeremy Duncan, Dominique Crapon de Caprona, and Bernd Fritzsch review the development of the vestibulocochlear efferent system in Chap. 7 and highlight some pertinent and surprising findings that have helped to shape our current understanding about mechanisms of neuronal development in general and efferent function in particular. There is a working hypothesis that "ontogeny recapitulates phylogeny" where the stages in embryonic development and differentiation approximate the evolutionary history of the species. These stages are hypothesized to resemble the adult phase of ancient ancestors. Researchers have been able to examine neurons during the period when structures were immature (and therefore less complicated) and infer which groups of neurons were conserved through evolution. This strategy generated our basic knowledge about the development of specific fiber tracts and the cellular organization of the spinal cord, brainstem and cerebral hemispheres (Ramón y Cajal 1909). In addition, because the lengths and numbers of cell processes are reduced in the immature brain, details about cell-to-cell connections are more easily revealed. Combining developmental and comparative anatomical studies revealed that vestibulocochlear efferents differ embryologically from motor efferents in terms of their exclusive projection to placodally derived targets, the laterality of these projections, and the variety of neurotransmitters used (Fritzsch et al. 1999; Simmons 2002). Such data led to ideas about which cell masses were associated with which types of functions and behaviors.

The application of principles of comparative anatomy and evolutionary biology has been used to characterize and understand functional changes of complex structures. The structures involved in vertebrate hearing, for example, reflect shifts and specializations that occur over long periods of time in response to environmental pressures and physical requirements. The ecological demands establish a framework in which to consider functional morphology of the system and behavior of the organism. Individual phenotypes are hypothesized to represent a synergy of genotypes and environment, necessary for the organism to perform its biological roles. The presence of a general blueprint for the vertebrate nervous system suggests that a basic plan was established in an early common ancestor, and that with the evolution of "higher" animals, ancient structures were elaborated and/or new structures emerged that expanded behavioral and survival capabilities. Since there are no fossil records of hair cells or efferent neural systems, one strategy for studying the evolution of efferent systems is to compare basic features of hair cell sensory organs in "living fossils." The study of such relicts, proposed to be extant examples of a distant ancestor, may reveal the adaptive significance of features and structures that accompany the behavioral requirements in modern organisms. This premise, however, is complicated by the fact that such species have a long and independent history during which time they adapted to their own local environments and evolved their own specialized lifestyles. It remains to be seen if they truly represent an unchanged ancestral form.

Chapter 8 by Christine Köppl addresses the structural and functional diversity of efferent systems in terms of phylogenetic trends. The preservation, and indeed the variations of features over time that contribute to an improved "fit" of a species within its environment, are discussed as an approximation of natural selection for hearing. It is argued that inner ear efferents together with the lateral line form a coherent and whole octavolateralis system. This system is unified by the presence of hair cells, afferents, and efferents. The efferents across vertebrates are cholinergic and stain for either choline acetyltransferase or acetylcholinesterase (Roberts and Meredith 1992). Intriguingly, a group of efferent cells in the diencephalon of some otophysan fish immunostain for tyrosine hydroxylase (Bricaud et al. 2001). Could these cells be the ancestors of the dopaminergic efferent cells of the lateral olivocochlear system in mammals (Ruel et al. 2001, 2006)? Regardless of the answer to the question, efferent innervation of hair cells in the octavolateralis system has been observed in every vertebrate examined, suggesting an ancient and highly specialized feature of hearing.

During the course of vertebrate evolution, while auditory endorgans became more specialized, there were parallel changes in vertebrate brains. The correlation of structure and function has been important for evolutionary concepts because behavioral potential has been inferred from the presence and relative size of various brain structures (Sarnat and Netsky 1974; Webster et al. 1992). The growth of the cerebral cortex (encephalization) is one key to the evolutionary success of a species. The cerebral cortex and its interconnected structures enlarge substantially in size as one ascends the phylogenetic tree (see Fig. 1 of Meltzer and Ryugo 2006). With this "encephalization" has come an expansion in behavioral capacity that has culminated in the richness of human culture – ethics, art, science, philosophy, and so on. The corticofugal pathway – descending projections from cerebral cortex to lower brain centers – was elaborated along with growth of the forebrain and is one component to the efferent system (Malmierca and Ryugo 2010). Its prominence in mammals presumably contributes to the refinement and enhancement of auditory processing (Winer and Lee 2007).

Chapter 9 by Brett Schofield describes in detail the long descending "chains" of projections as well as feedback "loops" that comprise the descending auditory circuitry. These descending systems extend throughout the brain and their distribution emphasizes that the modulation of ascending information occurs beyond that observed at the auditory end organ. The complexity of the circuitry underlies the variety of auditory functions that could be influenced.

Chapter 10 by Donald Robertson and Wilhelmina Mulders explores electrophysiological data that address the central effects of efferent activation. If the main effect of olivocochlear efferents is to reduce the gain of cochlear sensitivity, what is the function of the descending circuits that terminate in central structures? With the enormous and complex ascending auditory pathways, olivocochlear anti-masking does not by itself seem sufficient to account for the diverse neuronal properties involved with signal processing (Mulders et al. 2009). The chapter provides a thorough discussion of technical considerations for studying this question and a constructive critique of some of the seminal research on the topic, and includes a review of the effects on single cell responses in the cochlear nucleus and inferior colliculus by olivocochlear activation. The authors report that approximately half the neurons show olivocochlear effects similar to those described in primary afferents, whereas the others exhibit a variety of novel effects. There is definitely more research to be done on this issue.

Nobuo Suga, Weiqing Ji, Xiaofeng Ma, Jie Tang, Zhongju Xiao, and Jun Yan highlight our topic in Chap. 11 by reviewing some of the features that unify the function of descending projections. By virtue of the sensory maps, the interactions between ascending and descending pathways may be "matched" or "unmatched" in terms of shared physiological properties and receptive field responsiveness. When activation of a cortical region that is matched to a subcortical site, for example, in terms of best frequency sensitivity, visual field location or body surface, the resulting response tends to be amplified or augmented. The cortical function that mediates adjustment and enhanced sensory processing has been termed "egocentric" selection (Gao and Suga 1998). In contrast, when the sites are unmatched, the resulting response can be a shift in properties of the recipient neurons toward that of the activated source neurons, or the recipient neurons can be unaffected or even suppressed by the phenomenon of lateral inhibition. The necessary and sufficient parameters for these concepts are expertly developed.

The precise organization of sensory systems in terms of topographic maps of their respective receptor epithelia creates a powerful template with which to study development and plasticity in the central pathways. Perturbations in this organization can therefore be readily detected when the organization is modified by selective activation, sensory deprivation, injury, or experience (Parks et al. 2004). In the bat, electrical stimulation of the highly tonotopic auditory cortex results in an upward or downward shift of the preferred frequencies of collicular neurons toward that of the stimulated cortical neurons. Moreover, centrifugal shifts in tuning curves can be manipulated by experimental conditions toward or away from the neuron's preferred frequency (Suga et al. 2002). In short, response plasticity is evident in terms of changes in single neuron sensitivity as well as in expanded or compressed reorganizations of sensory maps. In both young and adult animals, response changes have been attributed to alterations in the divergent and convergent projections of the ascending projections (van der Loos and Woolsey 1973; Katz and Shatz 1996; Antonini et al. 1999; Parks et al. 2004; Sato and Stryker 2008) as well as through descending corticofugal pathways (Yan and Suga 1998; Yan et al. 2005).

1.3 Comparison with Other Sensory Systems

It is worth noting that the organization of the auditory, somatosensory, and visual pathways enjoys structural similarities but they are not identical. Each pathway exhibits a cochleotopic, somatotopic, and retinotopic structure that comprises ascending projections as well as descending loops and projections. In the somatosensory system, the processing of touch sensations begins with peripheral inputs from the body surface and continues along the ascending somatosensory pathway that includes the dorsal column nuclei (nucleus gracilis and cuneatus), the thalamic ventroposterior nucleus, and multiple neocortical regions.

Cortical control of sensory information is highlighted by descending corticofugal projections that converge on the various ascending pathways (Jabbur and Towe 1961; Landry and Dykes 1985; Weisberg and Rustioni 1976, 1977, 1979; Martinez-Lorenzana et al. 2001). Descending corticothalamic projections exhibit a dual physiological effect on neurons of the ventroposterior nucleus. Glutamatergic projections mediate tonic inhibitory actions via GABAergic neurons of the thalamic reticular nucleus as well as excitatory topographic effects on matched thalamic cells and inhibition of adjacent unmatched thalamic cells (Krupa et al. 1999). While the basic properties of the thalamic neurons are determined by the ascending feedforward projections, matched cortical activation enhances activity for discrete loci of the body and unmatched activation exerts a suppressive surround (Rapisarda et al. 1992; Ghazanfar et al. 2001; Wang et al. 2007). Neurons of pars interpolaris of the trigeminal nucleus that respond to peripheral mechanical stimulation of the face exhibit response enhancement when the facial cortical field is electrically stimulated. In contrast, when a region of cortex was stimulated where receptive fields did not include the face, responses were suppressed (Woolston et al. 1983). Functional and coherent correlated interactions within somatosensory pathways provide two complementary mechanisms for response enhancement: there is a type of "on-center" feedback excitation that augments the response to the main stimulus; and there is surround-inhibition where "off center" cortical influences reduce the response to (background) stimuli, which serves to enhance further the biologically significant stimulus.

There are significant corticofugal projections from visual cortex to the thalamus and midbrain (Guillery 1969; Van Horn et al. 2000). Elimination of visual cortex activity by cooling causes a decrease in activity of lateral geniculate cells that includes reduced spontaneous activity and peak response amplitudes (Waleszczyk et al. 2005) and loss of thalamo–cortico–thalamic synchronization (Sillito et al. 1994). The visual system has a highly topographic organization where retinal fields and binocular interactions are tightly mapped. Coherent and similar stimulation in terms of orientation, direction of movement, and contrast produced enhanced neuronal responses in the dorsal lateral geniculate nucleus, whereas dissimilar stimulation had a suppressive effect (Varela and Singer 1987). Ablation of visual cortex abolished these feature-dependent interactions. Such observations promoted the idea that corticothalamic projections are involved in the mediation of binocular interactions (Murphy and Sillito 1987; Marrocco et al. 1996). More importantly, they are consistent with the theme of sensory efferent action where descending feedback strengthens thalamic transmission when cortical activation patterns and retinal signals are congruent.

1.4 Summary

One challenge for sensory system researchers is to unravel how central efferent systems engineer the extraction of signals from noise under the wide and varied conditions of a natural environment. With respect to the auditory system, the relatively small number of efferent neurons coupled with the often small impact on gain control in the ear seems contradictory to the behavioral consequences of efferent activity. Subtle manipulations enacted by the olivocochlear system on responses from the inner ear appear to be amplified and implemented by complex brain processes. The vestibular system continues to hold many secrets regarding mechanisms of function as well. It has received somewhat less attention over the years, and its central pathways have yet to be fully described. Equally daunting is the rapid and enduring plasticity of the central vestibular system where neuronal responses exhibit remarkable compensatory adaptations to systemic perturbations. There are many avenues of research for the intrepid explorer.

In spite of our gaps in knowledge, it is evident that sensory systems distribute information over bidirectional divergent and convergent pathways. Moreover, their organization enables both parallel and serial processing at every synaptic junction. This arrangement promotes feedback modulation of signals as they pass from one structure to the next, and coding schemes can change as rapidly as required by environmental demands. If we accept the concept of "top down" influences for ongoing and continuous monitoring of our conscious experiences, we are faced with the question of what structure represents the "top"? Given that most brain structures give rise to both ascending and descending projections, perhaps there is no "top" but a continuous series of loops at every level that monitors what ascends and descends. This arrangement suggests that "on-the-fly" modifications of neural activity can be initiated at every level of the relevant pathway, emphasizing that the underlying mechanisms of efferent action are just beginning to be understood.

Acknowledgments The author was supported in part by NIH grants DC004395 and DC000232, a grant from Advanced Bionics Corporation, and a Life Science Research Award from the Office of Science and Medical Research, New South Wales, Australia.

References

Aitkin LM, Irvine DRF, Webster WR (1984) Central neural mechanisms of hearing. In: Brookhart JM, Mountcastle VB (eds) Handbook of physiology – the nervous system. American Physiological Society, Bethesda, pp 675–737

Antonini A, Fagiolini M, Stryker MP (1999) Anatomical correlates of functional plasticity in mouse visual cortex. J Neurosci 19:4388–4406

Barmack NH (2003) Central vestibular system: vestibular nuclei and posterior cerebellum. Brain Res Bull 60:511–541

Bricaud O, Chaar V, Dambly-Chaudiere C, Ghysen A (2001) Early efferent innervation of the zebrafish lateral line. J Comp Neurol 434:253–261

Brichta AM, Goldberg JM (2000) Responses to efferent activation and excitatory response-intensity relations of turtle posterior-crista afferents. J Neurophysiol 83:1224–1242

Brown MC (1987) Morphology of labeled efferent fibers in the guinea pig cochlea. J Comp Neurol 260:605–618

Canedo A (1997) Primary motor cortex influences on the descending and ascending systems. Prog Neurobiol 51:287–335

Card JP, Sved JC, Craig B, Raizada M, Vazquez J, Sved AF (2006) Efferent projections of rat rostroventrolateral medulla C1 catecholamine neurons: implications for the central control of cardiovascular regulation. J Comp Neurol 499:840–859

Cooper NP, Guinan JJ Jr (2006) Efferent-mediated control of basilar membrane motion. J Physiol 576:49–54

Cullen KE, Minor LB (2002) Semicircular canal afferents similarly encode active and passive head-on-body rotations: implications for the role of vestibular efference. J Neurosci 22:RC226

Dallos P (1997) Outer hair cells: the inside story. Ann Otol Rhinol Laryngol Suppl 168:16–22

Darrow KN, Maison SF, Liberman MC (2006) Cochlear efferent feedback balances interaural sensitivity. Nat Neurosci 9:1474–1476

Dewson JH III (1967) Efferent olivocochlear bundle: some relationships to noise masking and to stimulus attenuation. J Neurophysiol 30:817–832

Evans EF (1975) Cochlear nerve and cochlear nucleus. In: Keidel WD, Neff WD (eds) Handbook of sensory physiology, vol 5/2. Springer, Berlin, pp 1–108

Flock A, Russell I (1976) Inhibition by efferent nerve fibres: action on hair cells and afferent synaptic transmission in the lateral line canal organ of the burbot *Lota lota*. J Physiol 257:45–62

Fritzsch B, Pirvola U, Ylikoski J (1999) Making and breaking the innervation of the ear: neurotrophic support during ear development and its clinical implications. Cell Tissue Res 295:369–382

Fuchs PA, Murrow BW (1992) A novel cholinergic receptor mediates inhibition of chick cochlear hair cells. Proc Biol Sci 248:35–40

Galambos R (1956) Suppression of auditory nerve activity by stimulation of efferent fibers to cochlea. J Neurophysiol 19:424–437
Gao E, Suga N (1998) Experience-dependent corticofugal adjustment of midbrain frequency map in bat auditory system. Proc Natl Acad Sci USA 95:12663–12670
Ghazanfar AA, Krupa DJ, Nicolelis MA (2001) Role of cortical feedback in the receptive field structure and nonlinear response properties of somatosensory thalamic neurons. Exp Brain Res 141:88–100
Goldberg JM, Fernández C (1980) Efferent vestibular system in the squirrel monkey: anatomical location and influence on afferent activity. J Neurophysiol 43:986–1025
Granit R (1955) Centrifugal and antidromic effects on ganglion cells of retina. J Neurophysiol 18:388–411
Guillery RW (1969) The organization of synaptic interconnections in the laminae of the dorsal lateral geniculate nucleus of the cat. Z Zellforsch Mikrosk Anat 96:1–38
Guinan JJ Jr (1996) The physiology of olivocochlear efferents. In: Dallos PJ, Popper AN, Fay RR (eds) The cochlea. Springer, New York, pp 432–435
Hagbarth KE, Kerr DI (1954) Central influences on spinal afferent conduction. J Neurophysiol 17:295–307
Highstein SM (1992) The efferent control of the organs of balance and equilibrium in the toadfish, *Opsanus tau*. Ann NY Acad Sci 656:108–123
Jabbur SJ, Towe AL (1961) Cortical excitation of neurons in dorsal column nuclei of cat, including an analysis of pathways. J Neurophysiol 24:499–509
Jones EG (2002) Thalamic circuitry and thalamocortical synchrony. Philos Trans R Soc Lond B Biol Sci 357:1659–1673
Katz LC, Shatz CJ (1996) Synaptic activity and the construction of cortical circuits. Science 274:1133–1138
Kiang NY-S, Watanabe T, Thomas EC, Clark LF (1965) Discharge patterns of single fibers in the cat's auditory nerve. MIT, Cambridge
Krupa DJ, Ghazanfar AA, Nicolelis MA (1999) Immediate thalamic sensory plasticity depends on corticothalamic feedback. Proc Natl Acad Sci USA 96:8200–8205
Landry P, Dykes RW (1985) Identification of two populations of corticothalamic neurons in cat primary somatosensory cortex. Exp Brain Res 60:289–298
Mackay DM (1956) Towards an information-flow model of huan behaviour. Br J Psychol 47:30–43
Maison SF, Liberman MC (2000) Predicting vulnerability to acoustic injury with a noninvasive assay of olivocochlear reflex strength. J Neurosci 20:4701–4707
Malmierca MS, Ryugo DK (2010) Descending connections to the midbrain and brainstem. In: Winer JA, Schreiner CE (eds) The Auditory Cortex. Springer, New York (in press)
Marlinski V, Plotnik M, Goldberg JM (2004) Efferent actions in the chinchilla vestibular labyrinth. JARO 5:126–143
Marrocco RT, McClurkin JW, Alkire MT (1996) The influence of the visual cortex on the spatiotemporal response properties of lateral geniculate nucleus cells. Brain Res 737:110–118
Martinez-Lorenzana G, Machin R, Avendano C (2001) Definite segregation of cortical neurons projecting to the dorsal column nuclei in the rat. Neuroreport 12:413–416
Matesz C, Kulik A, Bacskai T (2002) Ascending and descending projections of the lateral vestibular nucleus in the frog *Rana esculenta*. J Comp Neurol 444:115–128
McCue MP, Guinan JJ Jr (1994) Influence of efferent stimulation on acoustically responsive vestibular afferents in the cat. J Neurosci 14:6071–6083
Meltzer NE, Ryugo DK (2006) Projections from auditory cortex to cochlear nucleus: a comparative analysis of rat and mouse. Anat Rec A Discov Mol Cell Evol Biol 288:397–408
Menétrey D, Basbaum AI (1987) Spinal and trigeminal projections to the nucleus of the solitary tract: a possible substrate for somatovisceral and viscerovisceral reflex activation. J Comp Neurol 255:439–450
Mulders WH, Paolini AG, Needham K, Robertson D (2009) Synaptic responses in cochlear nucleus neurons evoked by activation of the olivocochlear system. Hear Res 256:85–92

Murphy PC, Sillito AM (1987) Corticofugal feedback influences the generation of length tuning in the visual pathway. Nature 329:727–729

Oertel D, Young ED (2004) What's a cerebellar circuit doing in the auditory system. Trends Neurosci 27:104–110

Parks TN, Rubel EW, Fay RR, Popper AN (eds) (2004) Plasticity of the auditory system. Springer, New York

Popper AN, Fay RR (eds) (2005) Sound source localization. Springer, New York

Ramón y Cajal R (1909) Histologie du Système Nerveux de l'Homme et des Vertébrés. Instituto Ramón y Cajal, Madrid, pp 774–838

Rapisarda C, Palmeri A, Sapienza S (1992) Cortical modulation of thalamo-cortical neurons relaying exteroceptive information: a microstimulation study in the guinea pig. Exp Brain Res 88:140–150

Rasmussen GL (1946) The olivary peduncle and other fiber projections of the superior olivary complex. J Comp Neurol 84:141–219

Rasmussen GL (1953) Further observations of the efferent cochlear bundle. J Comp Neurol 99:61–74

Roberts BL, Meredith GE (1992) The efferent innervation of the ear: variations on an enigma. In: Webster DB, Fay RR, Popper AN (eds) The evolutionary biology of hearing. Springer, New York, pp 185–210

Ruel J, Nouvian R, Gervais d'Aldin C, Pujol R, Eybalin M, Puel JL (2001) Dopamine inhibition of auditory nerve activity in the adult mammalian cochlea. Eur J Neurosci 14:977–986

Ruel J, Wang J, Dememes D, Gobaille S, Puel JL, Rebillard G (2006) Dopamine transporter is essential for the maintenance of spontaneous activity of auditory nerve neurones and their responsiveness to sound stimulation. J Neurochem 97:190–200

Sarnat HB, Netsky MG (1974) Evolution of the nervous system. Oxford University Press, New York

Sato M, Stryker MP (2008) Distinctive features of adult ocular dominance plasticity. J Neurosci 28:10278–10286

Schrott-Fischer A, Kammen-Jolly K, Scholtz A, Rask-Andersen H, Glueckert R, Eybalin M (2007) Efferent neurotransmitters in the human cochlea and vestibule. Acta Otolaryngol 127:13–19

Sherman SM, Guillery RW (2002) The role of the thalamus in the flow of information to the cortex. Philos Trans R Soc Lond B Biol Sci 357:1695–1708

Sillito AM, Jones HE, Gerstein GL, West DC (1994) Feature-linked synchronization of thalamic relay cell firing induced by feedback from the visual cortex. Nature 369:479–482

Simmons DD (2002) Development of the inner ear efferent system across vertebrate species. J Neurobiol 53:228–250

Suga N, Xiao Z, Ma X, Ji W (2002) Plasticity and corticofugal modulation for hearing in adult animals. Neuron 36:9–18

Tsuchitani C (1977) Functional organization of lateral cell groups of the cat superior olivary complex. J Neurophysiol 40:296–318

Van der Loos H, Woolsey TA (1973) Somatosensory cortex: structural alterations following early injury to sense organs. Science 179:395–398

Van Horn SC, Erisir A, Sherman SM (2000) Relative distribution of synapses in the A-laminae of the lateral geniculate nucleus of the cat. J Comp Neurol 416:509–520

Varela FJ, Singer W (1987) Neuronal dynamics in the visual corticothalamic pathway revealed through binocular rivalry. Exp Brain Res 66:10–20

Waleszczyk WJ, Bekisz M, Wrobel A (2005) Cortical modulation of neuronal activity in the cat's lateral geniculate and perigeniculate nuclei. Exp Neurol 196:54–72

Wang X, Wall JT (2005) Cortical influences on sizes and rapid plasticity of tactile receptive fields in the dorsal column nuclei. J Comp Neurol 489:241–248

Wang JY, Chang JY, Woodward DJ, Baccala LA, Han JS, Luo F (2007) Corticofugal influences on thalamic neurons during nociceptive transmission in awake rats. Synapse 61:335–342

Warr WB, Guinan JJ (1979) Efferent innervation of the organ of Corti: two separate systems. Brain Res 173:152–155

Webster DB, Fay RR, Popper AN (1992) The evolutionary biology of hearing. Springer, New York

Weisberg JA, Rustioni A (1976) Cortical cells projecting to the dorsal column nuclei of cats. An anatomical study with the horseradish peroxidase technique. J Comp Neurol 168:425–437

Weisberg JA, Rustioni A (1977) Cortical cells projecting to the dorsal column nuclei of rhesus monkeys. Exp Brain Res 28:521–528

Weisberg JA, Rustioni A (1979) Differential projections of cortical sensorimotor areas upon the dorsal column nuclei of cats. J Comp Neurol 184:401–421

Wiederhold ML, Kiang NYS (1970) Effects of electrical stimulation of the crossed olivocochlear bundle on single auditory-nerve fibers in the cat. J Acoust Soc Am 48:950–965

Winer JA, Lee CC (2007) The distributed auditory cortex. Hear Res 229:3–13

Woolston DC, La Londe JR, Gibson JM (1983) Corticofugal influences in the rat on responses of neurons in the trigeminal nucleus interpolaris to mechanical stimulation. Neurosci Lett 36:43–48

Yan W, Suga N (1998) Corticofugal modulation of the midbrain frequency map in the bat auditory system. Nat Neurosci 1:54–58

Yan J, Zhang Y, Ehret G (2005) Corticofugal shaping of frequency tuning curves in the central nucleus of the inferior colliculus of mice. J Neurophysiol 93:71–83

Chapter 2
Anatomy of Olivocochlear Neurons

M. Christian Brown

2.1 Introduction

Hair cell receptors for the hearing and balance organs, and the lateral line, are unique among the senses by receiving an efferent innervation of the periphery. Olivocochlear (OC) neurons supply this efferent innervation, and they are the most peripheral of the many descending neural systems of the central auditory pathway (see Schofield, Chap. 9). OC neurons are named by their origins in the superior olivary complex and terminations in the cochlea (Fig. 2.1). In the cochlea, they innervate the hair cells and auditory-nerve fibers. This chapter mainly covers the new ground on OC anatomy in mammals since Warr's (1992) comprehensive chapter on this topic about 15 years ago. Since that time, there is even stronger evidence for the separate innervation of the periphery by the two major groups of OC neurons. It is also now clear that both of these groups consist of distinct subgroups. There is additional information on the reflex pathways leading up to OC neurons that enables their response to sound. Overall, this anatomy may help to define the functions that OC neurons perform in the sense of hearing.

2.2 OC Neurons in the Brain Stem

2.2.1 Distributions of Lateral vs. Medial Olivocochlear Neurons

OC neurons were discovered by Rasmussen (1946, 1953). Later studies confirmed the interpretation of these early fiber degeneration studies with the use of retrograde labeling that filled the OC neurons with tracer (Warr 1975). The separation of OC

M.C. Brown (✉)
Department of Otology and Laryngology, Harvard Medical School;
and
Eaton-Peabody Laboratory, Massachusetts Eye and Ear Infirmary, Boston, MA 02114, USA
e-mail: Chris_Brown@meei.harvard.edu

D.K. Ryugo et al. (eds.), *Auditory and Vestibular Efferents*,
Springer Handbook of Auditory Research 38, DOI 10.1007/978-1-4419-7070-1_2,

neurons into the two major groups originated with Warr and Guinan (1979; Fig. 2.1). Generally, lateral olivocochlear (LOC) neurons originate in and/or near the lateral superior olive (LSO), whereas medial olivocochlear (MOC) neurons generally originate in the medial periolivary regions. The exact location of OC neurons depends on species. For example, many LOC neurons in rodents are found within the LSO (Campbell and Henson 1988; Vetter and Mugnaini 1992; Sánchez-González et al. 2003), whereas in cats they are found mainly in the dorsal hilus of the LSO (Warr 1975) and in squirrel monkey they appear to reside between the LSO and the medial superior olive (Thompson and Thompson 1986). Some LOC neurons are located in the anterolateral and dorsolateral periolivary areas (Warr et al. 2002).

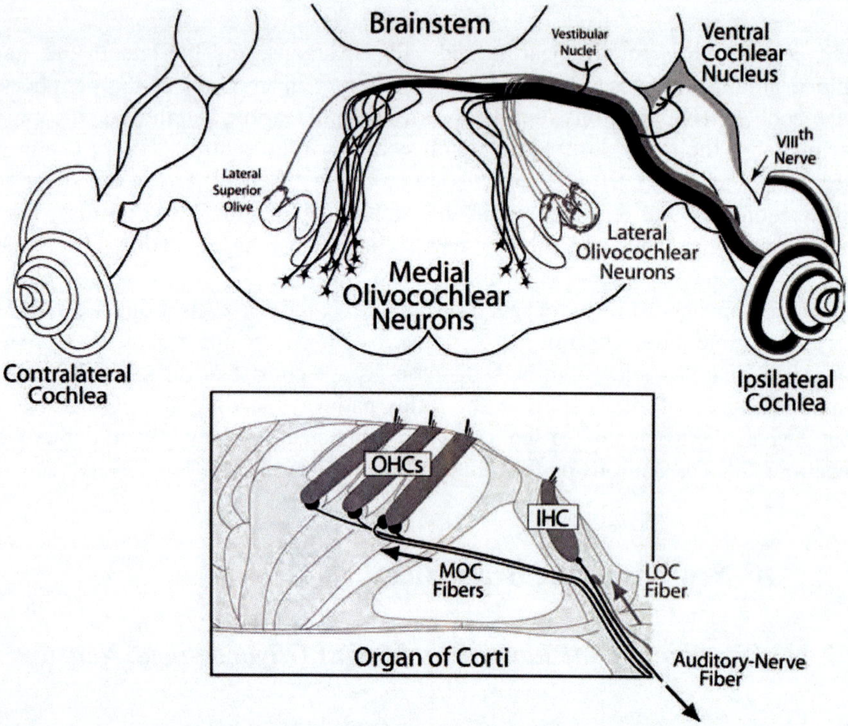

Fig. 2.1 Schematic showing OC neurons and the course of their axons to one cochlea, the cochlea on the *right* side of the figure designated "ipsilateral." Lateral olivocochlear (LOC) neurons have cell bodies located around and/or in the lateral superior olive (LSO), whereas medial olivocochlear (MOC) neurons have cell bodies located in the more medial parts of the superior olivary complex. Axons from LOC neurons (*gray lines* representing unmyelinated axons) and MOC neurons (*black lines* representing myelinated axons) coalesce into the OC bundle. As the bundle projects laterally, branches are given off to the vestibular and cochlear nuclei. *Inset below*: Schematic of peripheral terminations of LOC fibers on auditory-nerve dendrites below inner hair cells (IHCs) and terminations of MOC fibers on outer hair cells (OHCs). *Arrows* indicate direction of spike propagation (adapted from Warren and Liberman 1989)

Two subgroups of LOC neurons were discovered by Vetter and Mugnaini (1992). LOC "intrinsic" neurons are small and contained within the body of the LSO, whereas LOC "shell" neurons are larger and found on the margins of the LSO (Fig. 2.2). In cats, where all LOC neurons are outside the LSO, the subgroup distinction is made by proximity and dendritic association with the LSO (Warr et al. 2002). LOC intrinsic neurons in rodents have dendrites running across the LSO (Fig. 2.3), presumably within an isofrequency plane. LOC shell neurons are about twice the size of intrinsic neurons. Some of their dendrites enter the LSO and others run into the reticular formation.

MOC neurons in most species are found predominantly in the ventral nucleus of the trapezoid body (VNTB). Other nuclei containing MOC neurons include the dorsomedial preolivary nucleus, the medial nucleus of the trapezoid body, and the dorsal periolivary nucleus (Fig. 2.2; Warr 1975; Brown and Levine 2008). The distribution of MOC neurons extends more rostrally than LOC neurons, reaching the caudal end of the ventral nucleus of the lateral lemniscus. MOC neurons have large somata (Fig. 2.2). They give rise to dendrites that radiate in different directions, and the longest are directed medially (Brown and Levine 2008). The nonplanar arrangement of these dendrites seems incongruous with MOC responses, which are sharply tuned to sound frequency (Robertson and Gummer 1985; Liberman and Brown 1986).

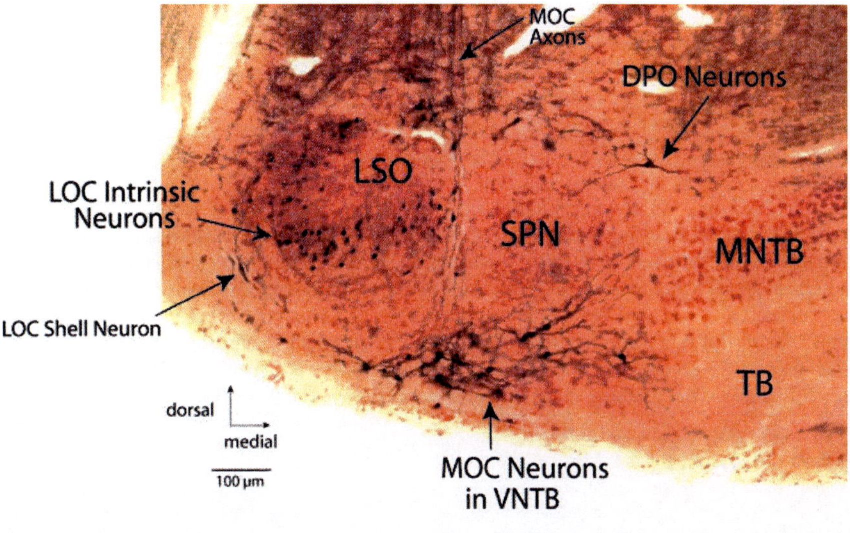

Fig. 2.2 Photomicrograph of presumed OC neurons in the brain stem of a mouse. In this transverse section through the left side, the stain for acetylcholinesterase appears as a *black color*. Darkly stained MOC neurons in the ventral nucleus of the trapezoid body (VNTB) have large cell bodies and long dendrites that extend medially from these somata toward the trapezoid body. A few large stained neurons are seen in the dorsal periolivary nucleus (DPO); these are probably MOC neurons. MOC axons project dorsally to eventually form the OC bundle. LOC intrinsic neurons with small somata are seen within the LSO and LOC shell neurons are seen on its margins. LOC dendrites and axons are not well stained by acetylcholinesterase (from Brown and Levine 2008)

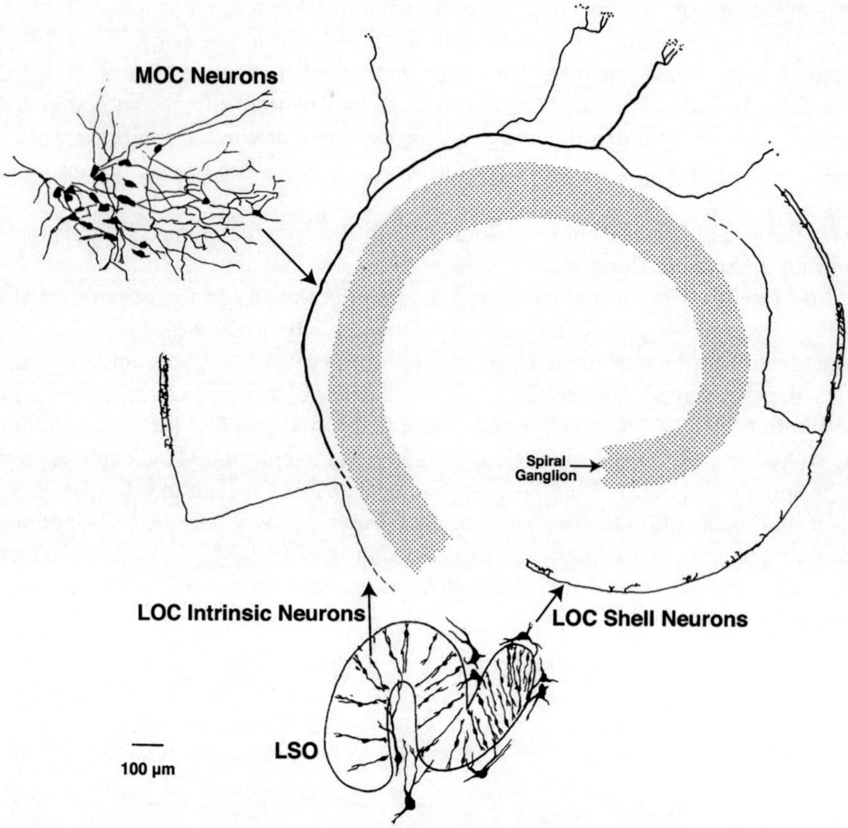

Fig. 2.3 Drawings of OC somata in the brain and their peripheral terminations in the organ of Corti. There are distinct types of LOC neurons. LOC intrinsic neurons have somata within the LSO and project peripherally to form "unidirectional" fibers that run in one direction in the inner and tunnel spiral bundles beneath IHCs. Their peripheral terminations have limited spans along the cochlea. LOC shell neurons on the margins of the LSO project peripherally to form "bidirectional" fibers that run both directions in the spiral bundles and terminate over extensive spans. MOC neurons project peripherally to innervate OHCs (*dots*) in a "patchy" pattern. The endings are distributed over substantial spans (modified from Brown 1987a, b; Warr et al. 1997)

2.2.2 Numbers of Neurons

Depending on the species, total numbers of OC neurons innervating one cochlea range between 341 (hamster) and 474 (mouse) on the lower end and 1,366 (cat) and 2,346 (guinea pig) on the higher end (reviews: Warr 1992; Sánchez-González et al. 2003). In humans, there are about 1,400 axons in the OC bundle (Arnesen 1984). In general, the smaller species have fewer neurons; thus, some of the variability in the number of OC neurons can be reduced by dividing by basilar membrane length, which is shorter in the smaller species (Bishop and Henson 1987). These numbers of OC neurons are dwarfed by the afferent neurons of the cochlea, which range from about 20,000 to 50,000 per cochlea in the various mammalian species (Nadol 1988).

The ratio of LOC to MOC neurons is also variable among species. In the extreme, a bat species similar to the horseshoe bat, *Rhinolophus rouxi*, lacks MOC neurons; in contrast, guinea pigs have approximately equal numbers of LOC and MOC neurons (Aschoff and Ostwald 1987). In cats and mice, the percentages are about 65% LOC and 35% MOC (Arnesen and Osen 1984; Campbell and Henson 1988; Warr et al. 2002). In humans, the percentages are about 70% LOC vs. 30% MOC (Arnesen 1984, counted as thin vs. thick axons in the OC bundle; see later). Shell neurons number about 15% of all LOC neurons in rats (Vetter and Mugnaini 1992).

OC neurons projecting to a single cochlea are distributed bilaterally in the brain stem (Fig. 2.1). LOC neurons are located predominantly on the same side of the brain as the cochlea that they innervate. For cats, the same side:opposite side ratio is about 3:1 (Warr et al. 2002), whereas for mice it is about 100:1 (Campbell and Henson 1988). MOC neurons are usually distributed unequally but their distribution is skewed toward more neurons on the side of the brain opposite to the innervated cochlea. For cats and most other species, there are about twice as many MOC neurons on the opposite side of the brain; for chinchillas the distribution is about even on the two sides (Azeredo et al. 1999). Opposite-side OC neurons have axons that cross the midline on their way to the innervated cochlea, and even uncrossed MOC axons can approach the midline (Fig. 2.1). About 5% of MOC neurons, but no LOC neurons, project bilaterally (Robertson et al. 1987a).

2.2.3 Axonal Characteristics

An important distinction between the groups of OC neurons is that LOC neurons have thin, unmyelinated axons and MOC neurons have thicker, myelinated axons (evidence reviewed by Guinan et al. 1983; Warr 1992). This distinction has important physiological implications because thin unmyelinated axons have much higher thresholds to electric stimulation. Thus, electric stimulation of the OC bundle is likely to activate the MOC axons exclusively (Guinan et al. 1983). Axons from both LOC shell neurons and intrinsic neurons are thin (Brown 1987a; Warr et al. 1997).

2.3 Peripheral Projections

2.3.1 Separate Terminations of LOC and MOC Neurons

The separate peripheral termination of LOC and MOC neurons was first demonstrated by Warr and Guinan (1979). Since then, the preponderance of the evidence suggests that LOC neurons project to inner hair cell (IHC)-associated targets, whereas MOC neurons project to outer hair cells (OHCs; Fig. 2.1, inset). Are there any OC neurons that innervate both IHCs and OHCs? Tracings of labeled fibers indicate separate innervation of the two types of hair cells, at least in the basal half of the cochlea (Fig. 2.3; Robertson and Gummer 1985; Liberman and Brown 1986; Brown 1987a). Also in support of separate innervation of the two types of hair cell, cuts of crossing MOC fibers reduce endings on OHCs, but have no effect on endings on IHCs in the mouse, where LOC neurons are almost entirely uncrossed (Maison et al. 2003). Separate innervations of the hair cells by the OC groups would parallel the separate innervation by afferent fibers, in which type I and type II auditory nerve fibers separately innervate IHC and OHC, respectively (Spoendlin 1971; Kiang et al. 1982).

2.3.2 *Terminations of LOC Fibers*

Both groups of OC neurons have fibers that branch extensively in the cochlea (Fig. 2.3). The end result of the branching is that a relatively small number of OC neurons gives rise to numerous synapses in the cochlea. LOC fibers synapse mainly on dendrites of auditory nerve fibers beneath IHCs. In the cat, those dendrites contacting the IHC on its modiolar side, which correspond to high-threshold, low-spontaneous rate fibers, receive an average of 15–25 synapses per fiber (Liberman et al. 1990). Those dendrites contacting the IHC on its pillar-cell side, which correspond to the low-threshold, high-spontaneous rate fibers, receive fewer synapses: an average of 5–10 synapses per fiber. Some LOC synapses are formed directly on the IHCs (Liberman 1980). In the inner spiral bundle, LOC synapses are formed by small en passant swellings (Fig. 2.4) and a few terminal branches. In some species, LOC swellings in the tunnel spiral bundle may contact MOC branches on their way to the OHCs (Iurato et al. 1978; Liberman 1980).

LOC intrinsic and shell neurons have very distinct terminal arbors (Fig. 2.3). Warr et al. (1997) studied this issue by injections of tracers either inside the LSO (to label intrinsic LOC neurons) or on its margins (to label shell neurons). Intrinsic neurons form peripheral fibers that turn one direction as they enter the inner spiral bundle. They terminate in arborizations of limited span along the organ of Corti. Shell neurons form peripheral fibers that bifurcate in the inner spiral bundle to run both apically and basally. They terminate in extensive arborizations over wide spans along the spiral bundles. Branching of axons giving rise to bidirectional fibers, but not unidirectional fibers, suggests that several bidirectional arborizations can arise from a single neuron (Brown 1987a).

The density of LOC terminals along the length of the cochlea is relatively even, with somewhat more innervation apically (Liberman et al. 1990). LOC neurons project to the cochlea in a way consistent with the known tonotopic mapping of the LSO (Guinan et al. 1984). For example, tracer injections into the lateral part of the rodent LSO (known to be a region of low characteristic frequency) result in labeling in the cochlear apex, whereas injections into the medial part of the LSO result in labeling the base (Stopp 1983; Robertson et al. 1987b). These results probably apply to intrinsic neurons; whether the mapping of shell neurons is tonotopic remains to be investigated.

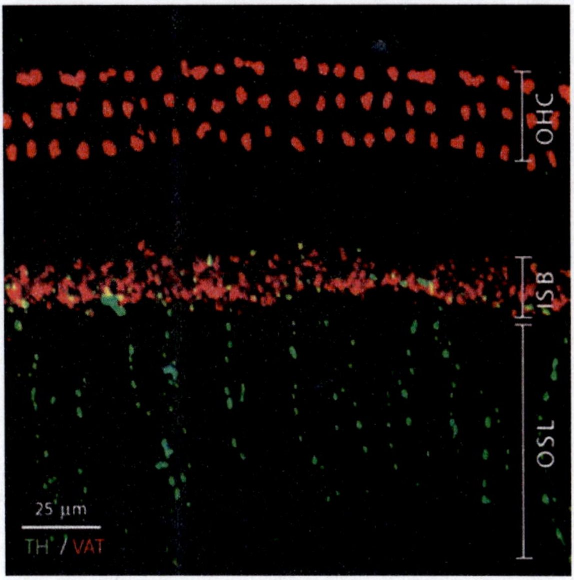

Fig. 2.4 Cochlear immunostains showing cholinergic MOC terminals on OHCs and mixed population of terminals in the inner spiral bundle (ISB) just beneath IHCs. This cochlear whole-mount from a mouse was stained for vesicular acetylcholine transporter (VAT, *red*) and for tyrosine hydroxylase (TH, *green*). Large, VAT-positive endings of MOC neurons are present on the OHC. Small, VAT-positive endings of LOC neurons are present in the ISB. Intermingled with these cholinergic endings are a few TH-positive LOC endings. There are also numerous TH-positive swellings in the osseous spiral lamina (OSL) (figure generously provided by M.C. Liberman)

2.3.3 Terminations of MOC Fibers

MOC endings on OHCs are larger than the LOC endings in the inner spiral bundle (Fig. 2.4). The density of these MOC terminals has a broad peak around the midpoint of the cochlea in mice (Fig. 2.5) or somewhat more basally in cats (Guinan et al. 1984). Compared to the more numerous terminals from crossed axons, terminals from uncrossed axons have a more even density and are distributed somewhat more apically (Guinan et al. 1984). In most species, as the terminations taper off apically, they do so initially for the third row of OHCs, next for the second row, and last for the first row (Liberman et al. 1990). However, in mouse, there is a relatively equal innervation of the three rows (Maison et al. 2003). The longitudinal distribution of endings is similar to the effects of stimulation in response to different frequencies (Fig. 2.5). The MOC terminals hyperpolarize OHCs, decrease the gain of the cochlear amplifier and the sensitivity of the IHCs, and finally reduce the responses of the auditory nerve fibers, which is the metric plotted in Fig. 2.5 (see Guinan, Chap. 3). OHCs in the cochlear base receive an average of nine MOC terminals (Liberman et al. 1990). Some MOC en passant swellings just below OHCs contact type II auditory nerve fibers (Thiers et al. 2002).

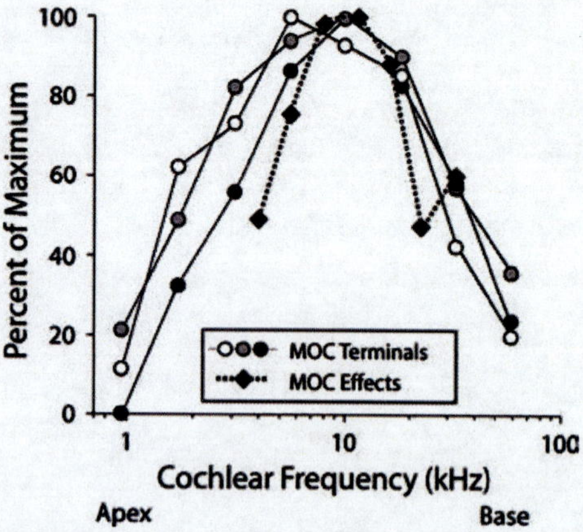

Fig. 2.5 Longitudinal distribution of cochlear MOC terminals in the mouse shows a broad peak in the middle of the cochlea. The *x*-axis is distance along the cochlear spiral plotted as cochlear frequency correlate. Equivalent counts of immunostaining for glutamic acid decarboxylase (GAD, *open circles*), calcitonin gene-related peptide, (CGRP, *shaded circles*), or vesicular acetylcholine transporter (VAT, *solid circles*) results from extensive co-localization. Presumably, the peak in distribution of terminals is responsible for the peak of MOC effect, which is the average suppression of the compound action potential of the cochlea expressed as effective attenuation in decibels as a result of electrical stimulation of the OC bundle. MOC terminal data from Maison et al. (2003); MOC effects adapted from Vetter et al. (1999)

Reconstructions of single MOC fibers in the cochlea indicate substantial spans of the endings along the organ of Corti. For example, the guinea pig fiber in Fig. 2.3 contacts OHCs over a span of about 1.5 mm. Cat MOC fibers can span up to 3.2 mm, corresponding to a cochlear distance of about an octave (Liberman and Brown 1986). The innervation pattern from a single fiber is "patchy," with clusters of OHCs innervated separated by long uninnervated regions; other fibers provide the innervation for those hair cells between the patches. MOC neurons project onto the cochlea in a mapping generally similar to the cochlear frequency mapping for auditory nerve fibers (Robertson and Gummer 1985; Liberman and Brown 1986; Brown 1989, 2002). For example, an MOC neuron with a characteristic frequency of 10 kHz projects to a region of the cochlea that has auditory-nerve fibers tuned to a similar characteristic frequency. MOC neurons thus affect the processing of information in the cochlea in a frequency-specific manner. Given the broad span of innervation it is expected that a single fiber would affect a band of frequencies rather than a single frequency (see Guinan, Chap. 3, for human data relevant to this issue).

MOC terminal anatomy is influenced by postsynaptic target. This has been demonstrated in mice lacking the α9 cholinergic receptor, a key component of the receptor for MOC action on OHCs (see Sewell, Chap. 4 and Katz et al., Chap. 5). In these mice, there is a single, large MOC terminal on each OHC rather than the multiple small terminals found in wild-type mice (Vetter et al. 1999). In these same mice, however, the central branches and somata of OC neurons have normal morphology (Brown and Vetter 2008), apparently because central receptors are different (see later) and are preserved in the knock-out mice.

2.4 Central Branches to the Cochlear and Vestibular Nuclei

MOC axons form branches to the cochlear nucleus en route to the periphery (Fig. 2.1). About two thirds of MOC axons in rodents form branches; crossing fibers as well as uncrossed fibers are branched (Brown 1993). There may be species differences in the number of branches formed. Mice and cats have many such branches (Osen et al. 1984), guinea pigs have fewer branches (Winter et al. 1989), and humans appear to lack them (Moore and Osen 1979). Some work (Brown et al. 1988) suggests that only the thick MOC axons form cochlear nucleus branches. Other studies (Ryan et al. 1990; Horvath et al. 2000), however, suggest branches from LOC neurons. Perhaps these differences arise from differences in the species used.

MOC branches terminate mainly in the edge regions and to a lesser extent the core of the cochlear nucleus (Osen et al. 1984; Ryan et al. 1990; Brown 1993; Brown and Vetter 2008). There are many branches at the medial edge (Fig. 2.6) and at the dorsal edges of the VCN where it abuts the granule cell lamina. Some branches are found in the core of the cochlear nucleus and on its superficial edge (Fig. 2.6), part of the shell of small and granule cells. Caudally and rostrally, the branches taper off. LOC processes are not well stained by acetylcholinesterase (Fig. 2.2), but LOC branches labeled by retrograde transport of amino acids terminate mainly in the core of the cochlear nucleus (Ryan et al. 1990).

The branches formed by the MOC give rise to numerous swellings, which are the site of synapses (Benson and Brown 1990). The synapses have round vesicles and are thus likely to be excitatory (Uchizono 1965). The most common targets of the synapses are large-diameter dendrites. Single dendrites receive multiple synapses from one branch, suggesting a powerful effect on the target. The target dendrites are likely to be from cochlear nucleus stellate/multipolar cells. In slice preparations, the type of neuron known as "T" stellate/multipolar cells is affected by cholinergic agonists (Fujino and Oertel 2001), indicating that it has the appropriate receptor for the acetylcholine released by OC branches. In vivo, electrical stimulation of the OC bundle causes excitation of neurons with an "onset chopper" response to sound (Mulders et al. 2007; see Robertson and Mulders, Chap. 10). Neurons with this response may correspond to a different subtype of stellate/multipolar cell and this difference remains to be resolved. The functional role of the OC branches to the cochlear nucleus also remains to be worked out. They may provide a form of "efferent copy" in which certain elements in the cochlear nucleus receive information about the type and amount of efferent feedback sent to the periphery (Benson and Brown 1990). For example, MOC neuron action on the periphery is inhibitory, in turn decreasing responses of cochlear nucleus neurons including those providing inputs to MOC neurons. This would cause MOC response to decrease,

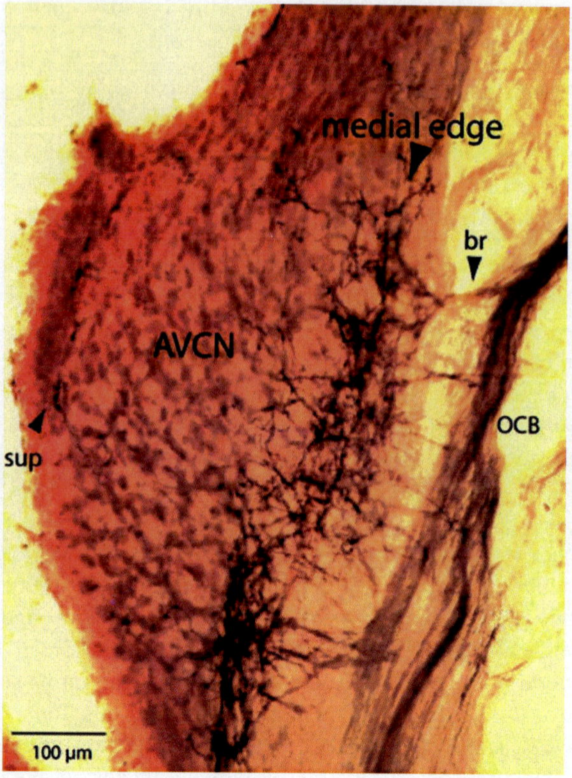

Fig. 2.6 OC branches to the cochlear nucleus. A mouse transverse section stained for acetylcholinesterase that shows the darkly stained olivocochlear bundle (OCB) giving off branches (br). In this section, most of the branches form endings at the medial edge (*arrow*), a few in the core of the anteroventral cochlear nucleus (AVCN), and one or two at its superficial edge (sup) (unpublished data from Brown and Levine)

which might compromise important functions such as providing protection from acoustic overstimulation. A scheme to maintain MOC response at a constant level would be to have MOC branches excite those neurons providing inputs to MOC neurons.

The cochlear nucleus contains neither the α9 nor α10 nicotinic receptor subunits that are present at MOC synapses in the cochlea. The central receptor for the MOC branches is most likely composed of the α7 nicotinic cholinergic receptor. Immunostaining (Yao and Godfrey 1999) and the presence of mRNA for this receptor (Happe and Morely 1998) overlap with the regional distributions of branches. The pharmacological profile of the effect of acetylcholine on cochlear nucleus stellate cells is also consistent with the involvement of an α7 receptor as well as the presence of other cholinergic receptors (Fujino and Oertel 2001).

MOC and LOC neurons also give off branches to some of the vestibular nuclei (Brown et al. 1988; Ryan et al. 1990; Brown 1993). These branches form swellings that, like those in the cochlear nucleus, produce synapses (Benson and Brown 1996). The postsynaptic targets in the inferior vestibular nucleus are thick dendrites and sometimes cell bodies. The function of these vestibular-nucleus branches is obscure.

2.5 Neurochemistry

Both groups of OC neurons are predominantly cholinergic, but the LOC group probably has additional neurons that are not cholinergic (see Sewell, Chap. 4 and Katz et al., Chap. 5). In immunohistochemical studies of the rat, about half of the LOC neurons are reported to stain for glutamic acid decarboxylase (GAD, an indicator of GABAergic neurotransmission), and the other half are reported to stain for calcitonin gene-related peptide (CGRP) (Vetter et al. 1991). In mouse, though, there is extensive co-localization of immunolabeling for these transmitter-related substances along with vesicular acetylcholine transporter (an indicator of cholinergic neurotransmission). This co-labeling suggests that LOC neurons are predominantly cholinergic. LOC shell neurons may be in large part dopaminergic because they immunostain for tyrosine hydroxylase (Darrow et al. 2006). These LOC neurons apparently account for the tyrosine hydroxylase-positive endings in the IHC area (Fig. 2.4). Some LOC shell neurons, however, may stain for acetylcholinesterase (Brown and Levine 2008). Clarification of the neurotransmitters of LOC neurons is an active area of research.

MOC neurons are cholinergic. At least in some species, there is co-localization of other transmitters in the terminals (Maison et al. 2003). For example, there are similar numbers of mouse OHC terminals stained for vesicular acetylcholine transporter (an indicator of cholinergic neurotransmission), GAD, and CGRP (Fig. 2.5). These numbers are similar to the numbers of OHC terminals stained for SNAP25, a marker of all vesiculated terminals, so there are apparently no other types of terminals. Consistent with their cholinergic neurotransmission, MOC neurons, their axons, and their processes all stain darkly for acetylcholinesterase (Fig. 2.2; Schuknecht and Nomura 1965; Osen et al. 1984). LOC neurons also stain, but their axons and dendrites are not as darkly stained (Fig. 2.2; Brown and Levine 2008). OC neurons also stain for the cholinergic marker, choline acetyltransferase (Thompson and Thompson 1986).

2.6 Ultrastructure of Synaptic Inputs to OC Neurons

MOC neurons receive several types of synaptic terminals (Fig. 2.7; White 1984, 1986; Spangler et al. 1986; Helfert et al. 1988; Benson and Brown 2006). One type contains large, round vesicles (Fig. 2.7a). These terminals are large in size and form up to seven synapses per terminal. Some of them are associated with spines of

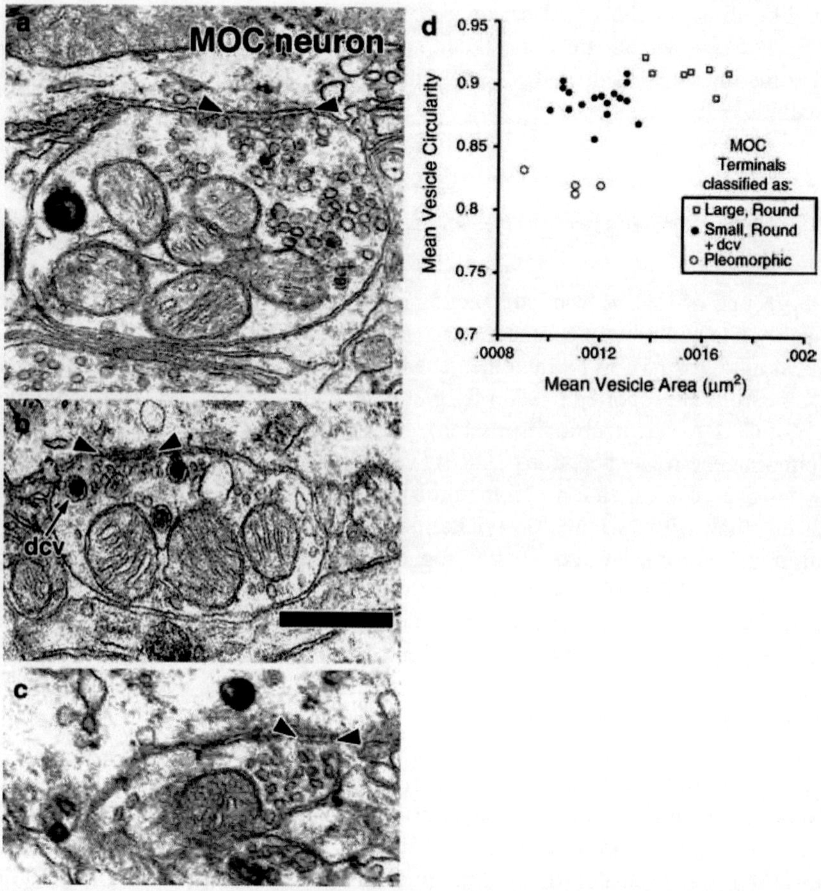

Fig. 2.7 Ultrastructure of synaptic input to MOC neurons, showing the three types of synaptic terminals (**a**–**c**) found on guinea pig MOC neurons. The MOC neuron is at the *top* in each panel, and the synaptic specialization is denoted by *arrowheads*. (**a**) Terminal with relatively large, round (spherical) vesicles. (**b**) Terminal with relatively smaller, round-to-oval vesicles and dense core vesicles (dcv, one indicated with *arrow*). (**c**) Terminal with pleomorphic or variously shaped vesicles. Scale bar = 1 μm. (**d**) Vesicle morphometry of synaptic terminals on MOC neurons. The terminals were first classified visually using size and shape of vesicles and presence of dense core vesicles (over all the serial sections containing the terminal). To make the measurements, all clear vesicles within 1 μm of a synaptic specialization in one or two sections from each terminal were measured using ImageJ, which defines circularity as $4\pi(\text{area/perimeter}^2)$. The number of vesicles measured per terminal ranged from 20 to 92; the means of these circularity/area measurements are plotted as a single point for each terminal. Dense core vesicles were not included. Data are from two guinea pigs ((**a**)–(**c**) from Benson and Brown 2006; (**d**) is unpublished data from Benson and Brown)

MOC neurons, which were first described by Mulders and Robertson (2000b). Both simple and "mushroom" spines are formed (Benson and Brown 2006). The type of terminal containing large, round vesicles is apparently lacking on MOC neurons from the cat (White 1984, 1986; Spangler et al. 1986). The second type (Fig. 2.7b) contains smaller oval-to-round vesicles as well as a few dense core vesicles. The presence of dense core vesicles is an important distinguishing feature, since morphometric measurement of vesicle area suggest a continuum in average vesicle size between the types containing "large, round" and "small, round" types (Fig. 2.7d). The small, round terminals are of moderate size and form up to four synapses per terminal. This type of terminal may originate from neurons in the cochlear nucleus and may excite the MOC neurons to respond to sound as part of the MOC reflex (Benson and Brown 2006). Interestingly, LOC neurons receive many such terminals (Helfert et al. 1988). A third type (Fig. 2.7c) contains flattened or pleomorphic vesicles and is thus likely to be inhibitory (Uchizono 1965). This type of terminal is small and forms a single synapse per terminal. In recordings of single MOC neurons, inhibition of activity has been reported for frequencies outside the excitatory area and sometimes for sounds in the nondominant ear (Liberman and Brown 1986; Brown 1989).

2.7 Neural Pathway of the Medial Olivocochlear Reflex

2.7.1 Direct Reflex Pathway

Medial OC neurons respond to sound as part of the MOC reflex (Fig. 2.8). The three neurons of the reflex pathway are auditory nerve fibers, cochlear nucleus neurons, and MOC neurons. Direct projections from the ventral cochlear nucleus to OC neurons have been demonstrated using neural tracers (Robertson and Winter 1988; Thompson and Thompson 1991; Ye et al. 2000). These direct projections are consistent with the short latency of the MOC neuron response to sound (Robertson and Gummer 1985; Liberman and Brown 1986; Brown et al. 2003). MOC neurons consist of two major subgroups defined on the basis of which ear excites their response (see Guinan, Chap. 3). Ipsi units respond to monaural sound in the ipsilateral ear whereas Contra units respond to monaural sound in the contralateral ear (Robertson and Gummer 1985; Liberman and Brown 1986). An additional minor subgroup responds to sound in either ear. Because of the distinct distribution of MOC neurons by the two major response classes, separate pathways can be drawn for the MOC reflex in response to ipsilateral vs. contralateral sound (black vs. gray pathways in Fig. 2.8).

The cochlear nucleus projections to OC neurons originate in two very different locations: the posteroventral subdivision (PVCN) (Thompson and Thompson 1991) and the shell of the anteroventral subdivision (AVCN) (Ye et al. 2000). The functional importance of the various projections has been explored by making small lesions using kainic acid. Lesions of the PVCN (Fig. 2.9), but not the AVCN, interrupt the ipsilateral MOC reflex (de Venecia et al. 2005), demonstrating that this cochlear nucleus subdivision contains the interneurons of the MOC reflex. Some data suggest a similar PVCN site for the neurons mediating the contralateral reflex.

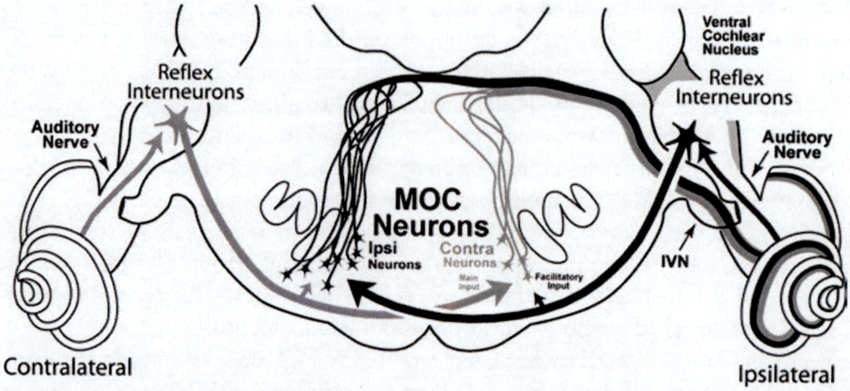

Fig. 2.8 Pathways of the sound-evoked MOC reflexes to one cochlea, the ipsilateral cochlea on the right of the figure. Indicated are the positions of MOC neurons that respond to sound in the ipsilateral ear (Ipsi Neurons) and those that respond to sound in the contralateral ear (Contra Neurons). The reflex pathway in response to ipsilateral sound begins in the ipsilateral cochlea with the responses of the hair cells and the auditory nerve. Nerve fibers project centrally into the cochlear nucleus. Here, MOC reflex interneurons from the cochlear nucleus send axons across the midline (*black* pathway) to innervate Ipsi neurons. These Ipsi neurons send axons back across the midline to innervate the ipsilateral cochlea. The reflex pathway in response to contralateral sound begins with hair cells and nerve fibers in the contralateral cochlea. From the contralateral cochlear nucleus, MOC reflex interneurons send axons that cross the midline (*gray* pathway) to innervate Contra neurons. Contra neurons in turn project without crossing to the ipsilateral cochlea. In addition to these dominant inputs, both types of MOC neurons receive inputs that facilitate the response to the dominant ear (*small arrows*, "Facilitatory Inputs") (modified from Liberman and Guinan 1998)

The PVCN contains projection neurons of three types: octopus cells, stellate/multipolar cells, and globular bushy cells (Osen 1969; Hackney et al. 1990). Octopus cells have onset responses to tone bursts at the characteristic frequency (Rhode et al. 1983; Rouiller and Ryugo 1984), but MOC neurons have very sustained responses (Brown 2001). Thus, octopus cells are unlikely to be the interneurons (Brown et al. 2003), leaving stellate/multipolar and bushy cells as possibilities.

Stellate/multipolar neurons have projections that are consistent with a role as intermediaries of the MOC reflex. At least some stellate/multipolar neurons project to the VNTB (Smith et al. 1993; Doucet and Ryugo 2003), the nucleus that contains most of the MOC neurons. The projection is likely formed by "planar" stellate/multipolar neurons. These neurons have dendrites confined to isofrequency laminae in the cochlear nucleus (Doucet and Ryugo 2003) and may, like MOC neurons, be sharply tuned to sound frequency. Another type of stellate/multipolar neuron, the radiate neuron, is inhibitory and thus cannot provide the excitation needed to drive the MOC response to sound. Which synaptic terminals on MOC neurons might correspond to those of the MOC reflex interneurons? Stellate/multipolar terminal ultrastructure in the inferior colliculus (Oliver 1987) and locally in the cochlear nucleus (Smith et al. 1993) is consistent with the type that has small, round vesicles and a few dense core vesicles (Fig. 2.7b), but this remains to be

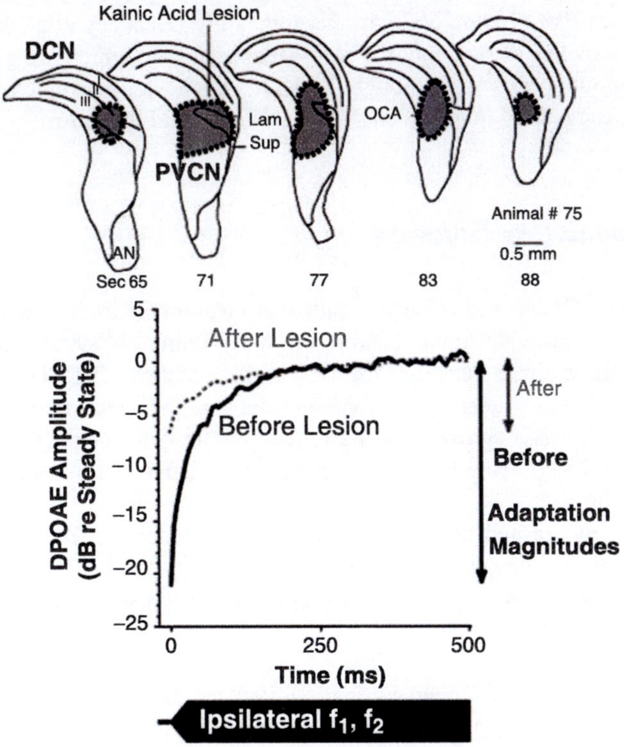

Fig. 2.9 Determination of the site of the MOC reflex interneurons in the cochlear nucleus. *Graph below*: Effect of a kainic-acid lesion on the adaptation of the DPOAE. Before the lesion, there was a large, positive-going adaptation of the DPOAE. Presumably, this takes place because the two primary tones (f_1, f_2 at bottom) activate Ipsi MOC neurons to respond to sound, and their influence on OHC alters the DPOAE. For particular frequency/level combinations of tones in which the DPOAE is composed of several components that are close to equal and opposite, a relatively small alteration of one can have a large effect on the sum. The DPOAE adaptation magnitude (*arrows at right*) was used as a metric for the MOC reflex because cuts of the OC bundle greatly reduce the adaptation (at least in the guinea pigs used here: Kujawa and Liberman 2001). After the kainic acid lesion, the adaptation was greatly reduced (*dashed line*), presumably because the reflex interneurons were lesioned. Post-experiment histology from the same experiment (*drawings above*), showing a region of complete cell loss (*gray shading*) centered in the dorsal and caudal portions of the posteroventral subdivision of the cochlear nucleus (PVCN), an area common to other reflex-interrupting lesions. In this particular lesion, there was also some involvement of the lamina (lam), the octopus cell area (oca), the superficial granular layer (sup), and the dorsal cochlear nucleus (DCN) (graph adapted from Brown et al. 2003; drawing from de Venecia et al. 2005)

directly demonstrated. Another type of cochlear nucleus neuron, the globular bushy cell, is less likely to play a direct role in the MOC reflex. Some bushy cells have projections to the VNTB (Spirou et al. 1990; Smith et al. 1991). However, the most effective reflex-interrupting lesions in the MOC reflex were in caudal PVCN locations (de Venecia et al. 2005), and the distribution of bushy cells is in restricted to more rostral PVCN and AVCN in the guinea pig (Hackney et al. 1990). Also, the waveform peaks of the auditory brain stem response, which are generated by bushy cells, could persist even though the MOC reflex was interrupted (de Venecia et al. 2005), suggesting independence of bushy cells and the reflex.

2.7.2 *Modulatory Pathways*

Both groups of OC neurons receive inputs that presumably modulate the response to sound. Serotoninergic inputs (Thompson and Thompson 1995) project to OC neurons and these inputs may arise from the midline raphe nuclei. Such neurons are labeled after cochlear injections of transneuronal labeling agents that first label OC neurons and then pass across synapses to neurons providing input to OC neurons (Horvath et al. 2003). Noradrenaline is present in varicosities surrounding OC neurons in the brain stem (Mulders and Robertson 2005); its likely source is the locus coeruleus. Serotonin and noradrenaline have been demonstrated to affect VNTB and OC neurons in slice preparations (Wang and Robertson 1997a, b).

MOC neurons receive inputs from higher centers that also presumably modulate the reflex. Descending inputs to MOC neurons arise from the inferior colliculus (Faye-Lund 1986; Thompson and Thompson 1993; Vetter et al. 1993) and the auditory cortex (Mulders and Robertson 2000a). Perhaps these higher centers mediate the trainable effects of the MOC reflex that have been documented in humans (de Boer and Thornton 2008). Such effects might be mediated by the synaptic terminals containing large, round vesicles (Fig. 2.7a), because these terminals are associated with spines. Spine-associated neural systems in the hippocampus mediate such plastic changes (reviews: Matsuzaki 2007; Bourne and Harris 2008). Working out the anatomical substrate for plasticity in the OC reflexes is a future goal.

2.8 Summary

Our knowledge of the anatomy of OC neurons has advanced, but is still incomplete. The separate innervations of the two hair cell groups by LOC and MOC neurons is clear, and it is apparent that MOC neuron action on OHCs decreases cochlear sensitivity. What separate role is played by LOC neurons is much less clear. By their limited dendritic extents and cochleotopic projections, the LOC intrinsic subgroup seems poised to operate within narrow bands of frequency information, whereas their radiating dendrites and extensive cochlear spans implies that the LOC shell subgroup operates in a much broader fashion. However, even whether the two subgroups differ in excitation vs. inhibition of the periphery has yet to be completely worked out. Our knowledge of MOC anatomy has advanced so that the basic reflex pathway is established although modulatory inputs and their roles are less well defined. The locations of MOC neurons responsive to sound in either ear is not indicated in Fig. 2.8 because they have not been established; similarly, how response characteristics of MOC neurons vary with brain stem location is not known. LOC neurons probably also participate in a sound-evoked LOC reflex, but the participating elements in cochlear nucleus and even whether it is a three-neuron or four-neuron pathway remains to be resolved. Further work on both groups of OC neurons will better establish their anatomical characteristics and roles in hearing.

Acknowledgments I thank Dr. M. Charles Liberman for comments on the manuscript and Ms. Marie Drottar for assistance with the figures. This work was Supported by NIH grant DCD 01089.

References

Arnesen AR (1984) Fibre population of the vestibulocochlear anastomosis in humans. Acta Otolaryngol 98:501–518

Arnesen AR, Osen KK (1984) Fibre spectrum of the vestibulo-cochlear anastomosis in the cat. Acta Otolaryngol 98:255–269

Aschoff A, Ostwald J (1987) Different origins of cochlear efferents in some bat species, rats, and guinea pigs. J Comp Neurol 264:56–72

Azeredo WJ, Kliment ML, Morley BJ, Relkin E, Slepecky NB, Sterns A, Warr WB, Weekly JM, Woods CI (1999) Olivocochlear neurons in the chinchilla: A retrograde fluorescent labelling study. Hear Res 134:57–70

Benson TE, Brown MC (1990) Synapses formed by olivocochlear axon branches in the mouse cochlear nucleus. J Comp Neurol 295:52–70

Benson TE, Brown MC (1996) Synapses from medial olivocochlear branches in the inferior vestibular nucleus. J Comp Neurol 372:176–188

Benson TE, Brown MC (2006) Ultrastructure of synaptic input to medial olivocochlear neurons. J Comp Neurol 499:244–257

Bishop AL, Henson OW Jr (1987) The efferent cochlear projections of the superior olivary complex in the mustached bat. Hear Res 31:175–182

Bourne JN, Harris KM (2008) Balancing structure and function at hippocampal dendritic spines. Annu Rev Neurosci 31:47–67

Brown MC (1987a) Morphology of labeled efferent fibers in the guinea pig cochlea. J Comp Neurol 260:605–618

Brown MC (1987b) Morphology of labeled afferent fibers in the guinea pig cochlea. J Comp Neurol 260:591–604

Brown MC (1989) Morphology and response properties of single olivocochlear fibers in the guinea pig. Hear Res 40:93–110

Brown MC (1993) Fiber pathways and branching patterns of biocytin-labeled olivocochlear neurons in the mouse brainstem. J Comp Neurol 337:600–613

Brown MC (2001) Response adaptation of medial olivocochlear neurons is minimal. J Neurophysiol 86:2381–2392

Brown MC (2002) Cochlear projections of single medial olivocochlear (MOC) axons in the guinea pig. ARO Absts 25:310

Brown MC, Levine JL (2008) Dendrites of medial olivocochlear (MOC) neurons in mouse. Neuroscience 154:147–159

Brown MC, Vetter DE (2008) Olivocochlear somata and central branches are normal in α9 knockout mice. J Assoc Res Otolaryngol 10:64–75

Brown MC, Liberman MC, Benson TE, Ryugo DK (1988) Brainstem branches from olivocochlear axons in cats and rodents. J Comp Neurol 278:591–603

Brown MC, de Venecia RK, Guinan JJ Jr (2003) Responses of medial olivocochlear (MOC) neurons: Specifying the central pathways of the MOC reflex. Exp Brain Res 153:491–498

Campbell JP, Henson MM (1988) Olivocochlear neurons in the brainstem of the mouse. Hear Res 35:271–274

Darrow KN, Simons EJ, Dodds L, Liberman MC (2006) Dopaminergic innervation of the mouse inner ear: Evidence for a separate cytochemical group of cochlear efferent fibers. J Comp Neurol 498:403–414

de Boer E, Thornton AR (2008) Neural correlates of perceptual learning in the auditory brainstem: efferent activity predicts and reflects improvement at a speech-in-noise discrimination task. J Neurosci 28:4929–4937

de Venecia RK, Liberman MC, Guinan JJ Jr, Brown MC (2005) Medial olivocochlear reflex interneurons are located in the posteroventral cochlear nucleus. J Comp Neurol 487: 345–360

Doucet JR, Ryugo DK (2003) Axonal pathways to the lateral superior olive labeled with biotinylated dextran amine injections in the dorsal cochlear nucleus in rats. J Comp Neurol 461:452–465

Faye-Lund H (1986) Projection from the inferior colliculus to the superior olivary complex in the albino rat. Anat Embryol (Berl) 175:35–52

Fujino K, Oertel D (2001) Cholinergic modulation of stellate cells in the mammalian ventral cochlear nucleus. J Neurosci 21:7372–7383

Guinan JJ Jr, Warr WB, Norris BE (1983) Differential olivocochlear projections from lateral vs. medial zones of the superior olivary complex. J Comp Neurol 221:358–370

Guinan JJ Jr, Warr WB, Norris BE (1984) Topographic organization of the olivocochlear projections from the lateral and medial zones of the superior olivary complex. J Comp Neurol 226:21–27

Hackney CM, Osen KK, Kolston J (1990) Anatomy of the cochlear nuclear complex of the guinea pig. Anat Embryol 182:123–149

Happe HK, Morely BJ (1998) Nicotinic acetylcholine receptors in rat cochlear nucleus: [^{125}I]-α-bungarotoxin receptor autoradiography and in situ receptor autoradiography of α7 nAChR subunit mRNA. J Comp Neurol 397:163–180

Helfert RH, Schwartz IR, Ryan AF (1988) Ultrastructural characterization of gerbil olivocochlear neurons based on differential uptake of ^{3}H-D-aspartic acid and a wheatgerm agglutinin-horseradish peroxidase conjugate from the cochlea. J Neurosci 8:3111–3123

Horvath M, Kraus KS, Illing R-B (2000) Olivocochlear neurons sending axon collaterals into the ventral cochlear nucleus of the rat. J Comp Neurol 422:95–105

Horvath M, Ribari O, Repassy G, Toth IE, Boldogkoi Z, Palkovits M (2003) Intracochlear injection of pseudorabies virus labels descending auditory and monoaminerg projections to olivocochlear cells in guinea pig. Eur J Neurosci 18:1439–1447

Iurato S, Smith CA, Eldredge DH, Henderson D, Carr C, Ueno Y, Cameron S, Richter R (1978) Distribution of the crossed olivocochlear bundle in the chinchilla's cochlea. J Comp Neurol 182:57–76

Kiang NYS, Rho JM, Northrop CC, Liberman MC, Ryugo DK (1982) Hair-cell innervation by spiral ganglion cells in adult cats. Science 217:175–177

Kujawa S, Liberman MC (2001) Effects of olivocochlear feedback on distortion product otoacoustic emissions in guinea pig. J Assoc Res Otolaryngol 2:268–278

Liberman MC (1980) Efferent synapses in the inner hair cell area of the cat cochlea: An electron microscopic study of serial sections. Hear Res 3:189–204

Liberman MC, Brown MC (1986) Physiology and anatomy of single olivocochlear neurons in the cat. Hear Res 24:17–36

Liberman MC, Guinan JJ Jr (1998) Feedback control of the auditory periphery: Anti-masking effects of middle ear muscles vs. olivocochlear efferents. J Commun Disord 31:471–483

Liberman MC, Dodds LW, Pierce S (1990) Afferent and efferent innervation of the cat cochlea: Quantitative analysis with light and electron microscopy. J Comp Neurol 301:443–460

Maison SF, Adams JC, Liberman MC (2003) Olivocochlear innervation in the mouse: Immunocytochemical maps, crossed versus uncrossed contributions, and transmitter colocalization. J Comp Neurol 455:406–416
Matsuzaki M (2007) Factors critical for the plasticity of dendritic spines and memory storage. Neurosci Rev 57:1–9
Moore JK, Osen KK (1979) The cochlear nuclei in man. Am J Anat 154:393–418
Mulders WHAM, Robertson D (2000a) Evidence for direct cortical innervation of medial olivocochlear neurones in rats. Hear Res 144:65–72
Mulders WHAM, Robertson D (2000b) Morphological relationships of peptidergic and noradrenergic nerve terminals to olivocochlear neurones in the rat. Hear Res 144:53–64
Mulders WHAM, Robertson D (2005) Catecholaminergic innervation of guinea pig superior olivary complex. J Chem Neuroanat 30:230–242
Mulders WHAM, Harvey AR, Robertson D (2007) Electrically-evoked responses in onset chopper neurons in guinea pig cochlear nucleus. J Neurophysiol 97:3288–3297
Nadol JB Jr (1988) Comparative anatomy of the cochlea and auditory nerve in mammals. Hear Res 34:253–266
Oliver DL (1987) Projections to the inferior colliculus from the anteroventral cochlear nucleus in the cat: Possible substrates for binaural interaction. J Comp Neurol 264:24–46
Osen KK (1969) Cytoarchitecture of the cochlear nuclei in the cat. J Comp Neurol 136:453–484
Osen KK, Mugnaini E, Dahl A-L, Christiansen AH (1984) Histochemical localization of acetylcholinesterase in the cochlear and superior olivary nuclei. A reappraisal with emphasis on the cochlear granule cell system. Arch Ital Biol 122:169–212
Rasmussen GL (1946) The olivary peduncle and other fiber connections of the superior olivary complex. J Comp Neurol 84:141–219
Rasmussen GL (1953) Further observations of the efferent cochlear bundle. J Comp Neurol 99:61–94
Rhode WS, Oertel D, Smith PH (1983) Physiological response properties of cells labeled intracellularly with horseradish peroxidase in cat ventral cochlear nucleus. J Comp Neurol 213:448–463
Robertson D, Gummer M (1985) Physiological and morphological characterization of efferent neurons in the guinea pig cochlea. Hear Res 20:63–77
Robertson D, Winter IM (1988) Cochlear nucleus inputs to olivocochlear neurones revealed by combined anterograde and retrograde labelling in the guinea pig. Brain Res 462:47–55
Robertson D, Cole KS, Corbett K (1987a) Quantitative estimate of bilaterally projecting medial olivocochlear neurons in the guinea pig brainstem. Hear Res 27:177–181
Robertson D, Anderson C-J, Cole KS (1987b) Segregation of efferent projections to different turns of the guinea pig cochlea. Hear Res 25:69–76
Rouiller EM, Ryugo DK (1984) Intracellular marking of physiologically characterized cells in the ventral cochlear nucleus of the cat. J Comp Neurol 225:167–186
Ryan AF, Keithley EM, Wang Z-X, Schwartz IR (1990) Collaterals from lateral and medial olivocochlear efferent neurons innervate different regions of the cochlear nucleus and adjacent brainstem. J Comp Neurol 300:572–582
Sánchez-González MA, Warr WB, López DE (2003) Anatomy of olivocochlear neurons in the hamster studied with FluoroGold. Hear Res 185:65–76
Schuknecht HF, Nomura Y (1965) The efferent fibers in the cochlea. Ann Otol Rhinol Laryngol 74:289–303
Smith PH, Joris PX, Carney LH, Yin TCT (1991) Projections of physiologically characterized globular bushy cell axons from the cochlear nucleus of the cat. J Comp Neurol 304:387–407
Smith PH, Joris PX, Banks MI, Yin TCT (1993) Responses of cochlear nucleus cells and projections of their axons. In: Merchan MA, Juiz JM, Godfrey DA, Mugnaini E (eds) The mammalian cochlear nuclei: Organization and function. Plenum Press, New York, pp 349–360
Spangler KM, White JS, Warr WB (1986) Electron microscopic features of axon terminals on olivocochlear neurons in the cat. Assoc Res Otolaryngol Abstr 9:37–38

Spirou GA, Brownell WE, Zidanic M (1990) Recordings from cat trapezoid body and HRP labeling of globular bushy cell axons. J Neurophysiol 63:1169–1190

Spoendlin H (1971) Degeneration behaviour of the cochlear nerve. Arch Klin Exp Ohren Nasen Kehlkopfheilkd 200:275–291

Stopp PE (1983) The distribution of the olivocochlear bundle and its possible role in frequency/intensity coding. In: Klinke R, Hartmann R (eds) Hearing-physiological bases and physchophysics. Springer, Berlin, pp 176–179

Thiers FA, Burgess BJ, Nadol JB Jr (2002) Axodendritic and dendrodendritic synapses within outer spiral bundles in a human. Hear Res 164:97–104

Thompson GC, Thompson AM (1986) Olivocochlear neurons in the squirrel monkey brainstem. J Comp Neurol 254:246–258

Thompson AM, Thompson GC (1991) Posteroventral cochlear nucleus projections to olivocochlear neurons. J Comp Neurol 303:267–285

Thompson AM, Thompson GC (1993) Relationship of descending inferior colliculus projections to olivocochlear neurons. J Comp Neurol 335:402–412

Thompson AM, Thompson GC (1995) Light microscopic evidence of serotoninergic projections to olivocochlear neurons in the bush baby (*Otolemur garnettii*). Brain Res 695:263–266

Uchizono K (1965) Characteristics of excitatory and inhibitory synapses in the central nervous system of the cat. Nature (Lond) 207:642–643

Vetter DE, Mugnaini E (1992) Distribution and dendritic features of three groups of rat olivocochlear neurons. A study with two retrograde cholera toxin tracers. Anat Embryol (Berl) 185:1–16

Vetter DE, Adams JC, Mugnaini E (1991) Chemically distinct rat olivocochlear neurons. Synapse 7:21–43

Vetter DE, Saldana E, Mugnaini E (1993) Input from the inferior colliculus to medial olivocochlear neurons in the rat: A double label study with PHA-L and cholera toxin. Hear Res 70:173–186

Vetter DE, Liberman MC, Mann J, Barhanin J, Boulter J, Brown MC, Saffiote-Kolman J, Heinemann SF, Elgoyhen AB (1999) Role of a9 nicotinic ACh receptor subunits in the development and function of cochlear efferent innervation. Neuron 23:93–103

Wang X, Robertson D (1997a) Effects of bioamines and peptides on neurones in the ventral nucleus of the trapezoid body and rostral periolivary regions of the rat superior olivary complex: An in vitro investigation. Hear Res 106:20–28

Wang X, Robertson D (1997b) Two types of actions of noradrenaline on identified auditory efferent neurons in rat brainstem slices. J Neurophysiol 78:1800–1810

Warr WB (1975) Olivocochlear and vestibular efferent neurons of the feline brainstem: Their location, morphology, and number determined by retrograde axonal transport and acetylcholinesterase histochemistry. J Comp Neurol 161:159–182

Warr WB (1992) Organization of olivocochlear efferent systems in mammals. In: Webster DB, Popper AN, Fay RR (eds) The mammalian auditory pathway: Neuroanatomy. Springer, New York, pp 410–448

Warr WB, Guinan JJ Jr (1979) Efferent innervation of the organ of Corti: Two separate systems. Brain Res 173:152–155

Warr WB, Beck Boche JE, Neely ST (1997) Efferent innervation of the inner hair cell region: Origins and terminations of two lateral olivocochlear systems. Hear Res 108:89–111

Warr WB, Boche JEB, Ye Y, Kim DO (2002) Organization of olivocochlear neurons in the cat studied with the retrograde tracer cholera toxin-B. J Assoc Res Otolaryngol 3:457–478

Warren EH III, Liberman MC (1989) Effects of contralateral sound on auditory-nerve responses I. Contributions of cochlear efferents. Hear Res 37:89–104

White JS (1984) Fine structure of medial olivocochlear neurons in the rat. Soc Neurosci Abstr 10:393

White JS (1986) Differences in the ultrastructure of labyrinthine efferent neurons in the albino rat. ARO Abstr 9:34–35

Winter IM, Robertson D, Cole KS (1989) Descending projections from auditory brainstem nuclei to the cochlea and cochlear nucleus of the guinea pig. J Comp Neurol 280:143–157
Yao W, Godfrey DA (1999) Immunolocalization of alpha4 and alpha7 subunits of nicotinic receptor in rat cochlear nucleus. Hear Res 128:97–102
Ye Y, Machado DG, Kim DO (2000) Projection of the marginal shell of the anteroventral cochlear nucleus to olivocochlear neurons in the cat. J Comp Neurol 420:127–138

Chapter 3
Physiology of the Medial and Lateral Olivocochlear Systems

John J. Guinan

3.1 Introduction

In this chapter we deal with the ways in which the two olivocochlear (OC) efferent systems, the medial (MOC) and lateral (LOC) systems, change the operation of the cochlea and how these changes may benefit hearing. To understand these changes, it is necessary to understand OC anatomy. OC anatomy is dealt with extensively in Brown (Chap. 2). Here we present the anatomy necessary for understanding OC physiology and function.

An important anatomical difference between MOC and LOC fibers is that MOC fibers are myelinated and can be recorded from and electrically stimulated, whereas LOC fibers are unmyelinated and have not been recorded from or directly stimulated. In addition, MOC fibers can be readily excited by acoustic stimulation, but it is not well established whether LOC fibers are acoustically excited, or not. The result is that we know a lot about the MOC system and very little about the LOC system.

Our focus is on OC function at the systems level of cochlear operation, covering important topics studied primarily since 1996. For a more elaborated review of older work, see the Guinan (1996) SHAR chapter. Sewell (Chap. 4) and Katz et al. (Chap. 5) present more detailed aspects of OC pharmacology, neurochemistry and the mechanisms of cholinergic inhibition. We will present enough of these topics for the reader to understand the material of this chapter.

The most important features of OC anatomy are shown in Fig. 3.1 (Smith 1961; Kimura and Wersäll 1962; Warr and Guinan 1979; Liberman 1980). Medial olivocochlear (MOC) fibers synapse on outer hair cells (OHCs), whereas lateral olivocochlear

J.J. Guinan (✉)
Eaton Peabody Laboratory, Massachusetts Eye and Ear Infirmary, Boston, MA 02114, USA
Department of Otology and Laryngology, Harvard Medical School, Boston, MA, USA
and
Speech and Hearing Bioscience and Technology Program, Harvard-MIT Division of Health Sciences and Technology, Boston, MA, USA
e-mail: jjg@epl.meei.harvard.edu

D.K. Ryugo et al. (eds.), *Auditory and Vestibular Efferents*,
Springer Handbook of Auditory Research 38, DOI 10.1007/978-1-4419-7070-1_3,

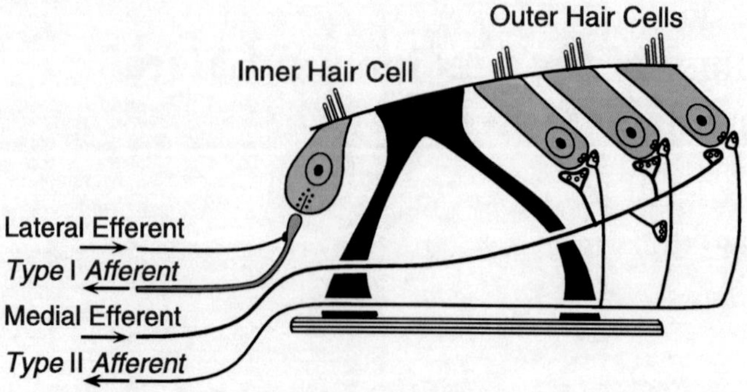

Fig. 3.1 Afferent and efferent innervation of the cochlea. A schematic of the organ of Corti showing (1) radial, type I, afferent auditory-nerve (AN) fibers that innervate inner hair cells (IHCs), (2) lateral olivocochlear (LOC) efferent fibers that synapse on these radial afferent fibers, (3) medial olivocochlear (MOC) efferent innervation of outer hair cells (OHCs), and (4) spiral (spiraling not shown), type II afferent AN fibers that form reciprocal synapses on OHCs and receive synapses from MOC fibers. Omitted are synapses between LOC and MOC efferents in the tunnel of Corti, and MOC synapses onto supporting cells

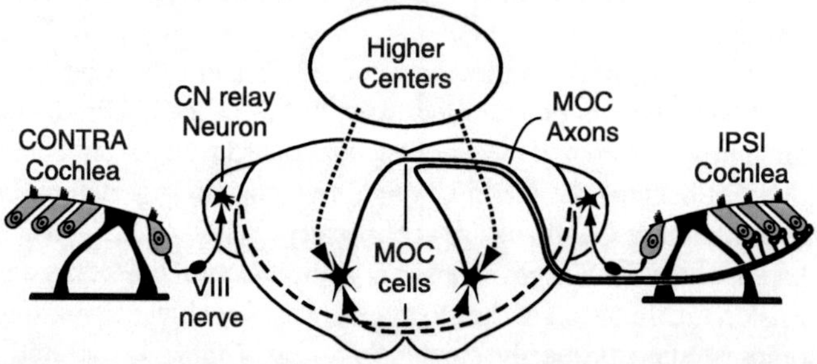

Fig. 3.2 MOC acoustic-reflex pathways to the right (ipsilateral) cochlea. Schematic transverse brain stem section and cochlear cross sections showing the three-neuron contralateral and ipsilateral MOC acoustic reflexes: (1) Auditory nerve (VIIIth nerve) fibers (*solid lines*) to the cochlear nucleus (CN), (2) CN relay neurons with their crossed projections (*dashed lines*) to MOC cell bodies, and (3) MOC neuron axons to the cochlea (*solid lines*). Also shown are descending inputs to MOC cells (*dotted lines*)

(LOC) fibers synapse on the dendrites of the type I afferent fibers, the fibers that make up the bulk of the auditory nerve (AN). Type II afferents send unmyelinated fibers to the brain and have much thicker processes that innervate OHCs with reciprocal synapses (i.e., the synapses go in both directions) (Thiers et al. 2002, 2008). These type II fibers spiral along the cochlea and receive innervation from MOC fibers at several places along their route (only one is shown in Fig. 3.1). The type II afferents may provide an additional way for MOC fibers to affect OHCs.

The main brain stem pathways of the MOC acoustic reflexes are shown in Fig. 3.2 (Guinan et al. 1983; Thompson and Thompson 1991; de Venecia et al. 2005). Inner hair cells (IHCs) sense cochlear mechanical movements and excite AN fibers. The AN fibers innervate neurons in the cochlear nucleus (CN). Reflex interneurons in the posteroventral cochlear nucleus send axons across the midline to innervate MOC neurons. These MOC neurons then send crossed or uncrossed axons that innervate OHCs. Note that the reflex signal crosses in the axons of the CN interneurons so that the contralateral MOC reflex (the MOC elicitor sound is contralateral to the side of the measurement) is mediated by the uncrossed MOC axons. Similarly, the ipsilateral reflex is mediated by the crossed MOC axons (it is a double-crossed reflex).

In most laboratory animals there are twice as many crossed as uncrossed MOC fibers (review: Warr 1992) which is consistent with the ipsilateral reflex being twice as strong as the contralateral reflex (see later). In the squirrel monkey, there are 1.5 times as many crossed as uncrossed MOC fibers. In humans the ratio is unknown. Anatomical data indicate that the LOC projection to the cochlea is tonotopic but anatomical data are inconclusive for the MOC reflex (Guinan et al. 1984; Robertson et al. 1987). However, single-fiber labeling, which combines anatomical and physiological data, shows that the MOC reflex projection to the cochlea is also tonotopic.

3.2 MOC Effects in the Cochlea: Overview

MOC activation produces effects in the cochlea on two time scales, fast (~100 ms) and slow (10s of seconds) (Sridhar et al. 1995; Cooper and Guinan 2003). The most common MOC effects are fast effects. Classic fast effects are due to turning down the gain of the cochlear amplifier, and this has different consequences depending on whether there is a background noise, or not. These classic fast effects are reviewed in Sects. 3.3 and 3.4. There are several MOC effects that are not due simply to turning down the gain of a cochlear amplified traveling wave. These effects appear to involve other vibrational motions of the organ of Corti in addition to motion that strictly follows basilar membrane (BM) motion. These are termed nonclassic MOC effects and are reviewed in Sect. 3.5. The MOC slow effect is reviewed in Sect. 3.6. The focus then changes, and Sects. 3.7 and 3.8 review MOC fiber responses to sound and MOC acoustic reflexes, followed in Sect. 3.9 by a discussion of MOC function in hearing. LOC physiology is reviewed in Sect. 3.10. This is short because we know little about LOC physiology. Finally, Sect. 3.11 summarizes the highlights of OC efferent physiology and makes suggestions for future research directions. But first, the next part of the present section provides some background and reviews two minor fast MOC electrical effects, an increase in cochlear microphonic (CM) and a decrease in endocochlear potential (EP).

3.2.1 MOC Activation Increases CM

The electrical effects produced by MOC efferents originate in the MOC synapses on OHCs. The neurochemistry and cellular physiology of these synapses are reviewed in detail in Sewell (Chap. 4) and Katz et al. (Chap. 5). Briefly stated, acetylcholine (ACh) released by the MOC presynaptic terminal acts on an unusual kind of ACh receptor that has a high calcium conductance (Elgoyhen et al. 1994, 2001). Activation of the ACh-receptors allows calcium (Ca^{2+}) ions to flow into the OHC and these Ca^{2+} ions turn on nearby Ca^{2+}-activated potassium (K^+) channels (Housley and Ashmore 1991; Fuchs 1996). The resulting outflow of K^+ ions overwhelms the smaller inflow of Ca^{2+} so that the net effect is a hyperpolarization of the OHC.

MOC stimulation *increases* CM (Fex 1959). This effect is produced primarily by the increase in OHC basolateral conductance brought about by the opening of both the K^+ and the ACh channels. This increase in OHC synaptic conductance increases the receptor current through OHC stereocilia by increasing the conductance of the current return path. The receptor current is also increased slightly by the OHC hyperpolarization which slightly increases the voltage that drives current through the OHC stereocilia (this voltage is the difference between the ~+100 mV EP and the ~–60 mV OHC intracellular potential). The OHC receptor current flowing through the resistance of the surrounding tissue produces the CM voltage, and the MOC-induced increase in the receptor current increases this CM. Although MOC activation reduces basilar-membrane (BM) motion and the resulting receptor currents near the best-frequency place of a tone, this region contributes little to the externally measured CM because the phase of BM motion, and the resulting current flow, changes by more than 360° across this region and mostly cancels out when measured at a distant electrode. A distant electrode records primarily CM from the in-phase current sources basal to the peak of the traveling wave. Since there is no cochlear amplification in this region to be decreased, the MOC effect in this region is to increase the receptor current resulting in an increased CM seen from a distant electrode.

3.2.2 MOC Activation Decreases EP and Has Other Related Effects

The MOC-induced increase in OHC receptor current produces a small (a few millivolts) decrease in EP (Fex 1959; Gifford and Guinan 1987). The EP source in stria vascularis, like any battery, has an associated resistance, and the reduction in EP is due to a voltage drop in this resistance produced by the receptor current. The MOC-induced increase in receptor current increases the voltage drop resulting in a decreased EP.

The decrease in EP produces several other effects. A large part of the driving voltage at both IHC and OHC stereocilia comes from EP. In the OHCs, the MOC reduction in EP contributes to the reduction in cochlear amplification. In the IHC, the MOC reduction in EP produces a reduction in IHC receptor potentials and reduces AN responses. This may account for part of the MOC-induced reduction of AN responses at high sound levels where there is little cochlear amplification (Guinan and Stankovic 1996). The MOC reduction in EP also lowers the resting IHC potential and thereby reduces AN spontaneous activity (Guinan and Gifford 1988b). More detail and a circuit diagram illustrating these concepts are given in the Guinan (1996) SHAR chapter.

3.3 Classic MOC Fast Effects in a Silent Background

Before we consider MOC effects on cochlear amplification we must understand the basics of cochlear amplification. "Cochlear amplification" is the name given to the process by which OHCs increase the amplitude of BM responses to sound. Cochlear amplification comes about by a process in which BM motion bends OHC stereocilia thereby causing OHC motion that feeds energy back into BM motion. The gain of cochlear amplification comes from the interplay of all parts of this cycle, but two aspects are particularly important and reasonably well understood. The gain of forward transduction (stereocilia motion to receptor current) is set by the *slope* of the OHC-receptor-current vs. stereocilia-angle curve and is reduced during two-tone suppression (e.g., Geisler et al. 1990; Geisler 1992). The gain of backward transduction (OHC voltage to OHC motion) is set by the characteristics of the protein prestin, which produces OHC somatic motility by voltage-controlled changes in molecular conformation that cause OHC elongation and contraction (Santos-Sacchi 1991; Dallos et al. 2008). The least well understood part of cochlear amplification is the micromechanical motions involved, both the coupling of BM motion to the bending of OHC stereocilia (including tilting of the reticular lamina – Nowotny and Gummer (2006), and possible traveling waves in the tectorial membrane – Ghaffari et al. (2007), and the coupling of OHC elongation back into BM motion (see Cooper and Kemp 2009). Another poorly understood area is how fluctuations in OHC receptor current produce an adequate change in OHC voltage at frequencies far above the OHC membrane low-pass frequency of ~1 kHz (see Lu et al. 2006 for one answer). Also poorly understood is the role of OHC stereocilia motility in mammalian cochlear amplification; its lack of a clear role is surprising considering that it is the main source of cochlear amplification in non-mammalian vertebrates (Hudspeth 2008).

Cochlear amplifier gain is changed during the MOC fast effect by two mechanisms: shunting and hyperpolarization. First, the MOC-induced increase in OHC synaptic conductance shunts the OHC receptor current thereby producing smaller changes in OHC voltage. The effect of OHC shunting should be less at frequencies above the OHC membrane low-pass frequency where the OHC capacitance dominates the OHC impedance (Guinan 1997). Second, the MOC-induced hyperpolarization of OHCs moves the operating point of reverse transduction away from its optimum point which reduces the OHC motion produced by a given OHC voltage change (Santos-Sacchi 1991). This effect should be independent of sound frequency. Finally, MOC fibers also synapse on type II afferent fibers and may indirectly exert effects on OHCs and cochlear amplification through the type II reciprocal synapses on OHCs (see Fig. 3.1).

3.3.1 Classic MOC Fast Effects on Basilar-Membrane Motion

MOC activation turns down the gain of cochlear amplification, and since amplification is greatest at low sound levels and at the best frequency, MOC activation has the largest effect at low sound levels and at the best frequency. In BM level functions, MOC activation shifts the response curve to higher levels and this shift is largest at low sound levels (Fig. 3.3a) (Murugasu and Russell 1996; Dolan et al. 1997; Cooper and Guinan 2003, 2006a; but see Russell and Murugasu 1997 with Ren and Nuttall 2001). Since MOC activity turns down the gain, to get the same BM response the sound level must be turned up. The shift to higher sound levels (called the "level shift") is a measure of how much cochlear amplification has been reduced. The level shift is greatest at the tip of the BM tuning curve (TC) and is near zero at much lower frequencies (Fig. 3.3b). This is consistent with the pattern of cochlear amplification deduced from measurements of BM motion in sensitive vs. damaged preparations (Robles and Ruggero 2001). Since MOC stimulation produces larger shifts near the best frequency (BF) than at the edges of the TC tip, the TC width is increased by MOC stimulation (Fig. 3.3b) (Murugasu and Russell 1996; Cooper and Guinan 2006a). MOC stimulation also produce phase leads in BM motion in the low-level tip portion of the response (Guinan and Cooper 2003).

MOC effects on BM responses to clicks follow what would be expected from the MOC effects on BM responses to tones (Guinan and Cooper 2008). MOC inhibition, as a percentage of the response, is strongest at low click levels and becomes very small at high click levels (Fig. 3.4a). There is no MOC inhibition on the first half cycle of the BM response at any level, consistent with this initial response being passive. After the first half cycle, inhibition builds up gradually, and ultimately increases the decay rate of the BM click response. BM click responses in sensitive preparations show prominent waxing and waning, presumably from the interaction of two resonances (Recio et al. 1998; Guinan and Cooper 2008). MOC stimulation had little effect on the waxing and waning of the responses or response instantaneous frequency. MOC stimulation also produced small phase leads in the response waveforms.

3.3.2 *Classic MOC Fast Effects on Otoacoustic Emissions*

Otoacoustic emissions (OAEs) are sounds that originate from within the cochlea and travel backward through the middle ear into the ear canal. They are useful because they can be measured noninvasively and provide a window into the mechanical response of the cochlea. They have been most valuable when used in humans, and here they are paired with noninvasive eliciting of MOC activity by sound. Most of the new things learned using OAEs will be considered in Sect. 3.8.2 (the MOC acoustic reflexes). The present section considers how well MOC effects on OAEs fit with what has been learned from more direct measurement techniques, and how well the various OAE types serve as metrics for MOC effects.

For a detailed presentation of the three types of evoked OAEs: transient evoked, stimulus frequency, and distortion product (TEOAEs, SFOAEs, and DPOAEs), see Shera and Guinan (2007). At the sound levels usually used, TEOAEs and SFOAEs are due to coherent reflection, whereas DPOAEs are a combination of components

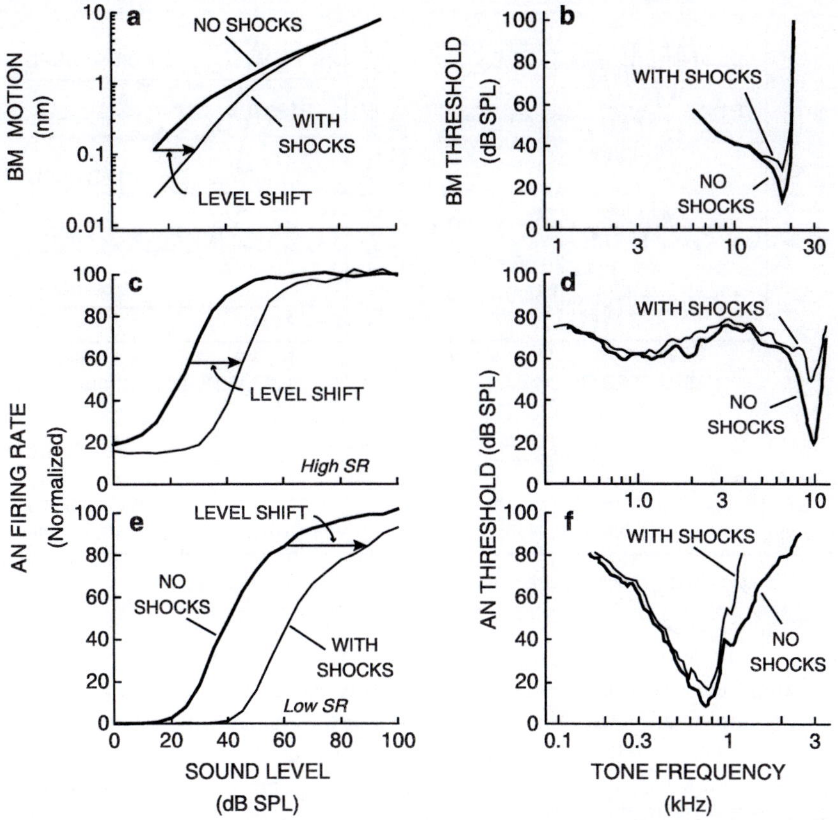

Fig. 3.3 Fast MOC effects on basilar membrane (BM) motion and auditory nerve (AN) firing. *Left*: BM motion (**a**) and AN firing rate (**c**, **e**) vs. sound level. *Right*: Tuning curves for BM motion (**b**) and AN firing (**d**, **f**) vs. tone frequency. *Thin* and *thick lines* are with and without MOC activity elicited by brain stem shocks. ((**a**, **b**) Adapted with permission from Cooper and Guinan 2006a, (**c**, **e**) Adapted with permission from Guinan and Stankovic 1996, (**d**) Adapted with permission from Guinan and Gifford 1988c, and (**f**) from unpublished data of Guinan and Gifford)

from a distortion source and a coherent reflection source[1] (Shera and Guinan 1999, 2007). The two DPOAE sources have different phase properties and they interfere in the ear canal, that is, they can be in-phase and add, or out-of-phase and cancel.

The principal on which OAE measurements of MOC effects are based is that OAEs get most of their energy from cochlear amplification and MOC activity turns down the gain of the cochlear amplifier thereby lowering OAE amplitudes. MOC effects are obtained by first measuring the OAE without MOC activation and then with MOC activation; the difference is the MOC effect. MOC effects on OAEs are

[1]In small animals such as guinea pigs, coherent reflection emissions are relatively weak (Zurek 1985) with the result that DPOAEs are primarily from the distortion component (see Fahey et al. 2008).

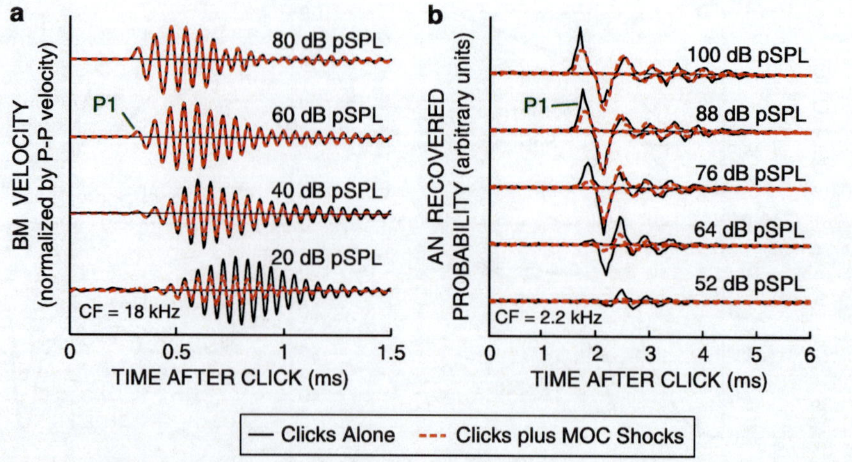

Fig. 3.4 MOC effects on BM motion (**a**) and AN firing (**b**) in response to clicks. (**b**) Compound histograms with the recovered-probability post-stimulus-time histogram from rarefaction plotted upwards and from condensation plotted downward. *Black solid line*=clicks alone; *red dashed line*=clicks plus MOC shocks (adapted with permission from Guinan et al. 2006)

usually expressed in a normalized form as the decibel change from the original OAE amplitude. This dB change is more closely related to the decibel change in cochlear amplification than the absolute value of the change which varies with the OAE amplitude as well as with changes in cochlear amplification. An even better measure is the MOC-induced level shift. The relationship between MOC-induced changes in OAEs and in cochlear output (i.e., in AN responses), is unknown. In a study using DPOAEs, the relationship varied widely from nearly equal changes to much smaller OAE changes than N_1 changes (Puria et al. 1996).

Since TEOAEs and SFOAEs originate from a single mechanism, linear coherent reflection, they show a simple pattern of MOC effects. The MOC effect on these emissions is almost always a reduction of their amplitude (e.g., Guinan 1990; Collet et al. 1990; Veuillet et al. 1991, 1996; Ryan and Kemp 1996; Guinan et al. 2003; Backus and Guinan 2006). In contrast, since the two DPOAE sources interfere, the MOC effect on DPOAEs is very complex and can even be an increased DPOAE (Siegel and Kim 1982; Moulin et al. 1993; Muller et al. 2005; Wagner et al. 2007). If the two DPOAE components normally cancel, and MOC stimulation inhibits one component more than the other, this inhibition reduces the cancelation and increases the DPOAE. Thus, the relative phases of the two DPOAE components greatly influences the DPOAE change measured. Since this phase relationship is unrelated to the MOC effect but strongly influences the result, the MOC change in DPOAE amplitude is not an accurate way of measuring MOC effects. Measuring MOC effects at response dips makes the value obtained larger, but, in humans, the result depends on the phase relationship more than on MOC strength.

A different kind of OAE paradigm is "DPOAE adaptation" which uses the DPOAE time course to measure the MOC effect (Liberman et al. 1996). The DPOAE primary tones are turned on abruptly and evoke MOC activity that builds up with a time constant of ~100 ms. The difference between the DPOAE just after the onset (which is not yet affected by MOC activity) and the DPOAE after a few hundred ms (which is affected by the MOC activity elicited by the primary tones) provides a metric of the MOC effect. This method measures effects of the ipsilateral MOC reflex, which is an advantage. A disadvantage is that the primary tones also produce a slower DPOAE change (time constant ~1 s) that remains after all efferents are cut and therefore is due to effects intrinsic to the cochlea (these intrinsic effects might be due to a build up of potassium around the OHC or an effect of the afferent type II neural network). The DPOAE adaptation technique has been applied in humans (e.g., Kim et al. 2001; Bassim et al. 2003; Muller et al. 2005) but, in humans, the results cannot be reliably interpreted because efferents cannot be cut and MOC effects cannot be separated from cochlear intrinsic effects.

In summary, MOC effects on TEOAEs and SFOAEs can be interpreted in a straightforward manner, but DPOAE measurements are complex because they originate from two separate cochlear mechanisms and places. DPOAE measurements of MOC effects can be made more accurate by separating the DPOAE into its two source components (e.g., Thompson et al. 2009; Abdala et al. 2009). Overall, OAEs provide imperfect measures of MOC effects on cochlear mechanical changes, but have the great advantage of being noninvasive.

3.3.3 Classic MOC Fast Effects on IHC and AN Responses

MOC effects were first observed on AN compound action potential (CAP) responses (N_1) to clicks. In N_1 responses evoked by clicks or tone-pips in a silent background, there are MOC level shifts as large as 20–30 dB at low sound levels (Galambos 1956; Desmedt 1962; Wiederhold and Peake 1966; Gifford and Guinan 1987). These N_1 level shifts decrease to near zero at high sound levels, consistent with the MOC effect being to turn down the gain of the cochlear amplifier.

The dependence of MOC inhibition on MOC firing rate has been determined from MOC effects on AN N_1 responses. The greatest inhibition of N_1 is produced by shock rates of 200–400/s with little change in inhibition across this range. Inhibition diminishes rapidly as shock rates are lowered below 200/s, for example, at 60/s shocks the inhibition is about ¼ of the maximum value (Gifford and Guinan 1987). Existing data suggest that all MOC effects have a similar dependence on shock rate (Desmedt 1962; Konishi and Slepian 1971; Gifford and Guinan 1987; Rajan 1988). MOC fibers follow MOC shocks one-for-one for shock rates up to rates of 200/s, but not at 400/s (McCue and Guinan, unpublished). Thus, the effects shown by shocks at rates of 200/s, or less, indicate the effects that would be produced by sound-evoked MOC activity at the same rate. The effects produced by trains with different numbers shocks indicate that the MOC synapse shows strong, time-dependent facilitation, and that this facilitation is the origin of the dependence of MOC effects on shock rate (Cooper and Guinan 2006b).

For low-level sounds, MOC stimulation shifts sound level functions of IHC receptor potential and AN-fiber firing rate toward higher sound levels (Fig. 3.3c) and this shift is greatest for tones at the characteristic frequency (CF) (Fig. 3.3d), similar to the MOC effect on BM motion (Wiederhold 1970; Teas et al. 1972; Guinan and Gifford 1988a; Guinan and Stankovic 1996). The pattern of MOC-induced AN rate shifts across fiber CFs closely matches the pattern of MOC innervation of OHCs along the cochlea (Guinan and Gifford 1988c; Liberman et al. 1990; Maison et al. 2003). The characteristics of these cochlear responses to low level sound are readily explained by the classic MOC effect of turning down the gain of the cochlear amplifier.

MOC inhibition makes TCs wider for fibers with CFs >3 kHz and for most fibers with lower CFs. Again, this is what is expected from the pattern of cochlear amplification relative to CF and efferents turning down the gain of the cochlear amplifier by a fixed ratio at each frequency. However, in some low CF fibers, MOC stimulation narrows the TC (e.g., Fig. 3.3f); this is not a classic MOC effect and will be dealt with in Sect. 3.5.

3.4 Classic MOC Fast Effects in a Noisy Background

In the presence of a low-level background noise, MOC stimulation can increase the neural response to a brief sound (Winslow and Sachs 1987; Kawase et al. 1993). The mechanisms for this are illustrated in Fig. 3.5. The top panels show the MOC inhibition of AN fiber responses to short tone bursts at CF without a background noise, which is a shift of the rate-vs.-level function to higher levels. The bottom panels show responses to the same stimuli with an added background noise. The AN response with a background noise but no MOC activation (dashed line in Fig. 3.5c) shows an increased firing rate at very low tone-burst levels from excitation by the background noise. Because the noise is continuous, the increased AN firing is continuous. This continuous firing causes AN fiber adaptation, principally by using up vesicles at the IHC-AN synapse. The resulting vesicle depletion has the effect of lowering the AN rate to high-level tone bursts because there are fewer vesicles to release. The result of both effects is to reduce the dynamic range of the AN output, that is, the noise partially masks the response to the tone. MOC stimulation reduces cochlear amplifier gain, which reduces the response to the low-level background noise (Fig. 3.5d, bottom left). The noise then causes less adaptation so high-level tones can evoke higher firing rates. The net effect is to partially restore the output dynamic range of the auditory fiber (called MOC unmasking). Although there is little change in the threshold produced by the MOC activity, the increase in output dynamic range means that small changes in the tone are more robustly signaled to the central nervous system (CNS), which increases the discriminability of the tone in the noise. This increase in the in the discriminability of the tone, or of other brief signals such as the transitions in speech, is probably the most important function of MOC efferents in everyday hearing (see Sect. 3.9).

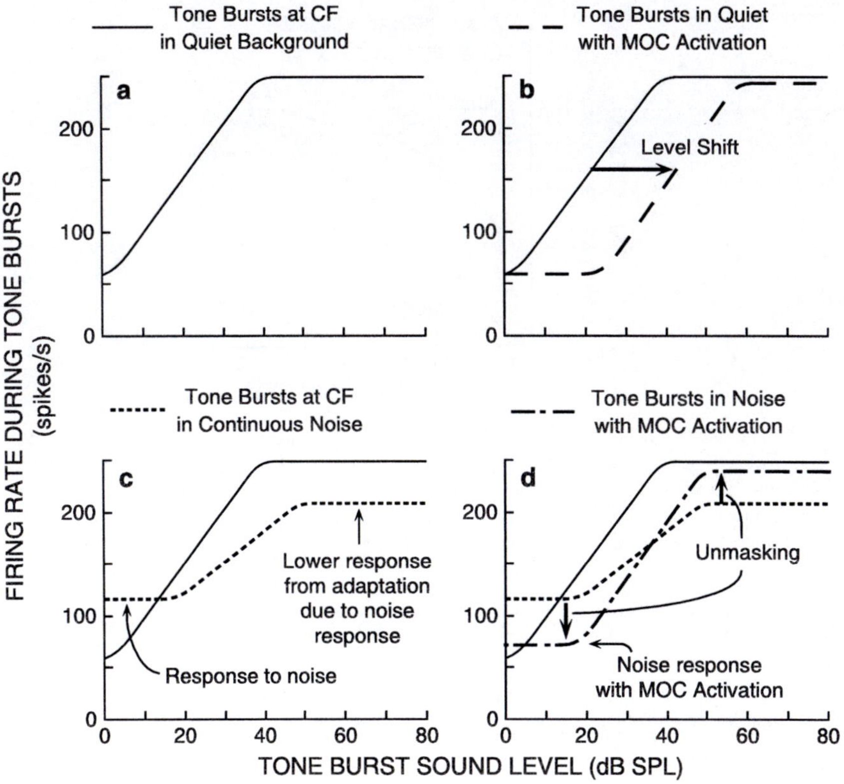

Fig. 3.5 Schematic of MOC unmasking of AN fiber responses. Each *panel* shows rate-vs.-level functions for a single AN fiber with different conditions coded by line styles (key at the *top* of *each panel*). Schematized from the results of Wiederhold (1970); Winslow and Sachs (1987); Guinan and Gifford (1988a); and Kawase et al. (1993) (adapted with permission from Guinan 1996)

Another circumstance in which MOC inhibition of the cochlear amplifier has been suggested to affect psychophysical performance is the signal-in-noise "temporal effect," also called "overshoot" (e.g., Zwicker 1965; Strickland and Krishnan 2005; Strickland 2008). The temporal effect is the phenomenon that a higher level of noise is needed to mask a brief tone when the tone is presented long after (>100 ms) the noise onset compared to just after the noise onset. One hypothesis is that the noise elicits efferent activity that turns down the gain of the cochlear amplifier thereby decreasing the response to the noise more than the response to the tone (Fig. 3.6). Since it takes 100 ms, or more, for MOC activity to reduce the gain, the S/N of the cochlear response is increased only for tones presented >100 ms after the onset of the noise. With SFOAEs as a monitor, the hypothesized increase in MOC activity was not seen even though the parameters used produced a substantial temporal effect (Keefe et al. 2009). However, the temporal effect shows a complicated dependence on the parameters of tone frequency and the noise spectrum

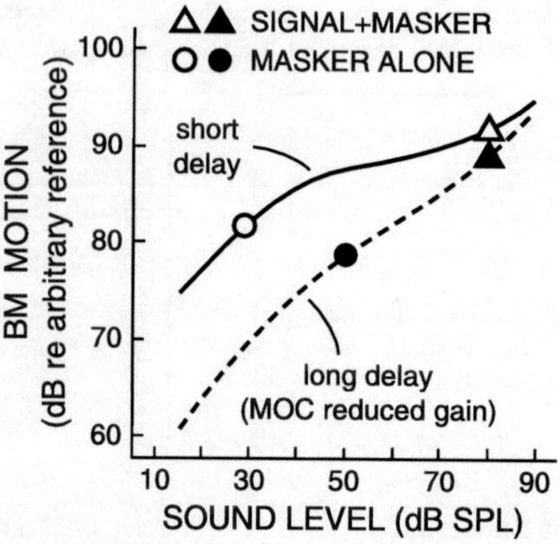

Fig. 3.6 The MOC-induced change in BM input–output function in the tone-detection-in-noise "temporal effect" (also called "overshoot"). The BM input–output function shows a large shift at low levels due to the noise-elicited MOC activity turning down the gain of cochlear amplification. The lowered gain means that a long-delayed tone in noise requires more noise to be just masked than a tone just after the noise onset (adapted with permission from Strickland and Krishnan 2005)

relative to the tone. It may be that the temporal effect is due both to MOC effects and to other, perhaps central, effects, depending on the parameters used. Overall, it seems likely that both Figs. 3.5 and 3.6 illustrate ways in which MOC inhibition can increase the ability to hear brief sounds in noise.

3.5 Nonclassic MOC Fast Effects in a Silent Background

Nonclassic MOC effects are effects that cannot be explained by the classic view that: (1) organ of Corti motion is a single vibration pattern that directly follows the BM motion of the traveling wave, (2) this motion is amplified by the cochlear amplifier, and (3) MOC effects are produced by turning down the gain of this amplifier. The part of this view that has the greatest need for revision is that organ of Corti motion is a single vibration pattern. Measurements in excised preparations show complex vibrational patterns of the organ of Corti consistent with the motion being the sum of motions from multiple vibrational modes (Mountain 1998). Further, some of these vibrational modes could lead to bending of IHC or OHC stereocilia without there being a direct coupling to BM motion (Nowotny and Gummer 2006; Karavitaki and Mountain 2007a, b; Ghaffari et al. 2007). It has not been possible to make micromechanical measurements in intact mammalian cochleas with demonstrated normal sensitivity. However, AN recording from intact cochleas with demonstrated normal sensitivity provide ample evidence that the motion that drives IHC stereocilia and leads to AN responses is due to multiple vibrational modes (e.g., Gifford and Guinan 1983; Liberman and Kiang 1984; Lin and Guinan 2000, 2004; Guinan et al. 2005).

MOC effects on BM motion have only been measured in the basal half of the cochlea (review: Cooper and Guinan 2006a). In fact, reliable measurements of mechanical responses of any kind in live preparations with thresholds shown to be normal (by tone-pip CAPs) are only from the basal half of the cochlea (Robles and Ruggero 2001). Measurements of cochlear motions have been made in the apex, but these have been without a good monitor of the preparation's sensitivity in the frequency region tested. Furthermore, many apical measurements are also contaminated by artifactual motion from the cochlear fast wave (Cooper and Rhode 1996). The result is that there is little direct knowledge of cochlear motions in the apical half of the cochlea. Motion in the apical half of the cochlea is often thought to be similar to motion in the base because of apex-to-base similarities in cochlear anatomy and in many aspects of AN responses. However, mechanical measurements in excised preparations show qualitative differences in the apex compared to the base (Nowotny and Gummer 2006) and there are many apex-to-base differences in AN response patterns. Because of this, nonclassic MOC effects in the base and apex are considered separately.

3.5.1 Nonclassic MOC Fast Effects in the Basal Half of the Cochlea

In the basal turn of guinea pigs, MOC stimulation produces an *increase* in BM motion in response to high-level tones at frequencies well above the local best frequency (Dolan et al. 1997; Guinan and Cooper 2003). In plots of BM motion vs. sound level, there is typically a dip in BM motion at these frequencies. Below the dip MOC stimulation decreases the BM response, above the dip MOC stimulation increases the response, and at the dip there is a phase change close to a reversal (Guinan and Cooper 2003). A hypothesis that fits the data is that BM motion is due to a cochlear-amplified component that is large at low sound levels and saturates at high levels, plus a passive component that grows linearly and is out-of-phase with the amplified component. When the two components are equal in amplitude they cancel. MOC stimulation inhibits the amplified component which reduces the cancelation so that the resulting BM motion increases (Guinan and Cooper 2003). The origin of the unamplified component and its vibration pattern in the organ of Corti is not known, but one possibility is that this motion is a direct mechanical response to the fast pressure wave (Rhode 2007). It is unknown whether there is a similar effect in AN responses in the cochlear base because there are no suitable measurements of MOC effects for tones at high levels and frequencies above CF. However, this effect has many similarities to the MOC effects in AN fibers with CFs near 1 kHz (Gifford and Guinan 1983) (see Sect. 3.5.2).

A second nonclassic MOC effect is that AN fibers with low spontaneous rates (SRs) have rate vs. level functions that show greater level shifts at moderate-to-high sound levels than at low sound levels (Fig. 3.3e) (Guinan and Stankovic 1996). At low sound levels, the level shift appears to be due to MOC activity turning down the gain of cochlear amplification. However, at moderate to high sound levels the level shifts are too large to be fully accounted for by a MOC reduction of cochlear amplifier gain. Some additional reduction in AN firing rate must come from the MOC-induced reduction of EP (Sect. 3.2.2), but it is not clear that this is enough to account for the large level shifts observed. Another possibility is that AN drive at high levels is a combination of two out-of-phase components (as suggested in the previous paragraph), and that the more linear component approaches the cochlear-amplified component but never becomes larger than it (in the motion of IHC stereocilia). With this scenario, at high levels these two components would partially cancel and MOC inhibition, by reducing the larger component, would increase the cancelation. This would make the resulting level shift greater than the reduction of cochlear amplification. A similar mechanism may explain the large two-tone suppression observed in low-SR AN fibers (Cai and Geisler 1996).

Another nonclassic MOC effect is that in cat AN fibers with high CFs (>10 kHz), MOC stimulation inhibits the response at frequencies much lower than the TC tip (called "tail" frequencies; Fig. 3.3d) (Stankovic and Guinan 1999). BM measurements indicate that there is no cochlear amplification at tail frequencies and there is no comparable MOC inhibition of BM motion at similar tail frequencies (Fig. 3.3b) (Murugasu and Russell 1996; Dolan et al. 1997; Guinan and Cooper 2003; Cooper and Guinan 2006a). MOC inhibition of ~1 dB at tail frequencies in AN fibers is attributable to the MOC-induced reduction of EP which reduces IHC receptor currents and the resulting AN response (Guinan and Gifford 1988b). However, the AN inhibition is particularly large (as much as 10 dB) near 2–3 kHz (Stankovic and Guinan 1999). It is noteworthy that AN response latencies (derived from phase-gradient group delays) at this 2–3 kHz region are slightly less than AN latencies at lower frequencies, despite the fact that the lowest frequency energy in the traditional traveling wave arrives first at any given cochlear location (Shera 2001; Rhode 2007). A hypothesis that fits these data is that: (1) there is a cochlear motion produced by sounds near 3 kHz that is inhibited by MOC efferents (which implies that the motion is derived from, or influenced by, OHCs), (2) this motion occurs slightly before the motion produced by low-frequency energy in the traveling wave, and (3) this motion produces bending of IHC stereocilia that excites AN fibers but with little or no associated BM motion (Guinan et al. 2005). The exact motion is unknown but one possibility is that the motion is a combination of fluid flow in the tunnel of Corti and an associated tilting of the reticular lamina (Karavitaki and Mountain 2007a; Nowotny and Gummer 2006).

3.5.2 *Nonclassic MOC Fast Effects in the Apical Half of the Cochlea*

Cat AN fibers with low CFs have rate and phase sound-level functions that show a sharp dip in rate at a high level (80–100 dB SPL) that is accompanied by an abrupt reversal of phase (Liberman and Kiang 1984). MOC stimulation reduces the firing rate at levels below the dip but not above the dip (Gifford and Guinan 1983). The reduction in rate below the dip follows the pattern of a classic MOC fast inhibition, but the whole phenomena of a rate dip and phase reversal does not. A hypothesis similar to the one for the MOC-induced enhancement of BM motion also fits these data (except that above the dip the AN response is not significantly enhanced), namely that the dip is due to the cancelation of two out-of-phase drives with the drive that is dominant at low levels being cochlear amplified and therefore inhibited by MOC activity. Presumably the AN fiber rate above the dip is not increased, whereas BM motion is, because the AN rate is limited by a separate saturation, for example, in the amount of transmitter that can be released. In both the high-CF BM enhancement and the low-CF AN dip phenomena, the low level component is presumably the cochlear amplified traveling wave drive. It is not known if the component that becomes dominant at high levels is the same for these two phenomena.

As noted earlier, for some low-CF AN fibers, MOC stimulation makes the TCs narrower instead of wider as expected. Guinan and Gifford (1988c) obtained some data that suggest an explanation. Low-CF AN fiber TCs, in addition to the basic V-shaped tuning, have side lobes. In at least some fibers, MOC produces a particularly large inhibition in the side-lobe region with level shifts far exceeding those seen at CF (Fig. 3.3f). A MOC-induced reduction of these side lobes might be the explanation for the MOC narrowing of TCs.

A particularly striking nonclassic MOC effect in the apical half of the cochlea is the MOC inhibition of the AN initial peak (ANIP) response to clicks (Guinan et al. 2005). In AN fibers with CFs <4 kHz, the synchronization of AN spike times is adequate to show individual peaks in post-stimulus-time histograms of responses to clicks. These peaks show the oscillations of the mechanical drive at frequencies near the local CF. AN fiber recordings with and without MOC stimulation show the MOC effects expected from the MOC effect seen in BM motion due to turning down the gain of the cochlear amplifier. That is, MOC stimulation completely inhibits AN click responses near threshold, and above threshold reduces the decaying part of the response consistent with the reduction in BM ringing due to widening cochlear tuning. However, in addition, MOC stimulation inhibits the initial peak of the AN response (the ANIP response). In BM motion (in the cochlear base) the first peak of the response is passive, grows linearly and is little changed by death (Recio et al. 1998; Guinan and Cooper 2008). The AN response was expected to have the same properties, so the inhibition of the first peak was unexpected. The ANIP inhibition cannot be explained by MOC stimulation turning down the gain of the classic cochlear amplifier because the BM first peak is not amplified. The ANIP inhibition implies that the first peak is produced by a motion that originates in (or is strongly modified by) OHCs and that the motion is distinct from the motion of the classic traveling wave which is not inhibited in the first peak. One possibility is that the presumed ANIP motion is due to tilting of the reticular lamina produced by OHC contractions (Nowotny and Gummer 2006; Karavitaki and Mountain 2007a) but another possibility is that it is a motion of the tectorial membrane (Ghaffari et al. 2007). The ANIP response might be the most behaviorally important response that shows nonclassic MOC effects because it is present at moderate to high sound levels and could have a large effect on the cue for the localization of transient sounds.

All of the nonclassic MOC effects imply the presence of two (or more) vibration modes of the organ of Corti. Most of these effects appear to be due to a classic MOC gain reduction in the amplification of the traveling wave vibrational mode which interacts with another, perhaps linear, response mode. The inhibition of the ANIP response, however, is different. It implies there is a nontraveling-wave mode that is MOC inhibited and therefore not linear and passive.

3.6 MOC Slow Effects

When MOC stimulation that produces a fast inhibition is continued for 10s of seconds, it produces an additional long-lasting inhibition called the MOC slow effect (Fig. 3.7). This effect was first seen in guinea-pig AN N_1 responses to clicks and high-frequency tone pips (Sridhar et al. 1995, 1997). Since the slow AN N_1 inhibition was accompanied by a slow CM increase, the slow inhibition was attributed to a slow increase in the MOC-OHC synaptic conductance. However, in BM motion, while fast and slow MOC effects both produce inhibition, there is a phase advance during the fast effect and a phase delay during the slow effect (Fig. 3.7b)

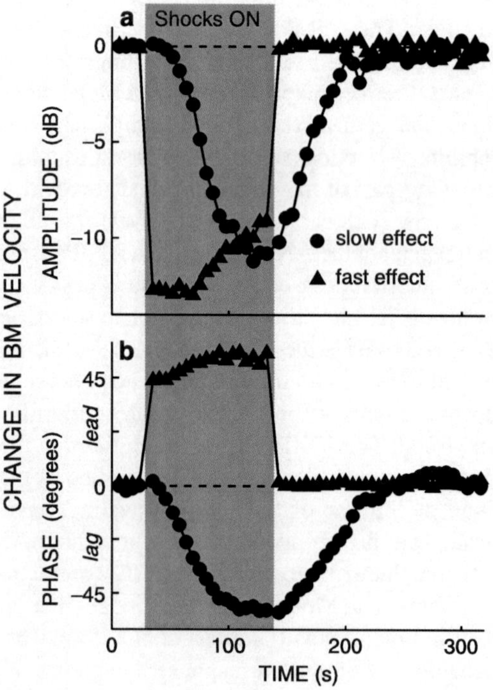

Fig. 3.7 MOC fast and slow effects both produce reductions of BM motion but with different phase changes. Variations in BM response amplitude (*top*) and phase (*bottom*) vs. time. The *gray bar* indicates the period with MOC stimulation (adapted from Cooper and Guinan 2003)

(Cooper and Guinan 2003). These phase changes in opposite directions mean that different underlying mechanisms in the OHC must be involved in producing MOC fast and slow effects. A related discovery is that application of the MOC neurotransmitter, ACh, to isolated OHCs produces a slow decrease in OHC stiffness (Dallos et al. 1997). A hypothesis that fits these data is that the MOC slow effect is due to the decrease in OHC stiffness, whereas the MOC fast effect is due to the OHC hyperpolarization and conductance change (Cooper and Guinan 2003). The increase in CM during the slow effect suggests there is also an OHC conductance change (Sridhar et al. 1995) but the BM phase change indicates the slow effect is dominated by the other change (e.g., stiffness). The ACh-induced decrease in OHC stiffness takes place in prestin molecules and the associated OHC cytoskeleton which are distant from the MOC synapse (He et al. 2003). One possibility is that there is a calcium action potential (a "calcium spark") that travels from the subsynaptic cistern at the OHC-MOC synapse to the OHC cytoskeleton and produces the slow change in stiffness and the MOC slow effect (see Sridhar et al. 1997). One reason for interest in MOC slow effects is that they may be involved in protecting the ear from acoustic trauma (Reiter and Liberman 1995). MOC slow effects are largest at high frequencies (>10 kHz) in guinea pigs, and it is not known if they are present in humans.

3.7 MOC-Fiber Responses to Sound

There are two ways that we can learn about MOC responses to sound: (1) by recording from MOC fibers and (2) by measuring sound-evoked MOC effects. In this section we review recordings from MOC fibers, and in the next section we review sound-evoked MOC effects. Most recordings from MOC fibers have been done in anesthetized animals. A variety of evidence indicates that anesthesia reduces sound-evoked MOC activity (Robertson and Gummer 1985; Liberman and Brown 1986; Brown 1989; Boyev et al. 2002) so results from anesthetized animals must be viewed with caution, especially MOC firing rates. Sound-evoked MOC effects can measured in awake humans but the sound levels have be kept low to avoid eliciting middle-ear-muscle reflexes. Overall, for awake intact animals, we do not have a good quantitative assessment of MOC firing rates or how big a neural change can be produced by sound-evoked MOC activity.

Recordings from single medial efferents come from two species: cats and guinea pigs (Fex 1962, 1965; Cody and Johnstone 1982; Robertson 1984; Robertson and Gummer 1985; Liberman and Brown 1986; Liberman 1988a, b; Gummer et al. 1988; Brown 1989, 2001; Brown et al. 1998a, b). MOC fibers have been contacted at two places: (1) in the vestibular–cochlear anastomosis – also called the "bundle of Oort" – where the MOC fibers pass from the vestibular nerve to the cochlear nerve and (2) in the intraganglionic spiral bundle within the cochlea. MOC fibers were distinguished from other nearby fibers by three criteria: (1) They respond to sound. If some MOC fibers do not respond to sound, they would not have been identified. (2) They have "regular" firing patterns (i.e., their spike-interval distributions

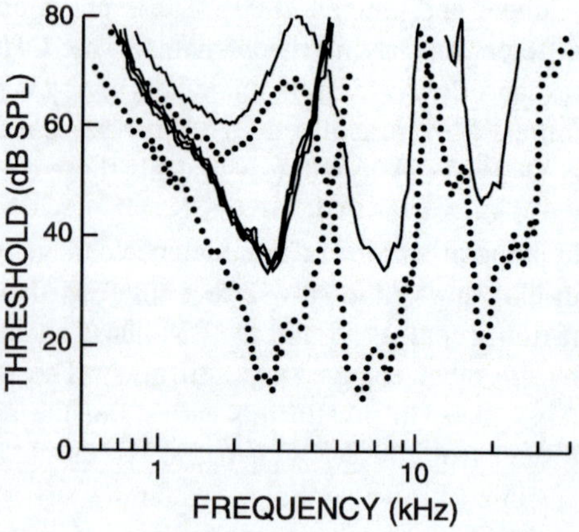

Fig. 3.8 TUNING CURVES (TCs) from cat MOC fibers. TCs measured in quiet (*solid*) compared with the tuning curves from the same fibers with broadband noise in the other ear (*dotted*) (adapted with permission from Liberman 1988a)

are approximately Gaussian) in both spontaneous activity (which is low or zero in most fibers) and responses to sound (Robertson and Gummer 1985; Liberman and Brown 1986), and (3) their latency to sound is 5 ms, or more.

The latency of the initial MOC response to sound is highly dependent on sound level. The first spike latency is tens of ms for sounds near threshold, but for high-level sounds it can be as short as 5 ms (Robertson and Gummer 1985; Liberman and Brown 1986; Brown 1989; Brown et al. 2003). This latency-vs.-level behavior suggests that MOC neurons integrate responses from their synaptic inputs and fire when threshold is reached. MOC response latency is more uniform, averaging 8.2 ± 1 ms, when measured as a modulation transfer function at moderate sound levels (Gummer et al. 1988). These data, particularly the 5 ms minimal latency, are consistent with the MOC reflex being the three-neuron arc shown in Fig. 3.2. However, additional contributions from longer pathways may also be present, for example, from the marginal shell of the anteroventral cochlear nucleus (Ye et al. 2000).

TCs from MOC fibers are similar to, or slightly wider than TCs from AN fibers, particularly at their tips (Fig. 3.8) (Cody and Johnstone 1982; Robertson 1984; Liberman and Brown 1986). Single MOC fibers labeled by dye injections show that each fiber innervates a cochlear frequency region close to the MOC fiber's best frequency (i.e., MOC projections are tonotopic) and most individual fibers innervate OHCs over a 0- to 1-octave range of cochlear length (Robertson 1984; Liberman and Brown 1986; Brown 1989, 2002) although some rat MOC fibers innervate OHCs over more than 40% of the length of the cochlea (Warr and Boche 2003). The TCs and cochlear innervation patterns of MOC fibers have lead to the idea that MOC fibers provide frequency-specific feedback to the cochlea (Winslow and Sachs 1987). The extent to which the MOC acoustic reflex actually produces narrow, tonotopic effects on cochlear responses will be considered later.

MOC fibers have been divided into three types based upon the ear that activates them using monaural sound; this ear is called the "main ear." The types are: Ipsi (these respond only to ipsilateral sound – ipsilateral re the ear innervated by the MOC fiber), Contra (these respond only to contralateral sound), and Either-Ear (these respond to sound in either ear). In cats and guinea pigs, most (~2/3) MOC efferents are Ipsi, some (~1/3) are Contra, and a few (4–11%) are Either-Ear. These percentages correspond well with the percentages of crossed and uncrossed MOC fibers (Warr 1975; Robertson 1985; Liberman 1988a; Brown 1989; Robertson and Gummer 1988) and the observation from a few labeled MOC fibers that crossed fibers respond to ipsilateral sound while uncrossed MOC fibers respond to contralateral sound (see Fig. 3.2). Although for monaural sound most MOC fibers only respond to one ear, almost all MOC fibers are binaural in that once they are activated by the main ear, sound in the opposite ear (re the main ear) can modulate their firing rate. Typically, sound in the opposite ear produces additional activation, although sometimes it inhibits the response (Robertson 1985; Liberman 1988a; Brown 1989; Brown et al. 1998a). Opposite-ear sound usually widens MOC TCs (Fig. 3.8).

The reported firing rates of MOC fibers in response to monaural sound are relatively low. In anesthetized animals, using monaural tones or noise at nontraumatic levels (<~90 dB SPL), the highest reported firing rate of a MOC fiber is ~60/s and for most MOC fibers the highest rate is one half to one third of that (Fex 1962; Robertson and Gummer 1985; Liberman 1988a; Brown 1989; Brown et al. 1998a). When using binaural sounds, or after priming by a previously traumatic sound, rates up to 134/s (cats) and 140/s (guinea pigs) have been found (Liberman 1988a; Brown et al. 1998a), however, on average, the highest MOC firing rate is much less.

These sound-evoked MOC firing rates are distinctly less than the 200/s shock rates typically used to excite MOC fibers. The sound/shock rate discrepancy has raised the question of whether sound-evoked MOC activity produces significant effects (Pfalz 1969). Sound-evoked MOC effects on OAEs in awake humans are typically only a few dB (see later), again raising the same question. There are several factors which may account for this apparent discrepancy. First, all early MOC recordings and most experiments using OAEs used monaural sound; binaural sounds evoke much higher rates. Second, anesthesia is likely to have reduced the measured firing rates, perhaps by as much as a factor of two considering that MOC effects decreased approximately by a factor of 2 when awake animals were anesthetized with a barbiturate (Boyev et al. 2002). Third, in many MOC fibers, especially fibers with BFs >2 kHz, the firing rate continues to increase as long as the sound level increased, that is, no rate saturation was found. This suggests that much higher firing rates would be produced by sounds at traumatic levels. Finally, only a fraction of MOC fibers are activated by shocks (as shown by recordings of MOC fibers at the bundle of Oort while stimulating at the floor of the fourth ventricle; McCue and Guinan, unpublished). In contrast, it is thought that all, or almost all, MOC fibers are activated by sound. Thus, to produce the same effect, shock activation of MOC fibers might require higher rates than sound activation.

MOC response properties vary with fiber BF. MOC fibers with low BFs have shorter latencies, lower thresholds and higher maximum rates than high-BF fibers (Liberman and Brown 1986). During the presentation of a BF ipsilateral tone, binaural facilitation by contralateral tones is largest for low BF MOC fibers, whereas binaural facilitation by contralateral noise is largest for high-BF MOC fibers (Liberman 1988a).

3.8 MOC Acoustic Reflexes

MOC fibers respond to sound and form ipsilateral, contralateral and bilateral MOC acoustic reflexes, terms that refer to the ear stimulated relative to the measurement ear. The MOC effects on cochlear responses presented in Sects 3.2–3.7 were mostly from shock-activation of MOC fibers. The same MOC effects are also produced by the MOC acoustic reflexes, but many aspects of the patterns of these effects may be altered by the brain stem control of MOC activation. For instance, shock activation of MOC fibers is not frequency specific, but sound activation may be frequency specific. In Sect. 3.8, first we consider the effects of sound-elicited MOC activation on AN responses. AN responses provide a direct measure of MOC effects on the output of the cochlea. Next, we consider MOC effects as seen by the changes induced in OAEs. OAEs are an indirect measure but have the advantage of being noninvasive so they can be measured in humans. Finally, we consider the influence of descending projections to MOC neurons.

3.8.1 Sound-Elicited MOC Effects on AN Fibers

Most MOC fast effects on AN responses produced by shocks are also produced by contralateral sound, but they are smaller when evoked by contralateral sound. Inhibition of AN N_1 has been reported by many papers (e.g., Buño 1978; Folsom and Owsley 1987; Liberman 1989; Warren and Liberman 1989a, b; Aran et al. 2000). In single AN fibers, contralateral sound shifts rate and synchrony sound-level functions to higher levels but has little effect on phase functions (Warren and Liberman 1989a) which is similar to findings with shocks (Gifford and Guinan 1983). In both of these studies, MOC effects were greatest with the probe sound at the CF of the AN fiber. In fibers that showed dips in rate-vs.-level functions, contralateral sound lowered the rate below the dip and produced little change above the dip, again, the same pattern as found with shock activation of MOC fibers (Gifford and Guinan 1983; Warren and Liberman 1989a). Both contralateral sound and shocks produced small decreases in AN spontaneous activity, presumably due to the small MOC-induced decrease in EP.

For AN fibers excited by an ipsilateral CF tone, the addition of a contralateral tone inhibited the AN response (Warren and Liberman 1989b). The inhibition had a complicated pattern across fiber CFs, but was always largest for contralateral tones at frequencies near the AN-fiber CF. Fibers with CFs of 2–5 kHz showed the largest inhibitions for contralateral tones 0.5–1 octave below the CF (Fig. 3.9a). In fibers with much lower CFs, the largest inhibitions were for contralateral tones at frequencies above CF, and in fibers with CFs higher than 5 kHz, the largest inhibitions were for tones near CF.

Several lines of evidence indicate that all of the sound-evoked efferent effects described above are due to MOC and not LOC activity. First, sound evokes firing

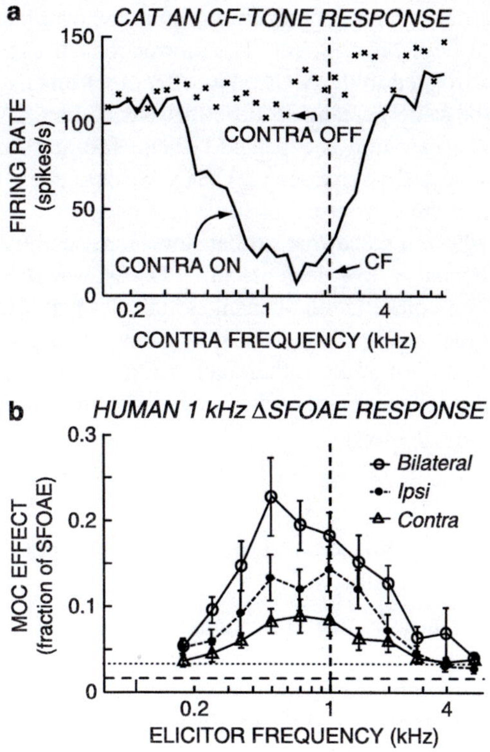

Fig. 3.9 Tuning of the MOC reflex is skewed when measured at midfrequencies by (**a**) AN responses, or (**b**) the change in stimulus frequency emissions (SFOAEs). (**a**) Responses to a characteristic frequency (CF) tone (2.14 kHz) from a cat AN fiber. *Solid line*: Inhibition from MOC activity elicited by a contralateral tone swept in frequency. *X's*: a sweep with the contra tone off. (**b**) The normalized SFOAE change from a 1-kHz probe tone vs. elicitor center frequency for 60 dB SPL half-octave noise-band elicitors. Compare the *solid curve* in (**a**) with the contralateral curve in (**b**) ((**a**) adapted with permission from Warren and Liberman 1989b, (**b**) adapted from Lilaonitkul and Guinan 2009a)

in MOC fibers that could produce these effects (see Sect. 3.7). Second, the fast time course and the patterns of the neural inhibition evoked by sound are qualitatively compatible with the inhibition produced by shocks to the olivocochlear bundle (OCB) (Warren and Liberman 1989a, b), and all of the effects elicited by OCB shocks are attributable to MOC fibers (review: Guinan 1996). Third, many studies show sound-evoked effects on OAEs that are similar to the effects on neural responses (see Sect. 3.8.2), and OAE changes are mechanical changes that cannot be produced by LOC synapses. Finally, the LOC effects that have been measured to date show very slow changes (Groff and Liberman 2003), and these are much slower than any of the sound-evoked fast effects reviewed above. Overall, it seems highly likely that the fast sound-evoked efferent effects described so far are all due to MOC efferents.

In addition to the fast effects, slower effects due to contralateral sound have been reported. Lima da Costa et al. (1997) found that a contralateral broad-band noise reduced ipsilateral round-window noise (RWN) with fast and slow time courses. RWN near 1 kHz is dominated by small contributions from far-field potentials from the random firing of AN fibers (Dolan et al. 1990). With no ipsilateral sound, RWN is due to spontaneous AN activity (or AN activity from animal noise) and a reduction of AN spontaneous activity can be produced by MOC activation (Guinan and Gifford 1988b). The fast reduction of RWN found by Lima da Costa et al. was blocked by a gentamicin injection, which blocks MOC synapses, and seems likely to be a MOC effect. The slower reduction of RWN by contralateral sound was attributed to the MOC slow effect by Lima da Costa et al., but was not blocked by the same gentamicin dose that blocked the fast effect. Blocking at the same ACh concentration is expected because fast and slow effects are both produced by the same ACh synapses on OHCs (Sridhar et al. 1995). An alternative explanation is that the slow effect of Lima da Costa et al. is due to LOC efferents (Yoshida et al. 1999). A somewhat similar slow increase in RWN along with a slow increase in AN N_1 and a slow decrease in DPOAEs, was found by Larsen and Liberman (2009). This constellation of changes indicates that MOC efferents are involved. Larsen and Liberman attributed the slow change to a centrally mediated slow increase in the MOC fast effect, rather than to a MOC slow effect. More work is needed to sort out the contributions of the various possible candidates for these slow sound-evoked efferent effects.

3.8.2 *Sound-Elicited MOC Effects on Otoacoustic Emissions*

The measurement of sound-elicited effects using OAEs has been one of the most productive ways of measuring MOC effects because it is noninvasive and can be done in humans. However, OAEs provide an indirect measure of cochlear mechanical responses and before considering results from them, we first consider the issues in their use.

Most measurements of MOC effects on OAEs used contralateral sound to elicit MOC activity because this is the easiest method. However, measurements can also be made with ipsilateral or bilateral elicitors if the two main problems with these are avoided. First, the high-level acoustic waveform of the ipsilateral elicitor can be canceled out by reversing sign of the elicitor on alternate presentations and averaging an even number of responses. The second problem, "two tone suppression"[2] produced by an ipsilateral elicitor, is more difficult. Two tone suppression is produced by energy in the elicitor that is near in frequency to the probe frequency and turns down the gain of the probe-frequency cochlear amplifier by bending OHC stereocilia into their nonlinear range (Geisler et al. 1990; Geisler 1992). There are two ways to get around this: (1) select ipsilateral elicitors that do not have energy near the probe frequency (e.g., notched noise, Backus and Guinan 2006; Lilaonitkul and Guinan 2009b), or (2) separate the effects in time using the difference in their decay rates, that is, two-tone suppression decays in a few milliseconds but MOC

[2] In the literature the term "contralateral suppression" is often used to mean the effect of MOC activity elicited by contralateral sound. We avoid this term because it does not distinguish between MOC inhibition elicited by contralateral sound and two-tone suppression produced by acoustic crosstalk from the contralateral to the ipsilateral ear. Instead, we use the term "contralateral inhibition" or "contralateral MOC inhibition".

inhibition decays with a time constant of ~100 ms (Guinan 1990). This second technique can be achieved by measuring in a "post elicitor window" that is after the two-tone suppression has died out but before the MOC inhibition has decayed away (Guinan et al. 2003). With the second technique, the MOC effect is measured while it is decaying and is not as large as it was during the elicitor. To make comparable measurements of ipsilateral and contralateral MOC effects, the same window has to be used for both.

There are several other issues that must be considered when using OAEs to measure MOC effects. First, the signal-to-noise ratio (S/N) must be adequate. Although this seems obvious, the S/N criterion has almost never been applied correctly in the literature. Most commonly, the S/N criterion has been applied to the OAE measurement. However, when measuring MOC effects, the "signal" of interest is the change in the OAE, not the OAE. To have an accurate measurement, the S/N of the change in the OAE must be adequate (e.g., 6 dB). A second consideration is to have an adequate number of alterations of elicitor-on vs. elicitor-off to remove any systematic drift. Third, there must be no middle-ear-muscle (MEM) contractions. MEM contractions interfere by changing sound transmission through the middle ear and by changing the impedance of the ear as seen by the acoustic source. Weak MEM contractions are not always shown by clinical MEM instruments (Feeney and Keefe 2001; Feeney et al. 2004). A more sensitive test for MEM contractions is the suppressed-OAE test (Lilaonitkul and Guinan 2009a, b). Finally, the sound used to evoke the OAE may also elicit unintentional MOC activity (Guinan et al. 2003). Such unintended MOC activity is certainly undesirable, but it is not known how much this changes measurements using contralateral elicitors.

MOC effects on OAEs in humans have been measured over a frequency range from ~0.5 to 5 kHz with the largest effects often at 1–2 kHz (e.g., Collet et al. 1990; Moulin et al. 1993; Lilaonitkul and Guinan 2009b). In animals, MOC effects on OAEs have been measured up to 30 kHz in cats and over 60 kHz in bats (Guinan 1986; Henson et al. 1995). Considering that MOC innervation peaks in the basal half of the cochlea in animals (the pattern in humans is assumed to be similar), it might seem surprising that the largest MOC effects are often at 1–2 kHz. However, contralateral sound evokes the highest MOC firing rates at these frequencies (Liberman 1988a). In addition, the change in OAEs may vary across frequency due to the processes by which OAEs are generated. Considering this, MOC effects vs. OAE probe frequency must be interpreted cautiously.

The threshold for MOC effects on OAEs is slightly (e.g., 10–15 dB) above the hearing threshold, and above that, MOC effects increase as the elicitor level increases (e.g., Collet et al. 1990; Ryan et al. 1991; Backus and Guinan 2006). Note, however, that the elicitor level range is limited by the need to avoid MEM activation, so it is not known if MOC effects saturate at higher elicitor levels or continue to increase. Presumably, the growth of MOC effects on OAEs with increases in sound level is due to, and provide a window on, the growth in MOC fiber activity.

The time course of MOC effects in humans is shown most clearly by ΔSFOAEs (Backus and Guinan 2006). For probes near 1 kHz, there is a delay of ~25 ms from sound onset, or offset, to the beginning of the change in the ΔSFOAE; ~20 ms of this is the delay to the beginning of the change in cochlear amplification, and ~5 ms is for this change to be carried backward in the cochlea to the ear canal. After the initial delay, there is a monotonic rise in the ΔSFOAE with an overall time constant of a few hundred ms, and after the offset delay there is a decrease in the ΔSFOAE that is typically faster than the rise. In some subjects the ΔSFOAE shows a small, short overshoot lasting a few tens of ms. The time course of MOC effects on TEOAEs has not been adequately demonstrated because of the noncontinuous nature of TEOAEs. However, this time course should be the same as for SFOAEs. In contrast to the above, MOC effects on DPOAEs can show very complicated time patterns, presumably because of the complicated mixing of the two different sources of DPOAEs (Muller et al. 2005). Overall, the OAE data indicate that the MOC fast effect is not fast enough to significantly change the hearing of speech on a syllable-by-syllable basis.

During continuous or intermittent contralateral noise, MOC effects are maintained for many minutes with little evidence of adaptation but with some evidence for a slow increase in the effect (Giraud et al. 1997a; Lima da Costa et al. 1997; Larsen and Liberman 2009; Zyl et al. 2009). After a long stimulation evoking MOC effects, there is sometimes a rebound enhancement lasting seconds (Zyl et al. 2009). This enhancement may be a rebound from a MOC slow effect, or may arise from an entirely different cause (see Maison et al. 2007).

3.8.2.1 MOC Reflex Tuning

OAE measurements in awake humans show there is tuning in the MOC reflex, but the largest MOC effects are not always centered on the elicitor frequency. Measurements of the MOC effects produced by 1/3 octave noise-band elicitors using DPOAEs or tone-pip TEOAEs show tuning in the MOC reflex for probes near 1 and 2 kHz but not 3 and 4 kHz (Veuillet et al. 1991; Chéry-Croze et al. 1993). Both studies concluded that the largest MOC effects were when the elicitor noise band was centered on the probe frequency, but both studies also show cases where elicitor bands below the probe frequency had larger effects than those at the probe frequency. Lilaonitkul and Guinan measured the MOC effects produced by 60 dB SPL tones or half-octave noise-bands using SFOAEs from probes at 0.5, 1, and 4 kHz. The change in SFOAEs (ΔSFOAE) for probes near 1 kHz show broad, skewed tuning with the most effective elicitor frequencies 0.5–1 octave below the probe frequency (Fig. 3.9b) (Lilaonitkul and Guinan 2009a). For 0.5-kHz probes, there was also broad tuning but with a skew in the opposite direction, whereas for 4-kHz probes, the tuning had a narrower peak and a broad low-frequency activation region (Lilaonitkul and Guinan, unpublished). These ΔSFOAE tuning patterns are similar to the effects of contralateral sound on AN fibers found by Warren and Liberman (1989b) (see Fig. 3.9 and Sect. 3.7) except that the human pattern is shifted down by about an octave from the cat pattern (presumably because the human hearing range is about an octave lower than cats). In contrast, a different SFOAE metric, the change in the magnitude of the SFOAE (SFOAEmoc) showed MOC effects that were much narrower and centered on the probe frequency

(Lilaonitkul and Guinan, unpublished). The correspondence between ΔSFOAE and the neural data gives support to the ΔSFOAE metric for MOC effects in humans. However, the more centered results found with TEOAEs, DPOAEs and SFOAEmoc indicates that we still do not fully understand the ways that MOC activity affects OAEs and the tuning of the MOC reflexes. Also, the above experiments were done with passive listening; perhaps in the context of a task such as identifying a 1-kHz signal in noise, the MOC effect might be more focused on 1 kHz.

3.8.2.2 MOC Reflex Amplitude as a Function of Elicitor Bandwidth

MOC effects on OAEs from elicitors of various bandwidths show little evidence of a frequency selective reflex. MOC effects measured with elicitors at a fixed level, but with increasing bandwidths, show that MOC activation increases as bandwidth increases up to 4–6.7 octaves (Fig. 3.10) (Maison et al. 2000; Lilaonitkul and Guinan 2009b). Because the elicitor SPL was held constant as bandwidth increased, the spectral level near the probe frequency decreased. However, despite the resulting decrease in MOC activation in frequency regions near the probe, the overall activation does not decrease, even at the widest bandwidths. This indicates that the increased activation from frequency regions remote from the probe frequency must have compensated for the activation lost near the probe frequency. These experiments indicate that the MOC reflex must integrate activation from practically the whole cochlea.

The wide frequency integration profile of the MOC reflex shown by the bandwidth experiments helps to explain why a wide variety of experiments found that the

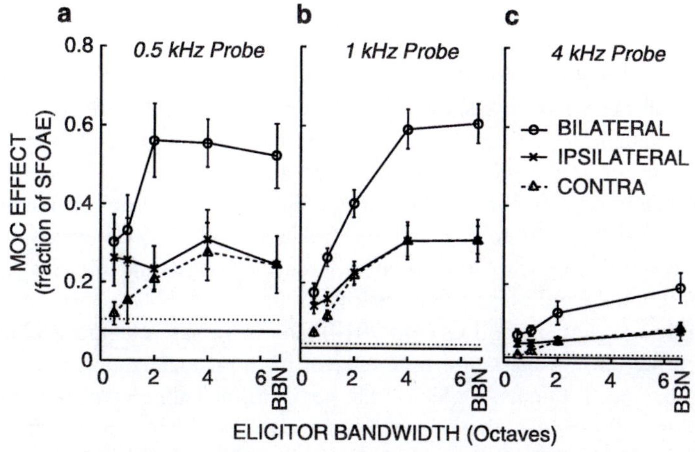

Fig. 3.10 MOC inhibition grows as elicitor noise bandwidth is increased despite the overall noise level being held constant at 60 dB SPL. MOC effect is the change in the SFOAE normalized by the magnitude of the SFOAE. *BBN* broad-band noise (0.1–10 kHz, or 6.67 octaves). Error bars are standard errors of the mean. *Horizontal lines*: *Solid*=noise-floor mean, *dotted*=1 std. above the noise mean (adapted with permission from Lilaonitkul and Guinan 2009b)

most potent elicitor of MOC effects is broad band noise (Ryan et al. 1991; Veuillet et al. 1991; Chéry-Croze et al. 1993; Norman and Thornton 1993; Maison et al. 2000; Guinan et al. 2003). Potency in eliciting MOC activity appears to increase in the following order: single tones, two tones (DPOAE primaries), repetitive tone pips, repetitive clicks and broad-band noise (Guinan et al. 2003). Note that the first four of these sounds are used to evoke OAEs, but they also elicit MOC activity.

White noise that is amplitude modulated (AM) at 100 Hz has been reported to produce larger MOC activations than unmodulated noise (Maison et al. 1999). Our experiments with AM noise show an increase for some subjects but not others, and it is unclear if these are two separate groups or a continuum (Backus and Guinan 2004). The extent to which AM might be used to increase MOC effects in humans is not yet clear.

3.8.2.3 MOC Reflex Laterality

The ratio of ipsilateral/contralateral MOC fibers is approximately 2:1 in small mammals but is unknown in humans. This ratio was derived largely from labeling the MOC neurons that project to one cochlea. Such data show that twice as many MOC fibers originate from the contralateral side as from the ipsilateral side (reviewed by Warr 1992) (keep in mind the innervation pattern of Fig. 3.2). Comparable experiments cannot be done on humans, but relevant information might be obtained by comparing the effects of ipsilateral vs. contralateral sound on OAEs. As shown in Fig. 3.10, binaural sound produces the largest MOC effects, often twice as large as the ipsilateral or contralateral effects alone. Monaural broad-band noise elicits ipsilateral and contralateral effects of very similar amplitudes (Fig. 3.10, the points labeled BBN). In contrast, for narrow-band noise, ipsilateral elicitors produce changes that are approximately twice as large as contralateral elicitors. The ratio of ipsilateral/ contralateral MOC fibers cannot change with elicitor bandwidth. These results show that the laterality of MOC effects is strongly influenced by central processes that change the activity in crossed vs. uncrossed MOC fibers according to the bandwidth of the stimulus.

One theory to explain the change in MOC laterality with elicitor bandwidth is that having the reflexes be equal when they produce large effects prevents the MOC effects from producing interaural time differences that disrupt binaural hearing. 60 dB SPL contralateral noise produces low-frequency cochlear phase advances in the 0.5 ms range (Francis and Guinan 2010), and this phase change would have a profound effect on the interaural time cue for binaural localization if it were not binaurally balanced. In cats, a species that hears at low enough frequencies to use interaural time differences, the crossed/uncrossed MOC fiber ratio is a function of the cochlear frequency region, that is, the overall ipsi/contra reflex ratio is 2/1, but at low frequencies the ratio is near 1:1 (Guinan et al. 1984). In contrast the ratio stays near 2:1 across frequency in the mouse, a species that does not have significant low-frequency hearing (Maison et al. 2003). These patterns are consistent with the hypothesis that MOC reflex equality at low frequencies evolved to enable good binaural localization at low frequencies.

3.8.2.4 MOC Reflex Strength

Most studies of MOC effects in humans used averages across groups of subjects. Such studies show that MOC efferents produced a change across the group as a whole, but not whether the change was present in each individual. To date, only one study applied an adequate S/N criterion to insure that the MOC effect on each subject was significantly different from the measurement noise (Backus and Guinan 2007). This study measured normalized ΔSFOAEs elicited by 60 dB SPL contralateral sound and found more variation between measurements at nearby frequencies than could be accounted for by the expected variation across frequency of the MOC effect. One explanation is that the MOC effect changes the weighting of the cochlear irregularities involved in the coherent reflection that produces SFOAEs (Backus and Guinan 2007). To get consistent measures of MOC strength, MOC effects had to be averaged across measurements at several nearby frequencies. The resulting MOC strength distribution is shown in Fig. 3.11. The distribution is approximately Gaussian with a mean of 36% (~3 dB) and a range of a factor of 5 between subjects with weak effects and subjects with strong effects. Thus, whatever benefits MOC reflexes provide, these benefits must vary considerably across individuals.

3.8.3 Descending Influences on MOC Acoustic Reflex Properties in Humans

Reviews of the anatomy of the descending auditory system and its influences within the CNS are given in Schofield (Chap. 9), Robertson and Mulders (Chap. 10), and Suga et al. (Chap. 11). Here we concentrate on descending influences on MOC neurons, particularly in humans. There are three areas that appear to show descending

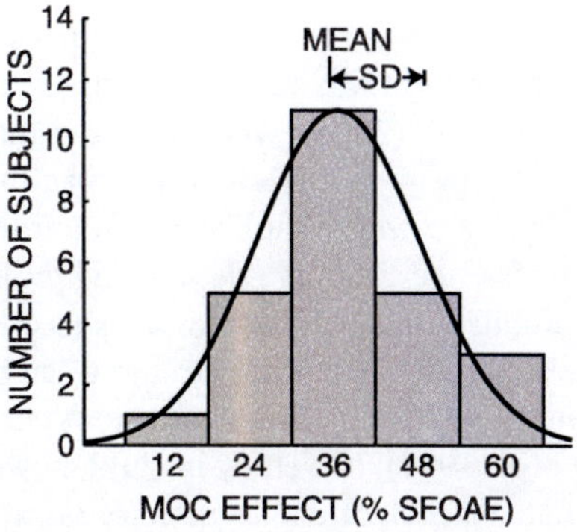

Fig. 3.11 A histogram of the MOC strengths at 1 kHz for 25 subjects (mean = 36.6%, SD = 11.7%). MOC activation from 60 dB SPL contralateral wide-band noise. MOC effect was measured by the SFOAE change. The curve is a Gaussian fit to the data (adapted with permission from Backus and Guinan 2007)

influences: (1) MOC reflex modulation by attention, (2) right-vs.-left ear differences, and (3) auditory learning.

There is an extensive literature on the effects of attention on cochlear processes. In one class of experiments, attention to a visual task is alternated with attention to an auditory task (reviews: Guinan 1996; Delano et al. 2007). Numerous experiments of this kind show changes in cochlear potentials and/or OAEs that indicate there is increased MOC activation during the visual task. However, such changes were not seen in all cases so perhaps the MOC activation is present in some subjects but not others. In other experiments, comparisons were made across auditory tasks, or MOC effects were elicited by contralateral noise (e.g., Giard et al. 1994; Michie et al. 1996; Maison et al. 2001; de Boer and Thornton 2007). A theme that may explain the diverse results from these experiments is that MOC efferents are activated for tasks when the MOC activity produces a benefit (e.g., reducing a distracting sound during a visual task, or aiding in difficult signal-in-noise tasks) but not when there is no benefit (e.g., doing an easy auditory task while ignoring a visual stimulus, or counting tone pips embedded in a click train).

An interesting attentional experiment was conducted by Scharf et al. (1997) using the probe-signal method, which focuses on subject detection of an "odd-ball" stimulus presented in a small fraction of the trials. Using a signal just above threshold in a background noise, Scharf et al. found that on-frequency signals were heard, but odd-ball signals that differed in frequency by 5% were not heard. Scharf et al. concluded that the off-frequency tone was MOC inhibited because the effect was not present in subjects with efferents cut for medical reasons. More recently, Tan et al. (2008) performed a set of similar experiments and drew an opposite conclusion. Tan et al. concluded that MOC efferents are activated by the tone cue used in the probe-signal method and produce a MOC benefit at the cued on-frequency, whereas targets at nearby off-frequencies get little or no MOC benefit and so their perception is worse than tones at the cued frequency.

There are right–left asymmetries in various aspects of the peripheral auditory system. There are small handedness and gender differences in OAE amplitudes, and in MOC effects on OAEs (e.g., Aidan et al. 1997; Khalfa et al. 1998; Morlet et al. 1999; Sininger and Cone-Wesson 2004). Benzodiazepines reduce contralaterally elicited MOC inhibition in the right ear, but not the left, according to Morand et al. (2001); Morand-Villeneuve et al. (2005), who suggested that this is because there are more benzodiazepine receptors in the left cortex than the right. At the cortical level, right–left asymmetries are well established. An attractive hypothesis is that these cortical right–left differences, through descending projections, produce MOC asymmetries and these produce the OAE asymmetries (see Khalfa et al. 2001).

A variety of evidence suggests that auditory training can have an effect on MOC reflex strength. Musicians have stronger MOC reflexes than people who never had musical training (Perrot et al. 1999). In a study in children with reading disabilities, certain children showed an absence of the asymmetry favoring the right ear that is found in average-reading children (Veuillet et al. 2007). After auditory training that improved their score in a speech task, these children's MOC function changed and their asymmetry became closer to normal. In a study using adults, subjects were given a 5-day training regimen on a speech-discrimination-in-noise task, and their

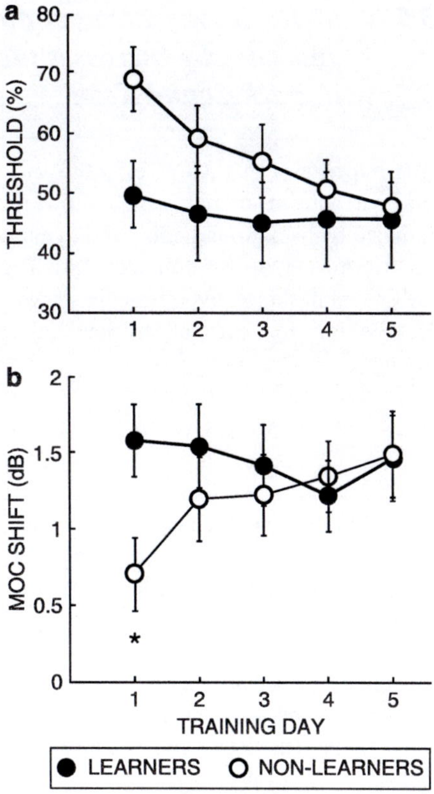

Fig. 3.12 MOC reflex activity and signal-in-noise detection change over 5 days of auditory training in some subjects, but not others. Subjects were divided into learners and nonlearners based their learning over 5 days. The discrimination thresholds for these two groups are shown in (**a**). On day 1, nonlearners had significantly lower MOC shifts than learners, but training erased this difference (**b**). MOC shift was induced by 40 dB SL contralateral noise. MOC shifts were measured by click-evoked TEOAEs (adapted with permission from de Boer and Thornton 2008)

MOC activation was measured on each day (de Boer and Thornton 2008). Subjects who originally had weaker MOC activation showed greater improvement in the speech-in-noise task and also showed increases in MOC activation so that after the training their MOC activation was similar to the subjects who originally had larger MOC activation (Fig. 3.12). More work is needed to show whether the improvement in perception is brought about by the increased MOC activity, or whether the increased MOC activity is simply a byproduct of other central changes. In either case, the work suggests that a test for MOC strength might predict subjects who would benefit from auditory training.

3.9 MOC Function in Hearing

There are two areas for which there is good evidence for a MOC function in hearing: aiding discrimination of signals in noise, and preventing or reducing acoustic trauma. In a third area, attention and learning, the evidence that the measured effects are primarily due to efferents is not clear, but there may be an important MOC function in this area as well.

3.9.1 MOC Activity Changes the Dynamic Range of Hearing and Thereby Increases the Discriminability of Transients in Background Noise

The hypothesis that MOC efferents aid in the discrimination of signals in noise originated from animal results showing that when AN output dynamic range is reduced by background noise, the dynamic range can be partially restored by MOC activation (see Fig. 3.5 and Sect. 3.4). There have been two main ways in which this "MOC unmasking" hypothesis has been tested: (1) by cutting MOC fibers and (2) by looking for correlations between psychophysical performance and MOC strength.

In many animal studies, MOC fibers were cut and performance deficits were investigated (review: Guinan 1996). In a noisy background, lesions of MOC efferents in cats reduce their performance in speech discrimination, high-frequency tone discrimination, and sound localization (May et al. 1995; Hienz et al. 1998; May et al. 2004). In contrast, functional MOC lesions in mice produced by altering the ACh receptor did not reveal any deficit in detecting signals in noise (May et al. 2002). This surprising result may be because the conditions tested were not right for mice. In humans, surgical cuts of the vestibular nerve (MOC fibers exit the brain in this nerve) revealed a MOC benefit in detecting speech in noise in some patients, but not others (Zeng and Shannon 1994; Zeng et al. 2000; Giraud et al. 1997b). Perhaps the inconsistent results were because these lesions interrupt MOC fibers to varying degrees (Zeng and Shannon 1994; Giraud et al. 1995; Chays et al. 2003). Scharf et al. (1997) performed signal-in-noise tests on humans with vestibular nerve cuts without finding a deficit, but the tests used conditions for which no MOC benefit is expected (tones and noise that were both continuous or both equal-duration bursts; see Sect. 3.4).

Measurements across subjects of psychophysical performance and MOC activation, with MOC activation sometimes increased by contralateral noise, have found correlations between performance and MOC activation that indicate both MOC benefits and handicaps (e.g., Micheyl and Collet 1996; Micheyl et al. 1997; Giraud et al. 1997b; Kumar and Vanaja 2004). Using the correct stimulus conditions appears to be the key to finding a MOC benefit. Subjects with strong MOC reflexes are better than those with weak reflexes in detecting signals in noise when the S/N is moderate (Fig. 3.13) but not near threshold or with high-level noise (Kumar and Vanaja 2004). In studies that found no correlation between MOC activation and subject ability to detect a signal in noise (e.g., Wagner et al. 2008; Mukari and Mamat 2008), the negative results may be due to the methods used (e.g., MOC-induced changes in human DPOAEs are poor MOC metrics; see earlier) and/or to using the wrong stimulus conditions (e.g., signals too close to threshold). Overall, the data support the hypothesis that one function of MOC efferents is to aid in the detection of signals in noise, although the conditions that produce this benefit are not well documented.

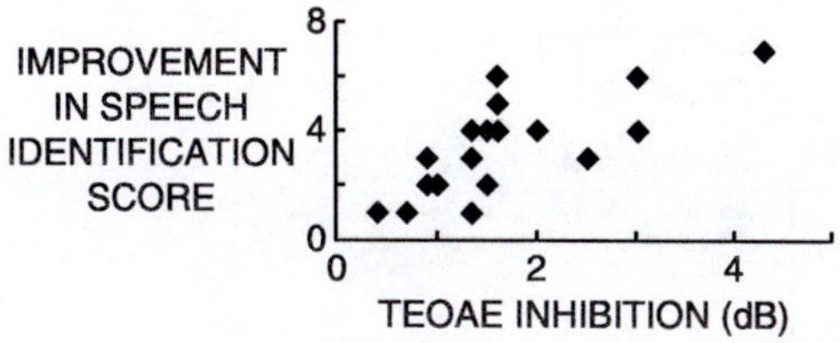

Fig. 3.13 A positive correlation across subjects between the improvement of speech perception in noise and MOC inhibition of transient evoked otoacoustic emissions (TEOAEs), which indicates that MOC activity provides a benefit in the detection of signals in noise. Correlation coefficient, $r=0.48$, $p=0.001$. MOC inhibition elicited by 30 dB SL contralateral noise. Speech at 50 dB HL and 10 dB S/N re ipsilateral noise (adapted with permission from Kumar and Vanaja 2004)

3.9.2 MOC Activity Helps to Protect Against Acoustic Trauma

Data from many animal experiments leave little doubt that MOC activity helps to prevent both temporary and permanent threshold shifts (TTS and PTS) due to traumatic sounds (review: Rajan 2000). We highlight one study, a prospective PTS study in which animals were exposed to traumatizing sounds after being classified into those with weak, average, and strong MOC reflexes using OAE measurements (Maison and Liberman 2000). The animals with strong MOC reflexes had the least PTS and the animals with weak MOC reflexes had the most PTS (Fig. 3.14). This result indicates that MOC activity reduces PTS, and also that an OAE-based test may show which subjects are susceptible to acoustic trauma.

3.9.3 Possible Roles of MOC Activity in Attention and Learning

Changes in MOC activation during attention and learning were shown in Sect. 3.8.3. At least for some subjects and conditions, MOC activation brought about by attention helps to reduce distracting sounds or aids in discriminating the attended sound when it is in a background noise. More data are needed to understand the extent to which attention modifies the function of the MOC acoustic reflex in everyday hearing.

Attention is important in learning and may bring about changes in MOC activity that are important for learning a task. The de Boer and Thornton (2008) results (see Sect. 3.8.3) suggest that some normal subjects had to learn to turn on their MOC efferents so that they could detect a signal in noise as well as was done by other subjects who required no training (Fig. 3.12). Learning to control MOC efferents seems to have been the key factor in this learning experiment. On the other hand,

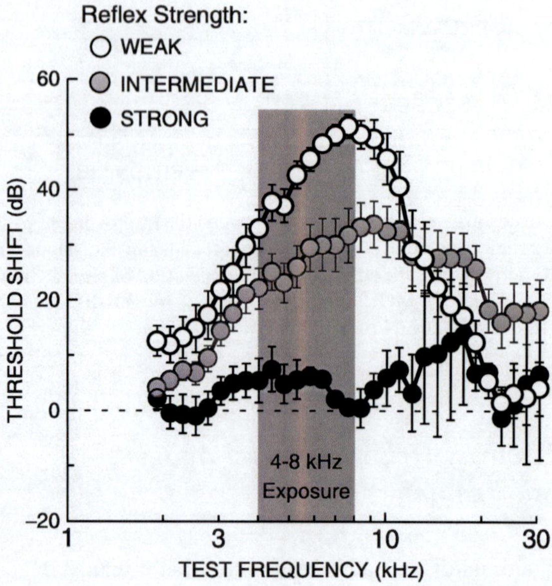

Fig. 3.14 MOC reflex strength predicts acoustic trauma. Twelve guinea pigs were grouped by the preexposure strength of their MOC reflexes. The noise-induced permanent threshold shifts (PTSs) measured from AN compound action potentials were least in the group with the strongest reflexes. The results imply that MOC reflexes reduce PTS. Error bars indicate SEM (adapted with permission from Maison and Liberman 2000)

the change in MOC activation and gaining of normal right–left differences in the reading impaired children of Veuillet et al. (1996) seems more likely to be a byproduct of some other central change. There may be an important MOC role in learning, but more work is needed before it is understood.

3.10 LOC Physiology and Function

Before we review LOC physiology, we briefly review LOC anatomy. There are two types of LOC neurons based on their position around the lateral superior olivary nucleus: intrinsic and shell neurons. In the cochlea, these appear to correspond to unidirectional fibers and bidirectional fibers, respectively (Brown 1987; Warr et al. 1997). In the mouse, intrinsic-unidirectional neurons are cholinergic and shell-bidirectional neurons are dopaminergic (Darrow et al. 2006b); whether this holds for other species is unknown. A wide variety of other neurotransmitters and neuroactive substances have been found in LOC neurons. For further details on LOC anatomy and neurochemistry, see Brown (Chap. 2) and Sewell (Chap. 4).

3.10.1 LOC Effects in the Cochlea

The fibers of both LOC groups are unmyelinated, and neither has been recorded from or stimulated. However, indirect activation of LOC neurons has been achieved by electrical stimulation in the inferior colliculus (IC). Depending on where the IC is stimulated, AN responses can be enhanced or reduced by effects that are attributable to LOC efferents (Groff and Liberman 2003). Considering that the two LOC groups have different neurotransmitter contents, an attractive hypothesis is that excitation is produced when one LOC group is activated and inhibition when the other group is activated. When LOC efferents are lesioned, AN activity is depressed (Liberman 1990; Le Prell et al. 2003) or enhanced (Darrow et al. 2006a). The difference in results might be explained by the lesions in different studies affecting the two LOC groups to different extents.

All of the effects attributable to LOC efferents are very slow (Groff and Liberman 2003). This is not surprising because these fibers are unmyelinated and have slow conduction velocities. In Groff and Liberman experiments, the LOC effects decayed with time constants (τ's) of minutes (the onset time courses of LOC effects were obscured by shock artifacts and MOC effects). Thus, efferent effects occur on three time scales: MOC fast effects with τ's of ~100 ms, MOC slow effects with τ's of 10's of seconds, and LOC effects with τ's of minutes.

3.10.2 LOC Response to Sound

LOC neurons receive innervation from the ipsilateral ventral cochlear nucleus and this innervation could form the basis of an ipsilateral LOC acoustic reflex (Thompson and Thompson 1991). There are no known inputs to LOC neurons from the contralateral side. However, most neurons in the vicinity of LOC neurons are excited by sound in the ipsilateral ear and inhibited by sound in the contralateral ear (Guinan et al. 1972), and MOC neurons might be similar. There are no recording from LOC neurons in intact preparations, so there are no definitive data on LOC responses to sound. LOC acoustic reflexes, if they exist, can be expected to act on a slow (minutes) time scale.

3.10.3 LOC Function in Hearing

LOC fibers synapse directly on AN fibers and can change their firing patterns. The wealth of neurotransmitters in the two types of LOC fibers suggests that LOC fibers have multiple functions; however, we know relatively little about these functions.

LOC fibers appear to reduce the acoustic-trauma-induced excitotoxic effect of the IHC neurotransmitter, that is, they reduce the swelling and bursting of AN fibers produced by over stimulating IHCs (e.g., Ruel et al. 2001). See Groff and Liberman (2003) for a review of the evidence that LOC fibers reduce acoustic trauma in this and other ways.

Guinan (1996) suggested that LOC neurons act to achieve balance in the outputs from the two ears to enable binaural localization based on interaural level differences. Support for this hypothesis comes from LSO lesions that produced an imbalance in ABR responses from the right ear vs. the left ear (Darrow et al. 2006a). However, more evidence is needed before this hypothesis can be considered as established.

3.11 Summary and Future Directions

Most MOC effects are classical effects owing to turning down the gain of cochlear amplified traveling waves. At high sound levels, the traveling wave interacts with a more linear mechanical response and produces most of the nonclassic MOC effects. Click responses in the apical half of the cochlea show there is an additional motion that is MOC inhibited and that produces the ANIP response. Work is needed to determine the exact nature of the ANIP motion. Another unknown is what tone response corresponds to the ANIP click response. One possibility is TC side lobes (Fig. 3.3f). Overall, progress in understanding MOC effects requires a better understanding of the multiple motions involved in cochlear mechanics.

In addition to MOC fast effects, there is a MOC slow effect that has been seen only at high frequencies in guinea pigs. Its presence and frequency range in other species remains to be determined. It is also unknown whether the MOC slow effect originates from an OHC stiffness change and whether it has any relationship to protection from TTS.

Sound in either ear evokes MOC activity and produces MOC effects. More work is needed to show the single-fiber MOC responses evoked by ipsilateral, contralateral and bilateral noise of various bandwidths. Measurements in awake animals are necessary for a full understanding of MOC responses to sound.

Aiding the detection and discrimination of signals in noise is probably the most important MOC function. Physiological mechanisms capable of producing this benefit have been demonstrated, but we do not yet understand how powerful these mechanisms are or the conditions under which MOC activity provides a benefit and when it does not. Although it seems likely that MOC activity plays an important role in the signal-detection-in-noise temporal effect (i.e., "overshoot"), the MOC role needs to be more clearly shown. For instance, the fact that narrow-band noise centered on the signal frequency produces little temporal effect may be because narrow-band noise elicits little MOC effect (Fig. 3.10).

A wide variety of data indicate that both MOC and LOC activity reduce acoustic trauma, but the mechanisms for this are largely unknown. Small reductions in sound level can greatly reduce both TTS and PTS, so one possibility is that a small MOC-induced reduction in cochlear mechanical motion may be involved, but such reductions have not yet been demonstrated at traumatic sound levels. Although dopamine released by LOC activity may reduce TTS by blocking excitotoxity from too much transmitter released by IHCs (Ruel et al. 2001), the role of the many other neuroactive substances released by LOC synapses needs to be elucidated, for TTS, PTS, and on neural signaling. Prospective tests are needed to determine whether a MOC strength test can determine whether a person is particularly susceptible to noise damage. MOC reflex testing needs to be done with an adequate S/N whenever the results are to apply to a single individual.

Finally, at least under some circumstances, MOC activation changes during learning. Whether the learning causes the change in MOC activation or is caused by MOC activation needs to be determined. Also, work is needed to show whether MOC activation tests can predict which subjects will be aided by auditory training.

Acknowledgments This work was supported by NIH NIDCD RO1 000235 and RO1 005977.

References

Abdala C, Mishra SK, Williams TL (2009) Considering distortion product otoacoustic emission fine structure in measurements of the medial olivocochlear reflex. J Acoust Soc Am 125:1584–1594

Aidan D, Lestang P, Avan P, Bonfils P (1997) Characteristics of transient-evoked otoacoustic emissions (TEOES) in neonates. Acta Otolaryngol 117:25–30

Aran JM, Pajor AM, de Sauvage RC, Erre JP (2000) Role of the efferent medial olivocochlear system in contralateral masking and binaural interactions: an electrophysiological study in guinea pigs. Audiology 39:311–321

Backus BC, Guinan JJ Jr (2004) The efficacy of AM noise for activating the human MOC reflex measured using SFOAEs. Assoc Res Otolaryngol Abstr 27:535

Backus BC, Guinan JJ Jr (2006) Time course of the human medial olivocochlear reflex. J Acoust Soc Am 119:2889–2904

Backus BC, Guinan JJ Jr (2007) Measurement of the distribution of medial olivocochlear acoustic reflex strengths across normal-hearing individuals via otoacoustic emissions. J Assoc Res Otolaryngol 8:484–496

Bassim MK, Miller RL, Buss E, Smith DW (2003) Rapid adaptation of the 2f1–f2 DPOAE in humans: binaural and contralateral stimulation effects. Hear Res 182:140–152

Boyev KP, Liberman MC, Brown MC (2002) Effects of anesthesia on efferent-mediated adaptation of the DPOAE. J Assoc Res Otolaryngol 3:362–373

Brown MC (1987) Morphology of labeled efferent fibers in the guinea pig cochlea. J Comp Neurol 260:605–618

Brown MC (1989) Morphology and response properties of single olivocochlear fibers in the guinea pig. Hear Res 40:93–110

Brown MC (2001) Response adaptation of medial olivocochlear neurons is minimal. J Neurophysiol 86:2381–2392

Brown MC (2002) Cochear projections of single medial olivocochlear (MOC) axons in the guinea pig. Asso Res Otolaryngol Abstr 25:310

Brown MC, Kujawa SG, Duca ML (1998a) Single olivocochlear neurons in the guinea pig. I. Binaural facilitation of responses to high-level noise. J Neurophysiol 79:3077–3087

Brown MC, Kujawa SG, Liberman MC (1998b) Single olivocochlear neurons in the guinea pig. II. Response plasticity due to noise conditioning. J Neurophysiol 79:3088–3097

Brown MC, de Venecia RK, Guinan JJ (2003) Responses of medial olivocochlear neurons: specifying the central pathways of the medial olivocochlear reflex. Exp Brain Res 153:491–498

Buño W Jr (1978) Auditory nerve fiber activity influenced by contralateral ear sound stimulation. Exp Neurol 59:62–74

Cai Y, Geisler CD (1996) Suppression in auditory-nerve fibers of cats using low-side suppressors. II. Effect of spontaneous rates. Hear Res 96:113–125

Chays A, Maison S, Robaglia-Schlupp A, Cau P, Broder L, Magnan J (2003) Are we sectioning the cochlear efferent system during vestibular neurotomy? Rev Laryngol Otol Rhinol (Bord) 124:53–58

Chéry-Croze A, Moulin A, Collet L (1993) Effect of contralateral sound stimulation on the distortion product 2f1–f2 in humans: evidence of a frequency specificity. Hear Res 68:53–58

Cody AR, Johnstone BM (1982) Acoustically evoked activity of single efferent neurons in the guinea pig cochlea. J Acoust Soc Am 72:280–282

Collet L, Kemp DT, Veuillet E, Duclaux R, Moulin A, Morgon A (1990) Effect of contralateral auditory stimuli on active cochlear micro-mechanical properties in human subjects. Hear Res 43:251–262

Cooper NP, Guinan JJ Jr (2003) Separate mechanical processes underlie fast and slow effects of medial olivocochlear efferent activity. J Physiol 548:307–312

Cooper NP, Guinan JJ Jr (2006a) Efferent-mediated control of basilar membrane motion. J Physiol 576:49–54

Cooper NP, Guinan JJ Jr (2006b) The dynamics of medial olivocochlear efferent fast effects on basilar membrane motion. Assoc Res Otolaryngol Abstr 30:273

Cooper NP, Kemp DT (2009) Concepts and challenges in the biophysics of hearing. World Scientific, Singapore

Cooper NP, Rhode WS (1996) Fast travelling waves, slow travelling waves and their interactions in experimental studies of apical cochlear mechanics. Audit Neurosci 2:289–299

Dallos P, He DZ, Lin X, Sziklai I, Mehta S, Evans BN (1997) Acetylcholine, outer hair cell electromotility, and the cochlear amplifier. J Neurosci 17:2212–2226

Dallos P, Wu X, Cheatham MA, Gao J, Zheng J, Anderson CT, Jia S, Wang X, Cheng WH, Sengupta S, He DZ, Zuo J (2008) Prestin-based outer hair cell motility is necessary for mammalian cochlear amplification. Neuron 58:333–339

Darrow KN, Maison SF, Liberman MC (2006a) Cochlear efferent feedback balances interaural sensitivity. Nat Neurosci 9:1474–1476

Darrow KN, Simons EJ, Dodds L, Liberman MC (2006b) Dopaminergic innervation of the mouse inner ear: evidence for a separate cytochemical group of cochlear efferent fibers. J Comp Neurol 498:403–414

de Boer J, Thornton AR (2007) Effect of subject task on contralateral suppression of click evoked otoacoustic emissions. Hear Res 233:117–123

de Boer J, Thornton AR (2008) Neural correlates of perceptual learning in the auditory brainstem: efferent activity predicts and reflects improvement at a speech-in-noise discrimination task. J Neurosci 28:4929–4937

de Venecia RK, Liberman MC, Guinan JJ Jr, Brown MC (2005) Medial olivocochlear reflex interneurons are located in the posteroventral cochlear nucleus: a kainic acid lesion study in guinea pigs. J Comp Neurol 487:345–360

Delano PH, Elgueda D, Hamame CM, Robles L (2007) Selective attention to visual stimuli reduces cochlear sensitivity in chinchillas. J Neurosci 27:4146–4153

Desmedt JE (1962) Auditory-evoked potentials from cochlea to cortex as influenced by activation of the efferent olivocochlear bundle. J Acoust Soc Am 34:1478–1496

Dolan DF, Nuttall AL, Avinash G (1990) Asynchronous neural activity recorded from the round window. J Acoust Soc Am 87:2621–2627

Dolan DF, Guo MH, Nuttall AL (1997) Frequency-dependent enhancement of basilar membrane velocity during olivocochlear bundle stimulation. J Acoust Soc Am 102:3587–3596

Elgoyhen AB, Johnson DS, Boulter J, Vetter DE, Heinemann S (1994) Alpha 9: an acetylcholine receptor with novel pharmacological properties expressed in rat cochlear hair cells. Cell 79:705–715

Elgoyhen AB, Vetter DE, Katz E, Rothlin CV, Heinemann SF, Boulter J (2001) Alpha10: a determinant of nicotinic cholinergic receptor function in mammalian vestibular and cochlear mechanosensory hair cells. Proc Natl Acad Sci USA 98:3501–3506

Fahey PF, Stagner BB, Martin GK (2008) Source of level dependent minima in rabbit distortion product otoacoustic emissions. J Acoust Soc Am 124:3694–3707

Feeney MP, Keefe DH (2001) Estimating the acoustic reflex threshold from wideband measures of reflectance, admittance, and power. Ear Hear 22:316–332

Feeney MP, Keefe DH, Sanford CA (2004) Wideband reflectance measures of the ipsilateral acoustic stapedius reflex threshold. Ear Hear 25:421–430

Fex J (1959) Augmentation of cochlear microphonic by stimulation of efferent fibers to the cochlea. Acta Otolaryngol 50:540–541

Fex J (1962) Auditory activity in centrifugal and centripetal cochlear fibers in cat. Acta Physiol Scand 55:2–68

Fex J (1965) Auditory activity in the uncrossed centrifugal cochlear fibers in cat. A study of a feedback system, II. Acta Physiol Scand 64:43–57

Folsom RC, Owsley RM (1987) N1 action potentials in humans. Influence of simultaneous contralateral stimulation. Acta Otolaryngol (Stockh) 103:262–265

Francis NA, Guinan JJ Jr (2010) Acoustic stimulation of human medial olivocochlear efferents reduces stimulus-frequency and click-evoked otoacoustic emission delays: Implications for cochlear filter bandwidths. Hear Res 267:36–45

Fuchs PA (1996) Synaptic transmission at vertebrate hair cells. Curr Opin Neurobiol 6:514–519

Galambos R (1956) Suppression of auditory activity by stimulation of efferent fibers to the cochlea. J Neurophysiol 19:424–437

Geisler CD (1992) Two-tone suppression by a saturating feedback model of the cochlear partition. Hear Res 63:203–210

Geisler CD, Yates GK, Patuzzi RB, Johnston BM (1990) Saturation of outer hair cell receptor currents causes two-tone suppression. Hear Res 44:241–256

Ghaffari R, Aranyosi AJ, Freeman DM (2007) Longitudinally propagating traveling waves of the mammalian tectorial membrane. Proc Natl Acad Sci USA 104:16510–16515

Giard M-H, Collet L, Bouchet P, Pernier J (1994) Auditory selective attention in the human cochlea. Brain Res 633:353–356

Gifford ML, Guinan JJ Jr (1983) Effects of crossed-olivocochlear-bundle stimulation on cat auditory nerve fiber responses to tones. J Acoust Soc Am 74:115–123

Gifford ML, Guinan JJ Jr (1987) Effects of electrical stimulation of medial olivocochlear neurons on ipsilateral and contralateral cochlear responses. Hear Res 29:179–194

Giraud AL, Collet L, Chery-Croze S, Magnan J, Chays A (1995) Evidence of a medial olivocochlear involvement in contralateral suppression of otoacoustic emissions in humans. Brain Res 705:15–23

Giraud AL, Collet L, Chery-Croze S (1997a) Suppression of otoacoustic emission is unchanged after several minutes of contralateral acoustic stimulation. Hear Res 109:78–82

Giraud AL, Garnier S, Micheyl C, Lina G, Chays A, Chery Croze S (1997b) Auditory efferents involved in speech-in-noise intelligibility. Neuroreport 8:1779–1783

Groff JA, Liberman MC (2003) Modulation of cochlear afferent response by the lateral olivocochlear system: activation via electrical stimulation of the inferior colliculus. J Neurophysiol 90:3178–3200

Guinan JJ Jr (1986) Effect of efferent neural activity on cochlear mechanics. Scand Audiol Suppl 25:53–62

Guinan JJ Jr (1990) Changes in stimulus frequency otoacoustic emissions produced by two-tone suppression and efferent stimulation in cats. In: Dallos P, Geisler CD, Matthews JW, Steele CR (eds) Mechanics and biophysics of hearing. Springer, New York, pp 170–177

Guinan JJ Jr (1996) The physiology of olivocochlear efferents. In: Dallos PJ, Popper AN, Fay RR (eds) The cochlea. Springer, New York, pp 435–502

Guinan JJ Jr (1997) Efferent inhibition as a function of efferent stimulation parameters and sound frequency: testing the OHC-shunt hypothesis. In: Lewis ER, Long GR, Lyon RF, Narins PM, Steele CR, Hecht-Poinar E (eds) Diversity in auditory mechanics. World Scientific, Singapore, pp 368–375

Guinan JJ Jr, Cooper NP (2003) Fast effects of efferent stimulation on basilar membrane motion. In: Gummer AW, Dalhoff E, Nowotny M, Scherer MP (eds) The biophysics of the cochlea: molecules to models. World Scientific, Singapore, pp 245–251

Guinan JJ Jr, Cooper NP (2008) Medial olivocochlear efferent inhibition of basilar-membrane responses to clicks: evidence for two modes of cochlear mechanical excitation. J Acoust Soc Am 124:1080–1092

Guinan JJ Jr, Gifford ML (1988a) Effects of electrical stimulation of efferent olivocochlear neurons on cat auditory-nerve fibers. I. Rate-level functions. Hear Res 33:97–114

Guinan JJ Jr, Gifford ML (1988b) Effects of electrical stimulation of efferent olivocochlear neurons on cat auditory-nerve fibers. II. Spontaneous rate. Hear Res 33:115–128

Guinan JJ Jr, Gifford ML (1988c) Effects of electrical stimulation of efferent olivocochlear neurons on cat auditory-nerve fibers. III. Tuning curves and thresholds at CF. Hear Res 37:29–46

Guinan JJ Jr, Stankovic KM (1996) Medial efferent inhibition produces the largest equivalent attenuations at moderate to high sound levels in cat auditory-nerve fibers. J Acoust Soc Am 100:1680–1690

Guinan JJ Jr, Norris BE, Guinan SS (1972) Single auditory units in the superior olivary complex II: locations of unit categories and tonotopic organization. Int J Neurosci 4:147–166

Guinan JJ Jr, Warr WB, Norris BE (1983) Differential olivocochlear projections from lateral vs. medial zones of the superior olivary complex. J Comp Neurol 221:358–370

Guinan JJ Jr, Warr WB, Norris BE (1984) Topographic organization of the olivocochlear projections from the lateral and medial zones of the superior olivary complex. J Comp Neurol 226:21–27

Guinan JJ, Backus BC, Lilaonitkul W, Aharonson V (2003) Medial olivocochlear efferent reflex in humans: otoacoustic emission (OAE) measurement issues and the advantages of stimulus frequency OAEs. J Assoc Res Otolaryngol 4:521–540

Guinan JJ Jr, Lin T, Cheng H (2005) Medial-olivocochlear-efferent inhibition of the first peak of auditory-nerve responses: evidence for a new motion within the cochlea. J Acoust Soc Am 118:2421–2433

Guinan JJ Jr, Lin T, Cheng H, Cooper N (2006) Medial-Olivocochlear-Efferent Effects on Basilar-Membrane and Auditory-Nerve Responses to Clicks: Evidence for a New Motion within the Cochlea. In: Nuttall AL, Ren T, Gillespie PG, Grosh K, de Boer E, eds. Auditory Mechanisms: Processes and Models. World Scientific, Singapore, pp:1–9

Gummer M, Yates GK, Johnstone BM (1988) Modulation transfer function of efferent neurons in the guinea pig cochlea. Hear Res 36:41–52

He DZ, Jia S, Dallos P (2003) Prestin and the dynamic stiffness of cochlear outer hair cells. J Neurosci 23:9089–9096

Henson OW, Xie DH, Keating AW, Henson MM (1995) The effect of contralateral stimulation on cochlear resonance and damping in the mustached bat: the role of the medial efferent system. Hear Res 86:111–124

Hienz RD, Stiles P, May BJ (1998) Effects of bilateral olivocochlear lesions on vowel formant discrimination in cats. Hear Res 116:10–20

Housley GD, Ashmore JF (1991) Direct measurement of the action of acetylcholine on isolated outer hair cells of the guinea pig cochlea. Proc R Soc Lond B 244:161–167

Hudspeth AJ (2008) Making an effort to listen: mechanical amplification in the ear. Neuron 59:530–545

Karavitaki KD, Mountain DC (2007a) Evidence for outer hair cell driven oscillatory fluid flow in the tunnel of corti. Biophys J 92:3284–3293

Karavitaki KD, Mountain DC (2007b) Imaging electrically evoked micromechanical motion within the organ of Corti of the excised gerbil cochlea. Biophys J 92:3294–3316

Kawase T, Delgutte B, Liberman MC (1993) Anti-masking effects of the olivocochlear reflex, II: Enhancement of auditory-nerve response to masked tones. J Neurophysiol 70:2533–2549

Keefe DH, Schairer KS, Ellison JC, Fitzpatrick DF, Jesteadt W (2009) Use of stimulus-frequency otoacoustic emissions to investigate efferent and cochlear contributions to temporal overshoot. J Acoust Soc Am 125:1595–1604

Khalfa S, Veuillet E, Collet L (1998) Influence of handedness on peripheral auditory asymmetry. Eur J Neurosci 10:2731–2737

Khalfa S, Bougeard R, Morand N, Veuillet E, Isnard J, Guenot M, Ryvlin P, Fischer C, Collet L (2001) Evidence of peripheral auditory activity modulation by the auditory cortex in humans. Neuroscience 104:347–358

Kim DO, Dorn PA, Neely ST, Gorga MP (2001) Adaptation of distortion product otoacoustic emission in humans. J Assoc Res Otolaryngol 2:31–40

Kimura R, Wersäll J (1962) Termination of the olivocochlear bundle in relation to the outer hair cells of the organ of Corti in guinea pig. Acta Otolaryng (Stockh) 55:11–32

Konishi T, Slepian JZ (1971) Effects of the electrical stimulation of the crossed olivocochlear bundle on cochlear potentials recorded with intracochlear electrodes in guinea pigs. J Acoust Soc Am 49:1762–1769

Kumar UA, Vanaja CS (2004) Functioning of olivocochlear bundle and speech perception in noise. Ear Hear 25:142–146

Larsen E, Liberman MC (2009) Slow build-up of cochlear suppression during sustained contralateral noise: central modulation of olivocochlear efferents? Hear Res 256:1–10

Le Prell CG, Shore SE, Hughes LF, Bledsoe SC Jr (2003) Disruption of lateral efferent pathways: functional changes in auditory evoked responses. J Assoc Res Otolaryngol 4:276–290

Liberman MC (1980) Efferent synapses in the inner hair cell area of the cat cochlea: An electron microscopic study of serial sections. Hear Res 3:189–204

Liberman MC (1988a) Response properties of cochlear efferent neurons: monaural vs. binaural stimulation and the effects of noise. J Neurophysiol 60:1779–1798

Liberman MC (1988b) Physiology of cochlear efferent and afferent neurons: direct comparisons in the same animal. Hear Res 34:179–192

Liberman MC (1989) Rapid assessment of sound-evoked olivocochlear feedback: suppression of compound action potentials by contralateral sound. Hear Res 38:47–56

Liberman MC (1990) Effects of chronic cochlear de-efferentation on auditory-nerve response. Hear Res 49:209–224

Liberman MC, Brown MC (1986) Physiology and anatomy of single olivocochlear neurons in the cat. Hear Res 24:17–36

Liberman MC, Kiang NYS (1984) Single-neuron labeling and chronic cochlear pathology. IV. Stereocilia damage and alterations in rate- and phase-level functions. Hear Res 16:75–90

Liberman MC, Dodds LW, Pierce S (1990) Afferent and efferent innervation of the cat cochlea: quantitative analysis with light and electron microscopy. J Comp Neurol 301:443–460

Liberman MC, Puria S, Guinan JJ Jr (1996) The ipsilaterally evoked olivocochlear reflex causes rapid adaptation of the $2f_1–f_2$ distortion product otoacoustic emission. J Acoust Soc Am 99:3572–3584

Lilaonitkul W, Guinan JJ Jr (2009a) Reflex control of the human inner ear: a half-octave offset in medial efferent feedback that is consistent with an efferent role in the control of masking. J Neurophysiol 101:1394–1406

Lilaonitkul W, Guinan JJ Jr (2009b) Human medial olivocochlear reflex: effects as functions of contralateral, ipsilateral, and bilateral elicitor bandwidths. J Assoc Res Otolaryngol 10:459–470

Lima da Costa DL, Chibois A, Erre JP, Blanchet C, de Sauvage RC, Aran JM (1997) Fast, slow, and steady-state effects of contralateral acoustic activation of the medial olivocochlear efferent system in awake guinea pigs: action of gentamicin. J Neurophysiol 78:1826–1836

Lin T, Guinan JJ Jr (2000) Auditory-nerve-fiber responses to high-level clicks: interference patterns indicate that excitation is due to the combination of multiple drives. J Acoust Soc Am 107:2615–2630

Lin T, Guinan JJ Jr (2004) Time-frequency analysis of auditory-nerve-fiber and basilar-membrane click responses reveal glide irregularities and non-characteristic-frequency skirts. J Acoust Soc Am 116:405–416
Lu TK, Zhak S, Dallos P, Sarpeshkar R (2006) Fast cochlear amplification with slow outer hair cells. Hear Res 214:45–67
Maison SF, Liberman MC (2000) Predicting vulnerability to acoustic injury with a noninvasive assay of olivocochlear reflex strength. J Neurosci 20:4701–4707
Maison S, Micheyl C, Collet L (1999) Sinusoidal amplitude modulation alters contralateral noise suppression of evoked otoacoustic emissions in humans. Neuroscience 91:133–138
Maison S, Micheyl C, Andeol G, Gallego S, Collet L (2000) Activation of medial olivocochlear efferent system in humans: influence of stimulus bandwidth. Hear Res 140:111–125
Maison S, Micheyl C, Collet L (2001) Influence of focused auditory attention on cochlear activity in humans. Psychophysiology 38:35–40
Maison SF, Adams JC, Liberman MC (2003) Olivocochlear innervation in the mouse: immunocytochemical maps, crossed versus uncrossed contributions, and transmitter colocalization. J Comp Neurol 455:406–416
Maison SF, Vetter DE, Liberman MC (2007) A novel effect of cochlear efferents: in vivo response enhancement does not require alpha9 cholinergic receptors. J Neurophysiol 97:3269–3278
May BJ, McQuone SJ, Lavoie A (1995) Effects of olivocochlear lesions on intensity discrimination in cats. Assoc Res Otolaryngol Abstr 18:146
May BJ, Prosen CA, Weiss D, Vetter D (2002) Behavioral investigation of some possible effects of the central olivocochlear pathways in transgenic mice. Hear Res 171:142–157
May BJ, Budelis J, Niparko JK (2004) Behavioral studies of the olivocochlear efferent system: learning to listen in noise. Arch Otolaryngol Head Neck Surg 130:660–664
Micheyl C, Collet L (1996) Involvement of the olivocochlear bundle in the detection of tones in noise. J Acoust Soc Am 99:1064–1610
Micheyl C, Perrot X, Collet L (1997) Relationship between auditory intensity discrimination in noise and olivocochlear efferent system activity in humans. Behav Neurosci 111:801–807
Michie PT, LePage EL, Solowlij N, Haller M, Terry L (1996) Evoked otoacoustic emissions and auditory selective attention. Hear Res 98:54–67
Morand N, Bouvard S, Ryvlin P, Mauguiere F, Fischer C, Collet L, Veuillet E (2001) Asymmetrical localization of benzodiazepine receptors in the human auditory cortex. Acta Otolaryngol 121:293–296
Morand-Villeneuve N, Veuillet E, Perrot X, Lemoine P, Gagnieu MC, Sebert P, Durrant JD, Collet L (2005) Lateralization of the effects of the benzodiazepine drug oxazepam on medial olivocochlear system activity in humans. Hear Res 208:101–106
Morlet T, Goforth L, Hood LJ, Ferber C, Duclaux R, Berlin CI (1999) Development of human cochlear active mechanism asymmetry: involvement of the medial olivocochlear system? Hear Res 134:153–162
Moulin A, Collet L, Duclaux R (1993) Contralateral auditory stimulation alters acoustic distortion products in humans. Hear Res 65:193–210
Mountain DC (1998) Modal analysis: a new paradigm for cochlear mechanics. Assoc Res Otolaryngol Abstr 21:61
Mukari SZ, Mamat WH (2008) Medial olivocochlear functioning and speech perception in noise in older adults. Audiol Neurootol 13:328–334
Muller J, Janssen T, Heppelmann G, Wagner W (2005) Evidence for a bipolar change in distortion product otoacoustic emissions during contralateral acoustic stimulation in humans. J Acoust Soc Am 118:3747–3756
Murugasu E, Russell IJ (1996) The effect of efferent stimulation on basilar membrane displacement in the basal turn of the guinea pig cochlea. J Neurosci 16:325–332
Norman M, Thornton ARD (1993) Frequency analysis of the contralateral suppression of evoked otoacoustic emissions by narrow-band noise. Br J Audiol 27:281–289

Nowotny M, Gummer AW (2006) Nanomechanics of the subtectorial space caused by electromechanics of cochlear outer hair cells. Proc Natl Acad Sci USA 103:2120–2125
Perrot X, Micheyl C, Khalfa S, Collet L (1999) Stronger bilateral efferent influences on cochlear biomechanical activity in musicians than in non-musicians. Neurosci Lett 262:167–170
Pfalz RKJ (1969) Absence of a function for the crossed olivocochlear bundle under physiological conditions. Arch Klin Exp Ohren Nasen Kehlkopfheilkd 193:89–100
Puria S, Guinan JJ Jr, Liberman MC (1996) Olivocochlear reflex assays: effects of contralateral sound on compound action potentials vs. ear-canal distortion products. J Acoust Soc Am 99:500–507
Rajan R (1988) Effect of electrical stimulation of the crossed olivocochlear bundle on temporary threshold shifts in auditory sensitivity. I. Dependence on electrical stimulation parameters. J Neurophysiol 60:549–568
Rajan R (2000) Centrifugal pathways protect hearing sensitivity at the cochlea in noisy environments that exacerbate the damage induced by loud sound. J Neurosci 20:6684–6693
Recio A, Rich NC, Narayan SS, Ruggero MA (1998) Basilar-membrane responses to clicks at the base of the chinchilla cochlea. J Acoust Soc Am 103:1972–1989
Reiter ER, Liberman MC (1995) Efferent-mediated protection from acoustic overexposure: relation to slow effects of olivocochlear stimulation. J Neurophysiol 73:506–514
Ren T, Nuttall AL (2001) Recording depth of the heterodyne laser interferometer for cochlear vibration measurement. J Acoust Soc Am 109:826–829
Rhode WS (2007) Basilar membrane mechanics in the 6–9 kHz region of sensitive chinchilla cochleae. J Acoust Soc Am 121:2792–2804
Robertson D (1984) Horseradish peroxidase injection of physiologically characterized afferent and efferent neurones in the guinea pig spiral ganglion. Hear Res 15:113–121
Robertson D (1985) Brainstem location of efferent neurons projecting to the guinea pig cochlea. Hear Res 20:79–84
Robertson D, Gummer M (1985) Physiological and morphological characterization of efferent neurons in the guinea pig cochlea. Hear Res 20:63–77
Robertson D, Gummer M (1988) Physiology of cochlear efferents in the mammal. In: Syka J, Masterton RB (eds) Auditory pathways: structure and function. Plenum, New York, pp 269–278
Robertson D, Anderson C-J, Cole KS (1987) Segregation of efferent projections to different turns of the guinea pig cochlea. Hear Res 25:69–76
Robles L, Ruggero MA (2001) Mechanics of the mammalian cochlea. Physiol Rev 81:1305–1352
Ruel J, Nouvian R, Gervais d'Aldin C, Pujol R, Eybalin M, Puel JL (2001) Dopamine inhibition of auditory nerve activity in the adult mammalian cochlea. Eur J Neurosci 14:977–986
Russell IJ, Murugasu E (1997) Medial efferent inhibition suppresses basilar membrane responses to near characteristic frequency tones of moderate to high intensities. J Acoust Soc Am 102:1734–1738
Ryan S, Kemp DT (1996) The influence of evoking stimulus level on the neural suppression of transient evoked otoacoustic emissions. Hear Res 94:140–147
Ryan S, Kemp DT, Hinchcliffe R (1991) The influence of contralateral acoustic stimulation on click-evoked otoacoustic emission in humans. Br J Audiol 25:391–397
Santos-Sacchi J (1991) Reversible inhibition of voltage-dependent outer hair cell motility and capacitance. J Neurosci 11:3096–3110
Scharf B, Magnan J, Chays A (1997) On the role of the olivocochlear bundle in hearing: 16 case studies. Hear Res 103:101–122
Shera CA (2001) Frequency glides in click responses of the basilar membrane and auditory nerve: their scaling behavior and origin in traveling-wave dispersion. J Acoust Soc Am 109:2023–2034
Shera CA, Guinan JJ Jr (1999) Evoked otoacoustic emissions arise by two fundamentally different mechanisms: a taxonomy for mammalian OAEs. J Acoust Soc Am 105:782–798
Shera CA, Guinan JJ Jr (2007) Cochlear traveling-wave amplification, suppression, and beamforming probed using noninvasive calibration of intracochlear distortion sources. J Acoust Soc Am 121:1003–1016

Siegel JH, Kim DO (1982) Efferent neural control of cochlear mechanics? Olivocochlear bundle stimulation affects cochlear biomechanical nonlinearity. Hear Res 6:171–182

Sininger YS, Cone-Wesson B (2004) Asymmetric cochlear processing mimics hemispheric specialization. Science 305:1581

Smith CA (1961) Innervation pattern of the cochlea. Ann Oto Rhinol Laryngol 70:504–527

Sridhar TS, Liberman MC, Brown MC, Sewell WF (1995) A novel cholinergic "slow effect" of olivocochlear stimulation on cochlear potentials in the guinea pig. J Neurosci 15:3667–3678

Sridhar TS, Brown MC, Sewell WF (1997) Unique post-synaptic signaling at the hair cell efferent synapse permits calcium to evoke changes on two different time scales. J Neurosci 17:428–437

Stankovic KM, Guinan JJ Jr (1999) Medial efferent effects on auditory-nerve responses to tail-frequency tones I: rate reduction. J Acoust Soc Am 106:857–869

Strickland EA (2008) The relationship between precursor level and the temporal effect. J Acoust Soc Am 123:946–954

Strickland EA, Krishnan LA (2005) The temporal effect in listeners with mild to moderate cochlear hearing impairment. J Acoust Soc Am 118:3211–3217

Tan MN, Robertson D, Hammond GR (2008) Separate contributions of enhanced and suppressed sensitivity to the auditory attentional filter. Hear Res 241:18–25

Teas DC, Konishi T, Nielsen DW (1972) Electrophysiological studies on the spatial distribution of the crossed olivocochlear bundle along the guinea pig cochlea. J Acoust Soc Am 51:1256–1264

Thiers FA, Burgess BJ, Nadol JB (2002) Reciprocal innervation of outer hair cells in a human infant. J Assoc Res Otolaryngol 3:269–278

Thiers FA, Nadol JB Jr, Liberman MC (2008) Reciprocal synapses between outer hair cells and their afferent terminals: evidence for a local neural network in the mammalian cochlea. J Assoc Res Otolaryngol 9:477–489

Thompson AM, Thompson GC (1991) Posteroventral cochlear nucleus projections to olivocochlear neurons. J Comp Neurol 303:267–285

Thompson S, Abdelrazeq S, Long GR, Henin S (2009) Differential effects of efferent stimulation by contralateral bandpass noise on the two major components of distortion product otoacoustic emissions. Assoc Res Otolaryngol Abstr 32:244

Veuillet E, Collet L, Duclaux R (1991) Effect of contralateral acoustic stimulation on active cochlear micromechanical properties in human subjects: dependence on stimulus variables. J Neurophysiol 65:724–735

Veuillet E, Duverdy-Bertholon F, Collet L (1996) Effect of contralateral acoustic stimulation on the growth of click-evoked otoacoustic emissions in humans. Hear Res 93:128–135

Veuillet E, Magnan A, Ecalle J, Thai-Van H, Collet L (2007) Auditory processing disorder in children with reading disabilities: effect of audiovisual training. Brain 130:2915–2928

Wagner W, Heppelmann G, Muller J, Janssen T, Zenner HP (2007) Olivocochlear reflex effect on human distortion product otoacoustic emissions is largest at frequencies with distinct fine structure dips. Hear Res 223:83–92

Wagner W, Frey K, Heppelmann G, Plontke SK, Zenner HP (2008) Speech-in-noise intelligibility does not correlate with efferent olivocochlear reflex in humans with normal hearing. Acta Otolaryngol 128:53–60

Warr WB (1975) Olivocochlear and vestibular efferent neurons of the feline brain stem: their location, morphology and number determined by retrograde axonal transport and acetylcholinesterase histochemistry. J Comp Neurol 161:159–182

Warr WB (1992) Organization of olivocochlear efferent systems in mammals. In: Webster DB, Popper AN, Fay RR (eds) Mammalian auditory pathway: neuroanatomy. Springer, New York, pp 410–448

Warr WB, Boche JE (2003) Diversity of axonal ramifications belonging to single lateral and medial olivocochlear neurons. Exp Brain Res 153:499–513

Warr WB, Guinan JJ Jr (1979) Efferent innervation of the organ of Corti: two separate systems. Brain Res 173:152–155
Warr WB, Beck Boche JE, Neely ST (1997) Efferent innervation of the inner hair cell region: origins and terminations of two lateral olivocochlear systems. Hear Res 108:89–111
Warren EH III, Liberman MC (1989a) Effects of contralateral sound on auditory-nerve responses. I. Contributions of cochlear efferents. Hear Res 37:89–104
Warren EH III, Liberman MC (1989b) Effects of contralateral sound on auditory-nerve responses. II. Dependence on stimulus variables. Hear Res 37:105–122
Wiederhold ML (1970) Variations in the effects of electric stimulation of the crossed olivocochlear bundle on cat single auditory-nerve-fiber responses to tone bursts. J Acoust Soc Am 48:966–977
Wiederhold ML, Peake WT (1966) Efferent inhibition of auditory nerve responses: dependence on acoustic stimulus parameters. J Acoust Soc Am 40:1427–1430
Winslow RL, Sachs MB (1987) Effect of electrical stimulation of the crossed olivocochlear bundle on auditory nerve response to tones in noise. J Neurophysiol 57:1002–1021
Ye Y, Machado DG, Kim DO (2000) Projection of the marginal shell of the anteroventral cochlear nucleus to olivocochlear neurons in the cat. J Comp Neurol 420:127–138
Yoshida N, Liberman MC, Brown MC, Sewell WF (1999) Gentamicin blocks both fast and slow effects of olivocochlear activation in anesthetized guinea pigs. J Neurophysiol 82:3168–3174
Zeng F-G, Shannon RV (1994) Loudness-coding mechanisms inferred from electric stimulation of the human auditory system. Science 264:564–566
Zeng F, Martino KM, Linthicum FH, Soli SD (2000) Auditory perception in vestibular neurectomy subjects. Hear Res 142:102–112
Zurek PM (1985) Acoustic emissions from the ear: a summary of results from humans and animals. J Acoust Soc Am 78:340–344
Zwicker E (1965) Temporal effects in simultaneous masking by white-noise bursts. J Acoust Soc Am 37:653–663
Zyl AV, Swanepoel DW, Hall JW III (2009) Effect of prolonged contralateral acoustic stimulation on transient evoked otoacoustic emissions. Hear Res 254:77–81

Chapter 4
Pharmacology and Neurochemistry of Olivocochlear Efferents

William F. Sewell

4.1 Introduction

This chapter covers the chemistry and pharmacology of efferent transmission in the cochlea, starting with an overview of the biochemical and biophysical steps following the arrival of an action potential at the peripheral efferent nerve terminal (Sect. 4.1.1). A brief history of advances in understanding the efferent system follows (Sect. 4.1.2). The description of the neurochemistry and pharmacology of efferent action is organized around the sequence of events beginning with the arrival of an action potential at the medial efferent peripheral terminal and ending with the activation of calcium-dependent potassium channels in the outer hair cell (OHC; Sect. 4.2). In Sect. 4.3, other efferent neurotransmitters are covered and, because most of these are associated with the lateral efferent system, it is in this section that much of the knowledge of lateral efferents is presented.

4.1.1 Overview of Biochemical and Biophysical Steps in Efferent Activation

Medial efferent fibers are large, myelinated nerves that primarily terminate on the OHCs. The arrival of an action potential at an efferent nerve ending induces release of the neurotransmitter, acetylcholine (ACh), onto the OHCs. Released ACh crosses the synaptic cleft to activate nicotinic cholinergic receptors on the OHC side of the synapse. The nicotinic receptors on OHCs are unusual, comprising $\alpha 9$ and $\alpha 10$ subunits. They are permeable to cations with a relatively high affinity for calcium

W.F. Sewell (✉)
Department of Otology and Laryngology, Harvard Medical School, Boston, MA 02114, USA
and
Eaton Peabody Laboratory, Massachusetts Eye and Ear Infirmary, Boston, MA 02114, USA
e-mail: wfs@epl.meei.harvard.edu

D.K. Ryugo et al. (eds.), *Auditory and Vestibular Efferents*,
Springer Handbook of Auditory Research 38, DOI 10.1007/978-1-4419-7070-1_4,

compared to that for sodium. Calcium entering the OHC through the nicotinic receptors can activate a class of potassium channels, called calcium-activated potassium (K_{Ca}) channels. Potassium leaving the OHC via the K_{Ca} channels hyperpolarizes the cell. The calcium entering is also thought to activate calcium channels in a structure known as the synaptic cistern, which is very closely apposed (within 30 nm) to the OHC's membrane. These calcium-activated calcium release channels add more calcium to the intracellular space near the synapse, effectively serving to amplify the effects of calcium entering via the nicotinic receptors. Calcium sparks may travel from the synaptic cisterna to the subsurface cisterna of the OHC to evoke effects along the lateral wall of the OHC. The neurochemical processes involved in medial efferent activation at the OHC are illustrated in Fig. 4.1.

Lateral efferent fibers are small unmyelinated fibers that terminate primarily on the dendrites of the spiral ganglion cells near the inner hair cells (IHCs). Considerably less is known about lateral efferent action than medial efferent action, as it has proven difficult to directly activate these neurons with electrical stimulation. The lateral efferents, like the medial efferents, use ACh as a neurotransmitter. There are, however, a broad array of neurotransmitters and neuromodulators often associated with the lateral efferents, including opioid peptides, calcitonin gene-related peptide (CGRP), γ-aminobutyric acid (GABA), and dopamine.

4.1.2 Historical Perspective of Issues in the Pharmacology of the Olivocochlear Efferents

The first pharmacological analysis of efferent transmission actually occurred shortly before the functional effects of efferent stimulation were understood. In the late 1940s ACh was becoming established as a neurotransmitter, and pharmacology was becoming a valuable tool to analyze cholinergic transmission. Acetylcholinesterase (AChE) had been identified as the enzyme that terminated the action of ACh by cleaving the ACh molecule into choline and acetate (Nachmansohn and Wilson 1951). Drugs such as physostigmine and neostigmine had been developed to block the action of ACh, and drugs that blocked the ACh receptor, such as atropine, were well studied. Assays for AchE were available and were being used to probe for the possibility of cholinergic transmission at many synapses (Augustinsson 1946).

Gisselsson (1950) found AChE in perilymph and then completed a series of pharmacological experiments from which he concluded that ACh was indeed likely to play a role in cochlear function, though not in afferent transmission. He did not ascribe the role to efferent function, possibly because the efferent pathways had been described only in 1946 (Rasmussen 1946), and it wasn't until 1956 that Galambos (1956) had published experiments that demonstrated that stimulation of efferents could inhibit cochlear responses. At this time Churchill et al. (1956) completed a histochemical assay for cholinesterase, finding high concentrations of this enzyme beneath each OHC and in a dense band beneath the IHCs. These findings led to the first suggestion that the olivocochlear efferents were cholinergic.

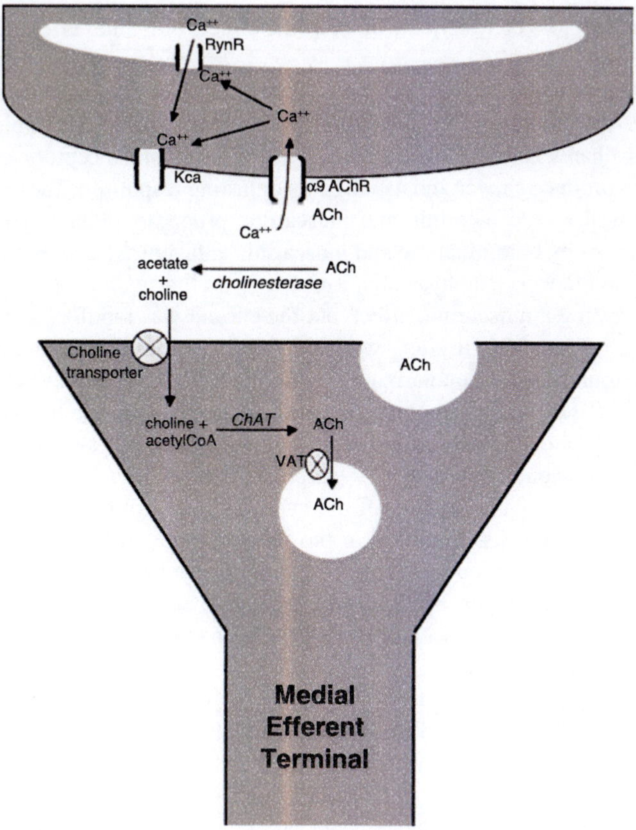

Fig. 4.1 Neurochemical processes involved in medial efferent activation at the outer hair cell are illustrated schematically. Acetylcholine (ACh) is synthesized in the medial efferent axon terminal from choline and acetyl-CoA by the enzyme choline acetyltransferase (ChAT), then packaged into synaptic vesicles by the vesicular acetylcholine transporter (VAT). Once released into the synaptic cleft, ACh is a target for degradation by the enzyme acetylcholinesterase (AChE), which cleaves it into acetate and choline. Choline is transported back into the efferent axonal terminal to be recycled into more ACh. Activation of the nicotinic ACh receptor (AChR) on the outer hair cell allows calcium to enter the cell and activate calcium-activated potassium channels (K_{Ca}), which hyperpolarize the hair cell. Calcium entering through the AChR can also activate ryanodine receptors (RynR) on the synaptic cistern to induce release of calcium from internal stores, which amplifies the calcium signal initiated by AChR activation

Thus Gisselson's deduction from pharmacological analysis that ACh played a role in cochlear responses proved to be essentially correct.

It was the discovery of enkephalins in efferent terminals (Fex and Altschuler 1981) that led to the idea that there might be fundamental differences in the pharmacology of the medial and lateral efferents. This was soon followed by a plethora of other neurotransmitter candidates in lateral efferents, including other opioid peptides, CGRP, dopamine, and GABA. All of these are now thought to be primarily (though not exclusively) associated with the lateral efferents, and the action of ACh on nicotinic receptors can account for all of the effects of medial efferents that have been described.

It was clear that the efferent effects were mediated by the ACh receptor. ACh receptors are classified as either nicotininic or muscarinic, with nicotinic receptors forming ion channels with fast responses and muscarinic receptors coupling to G-proteins to produce slower and usually longer lasting responses. The efferent ACh receptor showed mixed nicotinic and muscarinic properties. Pharmacologically it could be blocked by both nicotinic and muscarinic antagonists, as well as by strychnine, a glycine blocker. Biophysically, it was evident that ACh evoked a potassium current, suggesting a muscarinic effect, but the current was rapid, consistent with a nicotinic effect. Part of this mystery was solved in an analysis by Fuchs and Murrow (1992), who found a brief depolarizing current preceding the ACh evoked hyperpolarizing current. They deduced correctly that calcium entering the hair cell through a nicotinic receptor could activate K_{Ca} channels, whose activity "swamped that of the small amount of calcium entering, to hyperpolarize the cell."

One critical issue – the identity of the cholinergic receptor on the OHC – was completely confusing. As described in the preceding text, this obviously cholinergic receptor had both nicotinic and muscarinic pharmacological properties. The identity was resolved when Elgoyhen et al. (1994) discovered the α9 nicotinic receptor, whose pharmacological properties explained many of the pharmacological mysteries of medial efferents.

The discovery of two separate actions with very different time courses (fast vs. slow effects) (Sridhar et al. 1995) led to the idea that efferent activation may control different cellular mechanisms by different intracellular signal pathways. For example, protection from acoustic trauma and some actions on basilar membrane mechanics may arise from mechanisms different than those that attenuate cochlear responses to acoustic stimulation (Reiter and Liberman 1995; Dallos et al. 1997; Cooper and Guinan 2003).

Much of the historical scientific literature on efferent pharmacology and neurotransmitters is accessible in two comprehensive and now classic reviews. Paul Guth's comprehensive tome (Guth et al. 1976) published in 1976, covers virtually all work done up to that time. Eybalin's (1993) meticulous review paper took up the topic where Guth left off. The present chapter focuses on more recent work.

4.2 Cholinergic Medial Efferent Transmission

4.2.1 The Medial Efferent Synapse

Medial efferent fibers terminate at the base of the OHC in relatively large endings. The endings, described by Smith and Sjostrand (1961) in guinea pig, are 2–3 μm in length and less in diameter, with a synaptic cleft of around 20 nm. The endings are packed with clear-core vesicles about 28–35 nm in diameter. The poles of the terminals distal to the synapses are packed with mitochondria. The hair cell side of the synapse invariably demonstrates a pair of membranes forming a synaptic cistern about 18 nm in thickness. The synaptic cistern covers the synaptic area, is separated from the synaptic membrane by a narrow space of around 7–9 nm (Smith and Sjostrand 1961), and contains the ryanodine receptor, RyR1 (Lioudyno et al. 2004), suggesting a role in calcium-induced calcium release.

4.2.2 Events at the Efferent Terminal

Action potentials arriving at medial efferent terminals trigger release of synaptic vesicles containing ACh into the synaptic cleft. The fundamental molecular elements needed for vesicular release that are normally found in central nervous system synapses are also found in efferent terminals, including synapsin, SNAP-25, synaptophysin, and Cav1.2 (Gil-Loyzaga and Pujol 1988; Knipper et al. 1995; Kurc et al. 1998; Safieddine and Wenthold 1999; Waka et al. 2003; Bergeron et al. 2005).

4.2.3 ACh Metabolism

The metabolism of ACh is well understood. ACh is synthesized within the nerve terminal from choline and acetyl-CoA by the enzyme choline acetyltransferase (ChAT). Choline is taken up by efferent terminals. Acetyl-CoA is made in mitochondria, which are present in high density in cholinergic terminals. ChAT is synthesized in the cell body and transported to the nerve endings. Newly synthesized ACh is packaged into the synaptic vesicles via the vesicular ACh transporter (VAT). Once released into the synaptic cleft, ACh is broken down into choline and acetate by acetylcholinesterase (AChE), an enzyme excreted into the extracellular space in and around the synapse. The choline generated by AChE is taken up from extracellular fluid and pumped back into the efferent terminal by a high-affinity choline transporter (ChT1). The cycle of synthesis, packaging, and release then repeats itself.

There is an abundance of evidence to suggest ACh is synthesized as described in the preceding text in the cochlea. Jasser and Guth (1973) assayed the activity of the enzyme ChAT in the cat cochlea, demonstrating its presence and its disappearance when the efferents were lesioned. Similarly, Godfrey et al. (1976) demonstrated its presence, quantified enzymatic activity by cochlear turn, and compared ChAT levels to those in other cochlear tissues. In a tour de force of microchemical analysis, Godfrey and Ross (1985) quantified activities in discrete portions of the cochlea and were able to show much higher levels of activity in the IHC region than in the OHC region, and lower levels in the apical compared to middle or lower turns. Transection of the olivocochlear path on one side led to loss of ChAT activity in the ipsilateral organ of Corti. These biochemical data were confirmed by immunohistochemical analyses, indicating the presence of ChAT in efferent fibers beneath the OHC, below the OHCs, and in the inner spiral bundle (Altschuler et al. 1985b; Eybalin and Pujol 1987). These findings in the cochlea are supported by studies of the superior olivary complex where ChAT, VAT, and AChE were observed in both medial and lateral OC neurons (Yao and Godfrey 1998, 1999; Simmons et al. 1999; Bergeron et al. 2005).

The high-affinity choline transporter ChT1 can be blocked by the drug hemicholinium to reduce uptake of choline into the efferent nerve terminal. Because choline is required for ACh synthesis, new ACh cannot be synthesized. Thus hemicholinium induces a use-dependent depletion of ACh from the terminal and ultimately a block of cholinergic transmission. Medial efferent transmission can be blocked by hemicholinium (Galley et al. 1973; Comis and Guth 1974). ChT1 has been identified in lysates of the cochlea and, in the mature mouse, is found almost exclusively in the medial efferent terminals below the OHC (Bergeron et al. 2005).

After ACh is synthesized in the cytoplasm of the nerve terminal, it is transported into the synaptic vesicle by VAT, a process that can be blocked by the drug vesamicol. VAT is present in efferent cell bodies, as described previously, and has also been found in efferent terminals via an immunohistochemical approach (Maison et al. 2003).

AChE can be inhibited by cholinesterase inhibitors such as physostimine and neostigmine. Physostigmine can cross the blood–brain and blood–cochlear barriers, though an effect of efferent action with systemic injection has not been observed (see Guth et al. 1976 for an explanation). Cholinesterase inhibitors have been used in conjunction with ACh administration (to prolong the half-life of ACh) in studies of the effects of ACh on cochlear function (Kujawa et al. 1992).

4.2.4 Presynaptic Cholinergic Receptors

Numerous investigators have suggested that muscarinic receptors may play a presynaptic role at the medial efferent/OHC synapse. Bartolami et al. (1993) inferred the presence of muscarinic receptors in medial fibers by showing that a biochemical response (inositol phosphate formation) to carbachol (a cholinergic agonist) was eliminated by section of the olivocochlear bundle. Safieddine et al. (1996) followed up on this with a polymerase chain reaction analysis of M3 receptor RNA in regions of the cochlea and in the superior olivary complex. Colocalization of M3 receptor with ChAT mRNA confirmed a presynaptic role for muscarinic receptors in efferent transmission.

Kurc et al. (1998) analyzed the presence of heterotrimeric G-proteins (GTP-binding proteins) in the synaptic terminals of the guinea pig efferent system, and demonstrated the presence of Gq (Gα q/11) in regions of both the lateral and the medial efferent terminals where synaptic vesicles are most dense (as indicated by distribution of SNAP-25). These findings, together with the localization of M3 receptors in medial efferents by Safieddine et al. (1996) and Bartolami's et al. (1993)

finding of inositol phosphate formation in the cochlea associated with efferent innervation, are all consistent with the idea that presynaptic muscarinic receptors may be involved in regulating neurotransmitter release in efferent fibers. No pharmacological evidence exists yet to support this idea or to indicate a possible functional role for this process in efferent function.

4.2.5 Synaptic Facilitation of Efferent Effects

It has long been known that efferent effects are facilitated by multiple efferent shocks, and it has been suggested that this phenomenon might be due to a presynaptic mechanism to enhance transmitter release (Art and Fettiplace 1984). A quantal analysis of efferent facilitation in the neonatal rat supports the idea of a presynaptic facilitation of vesicle release from efferent fibers (Goutman et al. 2005).

4.2.6 Postsynaptic Cholinergic Receptor

4.2.6.1 Overview

The predominant effects of electrical stimulation of efferent fibers are mediated by activation of the α9/10 nicotinic cholinergic receptor. The identification of the α9 nicotinic receptor and determination of its unique pharmacological properties (Elgoyhen et al. 1994) explained several puzzling features of efferent pharmacology. Though it had been clear that the efferent neurotransmitter was ACh, the nature of the receptor was enigmatic. Electrical stimulation of efferent fibers could be blocked by a broad range of antagonists, including both muscarinic and nicotinic blockers, as well as bicuculline, a GABAergic blocker. Strychnine, classically thought of as a glycinergic blocker, was the most potent antagonist tested. Elgoyhen and colleagues demonstrated the α9 receptor is present on OHCs at the efferent synapse and that it possessed all of the unusual pharmacological characteristics demonstrated for medial efferent transmission. An α10 subunit was later identified as part of the cholinergic complex (Elgoyhen et al. 2001), which explained more of the biophysical behavior of the medial efferent receptor complex.

Work by Elgoyen et al. has extensively defined the biophysical and pharmacological profile of this receptor, a topic that is covered at length in Katz et al. (Chap. 5), so this topic will only be briefly reviewed in the present chapter. The cholinergic receptor on the OHC is a pentamer, comprising two α9 subunits and three α10 subunits. The receptor forms a nonspecific cation channel with a relatively high permeability to calcium. Activation of the receptor in hair cells produces a brief inward calcium current in the OHC that is quickly overwhelmed by a large outward potassium current. As potassium leaves the cell, the cell becomes hyperpolarized. The outward potassium current is mediated by calcium-activated potassium channels that are gated by calcium entering the cholinergic receptor. These have been identified as SK2 (small conductance potassium) channels. It is also suggested that the calcium entering the OHC interacts with calcium-induced-calcium-release channels in the synaptic cistern to release more calcium into the cytoplasm, a process that would amplify the effects of incoming calcium to activate potassium conductances (Sridhar et al. 1997; Lioudyno et al. 2004). This action is thought to be mediated via ryanodine receptors, so named for their sensitivity to the drug ryanodine. Ryanodine also has the ability to facilitate efferent transmission through a direct interaction with the cholinergic receptor (Zorrilla de San Martin et al. 2007), an action that must be considered in experiments on the role of ryanodine receptors in afferent transmission.

The synaptic cistern appears to be an endoplasmic reticulum-like structure. Activation of calcium-induced-calcium-release channels in the endoplasmic reticulum can induce calcium waves along that structure. It is hypothesized that a similar phenomenon occurs in the OHC, where the synaptic cistern becomes continuous with the subsurface cisternae, to produce slow effects of efferent stimulation (Sridhar et al. 1997). These slow effects build up and decay more slowly than the "classic fast effects" and behave in other ways different than the fast effects. For one, the slow effects desensitize over a period of tens of seconds, whereas fast effects do not. The slow effects can be enhanced with drugs that block the reuptake of cytosolic calcium, and the slow effects are observed over a different range of characteristic frequencies than the fast effects. Some of the protective effects of efferent stimulation are attributed to the slow effects (Reiter and Liberman 1995), as are components of the mechanical response of OHCs to efferent stimulation (Reiter and Liberman 1995; Dallos et al. 1997; Cooper and Guinan 2003).

4.2.6.2 Pharmacology of Medial Efferent Transmission

There are a number of steps susceptible to interference by drugs between the transmission of an action potential down the myelinated efferent fibers and the eventual hyperpolarization of the OHC. The synapses between the efferent terminals and the hair cells are readily accessible from the perilymph of the scala tympani, so that administration of agents into the scala tympani is a straightforward means of drug administration. The presence of the blood–cochlear barrier, similar to the blood–brain barrier, makes access from the systemic circulation difficult, if not impossible, for many drugs.

Several drugs are available to block efferent action at the presynaptic side of this synapse. Propagation of the action potential down the efferent fibers can be blocked by tetrodotoxin, an agent that blocks the regenerative sodium channels in neurons. Because tetrodotoxin will also block action potentials in the afferent fibers, any experiment using this agent would need to look at efferent effects on responses mediated by the OHCs, which are not sensitive to tetrodotoxin. Vesicular transmitter release by efferent fibers is mediated by the same molecular machinery involved in most synapses. Thus agents that block voltage-dependent calcium channels should block transmitter release. The synthesis of ACh requires uptake of choline from the extracellular fluid, which is mediated through the action of the high-affinity choline transporter CHT1. This transporter is blocked by hemicholinium. The action is not immediate, but depends upon a usage-dependent depletion of choline from the synaptic terminal. Packaging of ACh into the vesicle should be blocked with vesamicol, which inhibits VAT.

Once ACh is released from the terminal, it diffuses across the synaptic cleft to interact with nicotinic receptors on the OHC. ACh is cleared from the synapse by action of the enzyme AChE, which cleaves the ester linkage between acetate and choline in the ACh molecule. A number of drugs have been developed to block the activity of this enzyme; the most common cholinesterase inhibitors are neostigmine and physostigmine. Neostigmine is a charged molecule and thus cannot cross the blood–cochlear barrier. Physostigmine, on the other hand, is accessible from the systemic circulation. When ACh breakdown is blocked by these drugs, the neurotransmitter builds up, usually first producing a facilitation of cholinergic transmission, but ultimately blocks the receptor as ACh continually occupies the receptor to produce a desensitization.

The variety of drugs capable of blocking the α9/10 nicotinic receptor is astounding (Kujawa et al. 1993; Sridhar et al. 1995). It is blocked by agents traditionally thought of as muscarinic blockers and nicotinic blockers. These drugs include atropine, curare, and decamethonium. It is also blocked by strychnine, traditionally a glycinergic blocker, and bicuculline, traditionally thought of as a GABAergic blocker.

In addition, a broad array of other compounds used to analyze other aspects of efferent mechanisms have been shown to affect the α9/10 nicotine receptor. These include morphine (and the enkephalins) (Lioudyno et al. 2000, 2002); ryanodine, an agent used to analyze calcium-induced calcium release, which actually enhances affinity and efficacy of ACh at the receptor (Zorrilla de San Martin et al. 2007); linopirdine, an agent often used to silence contributions of KCQN channels (Gomez-Casati et al. 2004); aminoglycoside antibiotics (Rothlin et al. 2000); serotonin receptor ligands (Rothlin et al. 2003); GABA receptor ligands (Rothlin et al. 1999); and quinine drugs (Ballestero et al. 2005).

The medial efferent effects can be blocked by aminoglycoside antibiotics (Aran et al. 1994; Yoshida et al. 1999). Blocking of nicotinic receptors throughout the body by aminoglycosides is an action thought to be mediated in part by an interaction of the aminoglycoside with the cationic binding site for ACh on the nicotinic receptor. At the α9 nicotinic receptor, aminoglycosides produce a noncompetitive antagonism, suggesting a block of the cation channel in the receptor (Blanchet et al. 2000; Rothlin et al. 2000). A number of cationic agents, including the cationic styrl dyes (FM-143, etc.; Dawkins et al. 2005), can block efferent action, presumably by blocking calcium entry through the receptor. Other unusual blockers of efferents include memantine (an *N*-methyl-D-aspartate [glutamate] receptor blocker; Oliver et al. 2001), αRgIA (a conotoxin; Ellison et al. 2006), and quinine-like compounds (Ballestero et al. 2005).

As one can imagine, the sheer number of different types of ligands that interact with the α9/10 nicotinic receptor creates difficulties for taking a pharmacological approach to analyze the contributions of other neurotransmitter receptors (particularly those associated with lateral efferents) in efferent response. The take-home message of this extraordinary array of observations is that extreme caution is necessary in using pharmacological agents to infer cellular actions associated with efferent activation of this very unusual cholinergic receptor.

4.2.6.3 Pharmacology of K_{Ca} Channels

The pharmacology of the K_{Ca} channels that are thought to mediate efferent effects at the OHC is not clear cut. SK2 channels are present in the OHC (Nie et al. 2004) and are thought to mediate the effects of efferent stimulation. However, in vivo, medial efferent effects evoked by electrical stimulation are resistant to apamin, a K_{Ca} channel blocker derived from bee venom. Fast effects are unaltered by apamin and slow effects are blocked only at relatively high concentrations (Yoshida et al. 2001). However, in the chick (Yuhas and Fuchs 1999), the lateral line organ (Dawkins et al. 2005), and in isolated neonatal mammalian hair cells (Glowatzki and Fuchs 2000; Marcotti et al. 2004; Kong et al. 2006) apamin does block efferent effects.

Ryanodine is known to alter efferent transmission and it has been presumed to act via ryanodine receptors on the synaptic cisternae to enhance or block CICR. However, the finding of a direct action of ryanodine on the ACh receptor (Zorrilla de San Martin et al. 2007) calls into question whether ryanodine is indeed acting as presumed.

4.2.6.4 Medial Efferents: In Vivo vs. In Vitro Findings

Much of the work on understanding the efferent receptor has come from in vitro or in situ experiments, and most of these insights are consistent with observations from in vivo experiments. In general, experiments that require perfusion of the scala tympani require higher concentrations to produce effects. For example, EC_{50}s for a number of anticholinergic agents that block electrical stimulation of efferents are approximately tenfold higher than that required to block the action of ACh in isolated hairs cells. One inconsistency between in vitro and in vivo work is that it has been difficult to block the in vivo effects of efferent stimulation with apamin, a bee venom that is a potent blocker of the SK channels activated after efferent stimulation.

4.3 Other Efferent Neurotransmitters

4.3.1 Overview

A number of neuropeptides and other neurotransmitters have also been associated with the efferent innervation of the cochlea. These include opioid peptides, CGRP, dopamine, GABA, and serotonin. There has long been a general perception that the medial efferents are cholinergic and that the lateral efferents contain, in addition to ACh, an array of neuropeptides and other transmitters. Arguments have been made for subpopulations of lateral efferents based on differential distribution of these transmitters throughout the cochlea. However, the inconsistencies among reports on immunohistochemical localization of efferent neurotransmitters are too numerous to attempt to cover in a review. Explanations for these apparent inconsistencies range from species differences in localization to technical issues associated with a very difficult and intrinsically variable technique.

A viewpoint is evolving, however, to suggest that both lateral and medial efferents might contain multiple neurotransmitters. This is particularly true with CGRP, where some of the earliest electron microscopy showed immunolocalization beneath OHCs (Sliwinska-Kowalska et al. 1989). A more systematic analysis of CGRP, GABA, and cholinergic markers in the mouse indicated colocalization of these three candidates in both medial and lateral efferents (Maison et al. 2003). The recent finding that opioid receptors are distributed in the IHC and OCH regions of the guinea pig cochlea is also consistent with that idea (Jongkamonwiwat et al. 2006). There is a question of whether dopaminergic fibers may constitute a separate population of neurons intermingling with the lateral efferents, whose cell bodies may arise at the edges of the lateral superior olivary complexes (LSO). Darrow et al. (2006) suggested that these are the LOC shell neurons described by Warr et al. (1997). Safieddine et al. (1997), however, found that dopamine markers are colocalized with the others.

It is likely that these noncholinergic transmitters play a more subtle role in efferent transmission than that of ACh. This idea is especially evident for the medial efferents, where all of the effects of electrical stimulation can be accounted for by cholinergic activation of the α9/10 nicotinic receptor (Sridhar et al. 1995).

4.3.2 Lateral Efferent Origins

The lateral efferent fibers originate in cell bodies in and near the hilus of the LSO complex and generally terminate on the unmyelinated endings of the radial afferent fibers innervating the inner hair cells (though a fraction also innervate the IHCs). Though neurotransmitter chemistry of the lateral efferents represents an extraordinarily rich field, we know very little about what the lateral efferents do or even what role the various noncholinergic efferent transmitter candidates play in medial efferent function. This arises to a large extent from difficulty in electrically stimulating the lateral efferent fibers.

4.3.3 Acetylcholine

Acetylcholine is the major neurotransmitter of the lateral efferents and immunolabeling for cholinergic markers indicates a dense cholinergic innervation of the radial afferent fibers. The cholinergic receptor on auditory ganglion cells is not established. It is not, though, the α9/10 nicotinic receptor found on the OHC. A muscarinic receptor has been identified that can increase intracellular calcium independent of extracellular potassium (Rome et al. 1999).

4.3.4 Opioid Peptides

Met-enkephalin, an opioid peptide, was the first of the olivocochlear cotransmitters to be discovered (Fex and Altschuler 1981). Since then, members of all three opioid gene families have been described in efferents including Leu-enkephalin, dynorphin B, and proenkephalin B (Altschuler et al. 1985a; Eybalin et al. 1985; Hoffman et al. 1985). Opioid peptides are present both in lateral and in medial efferents (review: Eybalin 1993).

Opioid receptors are present in the inner ear and are differentially distributed in various cells of the cochlea including IHC, OHC, medial efferent terminals, and spiral ganglion cells. These receptors include mu, kappa, delta, and the orphanin/nociceptin receptor (Jongkamonwiwat et al. 2006; Kho et al. 2006). Pentazocine, an opioid agonist, can increase the amplitude of the auditory nerve responses near threshold (Sahley and Nodar 1994). Interpretation of the pharmacological results is complicated by the finding that opioid peptides and ligands for the opioid receptor are known to modulate the α9/10 nicotinic receptor (Lioudyno et al. 2002).

4.3.5 Calcitonin Gene-Related Peptide (CGRP)

CGRP was immunolocalized in the inner ear and presumed to be associated with efferent fibers in 1985 (Kitajiri et al. 1985). Since then a number of studies have shown CGRP to be present in efferent terminals both at the OHC and at the IHC region (e.g., Sliwinska-Kowalska et al. 1989; Cabanillas and Luebke 2002; Maison et al. 2003). Though CGRP is active in lateral line organ, where it increases afferent discharge rate and suppresses response to mechanical stimulation (Bailey and Sewell 2000), there is little direct evidence for an effect of CGRP in the cochlea. A CGRP knockout mouse displayed a small decrease in the growth of the auditory brain stem response amplitudes with stimulus intensity, but no change in responses to efferent stimulation (Maison et al. 2003).

4.3.6 GABA

GABA immunoreactivity has been reported in both medial and lateral efferent terminals. As with most of the other noncholinergic transmitter candidates, there is little evidence for an effect of GABA in the cochlea. Virtually all of the effects of medial olivocochlear electrical activation are accounted for by activation of the cholinergic receptor. Bobbin and Thompson (1978) found no effect of GABA perfused through the cochlea at concentrations of 10 mM, though GABA responses (hyperpolarization) are observed in isolated OHCs at micromolar concentrations (Gitter and Zenner 1992; Plinkert et al. 1993). Synthesis of GABA in the cochlea is very low: glutamate decarboxylase levels in the cochlea are 0.17 nM GABA formed /min/mg protein, or about 1/10 the level observed in the cochlear nucleus (Fex and Wenthold 1976). However, efferent terminals do demonstrate high-affinity uptake of GABA (Gulley et al. 1979; Schwartz and Ryan 1983), and GABA is released during intense acoustic stimulation (Drescher et al. 1983). GABA-A receptors (α 1–6, β 1–3, and γ 2) are present via PCR analysis (Drescher et al. 1993), and immunohistochemically (Plinkert et al. 1989). Pharmacological analysis of GABA contributions (Plinkert et al. 1993) suggests GABA may hyperpolarize OHC. Interpretations of putative GABAergic effects can be tricky, given the ability of bicuculline to block the $\alpha 9$ cholinergic receptor (Klinke and Oertel 1977; Rothlin et al. 1999).

4.3.7 Serotonin (5-Hydroxytryptamine)

Serotonin immunoreactive neurons were described in the organ of Corti with a distribution pattern similar to that of the lateral efferent system (Gil-Loyzaga et al. 1997). 5-HT receptors are present in the cochlea, with subtypes 1A, 1B, 2B, and 6 in the organ of Corti and 2C, 3, and 5B in the spiral ganglion (Oh et al. 1999). Serotonin receptor ligands are known to interact with the $\alpha 9$ cholinergic receptor (Rothlin et al. 2003).

4.3.8 Glycine

There is very little information about glycine as an efferent neurotransmitter in the cochlea. Glycine receptors (GlyRa3 and GlyRbeta) and gepheryn, though, have been found in the organ of Corti and in spiral ganglion neurons (Dlugaiczyk et al. 2008). Glycine receptors are located in the OHCs and beneath the IHCs, consistent with a role in efferent transmission.

4.3.9 Dopamine

The finding of tyrosine hydroxylase in the organ of Corti (Fex and Wenthold 1976) and immunoreactivity in efferent fibers (Jones et al. 1987) led to speculation that dopamine might be a neurotransmitter in olivocochlear efferents. These basic findings were confirmed and expanded by a number of investigators (Gil-Loyzaga and Pares-Herbute 1989; Eybalin et al. 1993; Mulders and Robertson 2004). Cochlear concentrations of dopamine are reduced with exposure to loud sound. Dopamine receptors (D2 long and D3) are detected in the cochlea by RT-PCR (Karadaghy et al. 1997), and Inoue et al. (2006) report immunolocalization of all receptor types in the spiral ganglion neurons.

Evidence that the dopaminergic neurons of the LSO may represent a subpopulation of lateral efferents comes from several sources. Mulders and Robertson (2004) describe a subpopulation of tyrosine hydroxylase–labeled neurons in the LSO region that were retrogradely labeled with tracers administered in the cochlea. Darrow et al. (2006) colabeled efferents for VAT and tyrosine hydroxylase, and found that dopaminergic neurons were generally not positive for VAT (though there were exceptions). Darrow et al. (2006) also found tyrosine hydroxylase–positive neurons in the LSO constituted around 5% of the total LSO efferent neurons.

The role for dopamine in cochlear function is not clear, though evidence is accumulating that dopaminergic activation may suppress responses of afferent fibers to transmitter released by the inner hair cell. Intracochlear application of dopamine suppressed both spontaneous and driven discharge rate in afferent fibers (d'Aldin et al. 1995). Dopamine antagonists SCH23390 and eticlopride decreased driven rates and improved thresholds (Ruel et al. 2001). Inhibitors of dopamine transport increase extracellular dopamine levels and suppress both spontaneous and evoked discharge rate (Ruel et al. 2006). Tyrosine hydroxylase is upregulated during stimuli that can condition the cochlea against acoustic trauma (Niu et al. 2007).

4.4 Summary

The acute effects of the electrical stimulation of efferent fibers on cochlear responses are now reasonably well understood and are likely mediated by the $\alpha 9/10$ nicotinic receptor on OHC. The challenges that remain are in understanding the biophysical and molecular mechanisms that account for those functional changes. A far greater challenge is in understanding the role and function of the lateral efferents. It has proven demanding to stimulate the lateral efferents, making a straightforward correlation between lateral efferent activation and changes in cochlear function difficult. This in turn precludes a simple pharmacological approach to dissect the roles of the myriad of neurotransmitter in efferent function. Most analyses of efferent function to date involve effects occurring between tens of milliseconds and tens of seconds. Another challenge is in understanding the role of efferents in cochlear function over much longer time courses, an area amenable to molecular biological approaches.

Acknowledgment This work was supported by a grant from the NIDCD (R01 DC000767).

References

Altschuler RA, Hoffman DW, Reeks KA, Fex J (1985a) Localization of dynorphin B-like and alpha-neoendorphin-like immunoreactivities in the guinea pig organ of Corti. Hear Res 17:249–258

Altschuler RA, Kachar B, Rubio JA, Parakkal MH, Fex J (1985b) Immunocytochemical localization of choline acetyltransferase-like immunoreactivity in the guinea pig cochlea. Brain Res 338:1–11

Aran JM, Erre JP, Avan P (1994) Contralateral suppression of transient evoked otoacoustic emissions in guinea-pigs: effects of gentamicin. Br J Audiol 28:267–271

Art JJ, Fettiplace R (1984) Efferent desensitization of auditory nerve fibre responses in the cochlea of the turtle *Pseudemys scripta elegans*. J Physiol 356:507–523

Augustinsson K-B (1946) Studies on the specificiy of choline esterase in *Helix pomatia*. Biochem J 40:343–349

Bailey GP, Sewell WF (2000) Calcitonin gene-related peptide suppresses hair cell responses to mechanical stimulation in the *Xenopus* lateral line organ. J Neurosci 20:5163–5169

Ballestero JA, Plazas PV, Kracun S, Gomez-Casati ME, Taranda J, Rothlin CV, Katz E, Millar NS, Elgoyhen AB (2005) Effects of quinine, quinidine, and chloroquine on alpha9alpha10 nicotinic cholinergic receptors. Mol Pharmacol 68:822–829

Bartolami S, Ripoll C, Planche M, Pujol R (1993) Localisation of functional muscarinic receptors in the rat cochlea: evidence for efferent presynaptic autoreceptors. Brain Res 626:200–209

Bergeron AL, Schrader A, Yang D, Osman AA, Simmons DD (2005) The final stage of cholinergic differentiation occurs below inner hair cells during development of the rodent cochlea. J Assoc Res Otolaryngol 6:401–415

Blanchet C, Erostegui C, Sugasawa M, Dulon D (2000) Gentamicin blocks ACh-evoked K^+ current in guinea-pig outer hair cells by impairing Ca^{2+} entry at the cholinergic receptor. J Physiol 525(Pt 3):641–654

Bobbin RP, Thompson MH (1978) Effects of putative transmitters on afferent cochlear transmission. Ann Otol Rhinol Laryngol 87:185–190

Cabanillas LA, Luebke AE (2002) CGRP- and cholinergic-containing fibers project to guinea pig outer hair cells. Hear Res 172:14–17

Churchill JA, Schuknecht HF, Doran R (1956) Acetylcholinesterase activity in the cochlea. Laryngoscope 66:1–15

Comis SD, Guth PS (1974) The release of acetylcholine from the cochlear nucleus upon stimulation of the crossed olivo-cochlear bundle. Neuropharmacology 13:633–641

Cooper NP, Guinan JJ Jr (2003) Separate mechanical processes underlie fast and slow effects of medial olivocochlear efferent activity. J Physiol 548:307–312

d'Aldin C, Puel JL, Leducq R, Crambes O, Eybalin M, Pujol R (1995) Effects of a dopaminergic agonist in the guinea pig cochlea. Hear Res 90:202–211

Dallos P, He DZ, Lin X, Sziklai I, Mehta S, Evans BN (1997) Acetylcholine, outer hair cell electromotility, and the cochlear amplifier. J Neurosci 17:2212–2226

Darrow KN, Simons EJ, Dodds L, Liberman MC (2006) Dopaminergic innervation of the mouse inner ear: evidence for a separate cytochemical group of cochlear efferent fibers. J Comp Neurol 498:403–414

Dawkins R, Keller SL, Sewell WF (2005) Pharmacology of acetylcholine-mediated cell signaling in the lateral line organ following efferent stimulation. J Neurophysiol 93:2541–2551

de San Z, Martin J, Ballestero J, Katz E, Elgoyhen AB, Fuchs PA (2007) Ryanodine is a positive modulator of acetylcholine receptor gating in cochlear hair cells. J Assoc Res Otolaryngol 8:474–483

Dlugaiczyk J, Singer W, Schick B, Iro H, Becker K, Becker CM, Zimmermann U, Rohbock K, Knipper M (2008) Expression of glycine receptors and gephyrin in the rat cochlea. Histochem Cell Biol 129:513–523

Drescher MJ, Drescher DG, Medina JE (1983) Effect of sound stimulation at several levels on concentrations of primary amines, including neurotransmitter candidates, in perilymph of the guinea pig inner ear. J Neurochem 41:309–320

Drescher DG, Green GE, Khan KM, Hajela K, Beisel KW, Morley BJ, Gupta AK (1993) Analysis of gamma-aminobutyric acid A receptor subunits in the mouse cochlea by means of the polymerase chain reaction. J Neurochem 61:1167–1170

Elgoyhen AB, Johnson DS, Boulter J, Vetter DE, Heinemann S (1994) Alpha 9: an acetylcholine receptor with novel pharmacological properties expressed in rat cochlear hair cells. Cell 79:705–715

Elgoyhen AB, Vetter DE, Katz E, Rothlin CV, Heinemann SF, Boulter J (2001) Alpha10: a determinant of nicotinic cholinergic receptor function in mammalian vestibular and cochlear mechanosensory hair cells. Proc Natl Acad Sci U S A 98:3501–3506

Ellison M, Haberlandt C, Gomez-Casati ME, Watkins M, Elgoyhen AB, McIntosh JM, Olivera BM (2006) Alpha-RgIA: a novel conotoxin that specifically and potently blocks the alpha9alpha10 nAChR. Biochemistry 45:1511–1517

Eybalin M (1993) Neurotransmitters and neuromodulators of the mammalian cochlea. Physiol Rev 73:309–373

Eybalin M, Pujol R (1987) Choline acetyltransferase (ChAT) immunoelectron microscopy distinguishes at least three types of efferent synapses in the organ of Corti. Exp Brain Res 65:261–270

Eybalin M, Cupo A, Pujol R (1985) Met-enkephalin-Arg6-Gly7-Leu8 in the organ of Corti: high performance liquid chromatography and immunoelectron microscopy. Brain Res 331:389–395

Eybalin M, Charachon G, Renard N (1993) Dopaminergic lateral efferent innervation of the guinea-pig cochlea: immunoelectron microscopy of catecholamine-synthesizing enzymes and effect of 6-hydroxydopamine. Neuroscience 54:133–142

Fex J, Altschuler RA (1981) Enkephalin-like immunoreactivity of olivocochlear nerve fibers in cochlea of guinea pig and cat. Proc Natl Acad Sci U S A 78:1255–1259

Fex J, Wenthold RJ (1976) Choline acetyltransferase, glutamate decarboxylase and tyrosine hydroxylase in the cochlea and cochlear nucleus of the guinea pig. Brain Res 109:575–585

Fuchs PA, Murrow BW (1992) Cholinergic inhibition of short (outer) hair cells of the chick's cochlea. J Neurosci 12:800–809

Galambos R (1956) Suppression of auditory nerve activity by stimulation of efferent fibers to cochlea. J Neurophysiol 19:424–437

Galley N, Klinke R, Oertel W, Pause M, Storch WH (1973) The effect of intracochlearly administered acetylcholine-blocking agents on the efferent synapses of the cochlea. Brain Res 64:55–63

Gil-Loyzaga P, Pares-Herbute N (1989) HPLC detection of dopamine and noradrenaline in the cochlea of adult and developing rats. Brain Res Dev Brain Res 48:157–160

Gil-Loyzaga P, Pujol R (1988) Synaptophysin in the developing cochlea. Int J Dev Neurosci 6:155–160

Gil-Loyzaga P, Bartolome MV, Vicente-Torres MA (1997) Serotonergic innervation of the organ of Corti of the cat cochlea. Neuroreport 8:3519–3522

Gisselsson L (1950) Experimental investigation into the problem of humoral transmission in the cochlea. Acta Otolaryngol 82:9–78

Gitter AH, Zenner HP (1992) Gamma-aminobutyric acid receptor activation of outer hair cells in the guinea pig cochlea. Eur Arch Otorhinolaryngol 249:62–65

Glowatzki E, Fuchs PA (2000) Cholinergic synaptic inhibition of inner hair cells in the neonatal mammalian cochlea. Science (NY) 288:2366–2368

Godfrey DA, Ross CD (1985) Enzymes of acetylcholine metabolism in the rat cochlea. Ann Otol Rhinol Laryngol 94:409–414

Godfrey DA, Krzanowski JJ Jr, Matschinsky FM (1976) Activities of enzymes of the cholinergic systems in the guinea pig cochlea. J Histochem Cytochem 24:470–472

Gomez-Casati ME, Katz E, Glowatzki E, Lioudyno MI, Fuchs P, Elgoyhen AB (2004) Linopirdine blocks alpha9alpha10-containing nicotinic cholinergic receptors of cochlear hair cells. J Assoc Res Otolaryngol 5:261–269

Goutman JD, Fuchs PA, Glowatzki E (2005) Facilitating efferent inhibition of inner hair cells in the cochlea of the neonatal rat. J Physiol 566:49–59

Gulley RL, Fex J, Wenthold RJ (1979) Uptake of putative neurotransmitters in the organ of Corti. Acta Otolaryngol 88:177–182

Guth PS, Norris CH, Bobbin RP (1976) The pharmacology of transmission in the peripheral auditory system. Pharmacol Rev 28:95–125

Hoffman DW, Zamir N, Rubio JA, Altschuler RA, Fex J (1985) Proenkephalin and prodynorphin related neuropeptides in the cochlea. Hear Res 17:47–50

Inoue T, Matsubara A, Maruya S, Yamamoto Y, Namba A, Sasaki A, Shinkawa H (2006) Localization of dopamine receptor subtypes in the rat spiral ganglion. Neurosci Lett 399: 226–229

Jasser A, Guth PS (1973) The synthesis of acetylcholine by the olivo-cochlear bundle. J Neurochem 20:45–53

Jones N, Fex J, Altschuler RA (1987) Tyrosine hydroxylase immunoreactivity identifies possible catecholaminergic fibers in the organ of Corti. Hear Res 30:33–38

Jongkamonwiwat N, Phansuwan-Pujito P, Casalotti SO, Forge A, Dodson H, Govitrapong P (2006) The existence of opioid receptors in the cochlea of guinea pigs. Eur J Neurosci 23: 2701–2711

Karadaghy AA, Lasak JM, Chomchai JS, Khan KM, Drescher MJ, Drescher DG (1997) Quantitative analysis of dopamine receptor messages in the mouse cochlea. Brain Res Mol Brain Res 44:151–156

Kho ST, Lopez IA, Evans C, Ishiyama A, Ishiyama G (2006) Immunolocalization of orphanin FQ in rat cochlea. Brain Res 1113:146–152

Kitajiri M, Yamashita T, Tohyama Y, Kumazawa T, Takeda N, Kawasaki Y, Matsunaga T, Girgis S, Hillyard CJ, MacIntyre I et al (1985) Localization of calcitonin gene-related peptide in the organ of Corti of the rat: an immunohistochemical study. Brain Res 358:394–397

Klinke R, Oertel W (1977) Evidence that GABA is not the afferent transmitter in the cochlea. Exp Brain Res 28:311–314

Knipper M, Zimmermann U, Rohbock K, Kopschall I, Zenner HP (1995) Synaptophysin and GAP-43 proteins in efferent fibers of the inner ear during postnatal development. Brain Res Dev Brain Res 89:73–86

Kong WJ, Guo CK, Zhang S, Zhang XW, Wang YJ, Li ZW (2006) Fast cholinergic efferent inhibition in guinea pig outer hair cells. Brain Res 1102:103–108

Kujawa SG, Glattke TJ, Fallon M, Bobbin RP (1992) Intracochlear application of acetylcholine alters sound-induced mechanical events within the cochlear partition. Hear Res 61: 106–116

Kujawa SG, Glattke TJ, Fallon M, Bobbin RP (1993) Contralateral sound suppresses distortion product otoacoustic emissions through cholinergic mechanisms. Hear Res 68:97–106

Kurc M, Dodane V, Pinto DS, Kachar B (1998) Presynaptic localization of G protein isoforms in the efferent nerve terminals of the mammalian cochlea. Hear Res 116:1–9

Lioudyno MI, Verbitsky M, Holt JC, Elgoyhen AB, Guth PS (2000) Morphine inhibits an alpha9-acetylcholine nicotinic receptor-mediated response by a mechanism which does not involve opioid receptors. Hear Res 149:167–177

Lioudyno MI, Verbitsky M, Glowatzki E, Holt JC, Boulter J, Zadina JE, Elgoyhen AB, Guth PS (2002) The alpha9/alpha10-containing nicotinic ACh receptor is directly modulated by opioid peptides, endomorphin-1, and dynorphin B, proposed efferent cotransmitters in the inner ear. Mol Cell Neurosci 20:695–711

Lioudyno M, Hiel H, Kong JH, Katz E, Waldman E, Parameshwaran-Iyer S, Glowatzki E, Fuchs PA (2004) A "synaptoplasmic cistern" mediates rapid inhibition of cochlear hair cells. J Neurosci 24:11160–11164

Maison SF, Emeson RB, Adams JC, Luebke AE, Liberman MC (2003) Loss of alpha CGRP reduces sound-evoked activity in the cochlear nerve. J Neurophysiol 90:2941–2949

Marcotti W, Johnson SL, Kros CJ (2004) A transiently expressed SK current sustains and modulates action potential activity in immature mouse inner hair cells. J Physiol 560:691–708

Mulders WH, Robertson D (2004) Dopaminergic olivocochlear neurons originate in the high frequency region of the lateral superior olive of guinea pigs. Hear Res 187:122–130

Nachmansohn D, Wilson IB (1951) The enzymic hydrolysis and synthesis of acetylcholine. Adv Enzymol Relat Subj Biochem 12:259–339

Nie L, Song H, Chen MF, Chiamvimonvat N, Beisel KW, Yamoah EN, Vazquez AE (2004) Cloning and expression of a small-conductance $Ca^{(2+)}$-activated K^{+} channel from the mouse cochlea: coexpression with alpha9/alpha10 acetylcholine receptors. J Neurophysiol 91:1536–1544

Niu X, Tahera Y, Canlon B (2007) Environmental enrichment to sound activates dopaminergic pathways in the auditory system. Physiol Behav 92:34–39

Oh CK, Drescher MJ, Hatfield JS, Drescher DG (1999) Selective expression of serotonin receptor transcripts in the mammalian cochlea and its subdivisions. Brain Res Mol Brain Res 70:135–140

Oliver D, Ludwig J, Reisinger E, Zoellner W, Ruppersberg JP, Fakler B (2001) Memantine inhibits efferent cholinergic transmission in the cochlea by blocking nicotinic acetylcholine receptors of outer hair cells. Mol Pharmacol 60:183–189

Plinkert PK, Mohler H, Zenner HP (1989) A subpopulation of outer hair cells possessing GABA receptors with tonotopic organization. Arch Otorhinolaryngol 246:417–422

Plinkert PK, Gitter AH, Mohler H, Zenner HP (1993) Structure, pharmacology and function of GABA-A receptors in cochlear outer hair cells. Eur Arch Otorhinolaryngol 250:351–357

Rasmussen GL (1946) The olivary peduncle and other fiber projections of the superior olivary complex. J Comp Neurol 84:141–219

Reiter ER, Liberman MC (1995) Efferent-mediated protection from acoustic overexposure: relation to slow effects of olivocochlear stimulation. J Neurophysiol 73:506–514

Rome C, Luo D, Dulon D (1999) Muscarinic receptor-mediated calcium signaling in spiral ganglion neurons of the mammalian cochlea. Brain Res 846:196–203

Rothlin CV, Katz E, Verbitsky M, Elgoyhen AB (1999) The alpha9 nicotinic acetylcholine receptor shares pharmacological properties with type A gamma-aminobutyric acid, glycine, and type 3 serotonin receptors. Mol Pharmacol 55:248–254

Rothlin CV, Katz E, Verbitsky M, Vetter DE, Heinemann SF, Elgoyhen AB (2000) Block of the alpha9 nicotinic receptor by ototoxic aminoglycosides. Neuropharmacology 39:2525–2532

Rothlin CV, Lioudyno MI, Silbering AF, Plazas PV, Casati ME, Katz E, Guth PS, Elgoyhen AB (2003) Direct interaction of serotonin type 3 receptor ligands with recombinant and native alpha 9 alpha 10-containing nicotinic cholinergic receptors. Mol Pharmacol 63:1067–1074

Ruel J, Nouvian R, Gervais d'Aldin C, Pujol R, Eybalin M, Puel JL (2001) Dopamine inhibition of auditory nerve activity in the adult mammalian cochlea. Eur J Neurosci 14:977–986

Ruel J, Wang J, Dememes D, Gobaille S, Puel JL, Rebillard G (2006) Dopamine transporter is essential for the maintenance of spontaneous activity of auditory nerve neurones and their responsiveness to sound stimulation. J Neurochem 97:190–200

Safieddine S, Wenthold RJ (1999) SNARE complex at the ribbon synapses of cochlear hair cells: analysis of synaptic vesicle- and synaptic membrane-associated proteins. Eur J Neurosci 11:803–812

Safieddine S, Bartolami S, Wenthold RJ, Eybalin M (1996) Pre- and postsynaptic M3 muscarinic receptor mRNAs in the rodent peripheral auditory system. Brain Res Mol Brain Res 40:127–135

Safieddine S, Prior AM, Eybalin M (1997) Choline acetyltransferase, glutamate decarboxylase, tyrosine hydroxylase, calcitonin gene-related peptide and opioid peptides coexist in lateral efferent neurons of rat and guinea-pig. Eur J Neurosci 9:356–367

Sahley TL, Nodar RH (1994) Improvement in auditory function following pentazocine suggests a role for dynorphins in auditory sensitivity. Ear Hear 15:422–431

Schwartz IR, Ryan AF (1983) Differential labeling of sensory cell and neural populations in the organ of Corti following amino acid incubations. Hear Res 9:185–200

Simmons DD, Bertolotto C, Typpo K, Clay A, Wu M (1999) Differential development of cholinergic-like neurons in the superior olive: a light microscopic study. Anat Embryol 200:585–595

Sliwinska-Kowalska M, Parakkal M, Schneider ME, Fex J (1989) CGRP-like immunoreactivity in the guinea pig organ of Corti: a light and electron microscopy study. Hear Res 42:83–95

Smith CA, Sjostrand FS (1961) Structure of the nerve endings on the external hair cells of the guinea pig cochlea as studied by serial sections. J Ultrastruct Res 5:523–556

Sridhar TS, Liberman MC, Brown MC, Sewell WF (1995) A novel cholinergic "slow effect" of efferent stimulation on cochlear potentials in the guinea pig. J Neurosci 15:3667–3678

Sridhar TS, Brown MC, Sewell WF (1997) Unique postsynaptic signaling at the hair cell efferent synapse permits calcium to evoke changes on two time scales. J Neurosci 17:428–437

Waka N, Knipper M, Engel J (2003) Localization of the calcium channel subunits Cav1.2 (alpha1C) and Cav2.3 (alpha1E) in the mouse organ of Corti. Histol Histopathol 18:1115–1123

Warr WB, Boche JB, Neely ST (1997) Efferent innervation of the inner hair cell region: origins and terminations of two lateral olivocochlear systems. Hear Res 108:89–111

Yao W, Godfrey DA (1998) Immunohistochemical evaluation of cholinergic neurons in the rat superior olivary complex. Microsc Res Tech 41:270–283

Yao W, Godfrey DA (1999) Vesicular acetylcholine transporter in the rat cochlear nucleus: an immunohistochemical study. J Histochem Cytochem 47:83–90

Yoshida N, Liberman MC, Brown MC, Sewell WF (1999) Gentamicin blocks both fast and slow effects of olivocochlear activation in anesthetized guinea pigs. J Neurophysiol 82:3168–3174

Yoshida N, Liberman MC, Brown MC, Sewell WF (2001) Fast, but not slow, effects of olivocochlear activation are resistant to apamin. J Neurophysiol 85:84–88

Yuhas WA, Fuchs PA (1999) Apamin-sensitive, small-conductance, calcium-activated potassium channels mediate cholinergic inhibition of chick auditory hair cells. J Comp Physiol 185:455–462

Chapter 5
Cholinergic Inhibition of Hair Cells

Eleonora Katz, Ana Belén Elgoyhen, and Paul Albert Fuchs

5.1 Introduction

In the inner ear, the activity of hair cells that transform sound into electrical signals is modulated by a descending efferent innervation from the brain. A major component of this feedback involves cholinergic inhibition of hair cells via an unusual ionic mechanism. It activates rapidly (on the order of milliseconds), but instead of being mediated by a hyperpolarizing conductance through γ-aminobutyric acid (GABA) and/or glycine receptors, it is served by nicotinic cholinergic receptors (nAChR), which usually mediate excitatory postsynaptic responses. How is fast inhibition accomplished if the activation of a cationic channel (the nAChR) at the resting membrane potential should depolarize the hair cell?

Current data show that this response occurs via the activation of a peculiar type of nAChR that allows calcium entry into the hair cell with the subsequent activation of calcium-activated potassium channels that hyperpolarize the cell membrane. This chapter focuses on the experimental evidence that gives support to this "two-channel hypothesis": electrophysiological experiments performed on hair cells from lower vertebrates, cloning and cellular localization of the nAChR subunits that make up the hair cell receptor, and electrophysiological recordings from inner and outer hair cells in a microdissected preparation of the mammalian organ of Corti. Finally, the focus will turn to the generation and analysis of mouse models with genetic modifications of the molecules that are key participants in this peculiar type of fast synaptic inhibition.

E. Katz (✉)
Instituto de Investigaciones en Ingeniería Genética y Biología Molecular (Consejo Nacional de Investigaciones Científicas y Técnicas), 1428 Buenos Aires, Argentina
e-mail: ekatz@dna.uba.ar

D.K. Ryugo et al. (eds.), *Auditory and Vestibular Efferents*,
Springer Handbook of Auditory Research 38, DOI 10.1007/978-1-4419-7070-1_5,

5.2 Historical Background

In vertebrates, mechanosensory hair cells of the inner ear convert sound, head position, and motion into electrical signals that are conveyed to the central nervous system (CNS) by peripheral afferent neurons. This afferent flow is regulated by the octavo-lateralis efferent system that receives input from ascending sensory fibers in the hindbrain and projects centrally to first-order sensory nuclei and peripherally to mechanoreceptive end organs of the inner ear (cochlea, otolith macula, and semicircular canal cristae) (see Chap. 2). Efferent feedback is provided by cholinergic efferent neurons present in the superior olivary complex of the brainstem that synapse with hair cells or primary afferent fibers in the inner ear. Degeneration studies by Rasmussen first identified this cochlear pathway (Rasmussen 1946), and since then, the pattern of efferent innervation in the mammalian ear has been studied by many, as exemplified by the contributions of Warr (1975; reviews: Warr 1992; Guinan 1996). Efferent activity inhibits auditory end-organs, but provides both excitation and inhibition to the vestibular periphery of most vertebrates. This chapter is focused on cholinergic inhibition of auditory hair cells; therefore, readers interested in vestibular end-organs are referred to Goldberg et al. (Chap. 6). Olivocochlear (OC) efferent neurons allow the CNS to control the transduction of sound in the auditory periphery, providing improved detection of signals in background noise, selective attention to particular signals, and protection of the periphery from damage caused by extremely loud sounds (see Chap. 3). A first anatomical subdivison of the efferent pathway was made by Rasmussen (1946), who noted that the efferent fibers formed uncrossed and crossed OC bundles, the latter crossing near the floor of the fourth ventricle. Warr and Guinan (1979) used tract-tracing to subdivide the OC efferents into the lateral and medial components. Neurons in the lateral superior olivary complex (LOCs) have unmyelinated fibers that synapse on the dendrites of type I afferents beneath inner hair cells (IHCs). Larger, myelinated medial OC (MOC) efferents are located near the medial superior olivary complex and synapse directly onto OHCs. In rodents, during the first 2 weeks of life, the MOC efferents temporarily contact IHCs directly (for an in depth description of the LOC and MOC systems, see Chap. 2).

Biochemical and immunohistochemical studies support the hypothesis that ACh is the main neurotransmitter of the MOC system (Eybalin 1993). These include the demonstration of choline acetyltransferase (ChAT) and acetylcholinesterase (AChE) in the cochlea of various species. Also, ChAT-immunoreactive neurons as well as AChE-positive neurons were detected in the nuclei containing the cell bodies of origin of the MOC axons. Similarly, both enzymes were localized to the organ of Corti itself. Both the ChAT-like immunolabel and the AChE histochemical reaction product are present in fibers crossing the tunnel of Corti in an upper tract to form patches below the OHCs. At the electron microscopic level, these ChAT-like immunolabeled patches were shown to correspond to large axosomatic synapses on the OHCs, nearly all of which were immunolabeled for

ChAT. Moreover, lesions of part or all of the efferent cochlear supply strongly decreased the immunoreactivity to ChAT (Eybalin 1993). Thus, many histological, biochemical and pharmacological studies have demonstrated that most of the efferent fibers to the inner ear are cholinergic (reviewed in Guth et al. 1998). There is evidence, however, indicating that a small fraction of efferent fibers may use GABA as their neurotransmitter (Eybalin et al. 1988; Vetter et al. 1991; Maison et al. 2003). There is also some evidence pointing to other neurotransmitters such as calcitonin-gene related peptide (CGRP) (Cabanillas and Luebke 2002), opioids, enkephalins (Fex and Altschuler 1981; Altschuler et al. 1983, 1984), and dynorphins (Altschuler et al. 1985) in efferent neurons (see Chap. 4 for a detailed review of the literature on cochlear efferent neurotransmitters).

Stimulation of the OC axons reduces the amplitude of the compound action potential (CAP) recorded extracellularly from the VIIIth nerve in response to an acoustic click or brief tone burst (Galambos 1956). In the mammalian cochlea this inhibitory effect is now known to result from suppression of the active electromotile response of outer hair cells (OHCs) that sustains cochlear amplification, sensitivity, and fine tuning (Guinan and Stankovic 1996). Owing to the predominance of cholinergic innervation to the cochlea, and the lower threshold to shock of MOC axons, it is likely that the effects of brain stem electrical stimulation are due to ACh released by medial efferents. The best known cochlear effects of activation of the OC bundle are the suppression of cochlear responses such as CAPs and distortion product otoacoustic emissions (DPOAEs). These classic OC effects have long been thought to arise from cholinergic effects of the MOC synapses on OHCs (see Chap. 3). This idea has been confirmed by the demonstration that all such effects of OC stimulation on CAPs and DPOAEs are absent in mice lacking the nicotinic cholinergic receptor subunits expressed by OHCs (Vetter et al. 1999, 2007). Whereas cochlear hair cells are differentially innervated by efferent and afferent neurons in adult mammals, it is interesting to note that before the onset of hearing (around postnatal day 12, P12 in rats and mice), IHCs, in addition to being innervated by the dendrites of spiral ganglion neurons, are also directly contacted by cholinergic efferent fibers from the OC system (Liberman et al. 1990; Simmons et al. 1996; Simmons 2002). As discussed in the text that follows, these contacts have served as accessible experimental targets, as well as raising interesting questions as to their role in cochlear maturation.

In the CNS, fast synaptic inhibition is mediated by a hyperpolarizing chloride conductance through GABA and/or glycine receptors (Alger 1991; Betz et al. 1999). These inhibitory postsynaptic currents (IPSCs) rise within a few milliseconds and decay with time constants of tens of milliseconds (Takahashi and Momiyama 1991; Jones and Westbrook 1996). Synaptic inhibition of hair cells by the MOC system also takes place in the order of milliseconds. However, it differs from typical inhibitory synapses because it is mediated by the activation of nicotinic cholinergic receptors (nAChR) that usually mediate excitatory postsynaptic responses. The cellular mechanisms of this nicotinic cholinergic inhibition and the molecular constituents involved are well conserved among vertebrates. Therefore comparative studies can be used to delve into the processes that account for this peculiar synaptic mechanism.

5.3 Cellular Physiology

5.3.1 *Intracellular Recordings from Hair Cells of the Fish Lateral Line*

Elucidation of the cellular mechanisms of efferent inhibition came initially from studies in hair cells of nonmammalian vertebrates. The first intracellular recordings were performed by Flock and Russell (1976) in fish lateral line hair cells. These authors showed that efferent activity caused long-lasting hyperpolarizing inhibitory postsynaptic potentials (IPSPs) and that during inhibition, the associated afferent fiber excitation was diminished. This result suggested a reduction in transmitter release from the hair cell (Fig. 5.1) and the IPSPs were shown to be sensitive to cholinergic antagonists (Flock and Russell 1976).

Hyperpolarizing IPSPs or responses to applied ACh were later observed in hair cells of frogs (Ashmore 1983; Sugai et al. 1992), reptiles (Art and Fettiplace 1984;

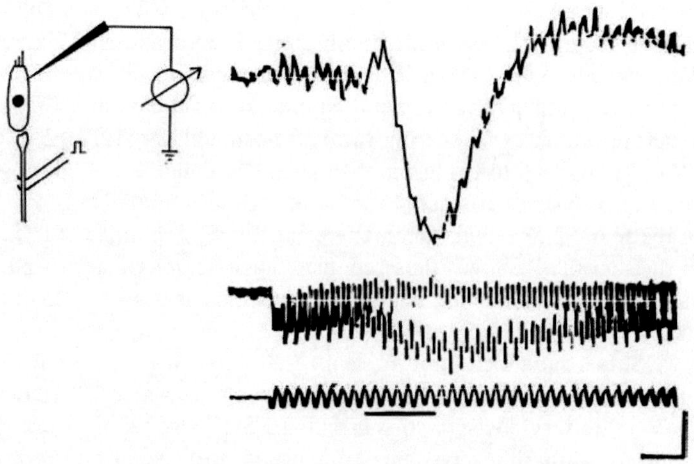

Fig. 5.1 Intracellular recording of an IPSP when the hair cell is mechanically stimulated. During the IPSP, the external double microphonic receptor potential is augmented (*middle trace*); the sinusoidal mechanical stimulus at 70 Hz is illustrated in the *lower trace*. The *bar* beneath the lower trace represents the period of electrical stimulation of the lateral line nerve at 200/s. *Vertical bar*: 1 mV for *upper trace* and 0.1 mV for *middle*. *Horizontal bar*: 50 ms (reproduced with permission from Fig. 3b in Flock and Russell 1976)

Art et al. 1984), and birds (Shigemoto and Ohmori 1991; Fuchs and Murrow 1992a, b), and in some instances could be seen to consist of a brief depolarization preceding the larger, longer-lasting hyperpolarization (see later).

5.3.2 *Details of Inhibitory Postsynaptic Potentials and Effect on Receptor Potentials in Turtle Hair Cells*

In the 1980s, Fettiplace and colleagues conducted an in-depth study of efferent inhibition in the turtle auditory papilla (Art and Fettiplace 1984; Art et al. 1984, 1985). First, by recording the response of single afferent fibers, they showed that electrical stimulation of efferent fibers could reduce the acoustic sensitivity of the auditory afferents up to a maximum of four orders of magnitude, depending on the pattern of stimulation. As also occurs in the mammalian cochlea (Kiang et al. 1970; Wiederhold and Kiang 1970), this desensitization was combined with reduced frequency selectivity (Art et al. 1984), requiring that inhibition somehow acted on the cochlear filter mechanism.

Then, intracellular recording in the turtle basilar papilla showed these inhibitory effects could be accounted for by a synaptic action on the hair cells. Electrical stimulation of the efferent axons evoked large hyperpolarizing synaptic potentials in the hair cells and a concomitant reduction in sensitivity to characteristic frequency tones (Art et al. 1982). These two effects were sufficient to account for the significant elevation in the acoustic threshold of the afferent axons. Stimulating the efferent fibers with a short train of shocks evoked maximal-amplitude hyperpolarizing IPSPs that in the different hair cells ranged from 12 to 30 mV, with a half-amplitude duration of 150–200 ms. The amplitude of IPSPs varied with the membrane potential and reversed in sign at –80 mV. The reversal potential was a function of the extracellular potassium concentration, thus leading to the conclusion that the hyperpolarizing phase of the IPSP was due to an increase in the potassium conductance of the hair cell. One important feature of these IPSPs was that a preceding short depolarizing phase was unmasked at the reversal potential of the longer lasting hyperpolarization. This early component also was evident in some cells at the resting potential. Application of either nicotinic or muscarinic cholinergic antagonists abolished the entire response (Art and Fettiplace 1984; Art et al. 1985).

One interesting feature is that single shocks given to the efferent axons generated small and infrequent responses, but with a train of shocks synaptic responses grew in a supralinear manner (Fig. 5.2). This increase in the probability and size of the response by high frequency stimulation was due to facilitation of transmitter release by successive action potentials. Functionally, the low probability of release to a single stimulus might be a protective mechanism to ensure that the tuning and sensitivity of the hair cells are not degraded by spontaneous impulses in the efferent fibers but rather depend on coherent, sustained synaptic drive of the efferent neurons (Art et al. 1984).

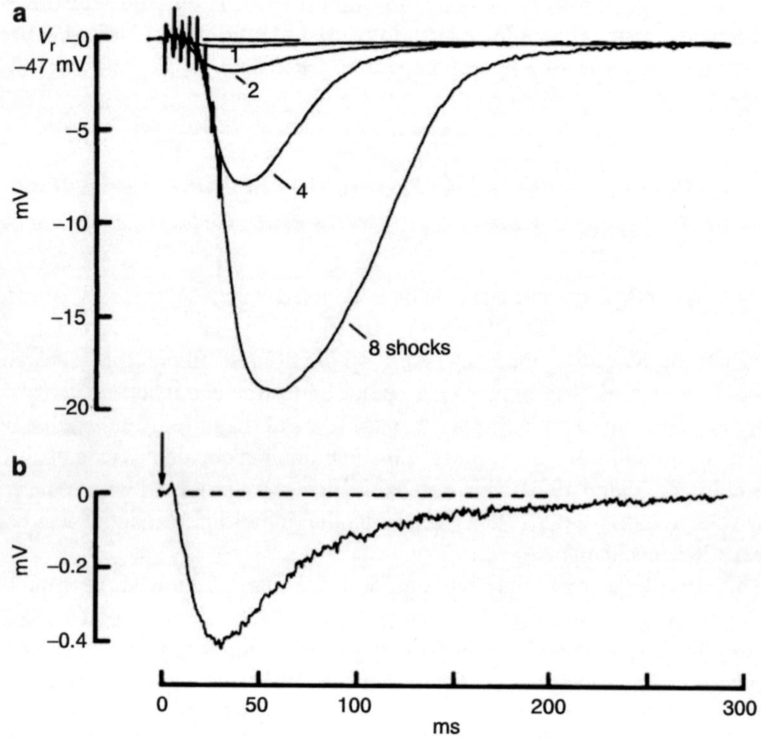

Fig. 5.2 Trains of efferent stimuli applied to an IHC. (**a**) Superimposed averages for one, two, four, and eight efferent shocks. Capacitive artifacts signal time of shock application; intershock interval 4 ms, repetition rate 2/s. (**b**) Single shock response at a higher gain. Ordinates are membrane potentials relative to the resting potential (V_r: –47 mV). Hair cell characteristic frequency, 325 Hz; temperature, 24.5°C (reproduced with permission from Fig. 1 in Art et al. 1984)

5.3.3 Application of ACh to Isolated OHCs

Several studies indicate that ACh is the main neurotransmitter released by the MOC efferents (e.g., Norris et al. 1988; Sewell 1996, see also Chap. 4). Confirming this view, the efferent inhibitory effects could be reproduced by applying ACh to hair cells isolated from the basilar papilla (auditory organ) of birds. The first experiment of this type was performed in hair cells isolated from the chicken's cochlea (Shigemoto and Ohmori 1990). By electrophysiological recording and calcium imaging, these authors showed that the application of ACh hyperpolarized the hair cells and also increased the internal Ca^{2+} concentration for several minutes. They concluded that the hyperpolarization was due to calcium-dependent potassium channels. In addition, these authors suggested that the effect of ACh was through a muscarinic cholinergic receptor causing the release of Ca^{2+} from intracellular stores. Voltage-clamp recordings in hair cells isolated from the chicken (Shigemoto and Ohmori 1991) and the guinea pig (Housley and Ashmore 1991) further supported the notion that a calcium-dependent potassium channel was involved in the cholinergic inhibition of cochlear hair cells. In contrast to the G-protein metabotropic effect proposed by Shigemoto and Ohmori, Housley and Ashmore (1991) suggested that the ACh receptor in guinea pig hair cells promoted Ca^{2+} influx from the extracellular space, which in turn activated the Ca-dependent K^+ current. The biphasic change in membrane potential, or in membrane current, was revealed by tight-seal recordings in chicken hair cells (Fuchs and Murrow 1992b), during brief (50–100 ms) application of ACh at a membrane potential of –40 mV (Fig. 5.3). ACh evoked a small inward current followed within milliseconds by a much larger and longer lasting outward K^+ current. This current ($IK_{(ACh)}$) was thought to flow through small-conductance, calcium activated potassium (SK) channels. This idea was later confirmed on mammalian (Nenov et al. 1996) and avian hair cells (Yuhas and Fuchs 1999) using specific blockers of the SK channel.

The calcium dependence of $IK_{(ACh)}$ is supported by the fact that the rapid calcium chelator BAPTA (10 mM) prevented the activation of the longer lasting outward current, leaving only the small inward current. Another indication of the calcium dependence of $IK_{(ACh)}$ was its steady-state voltage dependence, as observed in the "bell shaped" current–voltage relationship (Fig. 5.4).

This type of voltage dependence is typical of calcium-dependent potassium currents that rely on calcium influx. $IK_{(ACh)}$ is maximal around –40 mV but essentially disappears at more positive potentials, with the reduction of driving force and influx of calcium. The effect of membrane potential on calcium influx and accumulation can quantitatively explain the current–voltage relationship of $IK_{(ACh)}$ (Martin

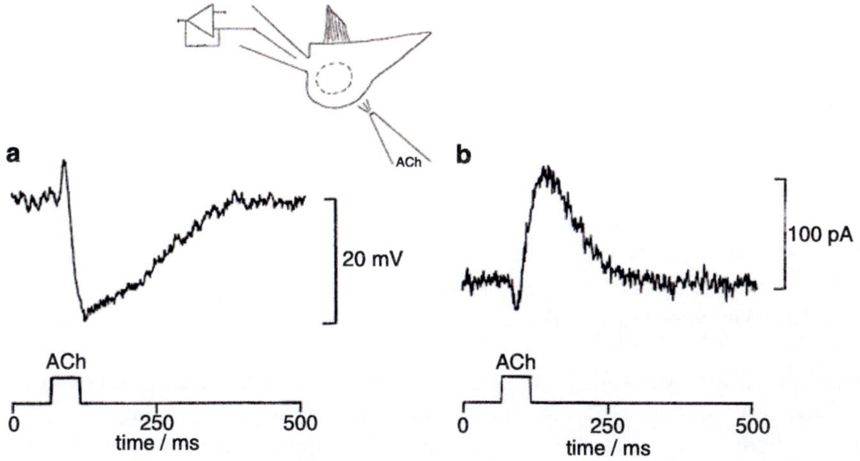

Fig. 5.3 ACh-evoked responses in chick short hair cells. (**a**) Effects of a 50-ms puff of 100 μM ACh on the voltage of an isolated hair cell (V_r was –50 mV before ACh application). Observe the large, long lasting hyperpolarization preceded by a brief depolarization caused by the application of ACh. (**b**) The same cell under voltage-clamp (V_{hold}: –54 mV). The *inset* drawing schematizes a whole cell recording from a short hair cell and the pipette used to apply ACh (reproduced with permission from Fig. 1 in Fuchs and Murrow 1992a)

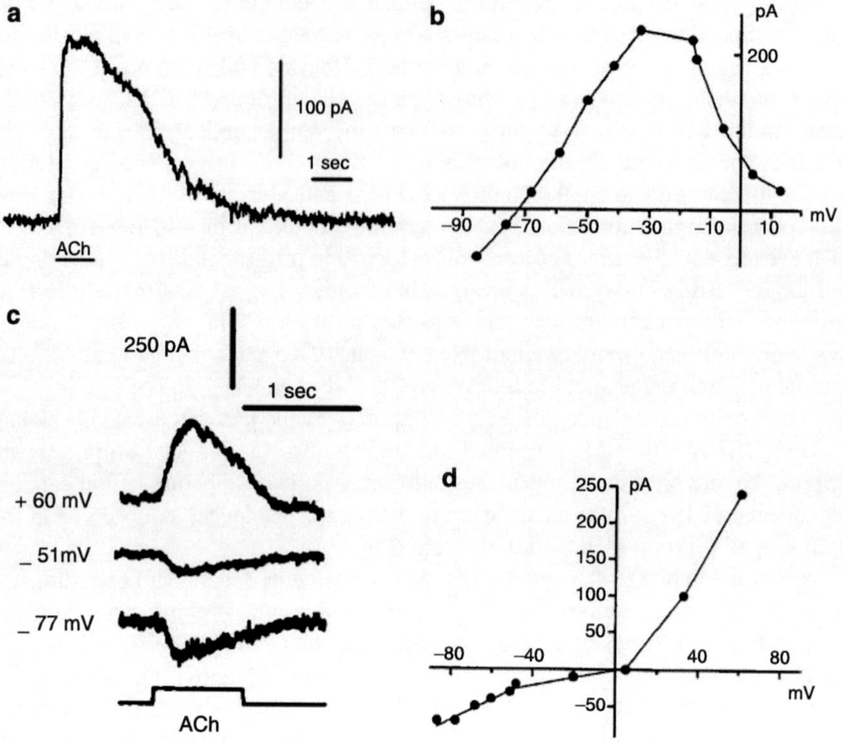

Fig. 5.4 Cholinergic responses in P12 OHCs. (**a**) Outward current evoked by ACh (100 μM ACh for 1 s) in a voltage-clamped OHC at −32 mV. (**b**) I–V curve of ACh-evoked currents. (**c**) Isolation of the "nicotinic-like" current in one OHC buffered with 10 mM BAPTA and stimulated with 100 μM ACh. Current traces at three potentials are shown on the *left*, and on the *right*, an I–V curve of ACh-evoked peak currents at different membrane potentials (reproduced with permission from Figs. 1a, c and 4a, b in Dulon and Lenoir 1996)

and Fuchs 1992). The nAChR current can be studied in hair cells buffered with BAPTA to prevent the activation of $IK_{(ACh)}$. Under these conditions the current–voltage relationship reverses near 0 mV, suggesting a Na^+/K^+ permeability ratio close to 1 (Dulon and Lenoir 1996; McNiven et al. 1996). The added fact that opening the nAChR gives rise to the activation of a calcium-dependent K^+-current indicates that Ca^{2+} also enters through the nAChR channel. However, the exact permeability ratio has been difficult to obtain because extracellular calcium is not only a charge carrier, but also is a required cofactor for nAChR gating (Weisstaub et al. 2002; Marcotti et al. 2004; Gomez-Casati et al. 2005). Therefore, cationic currents evoked by ACh in hair cells of chickens (McNiven et al. 1996) and guinea pigs (Blanchet et al. 1996; Evans 1996) completely disappeared when external calcium was removed. Moreover, at calcium concentrations above 10 mM, the current–voltage relationship strongly rectifies at negative potentials (McNiven et al. 1996). The permeability ratio of calcium and its effects on gating and conductance of the hair cell nAChR were better studied by functional expression of cloned receptors (Katz et al. 2000; Elgoyhen et al. 2001; Weisstaub et al. 2002) and later by isolated nAChR currents in IHCs present in the mammalian cochlear coil preparation (Gomez-Casati et al. 2005).

5.3.4 *Tight-Seal Recordings in the Mammalian Organ of Corti*

The introduction of the ex vivo organ of Corti preparation has made it possible to answer the question of whether both the biophysical and pharmacological characteristics of the cholinergic response of isolated hair cells could account for the synaptic efferent effects observed in the auditory periphery. Using this preparation, it was possible to show that the ionic mechanism and pharmacology of efferent synaptic effects were essentially identical to those obtained by application of ACh to isolated hair cells. Hair cells, supporting cells and neuronal synaptic contacts all remain functional for several hours after removal of the apical cochlear turn from young (0–21 days) mouse or rat cochleas. Therefore, one can study both the pre- and postsynaptic components of functioning OC synapses (Glowatzki and Fuchs 2000; Oliver et al. 2000; Katz et al. 2004; Lioudyno et al. 2004; Gomez-Casati et al. 2005; Goutman et al. 2005).

5.3.4.1 Responses to ACh in IHCs and OHCs

ACh causes a biphasic change in the membrane conductance of mammalian cochlear hair cells (see earlier). This response is observed either by spontaneous or electrically evoked release from efferent axons, or by exogenous application of this agonist. This has been shown for neonatal IHCs that are innervated by OC efferent fibers prior to the onset of hearing (Glowatzki and Fuchs 2000; Katz et al. 2004; Gomez-Casati et al. 2005; Goutman et al. 2005) and for both neonatal (Vetter et al. 2007; Taranda et al. 2009b; Taranda et al. 2009a) and older OHCs (Oliver et al. 2000; Lioudyno et al. 2004) of the mammalian cochlea. In all vertebrate hair cells examined to date, the efferent neurotransmitter ACh opens ligand-gated cation channels (the $\alpha9\alpha10$-containing nAChR, see later) through which calcium and sodium enter the hair cell, followed by activation of calcium-sensitive potassium channels. These are encoded by the SK2 (*KCNN2*) gene in mammals (Dulon et al. 1998; Marcotti et al. 2004) and birds (Yuhas and Fuchs 1999; Matthews et al. 2005). The relatively rapid functional coupling between these two ion channels, as well as the sensitivity to the fast calcium chelator, BAPTA, has led to the assumption that SK channels are directly activated by calcium influx through the nAChR (Fuchs and Evans 1990; Fuchs and Murrow 1992b; Martin and Fuchs 1992; Oliver et al. 2000). Other lines of evidence suggest that release of calcium from an internal store, perhaps the nearby synaptic cistern (Shigemoto and Ohmori 1991; Kakehata et al. 1993; Yoshida et al. 1994; Lioudyno et al. 2004), also contributes to activation of the SK channels.

Current through the nonselective nAChR can be carried by a combination of sodium, potassium, and calcium. At negative membrane potentials, inward current is carried by sodium and calcium. Measurements of the relative divalent permeability (P_{Ca}/P_{Na}) of α9α10-containing nAChRs expressed in *Xenopus* oocytes gave a permeability ratio of ~10.0 (Weisstaub et al. 2002) and ~8.0 in hair cells (Gomez-Casati et al. 2005). To date, only limited single channel data have been collected from α9α10 receptors expressed in *Xenopus* oocytes. These give a single channel conductance near 100 pS (with very low external calcium) and open time distributions with components at 90 and 320 μs at room temperature (Plazas et al. 2005b). The single-channel conductance in hair cells is probably much smaller owing to block by physiological levels of external calcium (Gomez-Casati et al. 2005). The SK2 channel has a small conductance (10 pS), is half-activated by 0.6 μM calcium and is voltage-insensitive (Kohler et al. 1996).

5.3.4.2 Spontaneous and Evoked Synaptic Currents in IHCs and OHCs

Spontaneous efferent synaptic currents have been observed in neonatal IHCs (Glowatzki and Fuchs 2000; Katz et al. 2004; Gomez-Casati et al. 2005; Goutman et al. 2005) and older OHCs (Oliver et al. 2000; Lioudyno et al. 2004). These synaptic currents are all relatively uniform. They are biphasic (fast inward current followed by a slower outward component) at membrane potentials between E_K and 0 mV (Fig. 5.5b). The nonspecific cationic current through the nAChR can be observed in isolation at the potassium equilibrium potential. Negative to E_K, the entire ACh activated current is inward, as both currents through the nAChR and the SK channel flow in the same direction (Fig. 5.5a, b). The kinetically dominant outward SK component has a decay time constant of 30–50 ms (at room temperature), while inward current through the nAChR (isolated by using the fast calcium chelator BAPTA) decays approximately three- to fivefold faster.

The rapid time course of IPSCs implies relatively tight coupling between the activation of cholinergic receptors and associated potassium channels. Other factors can come into play under physiological conditions. For example, repetitive activation of efferent fibers gives rise to strong facilitation of hair cell hyperpolarization (Art and Fettiplace 1984; Art et al. 1984). Hair cell hyperpolarization resulting from trains of shocks rises more gradually, and decays with significantly longer time constants (Goutman et al. 2005). The slowing of recovery suggests that additional cytoplasmic processes may contribute under these conditions (Fig. 5.6). For example, it is possible that calcium-induced calcium release from the associated synaptic cistern could prolong the efferent activation of calcium-dependent potassium channels. Given the strong conservation of efferent synaptic mechanisms among vertebrate hair cells, it seems likely that the behavior described for neonatal IHCs also will occur in OHCs of the mature mammalian cochlea.

5.3.4.3 Cholinergic Inhibition of IHC Action Potentials

Mice and rats are born deaf but start to hear at around P12. Before the onset of hearing, IHCs, the primary phonoreceptors, fire spontaneous or evoked action

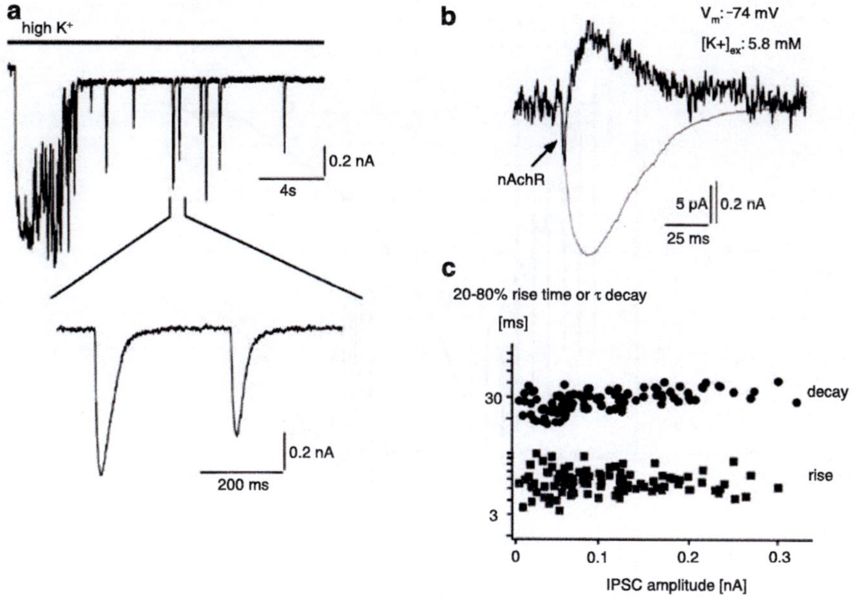

Fig. 5.5 IPSCs at the OHC synapse have a fast time course. (**a**) IPSCs recorded after bathing the cells with a saline containing 150 mM K^+. The same K^+-evoked IPSCs at an enlarged time scale (*lower panel*). (**b**) A spontaneous IPSCs recorded at 5.8 K^+ and at holding potential of −74 mV. For comparison, an IPSC from the experiment in (**a**) is illustrated. (**c**) Rise time (20–80%) and τ_{decay} of IPSCs recorded during a single application of high K^+ for 15 s plotted against IPSC amplitude (reproduced with permission from Fig. 2 in Oliver et al. 2000)

potentials resulting from the interplay of an inward Ca^{2+} current and the slowly activating delayed rectifier IK_{neo} (Kros et al. 1998; Marcotti et al. 2003). These Ca^{2+} action potentials have been shown to release transmitter at the first auditory synapse (Beutner and Moser 2001), thus driving activity in the immature auditory system, perhaps as a means to direct early stages of central synapse formation (Kotak and Sanes 1995; Kandler 2004; Leake et al. 2006). Coordinated activity among neighboring afferent neurons seems to result from calcium waves that spread among supporting cells in Kölliker's organ in the immature cochlea (Tritsch et al. 2007). This activity results in excitation of adjacent IHCs through the accompanying release of ATP, in turn releasing glutamate to establish "nearest-neighbor" activity patterns among afferent neurons.

As explained earlier, in the mature mammalian cochlea IHCs are mainly innervated by afferent fibers. However, during postnatal development, before the onset of hearing, a transient efferent innervation is found on IHCs even before OC fibers contact their final targets, the OHCs. This innervation, like that on mature OHCs, is cholinergic, inhibitory and is mediated by the same "two-channel" mechanism described in the preceding text (Glowatzki and Fuchs 2000; Elgoyhen et al. 2001; Sgard et al. 2002; Katz et al. 2004; Marcotti et al. 2004; Gomez-Casati et al. 2005). When ACh is applied, or the synapse is activated, either spontaneously (Glowatzki and Fuchs 2000) (Fig. 5.7a) or by electrical stimulation of the efferent fibers

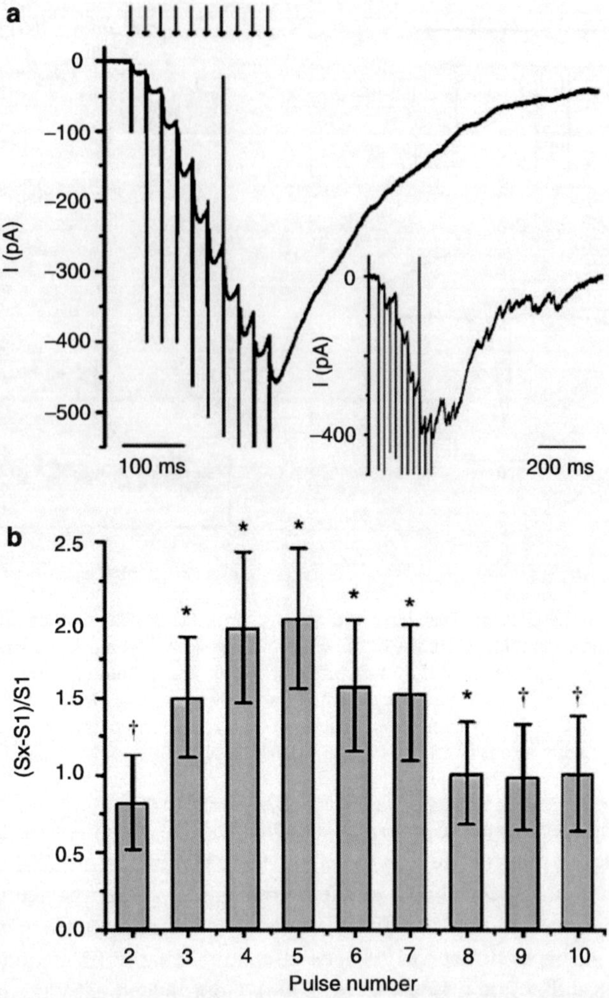

Fig. 5.6 Summation and facilitation gave rise to compound IPSCs upon repetitive electrical stimulation of efferent fibers contacting IHCs. (**a**) Response of an IHC at a holding voltage of –90 mV to a 10-ms pulse, 25-ms interval, train. The average compound IPSC elicited by a pulse train presented at a 20-s interval and repeated 30–50 times, is illustrated. *Inset*: marked fluctuations on the falling phase of the compound IPSC can be observed in an individual record. (**b**) Facilitation relative to the first response during a 10-ms pulse, 25-ms interval train (reproduced with permission from Fig. 4 in Goutman et al. 2005)

Fig. 5.7 (continued) generation of Ca^{2+} action potentials (*lower panel*). (**b**) Stimulation of efferent axons with a bipolar electrode positioned 10–20 μm below the IHC. Scale bar = 10 μm. (**c**) Current injections of 50 pA for 10 s induced repetitive firing of action potentials in IHCs. Efferent stimulation (*onset* indicated by *downward arrow*) at 2 Hz had little effect on IHC excitability. At, 5 Hz it caused a hyperpolarization of 5–10 mV and suppressed the generation of action potentials ((**a**) reproduced with permission from Fig. 4 in Glowatzki and Fuchs 2000; (**b**, **c**) reproduced from Figs. 1 and 5, respectively, in Goutman et al. 2005)

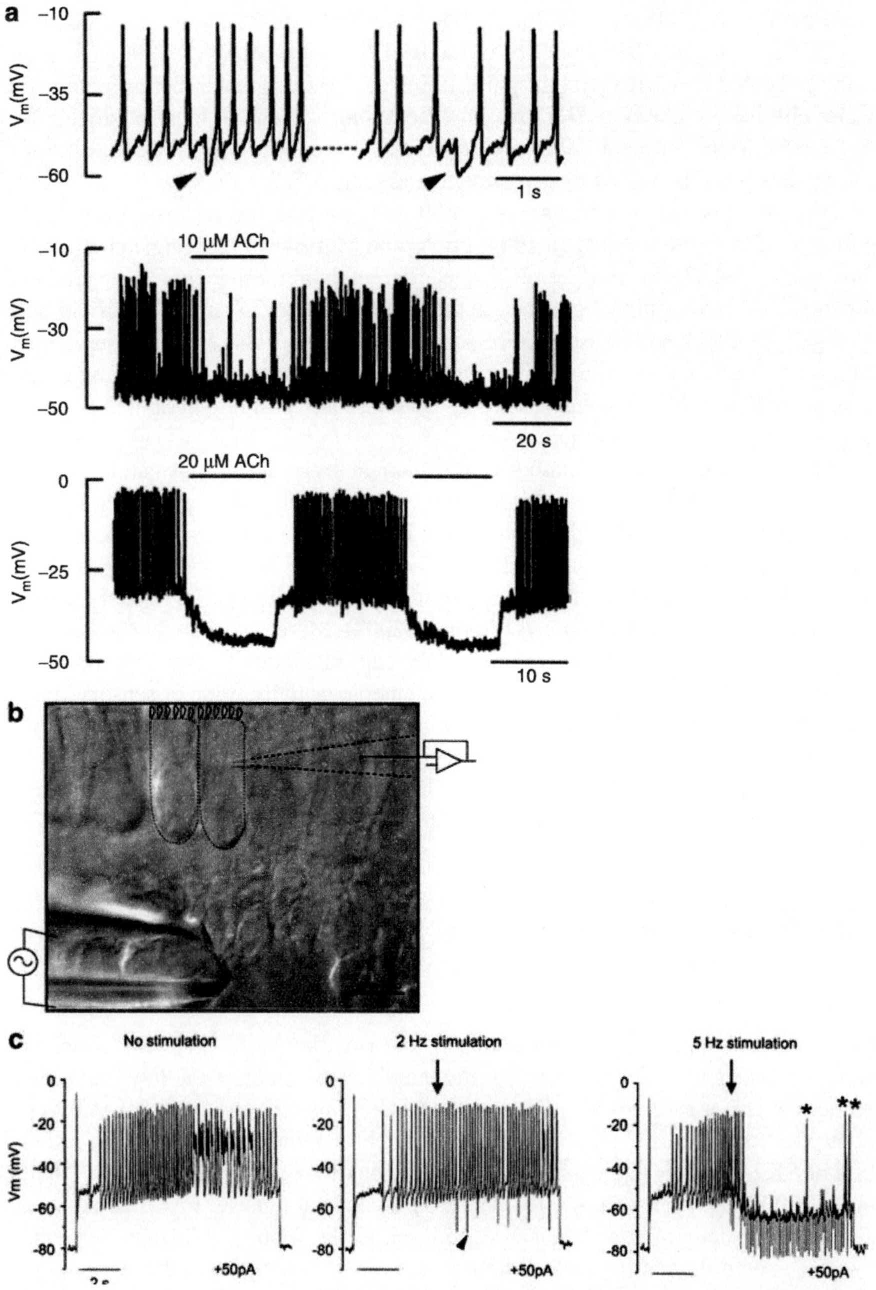

Fig. 5.7 Action potentials in IHCs from neonatal rats are inhibited by activation of cholinergic neurotransmission. (**a**) The firing frequency of Ca^{2+} action potentials was increased by injecting a constant current (100–120 pA) into IHCs. Spontaneously occurring currents (*arrowheads*) hyperpolarized the membrane potential (V_m) by –11 mV, delaying the generation of action potentials (*top panel*). At a V_m of –45 mV, 10 μM ACh reduced the firing rate (*middle panel*). At a higher concentration of ACh (20 μM), the V_m hyperpolarized from –30 to –45 mV, which abolished the

(Goutman et al. 2005) (Fig. 5.7b, c), IHCs are hyperpolarized and consequently, calcium action potential frequency is reduced or even abolished. Goutman et al. (2005) showed that with efferent pulse trains of 5 Hz or greater, action potentials were eliminated and the IHC resting membrane potential hyperpolarized by 5–10 mV. Thus, efferent inhibition becomes more effective when transmitter release has been facilitated by repetitive firing (Fig. 5.7c).

The "two-channel" mechanism of inhibition is well conserved among vertebrate hair cells (Flock 1983; Art et al. 1984; Fuchs and Murrow 1992b; Sugai et al. 1992; Goutman et al. 2005). Further, effective efferent inhibition seems to depend on repetitive, relatively high frequency activation of efferent axons (Wiederhold and Kiang 1970; Flock and Russell 1976; Art and Fettiplace 1984). These observations suggest that efferent synaptic inhibition normally requires facilitation to be effective, consistent with the suggestion that "intentional," programmatic activation is required to alter cochlear function.

What function does this inhibitory mechanism serve during postnatal development? It is clear that it must be interfering with transmitter release evoked by calcium action potentials (Beutner and Moser 2001) from ribbon synapses known to be functional at this stage (Glowatzki and Fuchs 2000). Several studies suggest that this transient efferent innervation may play a role in the ultimate functional maturation of cochlear hair cells (review: Simmons 2002). Moreover, a study showed that surgical lesion of the efferent nerve supply caused kittens to develop abnormal hearing (Walsh et al. 1998). Patterned spontaneous activity prior to sensory function is thought to be required for normal brain development (review: Katz and Shatz 1996). Thus it seems reasonable to suppose that transient efferent inhibition could direct both the functional maturation of IHCs, as well as linking together development in peripheral and central compartments of the auditory system.

5.4 Summary of "Two-Channel Hypothesis vs. Second-Messenger Mechanisms"

Current data support the notion that activation of the MOC pathway, either by sound or by shock trains delivered to the bundle at the floor of the IVth ventricle, reduces cochlear sensitivity by the action of ACh. This activates α9α10 nAChRs to allow Ca^{2+} entry, thus hyperpolarizing the hair cells through the subsequent activation of SK2 channels (Elgoyhen et al. 1994; Fuchs 1996; Dulon et al. 1998; Oliver et al. 2000; Elgoyhen et al. 2001). Thus, as in skeletal muscle, nAChRs trigger a subsequent calcium-dependent process in hair cells. Also, like in muscle, some evidence suggests that internal calcium stores might participate in this cholinergic response (Kakehata et al. 1993; Evans 1996; Lioudyno et al. 2004). Moreover, cochlear perfusion with agents that affect calcium stores alters efferent inhibition in vivo (Murugasu and Russell 1996; Sridhar et al. 1997). At the same time, the voltage dependence (Martin and Fuchs 1992) and time course of the cholinergic response (Oliver et al. 2000) imply that potassium channel gating depends on calcium influx through the nAChR. Calcium induced-calcium release (CIRC) from internal stores reconciles most of these observations.

Seconds-long application of ACh to isolated hair cells produced prolonged potassium currents, leading to the suggestion that internal calcium stores might participate in that process (Shigemoto and Ohmori 1990; Housley and Ashmore 1991; Shigemoto and Ohmori 1991; Kakehata et al. 1993; Yoshida et al. 1994). Subsequently it was found that SK currents arising from synaptic or exogenous application of ACh could be altered by agents acting on ryanodine-sensitive cytoplasmic calcium (Lioudyno et al. 2004). Further, a very important feature supporting this hypothesis is the physical presence at this synapse of the synaptic cistern, an endoplasmic reticulum closely apposed (within 20 nm) to the postsynaptic plasma membrane (Gulley and Reese 1977; Hirokawa 1978; Saito 1980). It was noted early on that these structures were similar in appearance to the sarcoplasmic reticulum of muscle (Gulley and Reese 1977). These various observations suggest that efferent inhibition may result both from calcium influx through the hair cell AChR, and by calcium-induced calcium release from the associated synaptic cistern. Studies of efferent inhibition in vivo also imply a role for cytoplasmic calcium stores (see Chap. 3).

5.5 Determination of Molecular Components

Acetylcholine is the main neurotransmitter released by medial OC (MOC) efferent axons. Throughout the nervous system ACh activates two pharmacologically, structurally, and genetically distinct receptor types, namely, the muscarinic and the nicotinic receptors. Metabotropic muscarinic receptors are linked to second-messenger systems, while the ionotropic nicotinic receptors are ligand-gated ion channels. Progress toward defining the molecular components that underlie mammalian hair cell hyperpolarization by the MOC system has been achieved via combined pharmacological, neuroanatomical, electrophysiological, and molecular studies.

While both muscarinic and nicotinic receptors have been proposed to mediate the effects of ACh in the cochlea, pharmacological and electrophysiological data suggest a central role for an atypical, nicotinic cholinergic receptor located at the synapse between efferent fibers and vertebrate OHCs and developing IHCs (Fuchs 1996). Current data support a model in which hair cell inhibition results from a small, transient, ACh-gated depolarization followed by activation of calcium-dependent potassium channels and consequent hair cell hyperpolarization (see earlier). A peculiar feature of the hair cell cholinergic receptor is that it exhibits a complex pharmacological profile, antagonized by atropine, nicotine, strychnine, α-bungarotoxin, *d*-tubocurarine, and bicuculline, but activated by very few compounds (carbachol, DMPP) other than ACh itself (Fuchs 1996).

5.5.1 Cloning of α9

The molecular composition of the hair cell ACh receptor was obtained via homology screening, and revealed to be related to nicotinic cholinergic receptors of nerve and muscle (Elgoyhen et al. 1994). The rat α9 nAChR subunit formed homomeric, ligand-gated cation channels in *Xenopus laevis* oocytes. These channels had high calcium permeability and their pharmacology was much like that of the native hair cell receptor. α9 has been shown to be expressed in cochlear and vestibular hair cells of several vertebrates by both in situ hybridization and reverse transcription-polymerase chain reaction (RT-PCR) (Elgoyhen et al. 1994; Anderson et al. 1997; Morley et al. 1998; Simmons et al. 1998; Simmons 2002; Kong et al. 2006).

Nicotinic acetylcholine receptors are formed by five homologous subunits oriented around a central ion-conducting pore, a structure shared by all members of the "Cys-loop" family of neurotransmitter-gated ion channels that also includes $GABA_A$, $GABA_C$, glycine, and 5-hydroxytryptamine-3 ($5\text{-}HT_3$) as well as some invertebrate anionic glutamate receptors (Karlin 2002). A pair of disulfide-linked cysteine residues in the ligand-binding amino portion is found in all members and gives this gene family its name. The canonical nicotinic receptor at the neuromuscular junction has a $(\alpha 1)_2\beta 1\gamma\delta$ stoichiometry and is a ligand-gated cation channel. α2–α8 and β2–β4 are thought to constitute neuronal nicotinic receptors by various heteropentameric combinations. The exceptions to this are the α7 and α8 subunits that can form homomeric receptors, and constitute the α-bungarotoxin-binding sites of the CNS (muscle receptors also bind α-bungarotoxin).

The amino acid sequence that α9 encodes is at best only 39% identical with that of other nicotinic receptors. Reflecting that sequence divergence, α9 breaks several pharmacological "rules" of the nicotinic receptor family, the most relevant being its failure to be activated by nicotine, which instead serves as a competitive antagonist (Elgoyhen et al. 1994; Verbitsky et al. 2000; Rothlin et al. 1999, 2003). α9 is thought to represent an early divergent branch of the cys-loop gene family, closer to the founder of the nicotinic gene family (Franchini and Elgoyhen 2006).

5.5.2 Cloning of α10

Although the α9 nicotinic subunit is capable of functioning as a homopentameric ACh-gated channel, several features differed from those known for the native hair cell AChR. Desensitization, the current-voltage relationship and the Ca^{2+} sensitivity, all differed from those obtained in isolated hair cells (Blanchet et al. 1996; Dulon and Lenoir 1996; Evans 1996). These discrepancies were resolved when the α10 nAChR was obtained from cochlear cDNA libraries.

Coexpression of α9 and α10 in *Xenopus laevis* oocytes gave functional receptors whose properties were essentially identical to those of the native hair cell AChRs (Elgoyhen et al. 2001). The native hair cell AChR probably consists of a hetero-pentamer with stoichiometry of $(\alpha 9)_2(\alpha 10)$ (Elgoyhen et al. 1994, 2001; Plazas et al. 2005a).

α9 and α10 share significant sequence identity. However, an examination of their sequences in different vertebrate species leads to the conclusion that α10 in particular has acquired nonsynonymous substitutions in the branch leading to the mammalian lineage (Franchini and Elgoyhen 2006). This observation takes additional significance from the fact that the outer hair cell "motor protein," prestin, likewise shows signs of positive selection pressure in mammals, fueling speculation that the hair cell AChR has acquired specific characteristics to conform with modulation of OHC electromotility and adaptations for higher frequency hearing in mammals.

5.6 Genetically Modified Mouse Models

5.6.1 α9 and α10 Knockouts

Identification of the α9 and α10 genes has enabled the generation of genetically modified mice that lack these receptor subunits. Thus, *Chrna9* has been shown to be required for effective efferent inhibition (Vetter et al. 1999), specifically, the suppression of sound-evoked CAPs and suppression of DPOAEs. Although efferent innervation patterns were generally intact, there was a significant change in the size and number of contacts on each OHC. The transgenic OHCs tended to have fewer, larger efferent boutons than did their wild-type littermates. Despite the demonstrable failure of efferent inhibition in the knockout mice, there were no other detectable behavioral consequences. Tone detection and intensity discrimination were normal, independent of background noise (May et al. 2002). As this is an obligatory knockout, it is thought that the animals develop central mechanisms to compensate for the loss of peripheral inhibition. Additional studies have confirmed that CAP thresholds, as well as OHC electromotility, are unaffected in the *Chrna9* knockouts (He et al. 2004a,b).

As predicted from oocyte expression studies, the hair cells of *Chrna10* knockout mice retain a slight cholinergic response (homomeric α9 forms functional channels in oocytes). Despite that residual cellular inhibition, distortion product suppression fails completely in the *Chrna10* knockout (Fig. 5.8). As in the *Chrna9* knockout, efferent terminals were also larger and fewer in number on the OHCs of *Chrna10* knockout mice (Vetter et al. 2007) (see Fig. 5.9c). It will be of interest to see what the respective roles of these two subunits are in acquisition and stabilization of efferent synaptic contacts on cochlear hair cells.

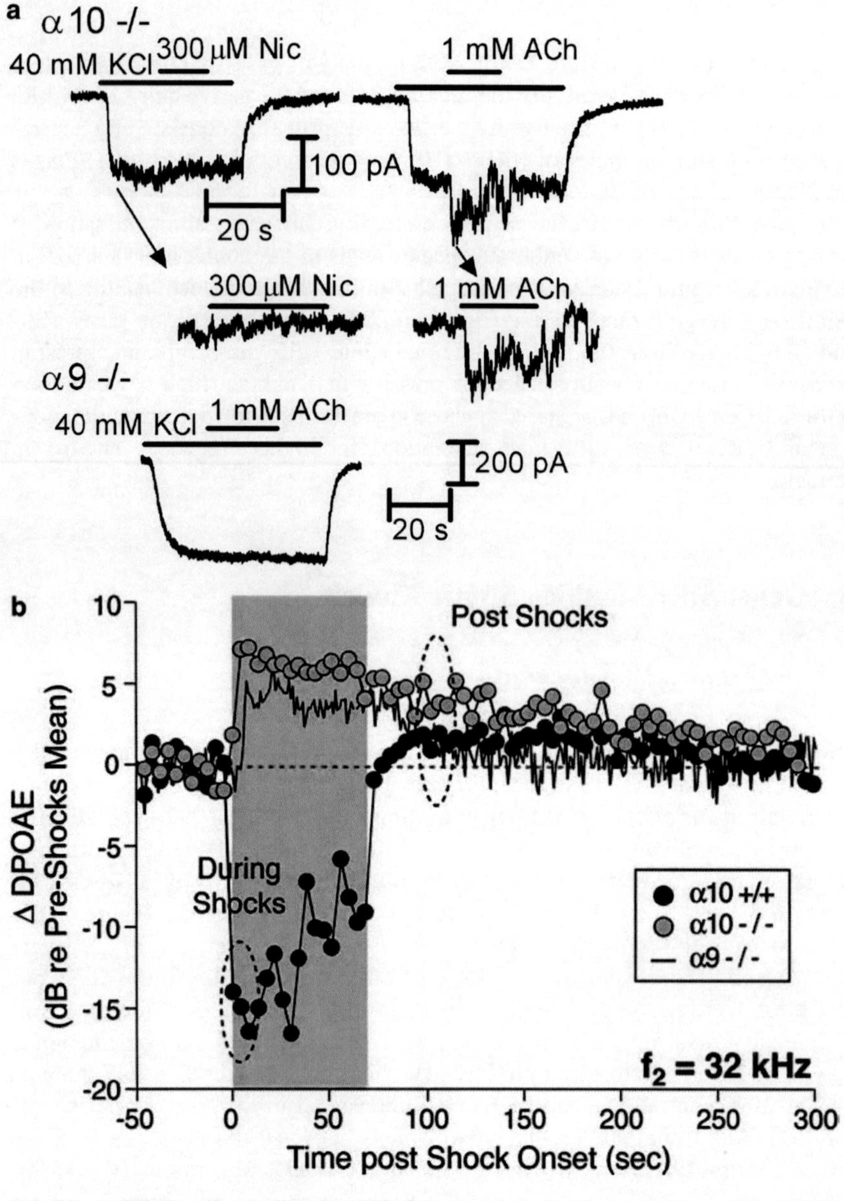

Fig. 5.8 (**a**) Whole cell recordings in OHCs from acutely isolated mouse organs of Corti (P10–13 $\alpha10^{-/-}$ and $\alpha9^{-/-}$). The figure illustrates the lack of effect of nicotine (300 μM, *left* and *inset*) in the same OHCs from $\alpha10^{-/-}$ mice in which 1 mM ACh could elicit an inward current (*middle* and *inset*). OHCs from $\alpha9^{-/-}$ mice were not sensitive to even 1 mM ACh applied in the presence of 40 mM KCl. Holding voltage was −90 mV. (**b**) Deletion of either the α9 or the α10 nAChR completely eliminates the inhibitory effect of the OC system. DPOAEs amplitudes (normalized to the mean preshock value in each case) are repeatedly measured before, during, and after a 70-s train of shocks to the OC bundle at the floor of the IVth ventricle (reproduced with permission from Figs. 3a, b and 4a in Vetter et al. 2007. Copyright National Academy of Sciences, USA 2007)

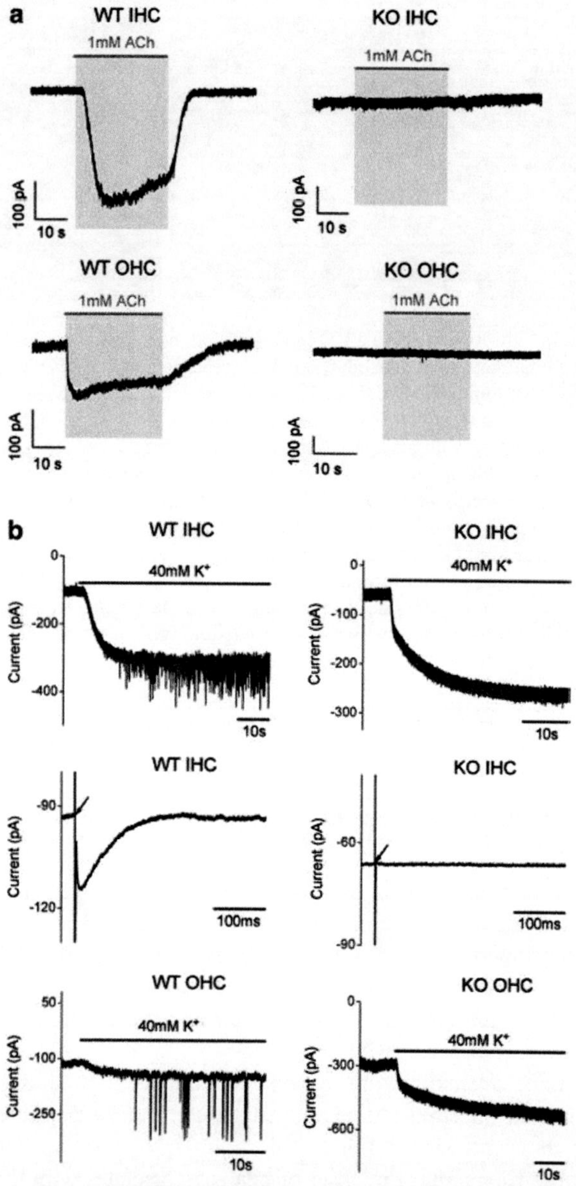

Fig. 5.9 Hair cells from SK2-knockout are insensitive to ACh and lack synaptic activity. (**a**) At a holding voltage of –94 mV, application of 1 mM ACh evoked an inward current in a wild-type (WT) IHC (P10) and in a WT OHC (P14); however, no changes in membrane current were observed either in a knockout (KO) IHC (P8) or in a KO OHC (P14). (**b**) In a WT IHC (P12) voltage-clamped at –84 mV, exposure to high K^+ saline caused steady inward current trough its effect on resting conductance, and transient inward currents due to release of ACh from depolarized efferent endings contacting these cells. Conversely, this same treatment caused a steady

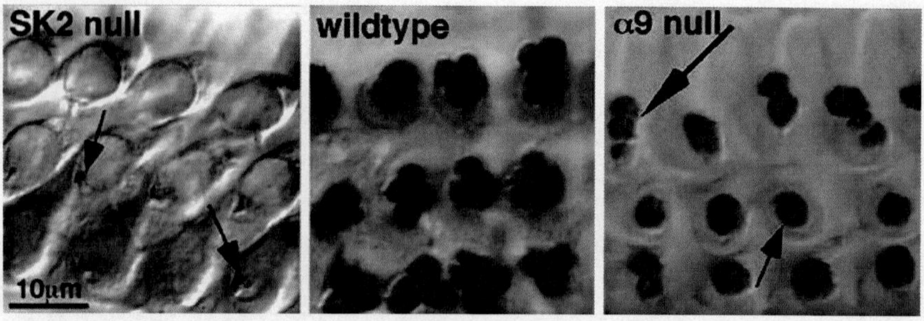

Fig. 5.9 (continued) inward current in a KO IHC (P16) but no synaptic currents. Efferent axons that were electrically stimulated produced synaptic currents in a WT IHC (P7), but not in a KO IHC (P6). As in IHCs, application of high K^+ saline in WT OHCs evoked synaptic currents whereas the same treatment in KO OHCs was ineffective. (**c**) In SK2$^{-/-}$ mice, OC synaptic boutons degenerate. Immunostaining with synaptophysin shows the efferent OC synaptic terminals contacting OHCs. OHCs in WT animals are innervated by two to five synaptic boutons (*left panel, arrows*). In α9$^{-/-}$ mice (*right panel*), OC synaptic contacts are less numerous (one or two per OHC) but are hypertrophied (*small arrow*). However, in some cases hair cells are contacted by two or three boutons (*large arrow*). In SK2$^{-/-}$ mice (*left panel*), terminals (*arrows*) degenerate in a progressive manner. All images are form mid-cochlea regions of adults (8–12 weeks old). Scale bar applies to all panels ((**a**, **b**) reproduced and modified with permission from Figs. 4a, b and 6a, b in Kong et al. 2008, (**c**) reproduced with permission from Fig. 1 in Murthy et al. 2009)

5.6.2 α9 and α10 Overexpressors

Transgenic mice overexpressing either the nAChR α9 or the α10 subunit also have been developed. Mice overexpressing the α9 subunit are significantly less sensitive to acoustic injury than their wild-type littermates (Maison et al. 2002). There was no difference in the number or size of efferent contacts onto OHCs, suggesting that the additional protection resulted from a greater density, or altered distribution of AChRs in the overexpressor OHCs.

Transgenic mice overexpressing the α10 subunit were generated to answer a completely different question. As explained earlier, before the onset of hearing, efferent fibers transiently make functional cholinergic synapses with IHCs. The retraction of these fibers after the onset of hearing correlates with the cessation of transcription of the *Chrna10* (but not the *Chrna9*) gene (Elgoyhen et al. 1994; Elgoyhen et al. 2001; Simmons 2002; Katz et al. 2004), with the disappearance of ACh-evoked currents and also with the down-regulation of the SK2 channel (Katz et al. 2004). Therefore, transgenic mice whose IHCs constitutively express α10 (*Pou4f3-α10* transgenic mice) into adulthood were generated to analyze this developmental change further (Taranda et al. 2009a). In situ hybridization showed that the α10 mRNA continues to be expressed by IHCs of 8-week-old transgenic mice (long after its down-regulation in wild-type littermates). In addition, this mRNA is translated into a functional protein because IHCs backcrossed to a *Chrna10*$^{-/-}$ background (whose IHCs completely lack cholinergic sensitivity) displayed normal synaptic ACh-evoked currents in patch-clamp recordings, showing that the transgene restored cholinergic function in these cells. However, the constitutive expression of the α10 subunit was not sufficient to maintain functional α9α10 receptors, leading to neither ACh responses nor efferent synaptic currents, after the onset of hearing. Therefore, the lack of cholinergic responses in IHCs goes beyond the transcription of the *Chrna10* gene. Presumably, after the onset of hearing, genes other than *Chrna10* also cease transcription and/or translation, as is the case of the *KCNN2* gene that codes for the SK2 channel. These genes might encode proteins that form a macromolecular synaptic complex that is necessary for assembly, trafficking, and/or anchorage of the nAChR to the plasma membrane of the IHC. For example, RIC3, a transmembrane protein that acts as a molecular chaperone, is required for efficient receptor folding, assembly, and functional expression of the α7 nAChR (Millar 2008). Even though chaperone proteins have not been described for the α9α10 nAChR, it is known that activation of this receptor leads to an increase in intracellular Ca^{2+} and the subsequent opening of closely associated SK2 channels (Housley and Ashmore 1991; Fuchs and Murrow 1992a, b; Dulon et al. 1998; Glowatzki and Fuchs 2000; Oliver et al. 2000; Katz et al. 2004; Gomez-Casati et al. 2005), implying a variety of intermolecular connections. The *Pou4f3-α10* transgenic mice also lack functional SK2 currents after the onset of hearing (Taranda et al. 2009b). Thus, these observations together with the fact that *KCNN2* knockout mice, as discussed below, totally lack ACh responses in hair cells (Johnson et al. 2007; Kong et al. 2008), might indicate the SK2 protein as fundamentally required for the assembly, trafficking, and/or anchorage of the nAChR macromolecular synaptic complex.

5.6.3 SK2 Knockout Mice

Neuronal firing patterns are shaped by the characteristics of the afterhyperpolarization (AHP) following action potentials (Kohler et al. 1996; Hallworth et al. 2003). The AHP is composed of fast and slow components, mediated by the large-conductance voltage- and calcium-activated potassium (BK) channels, and the voltage-insensitive, calcium-activated small conductance (SK) potassium channels, respectively.

As explained in the preceding text, cochlear hair cells employ what would ordinarily be an excitatory neurotransmitter to produce inhibition and they do this by means of the functional coupling of α9α10 nAChR to SK2 channels. By using SK2 knockout mice (Bond et al. 2004), this subtype of SK channel was shown to be solely responsible for encoding the calcium-activated potassium channel in cochlear hair cells (Johnson et al. 2007; Kong et al. 2008). When SK2 channels are blocked by potassium channel antagonists, the cholinergic response at the cellular level is excitatory (Marcotti et al. 2004) However, when SK2-knockout mice were examined for "reversed" cholinergic efferent effects in an intact animal model, the results were not as expected (Kong et al. 2008). Surprisingly, unlike voltage-gated conductances in hair cells, the ionotropic AChRs were profoundly affected by deletion of the SK2 gene. SK2-knockout OHCs were completely insensitive to exogenous ACh, implying absent or otherwise dysfunctional nAChRs (Fig. 5.9a). Likewise, spontaneous cholinergic synaptic currents were not seen in OHCs from these mice (Fig. 5.9b). In addition, neither efferent synaptic currents nor responses to exogenous ACh were seen in neonatal IHCs in the SK2-knockout mice (Fig. 5.9a, b).

Moreover, using this same animal model, SK2 channels were shown to be necessary for the long-term survival of OC fibers and synapses (Murthy et al. 2009). In distinct contrast with the hypertrophy seen at efferent-OHC synapses in $\alpha 9^{-/-}$ and $\alpha 10^{-/-}$ mice, the $SK2^{-/-}$ cochlea shows an age-related decline in number and size of OC terminals (Fig. 5.9c). This decline appears to be a specific effect on synaptic morphology because auditory brain stem response (ABR) and DPOAE thresholds are unaffected in the $SK2^{-/-}$ mice, indicating otherwise normal cochlear function. Perhaps surprisingly, given the lack of cholinergic response, mRNA for $\alpha 9$ and $\alpha 10$ remained near control levels (Murthy et al. 2009). Unfortunately, no effective antibodies yet exist for these proteins, precluding cellular localization of the encoded proteins.

These results suggest that the SK2 channel has a central position in OC innervation and that hair cell responses induced and/or modulated by OC activation are necessary for the survival of OC innervation. Moreover, the fact that cholinergic responses are completely absent in hair cells from these SK2 null mice, even though the amount of $\alpha 9$ and $\alpha 10$ mRNA, as evaluated by quantitative RT-PCR (Kong et al. 2008; Murthy et al. 2009), did not differ from those in wild-type animals, strongly suggests that the nAChR/SK2 channel complex is assembled prior to targeting and/or insertion into the membrane. This hypothesis is also supported by the lack of cholinergic responses in IHCs after the onset of hearing even in the case where the $\alpha 10$ gene is constitutively expressed (see preceding, $\alpha 10$ overexpressors). Thus, this suggests that the SK2 channel acts in a dominant manner important for transport of the entire postsynaptic receptor/channel complex.

In heart muscle, SK2 binds to α-actinin2 (Lu et al. 2007), and given that some neurotransmitter receptors are similarly bound to the membrane by cytoskeletal elements (Cabello et al. 2007), it may be that SK2 represents a component around which the SK2/nAChR multiprotein complex is transformed, via linker proteins unique to the SK2 channel, into a core complex ready to be inserted into the plasma membrane. The normal efferent innervation observed in SK2 heterozygotes and overexpressors (Maison et al. 2007) suggests that a lower level of SK2, as is presumed to be the case for heterozygotes, is sufficient for maintenance of OC synapses, and that overexpression of the gene has no deleterious effects on synaptic structure. Moreover, Murthy et al. (2009) studied the compound $SK2^{-/-}/\alpha 10^{-/-}$ mice and found that these animals showed synaptic degeneration similar to that found following the ablation of the SK2 gene. Therefore, SK2 channel activity is upstream of, and epistatic to, the role of nAChRs in synapse maintenance, as the hypertrophy of efferent synaptic terminals observed in the $\alpha 10^{-/-}$ mice is either masked or altered by the phenotype expressed in the SK2 nulls.

5.6.4 *α9 Knock-in Mice*

Although significant progress has been made in defining the cellular mechanisms of hair cell inhibition, the functional role(s) of the sound-evoked OC feedback system, including control of the dynamic range of hearing (Guinan 1996; see also Chap. 3), improvement of signal detection in background noise (Dolan and Nuttall 1988; Winslow and Sachs 1988), mediating selective attention (Oatman 1976; Delano et al. 2007), and protection from acoustic injury (Rajan 1988) remain controversial.

The α9 and α10 nAChR and the SK2 channel knockout mouse models are valuable tools to determine the molecular components of the OC synapse and to study their functional, biophysical, and pharmacological properties. However, because ablation of any of the three genes results in the disappearance of cholinergic sensitivity in cochlear hair cells (with the exception of small ACh-evoked currents in OHCs from α10 knockouts) and the complete loss of OC efferent function (see Sects. 5.6.1 and 5.6.3), it is not possible to explore the contribution of these genes in synaptic transmission at the OC-hair cell synapses.

Recently, a genetically modified mouse model in which the magnitude and duration of the MOC efferent effect is increased was generated to probe further the underlying molecular pathways, and to assess the consequences of this manipulation on auditory thresholds and susceptibility to noise-induced hearing loss (Taranda et al. 2009a). This mouse has a substitution of a threonine for a leucine at position 9′ (L9′T) of the second transmembrane domain of the α9 subunit. The (L9′T) mutation produced an increase in sensitivity to ACh and decreased rates of desensitization in both IHCs and OHCs (Fig. 5.10a). As was predicted, synaptic currents were dramatically prolonged in OHCs from $Chrna9^{L9'T/L9'T}$ mice (Fig. 5.10a). Consistent with these effects, $Chrna9^{L9'T/L9'T}$ mice had elevated acoustic thresholds, and shock-evoked MOC activation produced both enhanced and prolonged cochlear suppression in vivo (Fig. 5.10b).

In addition, this enhanced inhibitory effect attenuated sound-induced, permanent acoustic injury. This is consistent with the work showing that overexpression of wild-type α9 channels, which more modestly increased the MOC-mediated DPOAE suppression, also increased the resistance of the ear to acoustic injury (Maison et al. 2002). Thus, these animal models establish that the MOC efferent feedback system inhibits cochlear sensitivity and protects the inner ear from acoustic injury. Moreover, the cholinergic synapse between MOC terminals and cochlear hair cells provides a valuable model to study the phenotypic consequences of targeted mutations for both development and function of the auditory system.

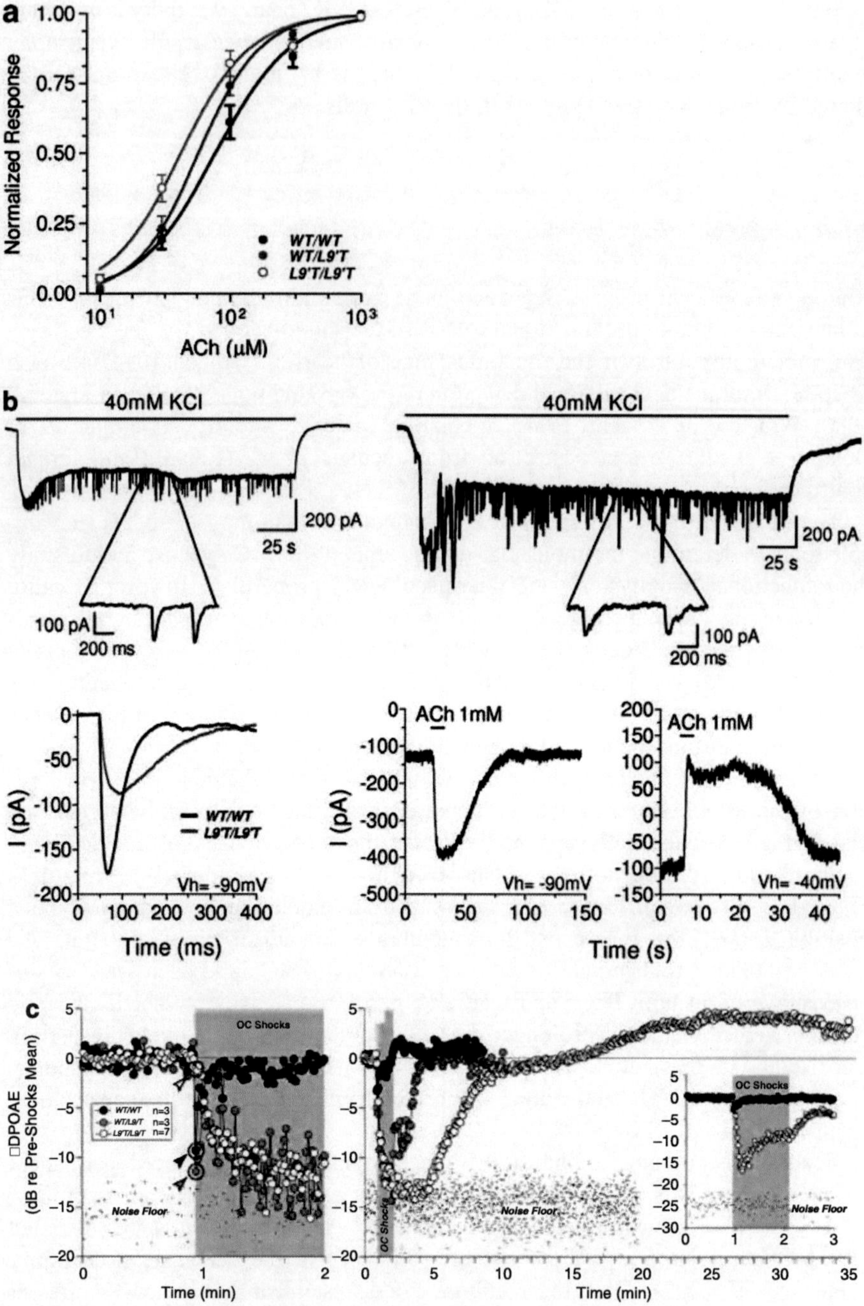

Fig. 5.10 (**a**) Concentration–response curves to ACh performed in P9–P10 IHCs from *Chrna9*$^{wt/wt}$ and *Chrna9*$^{L9'T/L9'T}$ mice. (**b**) Representative traces of IPSCs evoked by 40 mM K^+ in OHCs of P10–P11 *Chrna9*$^{wt/wt}$ at a V_{hold} of −90 mV (*top left panel*). *Top right panel* illustrates the same experiment but for *Chrna9*$^{L9'T/L9'T}$ mice. The *insets* show synaptic currents in an expanded time scale. *Lower left panel*: Superimposed representative traces of nAChR + SK2 synaptic currents

These mutations can be analyzed at the synaptic, whole-organ, and systems level. It is important to note that when analyzed at the level of nAChR function, the α9 L9′T mutant mice reproduce what has been previously described for the recombinant receptor expressed in *Xenopus laevis* oocytes, that is, a decrease in the EC_{50} for ACh and reduced desensitization kinetics (Plazas et al. 2005b). Both these effects probably derive from the effects of the L9′ on channel gating, and these alterations in gating properties translate into increased synaptic efficacy (Taranda et al. 2009a). Another relevant issue with this knock-in mouse is that the inhibitory sign of the efferent synapse is conserved as nAChR currents remain functionally coupled to SK2 channels. Interestingly, the L9′T mutation does not lead to cell death of cochlear hair cells. When similar mutations are introduced into α4 and α7 nAChR subunits (Labarca et al. 2001) they lead to death of dopaminergic neurons in substantia nigra and apoptotic cell death throughout the somatosensory cortex (Orr-Urtreger et al. 2000), respectively, probably due to Ca^{2+} excitotoxicity. The resilience of cochlear hair cells despite enhanced gating of α9α10 nAChRs with high Ca^{2+} permeability (Katz et al. 2000; Weisstaub et al. 2002; Gomez-Casati et al. 2005) likely results from their high levels of intrinsic calcium buffers (Hackney et al. 2005). Therefore, this animal model is useful to study additional physiological functions of MOC innervation. Moreover, this knock-in mouse is also an interesting model system to explore further all the components (those already established or those that are still unknown or putative) of this peculiar inhibitory synapse.

5.7 Summary and Conclusions

The experimental results described in this chapter support the hypothesis that cholinergic inhibition of cochlear hair cells is mediated by the α9α10-nAChR that allows Ca^{2+} into the cell and the subsequent activation of the voltage-independent, calcium-dependent SK2 potassium channel. This "two-channel" inhibitory mechanism appears to be conserved from fish to mammals. As discussed here, there also is evidence suggesting that calcium-induced calcium release from intracellular stores, particularly from the "synaptoplasmic cistern" coextensive with efferent synaptic contacts at the basal pole of the hair cell, enhances and sustains SK2 activation.

Fig. 5.10 (continued) from a *Chrna9*$^{wt/wt}$ (39 events) and a *Chrna9*$^{L9'T/L9'T}$ (77 events) mice. *Lower middle* and *right panels* show representative responses of OHCs from *Chrna9*$^{L9'T/L9'T}$ mice at a V_{hold} of −90 and −40 mV. (**c**) In mutant mice, OC-mediated suppression of cochlear DPOAEs is slowed, enhanced, and prolonged. DPOAE amplitudes measured before, during, and after a 70-s shock train to the OC bundle (*gray boxes*) are shown on two different time scales to emphasize the onset effects (*left*) and the offset effect (*middle*) of OC activation. *Arrowheads* indicate the first point after shock-train onset for each genotype. The *right panel* shows full suppression magnitude obtained by raising primary levels until preshock DPOAEs 25 dB above the noise floor: peak suppression in *Chrna9*$^{L9'T/L9'T}$ mice reached approximately 17 dB, whereas for *Chrna9*$^{wt/wt}$ mice, effects were less than approximately 5 dB at these higher stimulus levels. ((**a**) reproduced from Fig. 3a, (**b**) from Fig. 5, and (**c**) from Fig. 8 in Taranda et al. 2009a)

From the 1970s to 1990s, scientists in the field contributed a thorough and in-depth body of evidence supporting the notion that the inhibitory mechanism involved the activation of a cholinergic receptor, followed by entry of Ca^{2+} from the extracellular space and the subsequent activation of an SK channel. However, the molecular nature of the hair cell cholinergic receptor was still a puzzle. It was the cloning of the α9 subunit in 1994 that allowed its clear inclusion within the nicotinic family of cholinergic receptors. Then the cloning of the α10 subunit, together with the biophysical and pharmacological characterization of the recombinant α9α10 nAChR expressed in *Xenopus oocytes*, showed that this heteromeric receptor resembled more closely the characteristics of the one present at the hair cell synapse. After this, the introduction of the cochlear coil preparation fostered both the biophysical and pharmacological studies on the native nAChR receptor. They also allowed the direct stimulation of the efferent fibers contacting the hair cells to study the characteristics of efferent transmitter release and comparisons with in vivo functional measures of efferent inhibition. Both types of studies concluded that feedback from the central nervous system will have an effect only when it is strongly driven, preventing spontaneous activity from inadvertently altering cochlear function.

Finally, the genetic manipulation of the molecules involved in the cholinergic inhibition of hair cells confirmed that the α9 and α10 nAChR subunits are fundamental components of the native receptor. Both the α9 and α10 knockout mice fail to show suppression of cochlear responses (DPOAEs and CAP) during efferent fiber activation, demonstrating the key role the α9α10 nAChRs play in mediating the known effects of the OC system. In addition, the complete lack of cholinergic sensitivity of hair cells from SK2 knockout mice strongly suggests that this molecule has a central role and that it is necessary for the assembly, targeting, and/or insertion of the nAChR into the membrane. Further work will be necessary to elucidate how this nAChR/SK2 complex is assembled and targeted to the membrane, and to identify the chaperone molecules involved in this process.

References

Alger BE (1991) Gating of GABAergic inhibition in hippocampal pyramidal cells. Ann NY Acad Sci 627:249–263

Altschuler RA, Parakkal MH, Fex J (1983) Localization of enkephalin-like immunoreactivity in acetylcholinesterase-positive cells in the guinea-pig lateral superior olivary complex that project to the cochlea. Neuroscience 9:621–630

Altschuler RA, Fex J, Parakkal MH, Eckenstein F (1984) Colocalization of enkephalin-like and choline acetyltransferase-like immunoreactivities in olivochoclear neurons of the guinea pig. J Histochem Cytochem 32:839–843

Altschuler RA, Hoffman DW, Reeks KA, Fex J (1985) Localization of dynorphin B-like and alphaneoendorphin-like immunoreactivities in the guinea pig organ of Corti. Hear Res 17:249–258

Anderson AD, Troyanovskaya M, Wackym PA (1997) Differential expression of a2–7 and b2–4 nicotinic acetylcholine receptor subunit mRNA in the vestibular end-organs and Scarpa's ganglia of the rat. Brain Res 778:409–413

Art JJ, Fettiplace R (1984) Efferent desensitization of auditory nerve fibre responses in the cochlea of the turtle *Pseudemys scripta elegans*. J Physiol (Lond) 356:507–523
Art JJ, Crawford AC, Fettiplace R, Fuchs PA (1982) Efferent regulation of hair cells in the turtle cochlea. Proc R Soc Lond B Biol Sci 216:377–384
Art JJ, Fettiplace R, Fuchs PA (1984) Synaptic hyperpolarization and inhibition of turtle cochlear hair cells. J Physiol (Lond) 356:525–550
Art JJ, Crawford AC, Fettiplace R, Fuchs PA (1985) Efferent modulation of hair cell tuning in the cochlea of the turtle. J Physiol (Lond) 360:397–421
Ashmore JF (1983) Frequency tuning in a frog vestibular organ. Nature 304:536–538
Betz H, Kuhse J, Schmieden V, Laube B, Kirsch J, Harvey RJ (1999) Structure and functions of inhibitory and excitatory glycine receptors. Ann NY Acad Sci 868:667–676
Beutner D, Moser T (2001) The presynaptic function of mouse cochlear inner hair cells during development of hearing. J Neurosci 21:4593–4599
Blanchet C, Erostegui C, Sugasawa M, Dulon D (1996) Acetylcholine-induced potassium current of guinea pig outer hair cells: its dependence on a calcium influx through nicotinic-like receptors. J Neurosci 16:2574–2584
Bond CT, Herson PS, Strassmaier T, Hammond R, Stackman R, Maylie J, Adelman JP (2004) Small conductance Ca^{2+}-activated K^{+} channel knock-out mice reveal the identity of calcium-dependent afterhyperpolarization currents. J Neurosci 24:5301–5306
Cabanillas LA, Luebke AE (2002) CGRP- and cholinergic-containing fibers project to guinea pig outer hair cells. Hear Res 172:14–17
Cabello N, Remelli R, Canela L, Soriguera A, Mallol J, Canela EI, Robbins MJ, Lluis C, Franco R, McIlhinney RA, Ciruela F (2007) Actin-binding protein alpha-actinin-1 interacts with the metabotropic glutamate receptor type 5b and modulates the cell surface expression and function of the receptor. J Biol Chem 282:12143–12153
Delano PH, Elgueda D, Hamame CM, Robles L (2007) Selective attention to visual stimuli reduces cochlear sensitivity in chinchillas. J Neurosci 27:4146–4153
Dolan DF, Nuttall AL (1988) Masked cochlear whole-nerve response intensity functions altered by electrical stimulation of the crossed olivocochlear bundle. J Acoust Soc Am 83:1081–1086
Dulon D, Lenoir M (1996) Cholinergic responses in developing outer hair cells of the rat cochlea. Eur J Neurosci 8:1945–1952
Dulon D, Luo L, Zhang C, Ryan AF (1998) Expression of small-conductance calcium-activated potassium channels (SK) in outer hair cells of the rat cochlea. Eur J Neurosci 10:907–915
Elgoyhen AB, Johnson DS, Boulter J, Vetter DE, Heinemann S (1994) Alpha 9: an acetylcholine receptor with novel pharmacological properties expressed in rat cochlear hair cells. Cell 79:705–715
Elgoyhen A, Vetter D, Katz E, Rothlin C, Heinemann S, Boulter J (2001) Alpha 10: a determinant of nicotinic cholinergic receptor function in mammalian vestibular and cochlear mechanosensory hair cells. Proc Natl Acad Sci USA 98:3501–3506
Evans MG (1996) Acetylcholine activates two currents in guinea-pig outer hair cells. J Physiol (Lond) 491:563–578
Eybalin M (1993) Neurotransmitters and neuromodulators of the mammalian cochlea. Physiol Rev 73:309–373
Eybalin M, Parnaud C, Geffard M, Pujol R (1988) Immunoelectron microscopy identifies several types of GABA-containing efferent synapses in the guinea-pig organ of Corti. Neuroscience 24:29–38
Fex J, Altschuler RA (1981) Enkephalin-like immunoreactivity of olivochoclear nerve fibers in cochlea of guinea pig and cat. Proc Natl Acad Sci USA 78:1255–1259
Flock A (1983) Hair cells, receptors with a motor capacity? In: Klinke R, Hartmann R (eds) Hearing-physiological bases and psychophysics. Springer, Berlin, pp 2–8
Flock A, Russell I (1976) Inhibition by efferent nerve fibres: action on hair cells and afferent synaptic transmission in the lateral line canal organ of the burbot *Lota lota*. J Physiol 257:45–62

Franchini LF, Elgoyhen AB (2006) Adaptive evolution in mammalian proteins involved in cochlear outer hair cell electromotility. Mol Phylogenet Evol 41:622–635

Fuchs PA (1996) Synaptic transmission at vertebrate hair cells. Curr Opin Neurobiol 6:514–519

Fuchs PA, Evans MG (1990) Potassium currents in hair cells isolated from the cochlea of the chick. J Physiol (Lond) 429:529–551

Fuchs PA, Murrow BW (1992a) A novel cholinergic receptor mediates inhibition of chick cochlear hair cells. Proc R Soc Lond B Biol Sci 248:35–40

Fuchs PA, Murrow BW (1992b) Cholinergic inhibition of short (outer) hair cells of the chick's cochlea. J Neurosci 12:800–809

Galambos R (1956) Suppression of auditory nerve activity by stimulation of efferent fibers to cochlea. J Neurophysiol 19:424–437

Glowatzki E, Fuchs PA (2000) Cholinergic synaptic inhibition of inner hair cells in the neonatal mammalian cochlea. Science 288:2366–2368

Gomez-Casati ME, Fuchs PA, Elgoyhen AB, Katz E (2005) Biophysical and pharmacological characterization of nicotinic cholinergic receptors in rat cochlear inner hair cells. J Physiol 566:103–118

Goutman JD, Fuchs PA, Glowatzki E (2005) Facilitating efferent inhibition of inner hair cells in the cochlea of the neonatal rat. J Physiol 566:49–59

Guinan JJ (1996) Efferent physiology. In: Dallos P, Popper AN, Fay RR (eds) The cochlea. Springer, New York, pp 435–502

Guinan JJ Jr, Stankovic KM (1996) Medial efferent inhibition produces the largest equivalent attenuations at moderate to high sound levels in cat auditory-nerve fibers. J Acoust Soc Am 100:1680–1690

Gulley RL, Reese TS (1977) Regional specialization of the hair cell plasmalemma in the organ of corti. Anat Rec 189:109–123

Guth PS, Perin P, Norris CH, Valli P (1998) The vestibular hair cells: post-transductional signal processing. Prog Neurobiol 54:193–247

Hackney CM, Mahendrasingam S, Penn A, Fettiplace R (2005) The concentrations of calcium buffering proteins in mammalian cochlear hair cells. J Neurosci 25:7867–7875

Hallworth NE, Wilson CJ, Bevan MD (2003) Apamin-sensitive small conductance calcium-activated potassium channels, through their selective coupling to voltage-gated calcium channels, are critical determinants of the precision, pace, and pattern of action potential generation in rat subthalamic nucleus neurons in vitro. J Neurosci 23:7525–7542

He DZ, Jia S, Dallos P (2004a) Mechanoelectrical transduction of adult outer hair cells studied in a gerbil hemicochlea. Nature 429:766–770

He DZ, Cheatham MA, Pearce M, Vetter DE (2004b) Mouse outer hair cells lacking the alpha9 ACh receptor are motile. Brain Res Dev Brain Res 148:19–25

Hirokawa N (1978) The ultrastructure of the basilar papilla of the chick. J Comp Neurol 181:361–374

Housley GD, Ashmore JF (1991) Direct measurement of the action of acetylcholine on isolated outer hair cells of the guinea pig cochlea. Proc R Soc Lond B Biol Sci 244:161–167

Johnson SL, Adelman JP, Marcotti W (2007) Genetic deletion of SK2 channels in mouse inner hair cells prevents the developmental linearization in the Ca^{2+} dependence of exocytosis. J Physiol 583:631–646

Jones MV, Westbrook GL (1996) The impact of receptor desensitization on fast synaptic transmission. Trends Neurosci 19:96–101

Kakehata S, Nakagawa T, Takasaka T, Akaike N (1993) Cellular mechanism of acetylcholine-induced response in dissociated outer hair cells of guinea-pig cochlea. J Physiol (Lond) 463:227–244

Kandler K (2004) Activity-dependent organization of inhibitory circuits: lessons from the auditory system. Curr Opin Neurobiol 14:96–104

Karlin A (2002) Emerging structure of the nicotinic acetylcholine receptors. Nat Rev Neurosci 3:102–114

Katz LC, Shatz CJ (1996) Synaptic activity and the construction of cortical circuits. Science 274:1133–1138

Katz E, Verbitsky M, Rothlin CV, Vetter DE, Heinemann SF, Belen Elgoyhen A (2000) High calcium permeability and calcium block of the alpha9 nicotinic acetylcholine receptor. Hear Res 141:117–128

Katz E, Elgoyhen AB, Gomez-Casati ME, Knipper M, Vetter DE, Fuchs PA, Glowatzki E (2004) Developmental regulation of nicotinic synapses on cochlear inner hair cells. J Neurosci 24:7814–7820

Kiang NY, Moxon EC, Levine RA (1970) Auditory-nerve activity in cats with normal and abnormal cochleas. In: sensorineural hearing loss. Ciba Found Symp 241–273

Kohler M, Hirschberg B, Bond CT, Kinzie JM, Marrion NV, Maylie J, Adelman JP (1996) Small-conductance, calcium-activated potassium channels from mammalian brain. Science 273:1709–1714

Kong WJ, Cheng HM, van Cauwenberge P (2006) Expression of nicotinic acetylcholine receptor subunit alpha9 in type II vestibular hair cells of rats. Acta Pharmacol Sin 27:1509–1514

Kong JH, Adelman JP, Fuchs PA (2008) Expression of the SK2 calcium-activated potassium channel is required for cholinergic function in mouse cochlear hair cells. J Physiol 586:5471–5485

Kotak VC, Sanes DH (1995) Synaptically evoked prolonged depolarizations in the developing auditory system. J Neurophysiol 74:1611–1620

Kros CJ, Ruppersberg JP, Rusch A (1998) Expression of a potassium current in inner hair cells during development of hearing in mice. Nature 394:281–284

Labarca C, Schwarz J, Deshpande P, Schwarz S, Nowak M, Fonck C, Nashmi R, Kofuji P, Dang H, Shi W, Fidan M, Khakh B, Chen Z, Bowers B, Boulter J, Wehner J, Lester H (2001) Point mutant mice with hypersensitive alpha4 nicotinic receptors show dopaminergic deficits and increased anxiety. Proc Natl Acad Sci USA 98:2786–2791

Leake PA, Hradek GT, Chair L, Snyder RL (2006) Neonatal deafness results in degraded topographic specificity of auditory nerve projections to the cochlear nucleus in cats. J Comp Neurol 497:13–31

Liberman MC, Dodds LW, Pierce S (1990) Afferent and efferent innervation of the cat cochlea: quantitative analysis with light and electron microscopy. J Comp Neurol 301:443–460

Lioudyno M, Hiel H, Kong JH, Katz E, Waldman E, Parameshwaran-Iyer S, Glowatzki E, Fuchs PA (2004) A "synaptoplasmic cistern" mediates rapid inhibition of cochlear hair cells. J Neurosci 24:11160–11164

Lu L, Zhang Q, Timofeyev V, Zhang Z, Young JN, Shin HS, Knowlton AA, Chiamvimonvat N (2007) Molecular coupling of a Ca^{2+}-activated K^{+} channel to L-type Ca^{2+} channels via alpha-actinin2. Circ Res 100:112–120

Maison SF, Luebke AE, Liberman MC, Zuo J (2002) Efferent protection from acoustic injury is mediated via alpha9 nicotinic acetylcholine receptors on outer hair cells. J Neurosci 22:10838–10846

Maison SF, Adams JC, Liberman MC (2003) Olivocochlear innervation in the mouse: immunocytochemical maps, crossed versus uncrossed contributions, and transmitter colocalization. J Comp Neurol 455:406–416

Maison SF, Parker LL, Young L, Adelman JP, Zuo J, Liberman MC (2007) Overexpression of SK2 channels enhances efferent suppression of cochlear responses without enhancing noise resistance. J Neurophysiol 97:2930–2936

Marcotti W, Johnson SL, Holley MC, Kros CJ (2003) Developmental changes in the expression of potassium currents of embryonic, neonatal and mature mouse inner hair cells. J Physiol 548:383–400

Marcotti W, Johnson SL, Kros CJ (2004) A transiently expressed SK current sustains and modulates action potential activity in immature mouse inner hair cells. J Physiol 560:691–708

Martin AR, Fuchs PA (1992) The dependence of calcium-activated potassium currents on membrane potential. Proc R Soc Lond B Biol Sci 250:71–76

Matthews TM, Duncan RK, Zidanic M, Michael TH, Fuchs PA (2005) Cloning and characterization of SK2 channel from chicken short hair cells. J Comp Physiol A Neuroethol Sens Neural Behav Physiol 191:491–503

May BJ, Prosen CA, Weiss D, Vetter D (2002) Behavioral investigation of some possible effects of the central olivocochlear pathways in transgenic mice. Hear Res 171:142–157

McNiven AI, Yuhas WA, Fuchs PA (1996) Ionic dependence and agonist preference of an acetylcholine receptor in hair cells. Audit Neurosci 2:63–77

Millar NS (2008) RIC-3: a nicotinic acetylcholine receptor chaperone. Br J Pharmacol 153(suppl 1): S177–S183

Morley BJ, Li HS, Hiel H, Drescher DG, Elgoyhen AB (1998) Identification of the subunits of the nicotinic cholinergic receptors in the rat cochlea using RT-PCR and in situ hybridization. Brain Res Mol Brain Res 53:78–87

Murthy V, Maison SF, Taranda J, Haque N, Bond CT, Elgoyhen AB, Adelman JP, Liberman MC, Vetter DE (2009) SK2 channels are required for function and long-term survival of efferent synapses on mammalian outer hair cells. Mol Cell Neurosci 40:39–49

Murugasu E, Russell IJ (1996) The effect of efferent stimulation on basilar membrane displacement in the basal turn of the guinea pig cochlea. J Neurosci 16:325–332

Nenov AP, Norris C, Bobbin RP (1996) Acetylcholine responses in guinea pig outer hair cells. II Activation of a small conductance Ca^{2+}-activated K^+ channel. Hear Res 101:149–172

Norris CH, Housley GD, Williams WH, Guth SL, Guth PS (1988) The acetylcholine receptors of the semicircular canal of the frog (*Rana Pipiens*). Hear Res 32:197–206

Oatman LC (1976) Effects of visual attention on the intensity of auditory evoked potentials. Exp Neurol 51:41–53

Oliver D, Klocker N, Schuck J, Baukrowitz T, Ruppersberg JP, Fakler B (2000) Gating of Ca^{2+}-activated K^+ channels controls fast inhibitory synaptic transmission at auditory outer hair cells. Neuron 26:595–601

Orr-Urtreger A, Broide R, Kasten M, Dang H, Dani J, Beaudet A, Patrick J (2000) Mice homozygous for the L250T mutattion in the alpha7 nicotinic acetylcholine receptor show increased neuronal apoptosis and die within 1 day of birth. J Neurochem 74:2154–2166

Plazas PV, Katz E, Gomez-Casati ME, Bouzat C, Elgoyhen AB (2005a) Stoichiometry of the alpha9alpha10 nicotinic cholinergic receptor. J Neurosci 25:10905–10912

Plazas PV, De Rosa MJ, Gomez-Casati ME, Verbitsky M, Weisstaub N, Katz E, Bouzat C, Elgoyhen AB (2005b) Key roles of hydrophobic rings of TM2 in gating of the alpha9alpha10 nicotinic cholinergic receptor. Br J Pharmacol 145:963–974

Rajan R (1988) Effect of electrical stimulation of the crossed olivocochlear bundle on temporary threshold shifts in auditory sensitivity. I. Dependence on electrical stimulation parameters. J Neurophysiol 60:549–568

Rasmussen GL (1946) The olivary peduncle and other fiber projections of the superior olivary complex. J Comp Neurol 84:141–219

Rothlin CV, Katz E, Verbitsky M, Elgoyhen AB (1999) The alpha9 nicotinic acetylcholine receptor shares pharmacological properties with type A gamma-aminobutyric acid, glycine, and type 3 serotonin receptors. Mol Pharmacol 55:248–254

Rothlin CV, Lioudyno MI, Silbering AF, Plazas PV, Casati ME, Katz E, Guth PS, Elgoyhen AB (2003) Direct interaction of serotonin type 3 receptor ligands with recombinant and native alpha 9 alpha 10-containing nicotinic cholinergic receptors. Mol Pharmacol 63: 1067–1074

Saito K (1980) Fine structure of the sensory epithelium of the guinea pig organ of Corti: afferent and efferent synapses of hair cells. J Ultrastruct Res 71:222–232

Sewell W (1996) Neurotransmitters and synaptic transmission. In: Dallos P, Popper AN, Fay RR (eds) The cochlea. Springer, New York, pp 503–533

Sgard F, Charpentier E, Bertrand S, Walker N, Caput D, Graham D, Bertrand D, Besnard F (2002) A novel human nicotinic receptor subunit, $\alpha 10$, that confers functionality to the $\alpha 9$-subunit. Mol Pharmacol 61:150–159

Shigemoto T, Ohmori H (1990) Muscarinic agonists and ATP increase the intracellular Ca^{2+} concentration in chick cochlear hair cells. J Physiol (Lond) 420:127–148

Shigemoto T, Ohmori H (1991) Muscarinic receptor hyperpolarizes cochlear hair cells of chick by activating $Ca^{(2+)}$-activated K^+ channels. J Physiol 442:669–690

Simmons DD (2002) Development of the inner ear efferent system across vertebrate species. J Neurobiol 53:228–250

Simmons DD, Moulding HD, Zee D (1996) Olivocochlear innervation of inner and outer hair cells during postnatal maturation: an immunocytochemical study. Brain Res Dev Brain Res 95:213–226

Simmons DD, Bertolotto C, Kim J, Raji-Kubba J, Mansdorf N (1998) Choline acetyltransferase expression during a putative developmental waiting period. J Comp Neurol 397: 281–295

Sridhar TS, Brown MC, Sewell WF (1997) Unique postsynaptic signaling at the hair cell efferent synapse permits calcium to evoke changes on two time scales. J Neurosci 17:428–437

Sugai YJ, Sugitani M, Ooyama H (1992) Actions of cholinergic agonist and antagonists on the efferent synapse in frog sacculus. Hear Res 61:56–64

Takahashi T, Momiyama A (1991) Single-channel currents underlying glycinergic inhibitory postsynaptic responses in spinal neurons. Neuron 7:965–969

Taranda J, Maison SF, Ballestero JA, Katz E, Savino J, Vetter DE, Boulter J, Liberman MC, Fuchs PA, Elgoyhen AB (2009a) A point mutation in the hair cell nicotinic cholinergic receptor prolongs cochlear inhibition and enhances noise protection. PLoS Biol 7:e18

Taranda J, Ballestero JA, Hiel H, Souza FS, Wedemeyer C, Gomez-Casati ME, Lipovsek M, Vetter DE, Fuchs PA, Katz E, Elgoyhen AB (2009b) Constitutive expression of the alpha10 nicotinic acetylcholine receptor subunit fails to maintain cholinergic responses in inner hair cells after the onset of hearing. J Assoc Res Otolaryngol 10:397–406

Tritsch NX, Yi E, Gale JE, Glowatzki E, Bergles DE (2007) The origin of spontaneous activity in the developing auditory system. Nature 450:50–55

Verbitsky M, Rothlin C, Katz E, Elgoyhen AB (2000) Mixed nicotinic-muscarinic properties of the a9 nicotinic receptor. Neuropharmacology 39:2515–2524

Vetter DE, Adams JC, Mugnani E (1991) Chemically distinct rat olivocochlear neurons. Synapse 7:21–43

Vetter DE, Liberman MC, Mann J, Barhanin J, Boulter J, Brown MC, Saffiote-Kolman J, Heinemann SF, Elgoyhen AB (1999) Role of alpha9 nicotinic ACh receptor subunits in the development and function of cochlear efferent innervation. Neuron 23:93–103

Vetter DE, Katz E, Maison SF, Taranda J, Turcan S, Ballestero J, Liberman MC, Elgoyhen AB, Boulter J (2007) The alpha10 nicotinic acetylcholine receptor subunit is required for normal synaptic function and integrity of the olivocochlear system. Proc Natl Acad Sci USA 104:20594–20599

Walsh E, McGee J, McFadden S, Liberman M (1998) Long-term effects of sectioning the olivocochlear bundle in neonatal cats. J Neurosci 18:3859–3869

Warr WB (1975) Olivocochlear and vestibular efferent neurons of the feline brain stem: their location, morphology and number determined by retrograde axonal transport and acetylcholinesterase histochemistry. J Comp Neurol 161:159–181

Warr WB (1992) Organization of olivocochlear efferent systems in mammals. In: Douglas W, Popper AN, Fay RR (eds) The mammalian auditory pathway: neuroanatomy. Springer, New York, pp 410–448

Warr WB, Guinan JJ Jr (1979) Efferent innervation of the organ of Corti: two separate systems. Brain Res 173:152–155

Weisstaub N, Vetter DE, Elgoyhen AB, Katz E (2002) The alpha9alpha10 nicotinic acetylcholine receptor is permeable to and is modulated by divalent cations. Hear Res 167:122–135

Wiederhold ML, Kiang NYS (1970) Effects of electrical stimulation of the crossed olivocochlear bundle on cat single auditory nerve fibres. J Acoust Soc Am 48:950–965

Winslow RL, Sachs MB (1988) Single-tone intensity discrimination based on auditory-nerve rate responses in backgrounds of quiet, noise, and with stimulation of the crossed olivocochlear bundle. Hear Res 35:165–189

Yoshida N, Shigemoto T, Sugai T, Ohmori H (1994) The role of inositol triphosphate on ACh-induced outward currents in bullfrog saccular hair cells. Brain Res 644:90–100

Yuhas WA, Fuchs PA (1999) Apamin-sensitive, small-conductance, calcium-activated potassium channels mediate cholinergic inhibition of chick auditory hair cells. J Comp Physiol [A] 185:455–462

Chapter 6
The Efferent Vestibular System

Joseph C. Holt, Anna Lysakowski, and Jay M. Goldberg

6.1 Introduction

As is the case with most hair-cell organs, the vestibular labyrinth receives a dual innervation. Afferent nerve fibers arise from bipolar cells in the vestibular (Scarpa's) ganglion. The peripheral process of each ganglion cell gets synaptic inputs from hair cells in one of several discrete organs, and its central process conveys the resulting information, encoded in the spacing of action potentials, to the vestibular nuclei and the cerebellum. In addition, hair cells and afferent nerve terminals are innervated by efferent fibers originating in the brain stem and reaching the periphery by way of the vestibular nerve. This chapter reviews our understanding of the efferent vestibular system (EVS), including its neuroanatomical organization, candidate neurotransmitters, peripheral actions on afferent discharge as revealed by electrical stimulation of EVS pathways, and the underlying cellular (synaptic) and neurotransmitter mechanisms. To consider possible functions of the EVS, the chapter then describes the vestibular and nonvestibular signals carried by efferent neurons and how these signals might modify the information carried by afferents. Though our emphasis is on the mammalian EVS, results in nonmammalian species are also considered, as these provide insights into efferent function.

Efferent actions are related to the discharge properties of afferents, particularly to their regularity of discharge and their branching patterns and locations in the neuroepithelia of the various vestibular organs. Although these topics need to be reviewed, of necessity the treatment is brief. For readers wanting more detailed information, the following references can be consulted: anatomy of the peripheral vestibular organs (Lindeman 1969; Wersäll and Bagger-Sjöbäck 1974; Lysakowski and Goldberg 1997); functions of the semicircular canals and otolith organs (Wilson and Melvill Jones 1979; Lysakowski and Goldberg 2004); and relation of

J.M. Goldberg (✉)
Department of Neurobiology, Pharmacology, and Physiology,
University of Chicago, Chicago, IL 60637, USA
e-mail: jgoldber@bsd.uchicago.edu

D.K. Ryugo et al. (eds.), *Auditory and Vestibular Efferents*,
Springer Handbook of Auditory Research 38, DOI 10.1007/978-1-4419-7070-1_6,

discharge properties to peripheral terminations of afferents (Lysakowski and Goldberg 2004). Previous reviews of the EVS can also be recommended (Meredith 1988; Highstein 1991; Guth et al. 1998; Goldberg et al. 2000).

6.2 Afferents and Hair Cells

6.2.1 Afferent Discharge Properties

Vestibular-nerve afferents have a resting discharge in the absence of stimulation. This discharge can be quite high; for example, it averages ~100 spikes/s in monkeys (Fig. 6.1) (Goldberg and Fernández 1971; Hacque et al. 2004; Sadeghi et al. 2007). Rotations in the plane of a semicircular canal can, depending on their direction, result in an increase (excitation) or decrease (inhibition) in afferent discharge. Similarly, appropriately directed linear forces can excite or inhibit discharge in the utricular or saccular maculae.

Some vestibular afferents have a regular spacing of action potentials, whereas in others the spacing is irregular (Fig. 6.1) (Goldberg 2000). Discharge regularity is measured by a coefficient of variation (cv), the ratio of the standard deviation of intervals (s) to the mean interval ($\overline{x}$), or $cv = s / \overline{x}$. Because cv varies with $\overline{x}$, a normalized statistic (cv^*), the cv at a standard mean interval, is used. In mammals, cv^* at $\overline{x} = 15$ ms varies more than 20-fold, from <0.03 in the most regular units to >0.6 in the most irregular units. Discharge regularity has proved useful, as units first classified as regular or irregular differ in many of their other discharge characteristics (Table 6.1) (Goldberg 2000). A stochastic model of repetitive activity illustrates how discharge regularity impacts other neuronal properties (Smith and Goldberg 1986; Goldberg 2000).

As is generally the case (Connor 1978; Stocker 2004; Bean 2007), repetitive activity in the model reflects the interaction of an afterhyperpolarization (AHP) following each spike with synaptic and other depolarizing currents. In the model, regular units have slow, deep AHPs, whereas in irregular units, AHPs are fast and shallow. Of somewhat lesser importance in determining discharge regularity, miniature excitatory postsynaptic potentials (mEPSPs), which are the result of neurotransmission from hair cells to the afferent terminal, are smaller in regular units. Reflecting AHP differences, irregular units are much more sensitive to synaptic or externally applied currents than are regular units. Circumstantial evidence supporting the model comes from the responses of mammalian vestibular afferents to external currents (Goldberg et al. 1984). More direct evidence has been obtained from intracellular recordings of AHPs and mEPSPs in crista afferents from the red-eared turtle, *Trachemys scripta elegans* (Goldberg and Chatlani 2009).

The conclusion is that when the spike encoder, the mechanism that converts postsynaptic depolarizations into spike trains, has a much higher gain, the more irregular is the discharge. That is, a given depolarization results in a considerably larger increase in the discharge rate of irregular afferents. Sensitivity to externally

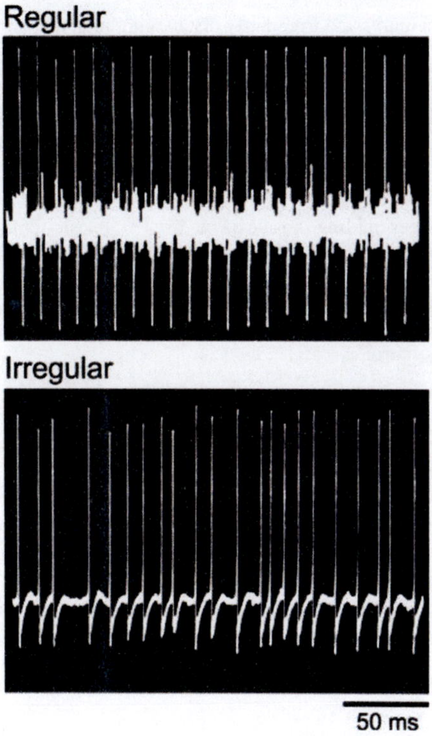

Fig. 6.1 Discharge regularity in vestibular-nerve afferents. Spike trains are shown during the resting discharge of two afferents, each innervating the superior semicircular canal in a squirrel monkey. Although both afferents have similar discharge rates of just under 100 spikes/s, they differ in the spacing of their action potentials, which is regular in the top afferent and irregular in the bottom afferent (Modified with permission from Goldberg and Fernández 1971. Copyright 1971, The American Physiological Society.)

applied galvanic currents provides a measure of encoder gain (Fig. 6.8c) (Goldberg et al. 1984). An inspection of Table 6.1 indicates that several of the differences between regular and irregular units, including the fact that efferent responses are much smaller in regular units, can be explained by differences in encoder gain (Goldberg 2000). In fact, the only item that cannot be so explained is the difference in response dynamics between the two afferent groups (Goldberg et al. 1982; Ezure et al. 1983).

6.2.2 *Hair Cells and Their Innervation*

In the vestibular organs of amniotes (reptiles, birds, and mammals), there are two kinds of hair cells (Fig. 6.2) (Wersäll and Bagger-Sjöbäck 1974; Lysakowski and

Table 6.1 Comparing regularly and irregularly discharging mammalian vestibular afferents

Regular	Irregular
Thin to medium-sized axons ending as bouton and dimorphic terminals in the peripheral (peripheral extrastriolar) zone (Goldberg and Fernández 1977; Yagi et al. 1977; Baird et al. 1988; Goldberg et al. 1990b; Lysakowski et al. 1995)	Medium to thick axons ending as calyx and dimorphic terminals in the central (striolar) zone
Tonic response dynamics, resembling those expected of end organ macromechanics (Goldberg and Fernández 1971; Fernández and Goldberg 1976; Schneider and Anderson 1976; Tomko et al. 1981; Curthoys 1982; Baird et al. 1988; Goldberg et al. 1990a; Lysakowski et al. 1995)	Phasic-tonic response dynamics, with sensitivity to the velocity of cupular (otolith) displacement
Low sensitivity to angular or linear forces (Goldberg and Fernández 1971; Fernández and Goldberg 1976; Schneider and Anderson 1976; Tomko et al. 1981; Curthoys 1982; Baird et al. 1988; Goldberg et al. 1990a; Lysakowski et al. 1995)	High sensitivity to angular or linear forces. Calyx afferents in cristae have unusually low sensitivity
Small responses to electrical stimulation of efferent fibers (Goldberg and Fernández 1980; McCue and Guinan 1994; Marlinski et al. 2004)	Large responses to electrical stimulation of efferent fibers
High thresholds and small responses to galvanic currents delivered to the perilymphatic space (Ezure et al. 1983; Goldberg et al. 1984; Brontë-Stewart and Lisberger 1994)	Low thresholds and large responses to galvanic currents delivered to the perilymphatic space

Goldberg 2004). Type I hair cells are distinctive in being amphora-shaped and having almost their entire basolateral surface contacted by a single calyx ending. Type II hair cells, also seen in fish and amphibians, are cylindrically shaped and receive an afferent innervation from several bud-shaped or bouton terminals. The two kinds of hair cells also differ in their efferent innervation. During development, both types of hair cells are contacted by efferent boutons, but the growth of calyces displaces these from type I hair cells (Favre and Sans 1979). With few exceptions (Wackym et al. 1991; Li et al. 2007), the efferent innervation in the adult terminates on calyx endings not on type I hair cells. Type II hair cells receive an efferent innervation, as do the bouton-shaped afferent terminals innervating these same hair cells.

Two methods have been used to characterize the afferent innervation of hair cells. First, fibers are dye-filled and their terminal fields reconstructed (Fernández et al. 1988, 1990, 1995). Second, different populations of afferents have distinctive molecular components, which can be discerned via immunohistochemistry (Desmadryl and Dechesne 1992; Lysakowski et al. 1999; Leonard and Kevetter 2002). There are three kinds of fibers in the cristae of the mammalian semicircular canals (Fernández et al. 1988, 1995; Desai et al. 2005a, b), organs that are involved in sensing angular head rotations (Fig. 6.3). *Calyx* fibers provide calyx endings to one or a few neighboring type I hair cells. *Bouton* fibers supply bud-shaped endings to several widely spaced type II hair cells. *Dimorphic* fibers provide a mixed innervation consisting of one or more calyx endings to type I hair cells and thin collaterals giving rise to bud-shaped endings to several type II hair cells. Calyx units are

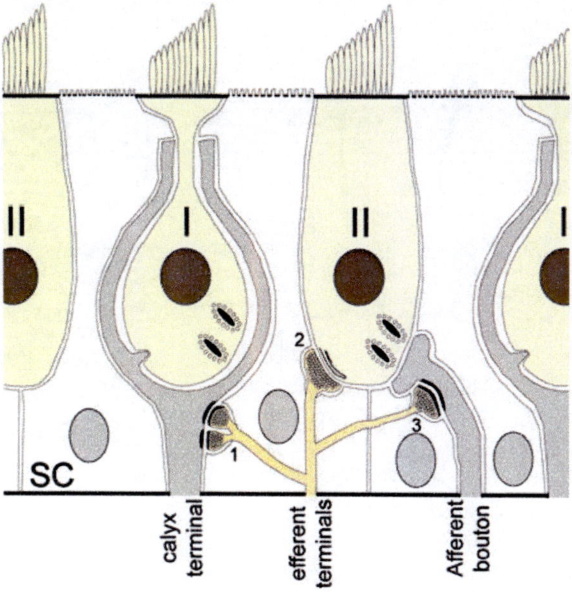

Fig. 6.2 Amphora-shaped type I hair cells are innervated by calyx endings derived from single afferent fibers, whereas cylindrically-shaped type II hair cells are innervated by bouton terminals from several afferents. In each instance, afferent synapses are marked by synaptic ribbons (a dense synaptic body surrounded by a halo of vesicles) in the hair cell. In addition to ribbon synapses, the calyx ending often invaginates into the type I hair cell. Single efferent fibers with highly vesiculated boutons terminate on the calyx ending (*1*) as well as on type II hairs (*2*) and their bouton terminals (*3*). Supporting cells (SC) span the width of the neuroepithelium and are recognized by their basally located nuclei and by microvilli on their apical surfaces

confined to a central zone (Fig. 6.3 right, CZ), whereas bouton units are confined to a peripheral zone (Fig. 6.3 right, PZ). Dimorphic units, which are the most numerous fiber type, are found throughout the neuroepithelium, as are type I and type II hair cells. Calyx fibers have the thickest axons and bouton fibers the thinnest axons. Dimorphic fibers are of intermediate caliber with those innervating the PZ being thinner and having more extensive terminal trees than those in the CZ.

The utricular and saccular maculae are sensors of linear forces acting on the head. In its afferent innervation, the utricular macula resembles the cristae (Fernández et al. 1990). There are three types of fibers, although bouton fibers are relatively infrequent. Calyx fibers are confined to the striola, a narrow stripe running throughout much of the length of the macula and separating the rest of the neuroepithelium into a medial and a lateral extrastriola. Dimorphic units are found in both the striola and the extrastriola. The relatively few bouton units are found only in the extrastriola at some distance from the striola (Fernández et al. 1990; Leonard and Kevetter 2002). Although a detailed description of the innervation patterns in the saccular macula is lacking, calyx fibers are confined to the striola (Leonard and Kevetter 2002; Desai et al. 2005a).

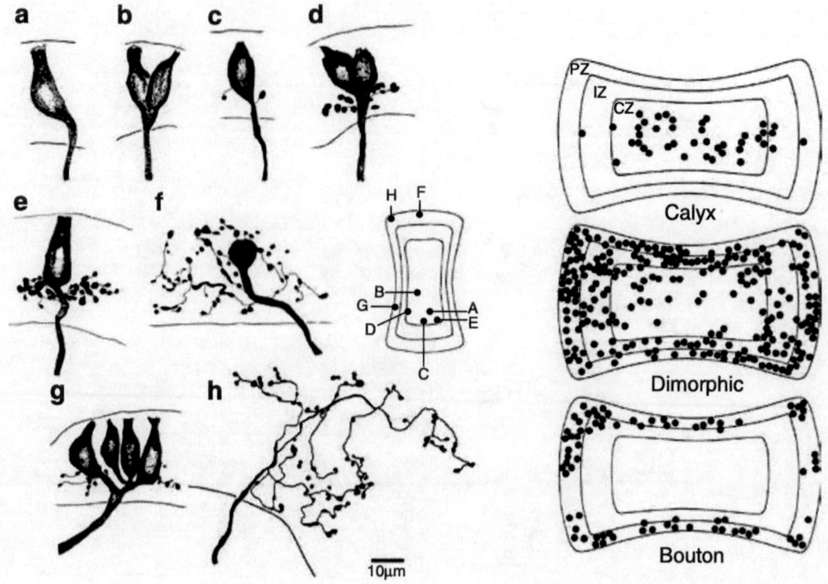

Fig. 6.3 (**a**–**h**) Branching patterns of individual semicircular canal afferents labeled by the extracellular deposit of horseradish peroxidase in the chinchilla vestibular nerve. There are three afferent classes: calyx (**a**, **b**), dimorphic (**c**–**g**), and bouton (**h**). Location of each unit is indicated on a flattened map of crista (*inset, middle right*). *Right column*: Distribution of calyx, dimorphic, and bouton units on a flattened reconstruction of the crista. In all maps, the crista is divided into central (CZ), intermediate (IZ), and peripheral (PZ) zones of equal areas (Modified with permission from Fernández et al. 1988. Copyright 1988, The American Physiological Society.)

6.2.3 Afferent Morphology and Physiology

Morphophysiological techniques can be used to relate terminal morphology and neuroepithelial location with the discharge properties of individual afferents. Here, an afferent fiber is impaled, its physiology characterized, after which an intracellular marker is injected into the fiber and allowed to diffuse to its peripheral termination (review: Lysakowski and Goldberg 2004). In the chinchilla cristae (Baird et al. 1988), it was found that dimorphic units in the CZ are irregularly discharging and have high rotational and galvanic gains, whereas those in the PZ are regularly discharging with low rotational and galvanic gains (Fig. 6.4). Compared to dimorphs also innervating the CZ, calyx units have a more irregular discharge, more phasic response dynamics, and larger responses to galvanic currents. This last observation implies that calyx fibers have especially sensitive spike encoders. Despite this, calyx units have considerably lower rotational gains than irregular dimorphs. In fact, a relatively low rotational gain has proved a reliable feature of calyx units in the cristae of several mammalian species (Lysakowski et al. 1995; Hullar et al. 2005; Sadeghi et al. 2007; Lasker et al. 2008). Bouton units, defined by their slow conduction velocities, resemble dimorphic afferents also localized in

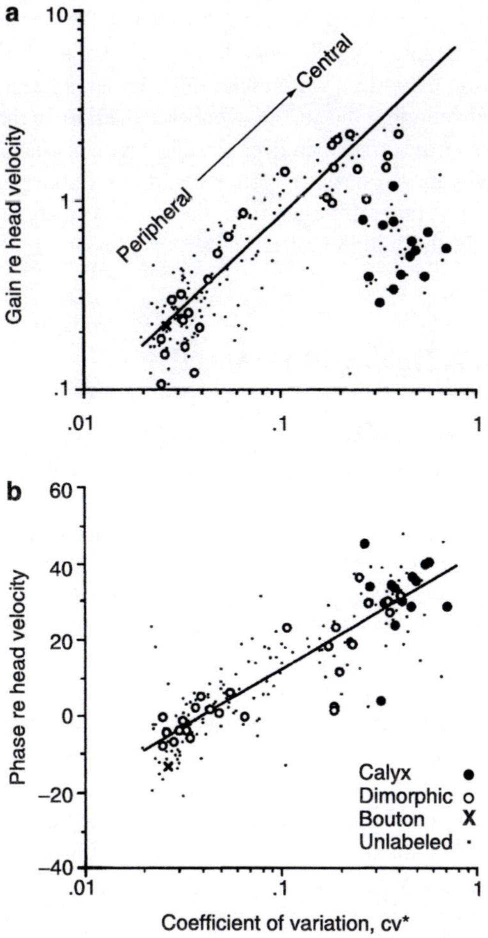

Fig. 6.4 Gains and phases vs. normalized coefficient of variation (*cv**) for labeled and unlabeled semicircular-canal afferents in the chinchilla responding to 2-Hz sinusoidal head rotations. Each point represents one unit. (**a**) Based on their gains, units fall into two groups. Straight line in (**a**) is best-fitting power law relation between gain and *cv** for one of the groups. (**b**) Sinusoidal phase re head velocity vs. *cv** for the same units shown in (**a**). Straight line in (**b**), best-fitting semi-logarithmic relationship between phase and *cv** for all units. Calyx units are the most irregularly discharging afferents, have the largest phase leads, and distinctively low gains. Gain and phase of dimorphic units increase with *cv**. Regular units are found in the peripheral zone; irregular units in the central zone. The one bouton unit was regular, had a low gain and phase, and was located in the peripheral zone (Modified with permission from Baird et al. 1988. Copyright 1988, The American Physiological Society.)

the PZ in having a regular discharge, tonic response dynamics, and low rotational and galvanic sensitivities (Lysakowski et al. 1995).

Morphophysiological experiments have also been done in the chinchilla utricular macula (Goldberg et al. 1990a). Recovered dye-filled fibers included calyx units in the striola and dimorphic units throughout the macula. Striolar units were irregular, with calyx units being more irregular than dimorphs. Extrastriolar dimorphs located at some distance from the striola were invariably regular, whereas those in the juxtastriola, a region surrounding the striola, were intermediate in their discharge regularity. As in the case of the cristae, an irregular discharge is associated with a higher galvanic sensitivity, a more phasic response dynamics, and a greater sensitivity to natural stimulation. The one difference from the cristae is that calyx units and striolar dimorphs have similarly high gains.

6.3 Efferents: A Historical Perspective

Once Rasmussen (1946, 1953) had demonstrated the presence of olivocochlear bundles, it seemed reasonable to suppose that vestibular and other hair-cell organs also received an efferent innervation. Ultrastructural evidence for this conjecture was provided by Engström (1958), who noted that there were two kinds of nerve endings in the cochlea and vestibular organs. One group was highly vesiculated, while the other group was poorly vesiculated. Based on observations at other synapses (de Robertis and Bennett 1955; Palay and Palade 1955), it was suggested that the highly vesiculated endings were efferents and the poorly vesiculated endings were afferents. Innervation of vestibular type I hair cells was consistent with the suggestion in that only the poorly vesiculated calyx endings were in contact with hair-cell ribbon synapses and, so, could be directly affected by sensory stimulation. In contrast, highly vesiculated endings contacted the calyx and thus were in a position to modulate afferent transmission. Histochemical studies in the cochlea (Churchill et al. 1956; Schuknecht et al. 1959) and vestibular labyrinth (Dohlman et al. 1958; Ireland and Farakashidy 1961) confirmed the dual innervation, as efferents, but not afferents, were acetylcholinesterase (AChE) positive. Hilding and Wersäll (1962) connected the ultrastructural and histochemical findings by demonstrating that it was the highly vesiculated endings that were AChE positive. Further proof that these endings were efferents was obtained by showing that they degenerated following central lesions (Smith and Rasmussen 1968; Iurato et al. 1972).

Rasmussen (1946) noted that the medial group of olivocochlear efferents crossed the midline immediately below the floor of the fourth ventricle. This observation provided a convenient site at which to stimulate efferent fibers while monitoring their influence on cochlear activity (Galambos 1956; Fex 1959; Desmedt and Monaco 1961). The disposition of peripheral nerves allowed similar studies to be done in lateral-line organs (Russell 1968; Flock and Russell 1973, 1976). Using tracing methods similar to those of Rasmussen, Gacek was able to define an efferent projection to the vestibular labyrinth, but could not specify the origin of the EVS or a convenient place to stimulate its fibers (Gacek 1960; Rasmussen and Gacek 1958). AChE histochemistry also failed to ascertain the locations of EVS neurons (Rossi and Cortesina 1965). For that reason, early attempts to characterize the peripheral actions of the EVS were unsuccessful (Sala 1965; Llinás and Precht 1969). It was only with the advent of modern retrograde tracer methods that the cells of origin of the EVS could be determined to lie bilaterally outside the confines of the vestibular nuclei (Gacek and Lyon 1974), at which point the trajectory of EVS fibers to the vestibular periphery could be traced by AChE histochemistry. This knowledge set the stage for studies of the anatomy, physiology, and pharmacology of the EVS.

6.4 Neuroanatomical Organization of the EVS

6.4.1 *Location of Cell Bodies and Their Dendritic Morphology*

There is an efferent innervation of the vestibular organs in every vertebrate class (Meredith 1988; Lysakowski 1996). The locations of the efferent cell bodies and their dendritic morphology in several representative species are summarized in Fig. 6.5. In mammals, separate groups of efferents innervate auditory and vestibular organs (Fig. 6.5a, cat). In all other vertebrate classes that have been examined, efferent neurons are contained in a single cell cluster, referred to in animals possessing lateral-line neuromasts as the octavolateralis efferent nucleus because cells of the nucleus innervate the lateral-line, as well as eighth-nerve derivatives. Branchiomotor facial motoneurons and octavolateralis efferents may arise embryologically from a common pool of neurons (see Simmons et al., Chap. 7). Possibly reflecting this origin, efferent neurons are found partly within the confines of the facial motor nucleus in eel and toad (Fig. 6.5a). In lizard and chicken, there is still a single efferent nucleus, but it is displaced from the facial nucleus. Retrograde tracer studies indicate that neurons destined for organs of different modalities (vestibular, vibratory, lateral line, and auditory) overlap within the efferent nucleus. In fact, some, but not all, efferent neurons can innervate the inner ear and lateral lines in fish, as well as in the African clawed frog (*Xenopus laevis*), an aquatic anuran still possessing a lateral line (Claas et al. 1981; Highstein and Baker 1986; Meredith and Roberts 1987). In reptiles (Strutz 1981, 1982a; Barbas-Henry and Lohman 1988) and birds (Whitehead and Morest 1981; Strutz and Schmidt 1982), there is a partial segregation of auditory and vestibular efferents in the efferent nucleus, with vestibular efferents being located more dorsally. Because the segregation is incomplete, it is conceivable that single efferent neurons could innervate both kinds of organs.

In mammals, vestibular efferents arise bilaterally in the brain stem from three collections of neurons (Fig. 6.6a, b). The first is a slender column of medium-sized multipolar neurons that extends rostrocaudally between the abducens and superior vestibular nuclei, just dorsal to the descending facial nerve (Gacek and Lyon 1974; Warr 1975; Goldberg and Fernández 1980; Perachio and Kevetter 1989; Marco et al. 1993). This group has been referred to as group *e* (Goldberg and Fernández 1980). A second, compact group of somewhat smaller fusiform neurons is situated dorsomedial to the facial genu (Goldberg and Fernández 1980; Marco et al. 1993). A third or ventral group of slightly larger neurons is scattered in the caudal pontine reticular formation (Strutz 1982b; Marco et al. 1993). By far, most efferent neurons are located in group *e*.

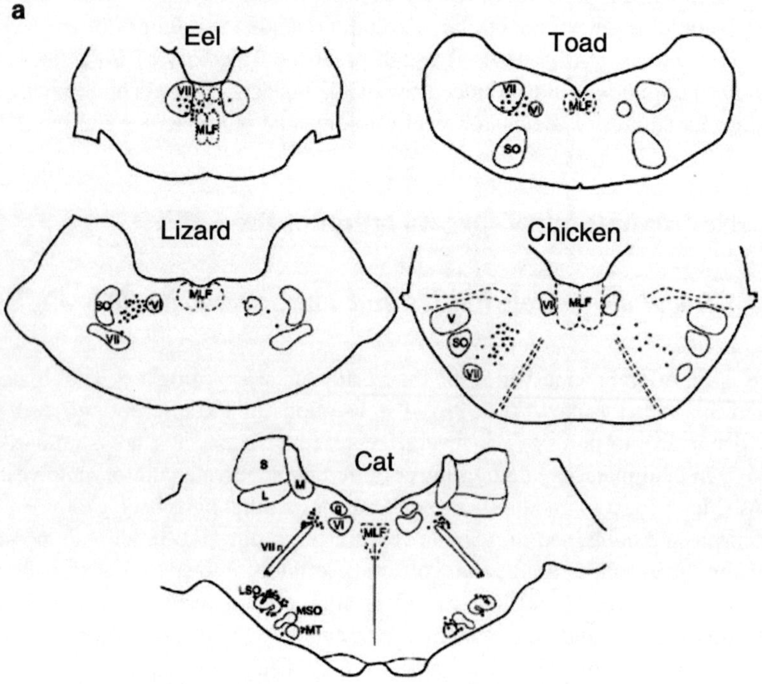

Fig. 6.5 The distribution of efferent neurons and their dendritic arborizations in the brain stem of representative vertebrates. (**a**) Distribution of efferent neurons. Locations of efferent neurons are indicated by *dots*. Examples include eel (Meredith and Roberts 1987), toad (Pellergrini et al. 1985), lizard (Barbas-Henry and Lohman 1988), chicken (Whitehead and Morest 1981), and cat (Warr 1975). (Data from eel, lizard, and cat adapted with permission from John Wiley & Sons; data from toad and chicken adapted with permission from Elsevier.)

The dendritic morphology, as well as the location of the efferent neurons, varies across the vertebrate scale. In animals ranging from the lamprey to amphibia, dendrites occupy much of the brain stem tegmentum (Fig. 6.5b, 1–4). In contrast, the dendritic arbors of the main mammalian efferent nucleus (group *e*) are quite restricted (Fig. 6.5b, 5). Another difference concerns the laterality of efferent projections. Roughly equal numbers of group *e* neurons project to the ipsilateral and contralateral labyrinths with possibly a slight contralateral preference; this is so in rats (Schwarz et al. 1986), guinea pigs (Strutz 1982b), cats (Gacek and Lyon 1974; Warr 1975; Dechesne et al. 1984), monkeys (Goldberg and Fernández 1980; Carpenter et al. 1987), chinchillas (Marco et al. 1993), and gerbils (Perachio and Kevetter 1989). There are a small number of efferent neurons projecting to both ears (Dechesne et al. 1984; Purcell and Perachio 1997). In contrast to the bilateral organization of efferent neurons in mammals, that in nonmammalian vertebrates is predominantly ipsilateral with some contralateral representation (Fig. 6.5a).

The differences in the central organization of the EVS suggest caution in extrapolating results from other vertebrate classes to mammals or vice versa.

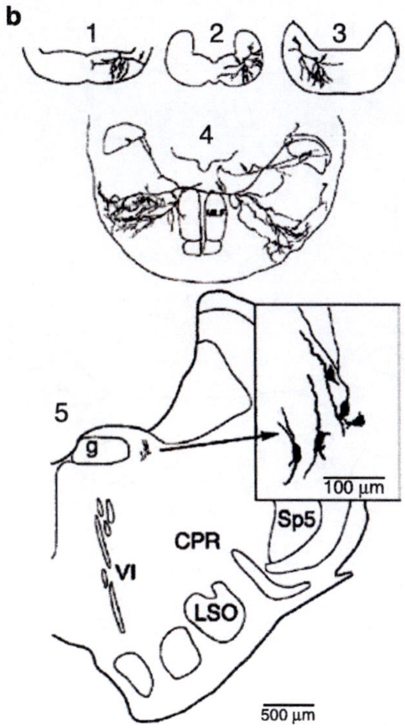

Fig. 6.5 (**b**) Efferent cell bodies and their dendritic arborizations in several vertebrates. Examples include: (1) lamprey (Fritzsch et al. 1989), (2) gymnophion (amphibian) (Fritzsch and Crapon de Caprona 1984), (3) salamander (Fritzsch 1981), (4) oyster toadfish (Highstein and Baker 1986), and (5) chinchilla (Lysakowski and Singer 2000). Scale bar found on the bottom right corner applies to all sections on the right. Efferent neurons in chinchilla shown at higher magnification in inset. (Figs. **b**1, **b**2 and **b**3 adapted with permission from Elsevier; Figs. **b**4 and **b**5 adapted with permission from John Wiley & Sons). CPR caudal pontine reticular formation; g genu of facial nerve; L, M, and S lateral, medial, and superior vestibular nuclei, respectively; LSO and MSO lateral and medial superior olive, respectively; MLF medial longitudinal fasciculus; MT medial trapezoid nucleus; SO superior olivary complex; V spinal trigeminal nucleus; VI, VII, and VIIn abducens nucleus, facial nucleus, and facial nerve, respectively.

As summarized in Sect. 6.6, peripheral efferent actions also differ, which reinforces the need for caution.

6.4.2 Axonal Pathways to the Periphery

The pathways leading from the efferent cell groups in mammals to the vestibular nerve are illustrated in Fig. 6.6c. Axons from the ipsilateral group *e* join fibers coming from the contralateral group *e*, as well as auditory (olivocochlear) efferents. All three contingents run together across the spinal trigeminal tract to join the vestibular

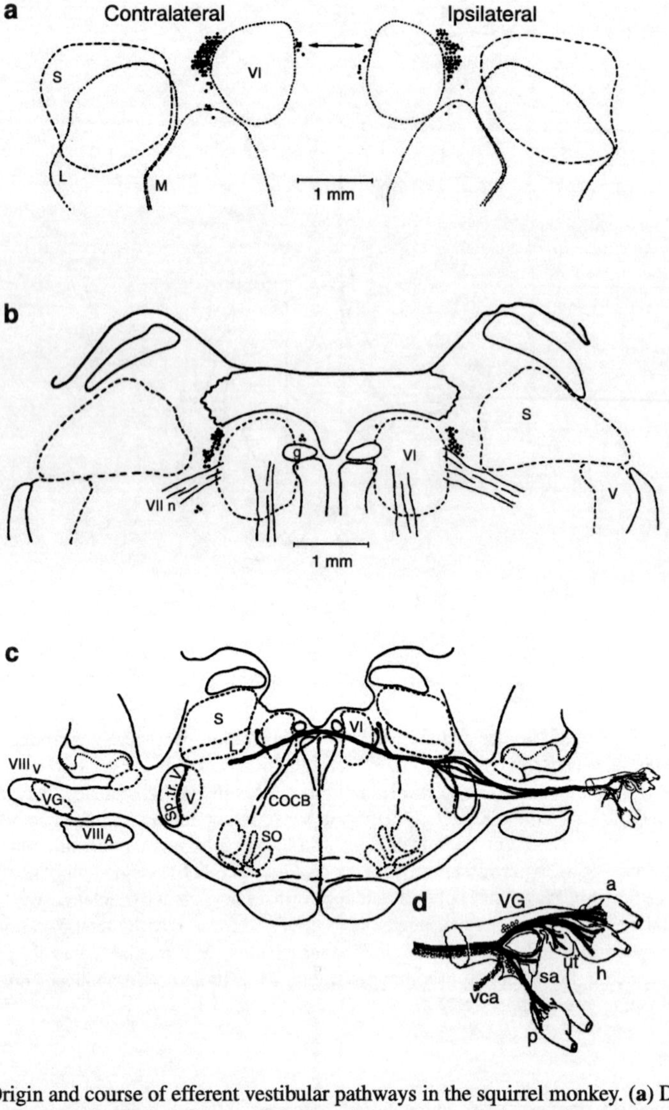

Fig. 6.6 Origin and course of efferent vestibular pathways in the squirrel monkey. (**a**) Distribution of neurons retrogradely labeled after an HRP deposit into the vestibule. Horizontal plan view. Each dot, approximately four neurons. (**b**) Same material as in (**a**). Frontal section at level indicated by *double arrow* in (**a**); each dot, one neuron. (**c**) Composite drawing of vestibular and auditory efferent pathways as revealed by AChE histochemistry. (**d**) Distribution of efferent fibers in the peripheral vestibular nerve to the different end organs. *a*, *h* and *p* anterior, horizontal, and posterior semicircular canals; *COCB* crossed olivocochlear bundle; *g* genu of facial nerve; *L*, *M*, and *S* lateral, medial, and superior vestibular nuclei, respectively; *sa* and *ut* saccular and utricular maculae; *SO* superior olivary complex; *Sp. tr. V* and *V* spinal trigeminal tract and nucleus, respectively; *vca* vestibulocochlear (Oort's) anastomosis; *VG* vestibular (Scarpa's) ganglion; *VI*, VII_n, VIII_a, and VIII_v abducens nucleus, facial, auditory and vestibular nerves, respectively (Modified with permission from Goldberg and Fernández 1980. Copyright 1980, The American Physiological Society.)

nerve. While still in the brain, contralateral and ipsilateral vestibular efferents send collateral projections to the cerebellar flocculus, ventral paraflocculus (Shinder et al. 2001), and the interstitial nucleus of the vestibular nerve (Perachio and Kevetter 1989). In the periphery, vestibular efferents distribute to all five end organs (Fig. 6.6d). Auditory efferents pass from the inferior vestibular nerve to the cochlear nerve via the vestibulo-cochlear (Oort's) anastomosis (*vca*, Fig. 6.6d).

6.4.3 Peripheral Branching Patterns

Counting ipsilateral and contralateral projections, 300–500 efferent neurons innervate each vestibular labyrinth in mammals (Goldberg and Fernández 1980; Marco et al. 1993). In contrast, more than 10,000 afferent nerve fibers innervate one or another of the five vestibular organs (Gacek and Rasmussen 1961; Hoffman and Honrubia 2002). Despite the 20:1 discrepancy in the numbers of parent axons, afferent boutons outnumber efferent boutons by only a 3:1 ratio (Goldberg et al. 1990b; Lysakowski and Goldberg 1997). The ratio suggests that efferent fibers branch more extensively than afferents. Two kinds of branching need to be considered. Individual efferent fibers could branch within a single organ or could innervate two or more organs.

Branching within individual organs has been studied by the anterograde labeling of efferent fibers in the gerbil brain stem and tracing their trajectories in the cristae (Purcell and Perachio 1997). Branching and the resulting terminal fields are much more extensive in efferent, as compared to afferent, fibers (cf. Figs. 6.3 and 6.7a). Most branching takes place after efferent fibers enter the neuroepithelium. Despite the large size of the efferent terminal fields, many of them are restricted to the CZ or to the PZ. Remarkably, efferent neurons arising on the contralateral side of the brain stem preferentially innervate the PZ. Ipsilateral efferents show less zonal selectivity with some projecting to the CZ and others to the PZ.

Electrophysiological techniques have been used in anurans to study branching to two or more organs; in particular, the entire nerve branch supplying one organ was electrically stimulated, while recordings were made from individual afferents in other organs (Rossi et al. 1980; Prigioni et al. 1983; Sugai et al. 1991). The recorded units showed typical efferent responses. A simple interpretation is that a parent efferent axon sends branches to both the stimulated and recorded organs. Presumably, responses are the result of axon reflexes involving both branches. All five organs tested were interconnected; this included the saccular macula, which in anurans monitors substrate-borne vibrations, rather than head movements (Narins and Lewis 1984). Evidence is lacking as to whether the branching of individual efferent fibers to multiple organs occurs in mammals (Purcell and Perachio 1997).

6.4.4 Synaptic Ultrastructure of Efferent Terminals

Upon reaching the vestibular periphery, efferent fibers end as boutons and can be readily distinguished from afferent endings by their highly vesiculated appearance

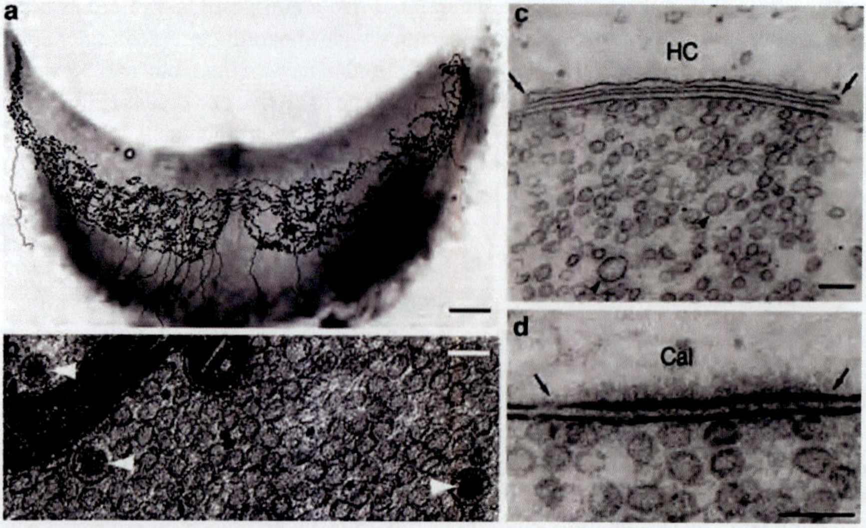

Fig. 6.7 Efferent neurons and their peripheral terminals. (**a**) Reconstructions of several efferent axons, labeled by extracellular injection of biocytin or biotinylated dextran amine in the contralateral brain stem and terminating in the posterior crista of a gerbil. These labeled efferent fibers are restricted in their innervation to the peripheral zone. Other fibers would be similarly restricted to the central zone. Each efferent fiber branches widely and gives rises to 100–300 bouton endings. (Modified with permission from Purcell and Perachio 1997. Copyright 1997, The American Physiological Society.) (**b–d**) Efferent terminals are recognized in electron micrographs by the large number of small, round synaptic vesicles they contain. (**b**) Electron micrograph of an efferent terminal from the red eared turtle (*Trachemys scripta elegans*) illustrating the presence of dense core vesicle (*arrowheads*) and several smaller, clear vesicles. (Unpublished micrograph, Dr. Anna Lysakowski) (**c**) An efferent synapse on a type II hair cell (HC) is marked by a subsynaptic cistern (*arrows*). *Arrowheads* point to larger vesicles that may contain neuropeptides. (**d**) Efferent ending on a calyx terminal (Cal) is marked by pre- and postsynaptic membrane thickenings delimited by *arrows*. The electron micrographs in **c** and **d** were taken from the chinchilla posterior crista. *Scale bars*: (**a**) 70 μm; (**b–d**) 100 μm (**c**, **d** courtesy of Dr. Anna Lysakowski, with permission)

(Engström and Wersäll 1958; Hilding and Wersäll 1962) (Fig. 6.7b–d). Efferent neurons can terminate on type II hair cells, the bouton afferents innervating type II hair cells, and the outer faces of calyx afferents (see Fig. 6.2). These terminations are characterized ultrastructurally by two types of contacts (Fig. 6.7c, d) (Iurato et al. 1972; Smith and Rasmussen 1968; Lysakowski and Goldberg 1997). Those on calyx endings and other afferent processes have asymmetric pre- and postsynaptic membrane thickenings, whereas those on type II hair cells are marked by subsynaptic cisterns, but have no obvious membrane specializations. It is thought that the cisterns are Ca^{2+} stores that might reinforce efferent synaptic actions (Sridhar et al. 1997; Lioudyno et al. 2004). Single efferent fibers can give rise to both kinds of endings (Smith and Rasmussen 1968; Lysakowski and Goldberg 1997).

6.5 Efferent Neurotransmitters and Receptors

This section considers the identity and summarizes the evidence for various efferent neurotransmitters and receptors based primarily on applying molecular biological techniques and immunohistochemistry to central neurons identified as efferents by retrograde labeling and to peripheral terminals presumed to be efferents by their being highly vesiculated. That highly vesiculated endings were AChE-positive suggested that the small clear vesicles in efferent terminals contain acetylcholine (ACh). Other substances commonly coexpressed in cholinergic neurons, such as adenosine-5′-triphosphate (ATP) and calcitonin gene-related peptide (CGRP), might also be present (Dowdall et al. 1974; Fontaine et al. 1986; New and Mudge 1986). Some context for studies of the EVS is provided by auditory efferents where there are several putative neurotransmitters other than ACh, including γ-aminobutyric acid (GABA), dopamine (DA), CGRP, and enkephalins (Eybalin 1993; see also Sewell, Chap. 4).

ACh has by far received the most attention and experimental support as the predominant efferent transmitter. Data regarding CGRP have also been reasonably compelling. The gaseous neurotransmitter nitric oxide (NO) has been implicated as a possible efferent signaling molecule. Of importance in evaluating a candidate efferent neurotransmitter are the effects of its agonists and antagonists on the responses to electrical stimulation of the EVS. Such evidence is available for ACh (Sect. 6.8), but is weak or nonexistent for most of the other candidates. For the latter, current information is based on the presence of synthesizing and degradative enzymes, of the neurotransmitter itself, and/or of various receptors. Determining the physiological effects of noncholinergic neurotransmitters remains important unfinished business. Even in the case of cholinergic mechanisms, much has to be learned about the roles of specific receptors.

6.5.1 *Acetylcholine*

Both the synthesizing (choline acetyltransferase [ChAT]) and the degradative enzyme (acetylcholinesterase [AChE]) are present in retrogradely labeled central EVS neurons (Schwarz et al. 1986; Perachio and Kevetter 1989; Ishiyama et al. 1994) and in highly vesiculated terminals in vestibular organs (Hilding and Wersäll 1962; Kong et al. 1994; Matsuda 1996). Clearly, ACh is a major neurotransmitter at peripheral efferent synapses. Yet, ChAT is not found in all brain stem EVS neurons, which has suggested that some of these neurons may not be cholinergic (Schwarz et al. 1986; Perachio and Kevetter 1989).

The two major classes of cholinergic receptors, nicotinic (nAChRs) and muscarinic (mAChRs), are found in vestibular neuroepithelia. nAChRs are pentameric, ligand-gated ion channels assembled from members of a currently identified family of 17 distinct subunits (α1–10, β1–4, δ, γ, and ε) (Millar and Gotti 2009;

Taly et al. 2009; Albuquerque et al. 2009). Native nAChRs expressed in the nervous system have one of the following compositions: homomeric, wherein all five subunits are identical (e.g., α7); heterodimeric, consisting of two distinct α subunits (e.g., α9α10) or of one α and one β subunit (e.g., α4β2); or multiple α and β subunits (e.g., α4α6β2β3). Differences in subunit composition and/or stoichiometry account for the diversity in the physiological and pharmacological properties of different nAChRs. mAChRs are heptahelical, G-protein–coupled receptors that are represented by five distinct subtypes (m1–5) where the odd-numbered mAChRs work predominantly through G_q and activation of phospholipase C pathways and the even-numbered mAChRs use $G_{i/o}$ to inhibit adenylyl cyclase (Caulfield and Birdsall 1998; Eglen 2005). Given their inherent differences in signaling schemes, nAChR-mediated responses should have faster kinetics than those involving mAChRs.

Molecular biological and immunohistochemical data have implicated both α9 and α10 subunits in vestibular hair cells of mammals (Elgoyhen et al. 1994, 2001; Hiel et al. 1996; Anderson et al. 1997; Luebke et al. 2005), chicken (Lustig et al. 1999), trout (Drescher et al. 2004), and frog (Holt et al. 2001). Studies in vestibular ganglia have suggested that nAChRs on afferent terminals contain the α4 and β2 nAChR subunits (Ohno et al. 1993; Wackym et al. 1995), but other subunits including α2, α3, α5–7, α9, and β3–4 have also been identified (Hiel et al. 1996; Anderson et al. 1997; Luebke et al. 2005). All five subtypes of mAChRs have been localized to the vestibular neuroepithelium in rats (Wackym et al. 1996) and pigeons (Li et al. 2007), whereas m1, m2, and m5 have been found in the human vestibular periphery (Wackym et al. 1996). Binding of the mAChR antagonist quinuclidinyl-benzilate (QNB) demonstrates that functional mAChRs are present in gerbil vestibular end organs (Drescher et al. 1999).

6.5.2 Adenosine 5′-Triphosphate

ATP is colocalized in and coreleased from synaptic vesicles in the company of a variety of classical neurotransmitters (ACh, noradrenaline, DA, GABA) (review: Abbracchio et al. 2009). ATP is likely to be a cotransmitter with ACh in vestibular efferent terminals, as it is in motor-nerve terminals (Dowdall et al. 1974; Schweitzer 1987; Silinsky 1975). ATP can activate P2X ionotropic, ligand-gated ion channels and P2Y G-protein linked receptors. Exogenous application of ATP has been shown to depolarize vestibular hair cells from several different species (Rennie and Ashmore 1993; Rossi et al. 1994; Aubert et al. 1994, 1995). Pharmacological evidence suggests that a P2Y receptor is responsible for the depolarization in *Rana* (Aubert et al. 1994, 1995). RT-PCR, Western blot, and immunohistochemical data have provided evidence for P2X receptors in the vestibular ganglia and end organs of mammals (Troyanovskaya and Wackym 1998; Syeda and Lysakowski 2001).

6.5.3 *Calcitonin Gene-Related Peptide*

In addition to the large numbers of small clear vesicles found in efferent terminals, the presence of occasional larger dense core vesicles (Fig. 6.7b) suggests that peptide neurotransmitters are also present. Among the various neuropeptides considered, CGRP is a 37-amino-acid peptide alternately spliced with calcitonin from the same gene transcript (Rosenfeld et al. 1983).

Cell bodies of mammalian group *e* neurons and their peripheral bouton terminals are immunoreactive (IR) for CGRP (Tanaka et al. 1989; Perachio and Kevetter 1989; Wackym et al. 1991). Consistent with the two kinds of vesicles found in individual efferent terminals, most CGRP-positive neurons also label for ChAT, suggesting that ACh and CGRP are colocalized in the same endings (Ohno et al. 1991). In the vestibular periphery, these CGRP-positive fibers terminate predominantly on calyx and bouton terminals (Tanaka et al. 1989; Perachio and Kevetter 1989; Wackym et al. 1991). Several studies have identified a possible role for CGRP and CGRP1 receptors in the lateral line (Sewell and Starr 1991; Bailey and Sewell 2000a, b), but comparable physiological studies are lacking in the peripheral vestibular system.

6.5.4 *Opioid Peptides*

Other efferent neuropeptide candidates likely to be present in vestibular efferents are those mediating opioid actions. Specifically, the majority of brain stem EVS neurons express preproenkephalin mRNA (Ryan et al. 1991) and show met-enkephalin-like IR (Perachio and Kevetter 1989). Endomorphin I, endomorphin II, and β-endorphin IR have been reported for ChAT-positive efferent terminals in the rat crista (Popper and Wackym 2001). Both μ and κ opioid receptors have been localized to vestibular-nerve afferents (Popper et al. 2004). Recordings in amphibian vestibular organs indicate that μ opioid receptors provide an excitatory postsynaptic modulatory input to afferent neurons (Andrianov and Ryzhova 1999; Vega and Soto 2003) and κ receptors mediate an inhibitory, presynaptic input to hair cells (Vega and Soto 2003). Both endomorphin I and dynorphin B, μ and κ receptor agonists, respectively, have also been shown to directly interact with and block nAChRs in the inner ear (Lioudyno et al. 2002).

6.5.5 *γ-Aminobutyric Acid*

This section considers the possibility that GABA is an afferent or an efferent neurotransmitter in vestibular organs. GABA is synthesized by glutamic acid decarboxylase (GAD) and inactivated by conversion to succinic semialdehyde by

GABA-transaminase (GABA-T). Receptors include ionotropic $GABA_A$ and $GABA_C$, as well as metabotropic $GABA_B$ varieties (Chebib and Johnston 1999; Bowery et al. 2002).

The first suggestion that GABA was an afferent neurotransmitter was provided by Flock and Lam (1974), who found that GABA was synthesized by hair-cell organs, even those lacking an efferent innervation; in addition, both spontaneous and evoked discharge were blocked by picrotoxin, a $GABA_A$ antagonist. These last results were not confirmed by Annoni et al. (1984), who found that neither GABA agonists nor antagonists had consistent effects on afferent transmission in the frog posterior canal. Despite these negative findings, there is evidence that GABA is an afferent neurotransmitter. GAD-IR has been localized to type I and type II hair cells, whereas GABA-T-IR is found postsynaptically (López et al. 1992; Usami et al. 1989). Recent studies have suggested that GABA is colocalized with glutamate in a distinctive set of hair cells in the horizontal canal of the oyster toadfish and may serve to modulate glutamatergic transmission (Holstein et al. 2004b, c).

There is, at best, mixed evidence that GABA is an EVS neurotransmitter. GAD-IR, which is an obligatory marker in GABAergic neurons, is apparently not present in brain stem EVS neurons (Perachio and Kevetter 1989) or in fibers innervating the vestibular organs (López et al. 1992; Usami et al. 1989). Attempts to localize GABA have used antibodies to its conjugation by way of aldehyde fixatives with serum albumin. Results have been variable. GABA-like-IR in vestibular organs has been described as being present only in efferent terminals (Usami et al. 1987; Kong et al. 1998); being present only in calyx endings (Didier et al. 1990); being present in calyx endings, fibers, and hair cells (López et al. 1990); or not being present (Matsubara et al. 1995).

6.5.6 Nitric Oxide

Nitric oxide (NO) is a gaseous neurotransmitter that acts by way of soluble guanylate cyclase (sGC) to activate cGMP-dependent protein kinase (reviews: Lincoln et al. 1997; Moncada et al. 1991; Garthwaite 2008). Synthesis of NO from arginine and O_2 is controlled by three isoforms of nitric oxide synthase, two of which (neuronal or nNOS, endothelial or eNOS) are constitutive, Ca^{2+}/calmodulin-dependent, and lead to a brief increase in NO in response to a transient rise in intracellular Ca^{2+}. A third or inducible form (iNOS) leads to a larger and more prolonged rise in NO, which is induced by cytokines rather than Ca^{2+}/calmodulin and is cytotoxic to invading microorganisms and tumor cells. Because NO is lipid-soluble and can pass though membranes, it can act as a paracrine agent, influencing neighboring cells, as well as the cells in which it is synthesized.

nNOS has been localized to brain stem EVS neurons and peripheral efferent boutons, hair cells, and afferent terminals (Lysakowski and Singer 2000; Takumida and Anniko 2002; Desai et al. 2004). Physiological studies have concentrated on hair cells. There is evidence that NO, likely through cGMP, inhibits $I_{K,L}$, the distinctive K^+ current in type I hair cells (Behrend et al. 1997; Chen and Eatock 2000; Rennie 2002). This inhibition should facilitate neurotransmitter release. At the same time, NO inhibits voltage-gated Ca^{2+} channels in hair cells, which should serve to reduce afferent neurotransmission (Almanza et al. 2007). The balance between these opposing effects can potentially be assessed at the level of afferent discharge. Afferent transmission in the cristae of the axolotl, *Ambystoma tigrinum*, was facilitated by NO (Flores et al. 2001). Using criteria described later (Sect. 6.7), the effects were likely targeted to hair cells. It is unclear whether the results can be generalized to mammals as axolotls do not posses type I hair cells or an $I_{K,L}$ current. The presence of NOS in EVS neurons and its effects on hair-cell ion channels suggest a role for NO in efferent neurotransmission, but definitive evidence will require a pharmacological analysis of efferent-mediated responses.

6.6 Afferent Responses to Electrical Activation of the EVS

Given their terminations on afferent processes and type II hair cells, efferent neurons are in a position to modulate the activity of vestibular afferents. Efferent peripheral actions have been characterized by recording single-unit afferent responses to electrical stimulation of EVS pathways. The following sections consider results in mammals and then summarize findings in other vertebrates.

6.6.1 Mammals

Electrical stimulation of the mammalian EVS centrally invariably results in excitation as reflected by an increase in afferent discharge (Goldberg and Fernández 1980; McCue and Guinan 1994; Marlinski et al. 2004). This excitation, which is best seen in response to high-frequency shock trains, is similar whether efferents on the ipsilateral or contralateral sides of the brain stem are stimulated separately or simultaneously (Goldberg and Fernández 1980; Marlinski et al. 2004). This similarity is difficult to reconcile with the reported zonal projections of contralateral efferents (Purcell and Perachio 1997). Specifically, the neuroanatomical results imply that only regular units should be affected by contralateral stimulation. To the contrary, such stimulation influences both regular and irregular units.

Excitation is much larger in irregular, than in regular afferents (Fig. 6.8a). This difference is best seen by plotting efferent response magnitude vs. cv^* (Fig. 6.8b), which results in a relationship paralleling that between galvanic sensitivity and discharge regularity (Fig. 6.8c). The similarity in slopes for the two relationships suggests that much of the variation in efferent responses with discharge regularity reflects the sensitivity of the postsynaptic spike encoder. Note that calyx afferents, recognized by their irregular discharge and relatively low rotational gains

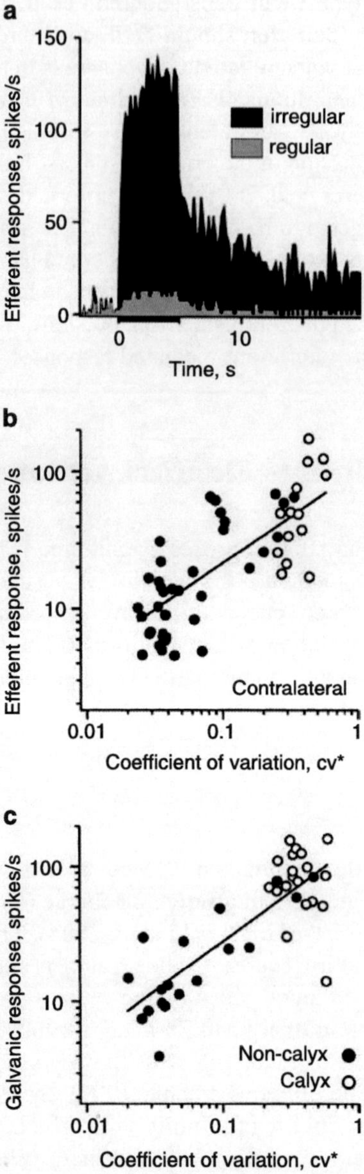

Fig. 6.8 Responses of vestibular-nerve afferents to electrical stimulation of the ipsilateral efferent nucleus in the chinchilla. (**a**) Responses of an irregular and a regular unit to 5-s shock trains, 333 shocks/s. (**b**) Relationship between efferent response magnitude and discharge regularity (*cv**). Stimulus parameters are the same as in (**a**). (**c**) Relationship between afferent response to perilymphatic galvanic polarization and discharge regularity (*cv**) (Adapted from Marlinski et al. 2004, with permission from the Association for Research in Otolaryngology.)

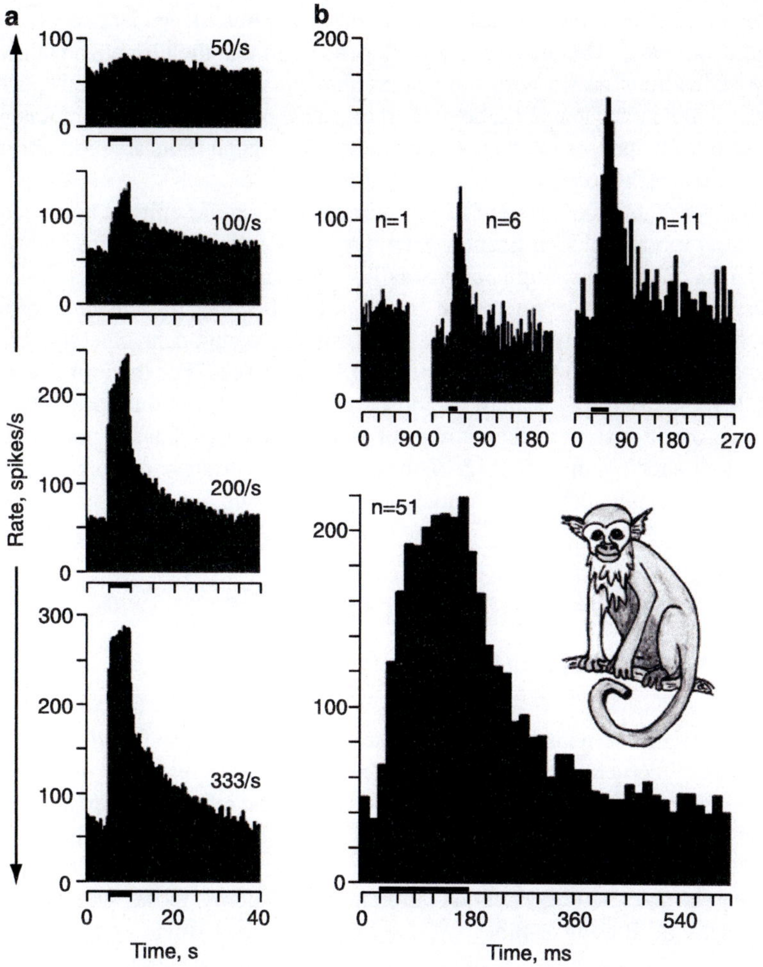

Fig. 6.9 Effects of shock-train parameters on the response of vestibular afferents in squirrel monkey to electrical stimulation of the ipsilateral efferent cell group. (**a**) Response of an irregular afferent to a 5-s efferent shock train (black bar) varying in shock frequency as stated on the *right* of each panel. (**b**) Response of another irregular afferent to efferent shock trains (*black bar*, 333 shocks/s) varying in the total number of shocks ($n = 1$–51) (Modified with permission from Goldberg and Fernández 1980. Copyright 1980, The American Physiological Society.)

(Fig. 6.4a), have efferent responses that are unexceptional in this regard (Fig. 6.8b). Consistent with their being the most irregular units in mammals, calyx units have the largest efferent responses.

For irregular units, the excitatory response can be decomposed into a fast component with kinetics of 10–100 ms and a slow response that builds up and declines with a time course of several seconds (Fig. 6.9a). The fast component is responsible for the abrupt transitions in discharge at the beginning and end of the shock train, with the difference in the two transitions reflecting a time-dependent adaptation of the fast component. The process is time dependent in that the discrepancy between the two transitions grows with train duration. A gradual buildup of the per-train response and its persistence in the post-train period reflect the slow component. In contrast to the responses of irregular units, those of regular units are predominantly small and slow (Fig. 6.8a).

Two other features of efferent responses seen in irregular units of mammals are of potential functional significance. First, large, fast responses require high shock rates. In Fig. 6.9a, for example, only small, slow responses are seen at a shock rate of 50/s. As shock rate is increased to 100/s and beyond, the ensuing response grows disproportionately as does the fast response component. Second, even at high shock rates, large responses require multiple shocks. For the unit illustrated in Fig. 6.9b, there is only a small response to single shocks. In this case, responses grow disproportionately as the number of shocks increases. Fast responses dominate at low shock numbers (Fig. 6.9b, $n=1$–6), but a slow response becomes evident at $n=11$ and continues to increase as shock number increases (Fig. 6.9b, $n=51$).

6.6.2 *Oyster Toadfish (Opsanus tau)*

Efferent-mediated responses in *Opsanus tau* are similar to those found in mammals (Fig. 6.10a). Electrical activation of the EVS almost always results in an increase in afferent discharge coupled with a small decrease in the afferent's response to head rotations or canal indentations (Boyle and Highstein 1990b; Boyle et al. 1991, 2009). Both fast and slow components are evident, with the largest responses appearing in so-called acceleration afferents, particularly those with low background activity. Based on morphophysiological studies (Boyle et al. 1991), the afferents showing large and small efferent responses are located, respectively, near the transverse center and edge of the crista. As compared to the latter units, the rotational responses of the former units have much higher gains and more phasic response dynamics.

6.6.3 *Anurans (Frogs and Toads, Rana and Bufo Species)*

Heterogeneous afferent responses to efferent stimulation are obtained in anurans (Rossi et al. 1980; Bernard et al. 1985; Sugai et al. 1991). Some irregular afferents are excited, whereas others are inhibited (Fig. 6.10b). As discussed in Sect. 6.7, both excitation and inhibition in anurans are the result of efferent actions on hair cells. Other discharge properties of excited and inhibited units appear similar. Efferent responses in regular afferents are small or nonexistent (Sugai et al. 1991).

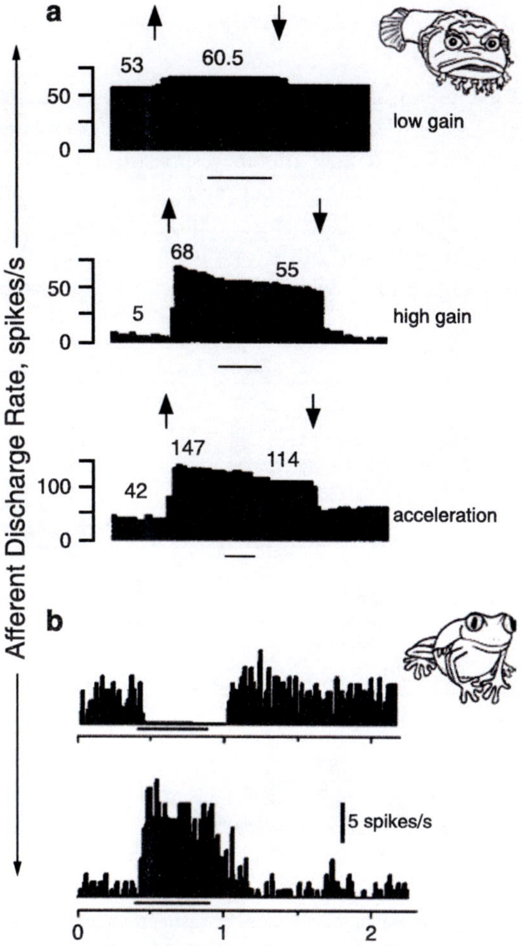

Fig. 6.10 (**a**) Firing rates of vestibular-nerve afferents during electrical stimulation of efferent pathways in the oyster toadfish, *Opsanus tau*. Three horizontal-canal afferents categorized as low-gain (*top*), high-gain (*middle*), or acceleration units (*bottom*). *Arrows* indicate the beginning and end of the efferent shock train, 100 shocks/s. *Bars*, 10 s in all cases. Numbers, firing rates at selected points. (Modified with permission from Boyle and Highstein 1990b. Copyright 1990, Society for Neuroscience.) (**b**) Responses of vestibular-nerve afferents in the Japanese toad, *Bufo vulgaris japonicus*, to electrical stimulation of efferent fibers by axon reflexes. A saccular afferent responds with inhibition to stimulation of the horizontal-canal nerve (*top*). An anterior-canal afferent is excited following stimulation of the posterior-canal nerve (*bottom*). Shock trains, 100 shocks/s, 0.5 s duration (indicated by *horizontal bar*) (Modified from Sugai et al. 1991. Copyright 1991, Physiological Society of Japan.)

6.6.4 Red-Eared Turtles (*Trachemys scripta elegans*)

Efferent-mediated excitation and inhibition are also observed in posterior crista afferents of the red-eared turtle. These efferent responses are related to the location of units in the neuroepithelium, to their afferent responses, and to the presence of efferent synapses on both hair cells and afferent processes (Brichta and Peterson 1994; Brichta and Goldberg 2000a, b; Holt et al. 2006). To understand the diversity of efferent responses in the red-eared turtle, one needs to review the organization of the posterior crista (Fig. 6.11a), which is made up of two triangularly shaped

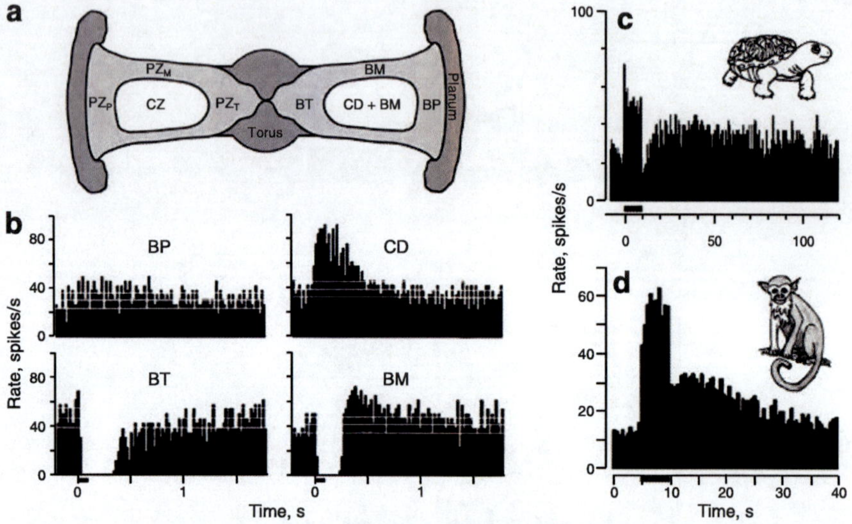

Fig. 6.11 (**a**) Regional organization of the posterior crista neuroepithelium of the red-eared turtle. The neuroepithelium consists of two hemicrista, each extending from the planum to the nonsensory torus. The hemicrista is divided into a central zone (CZ) and peripheral zone (PZ), which is further segmented as near the torus (PZ_T), near the planum (PZ_P), or between these two regions (PZ_M). Type I hair cells are located only in the CZ whereas type II hair cells are found throughout the neuroepithelium. Therefore, the PZ is innervated only by bouton afferents, further identified as those near the torus (BT), in intermediate regions (BM), or near the planum (BP). The CZ is innervated by calyx-bearing (CD) and bouton (BM) afferents. (Reproduced with permission from Holt et al. 2006. Copyright 2006, Society for Neuroscience.) (**b**) Responses of posterior-crista afferents to electrical stimulation of efferent fibers in the red-eared turtle. Shock trains, 200 shocks/s, 0.1 s duration (*bar*). Bouton afferents near the planum (BP) have a small excitatory response. Calyx-bearing (CD) units show a large excitatory efferent response. Bouton units in midportions (BM) of the hemicrista show inhibition followed by a postinhibitory excitation, whereas bouton afferents near the torus (BT) are inhibited by efferent activation without a postinhibitory excitation. (Reproduced with permission from Brichta and Goldberg 1996. Copyright 1996, John Wiley & Sons.) (**c**) Long-duration efferent stimulation (100 shocks/s, 10 s) for a CD unit results in a per-train excitation followed by a slowly developing, long-lasting (slow) excitation. (Modified with permission from Brichta and Goldberg 2000b. Copyright 2000, The American Physiological Society.) (**d**) Similar observations are seen in the squirrel monkey (333 shocks/s for 5 s) (Modified with permission from Goldberg and Fernández 1980. Copyright 1980, The American Physiological Society.)

hemicristae, each consisting of a central zone (CZ) surrounded on all sides by a peripheral zone (PZ). Calyx-bearing units are confined to the CZ, whereas bouton units are found in both the CZ and PZ. Four afferent classes need to be considered, including bouton units near the planum semilunatum (BP), at midportions (BM) of the hemicrista, and near the nonsensory torus (BT), as well as calyx-bearing (calyx and dimorphic, CD) units.

Responses of the various morphological classes to electrical stimulation of efferent fibers are illustrated in Fig. 6.11b. Resembling the situation in mammals, regularly discharging BP afferents and irregularly discharging CD units are excited by efferent stimulation with CD units having much larger responses. But unlike the situation in mammals, some afferents, including BT and BM units, are inhibited. BT/BM units are distinctive not only in their efferent responses. Their afferent responses are also unlike those seen in mammals, but are similar to units described in frogs (Blanks and Precht 1976) and the oyster toadfish (Boyle and Highstein 1990a; Boyle et al. 1991) in having distinctively high rotational gains and large phase leads re angular head velocity. Section 6.10.3 considers the possible functional implications of the high-gain (BT/BM) units being inhibited by efferent activation, while CD and BP units with more modest gains are excited.

The above responses in the red-eared turtle have relatively fast kinetics. However, slow responses, resembling those seen in mammals (Fig. 6.11d), are seen in turtle CD units in response to long-duration efferent shock trains (Fig. 6.11c) (Brichta and Goldberg 2000b). A second way to produce slow responses is to present several short efferent shock trains that are so closely spaced that the discharge does not relax to control values in the periods between trains, but continues to grow to a new asymptote (McCue and Guinan 1994; Marlinski et al. 2004; Holt and Goldberg, unpublished observations).

6.7 Sites of Efferent Actions: Hair Cells or Afferents

Resting activity is maintained by mEPSPs reflecting quantal neurotransmission to an afferent from the hair cells it innervates. Whether efferents target the hair cells or the afferent fibers can be deduced from intracellular recordings of synaptic activity in the afferent. Presynaptic actions on the hair cells should lead to a modulation of mEPSP rate, whereas postsynaptic actions on the afferent should be marked by direct efferent-mediated PSPs.

In the frog, efferent fibers contact hair cells, but not afferent terminals (Hillman 1969; Lysakowski 1996). Consistent with this innervation, efferent inhibition and excitation are associated, respectively, with decreases and increases in mEPSP rate (Rossi and Martini 1991; Bernard et al. 1985; Sugai et al. 1991). Efferent-mediated reduction in spike discharge is associated with a cessation of mEPSPs (Fig. 6.12a). When there is an increase in mEPSP traffic, these can summate to result in a depolarizing shift in membrane potential and an increased spike discharge (Fig. 6.12b).

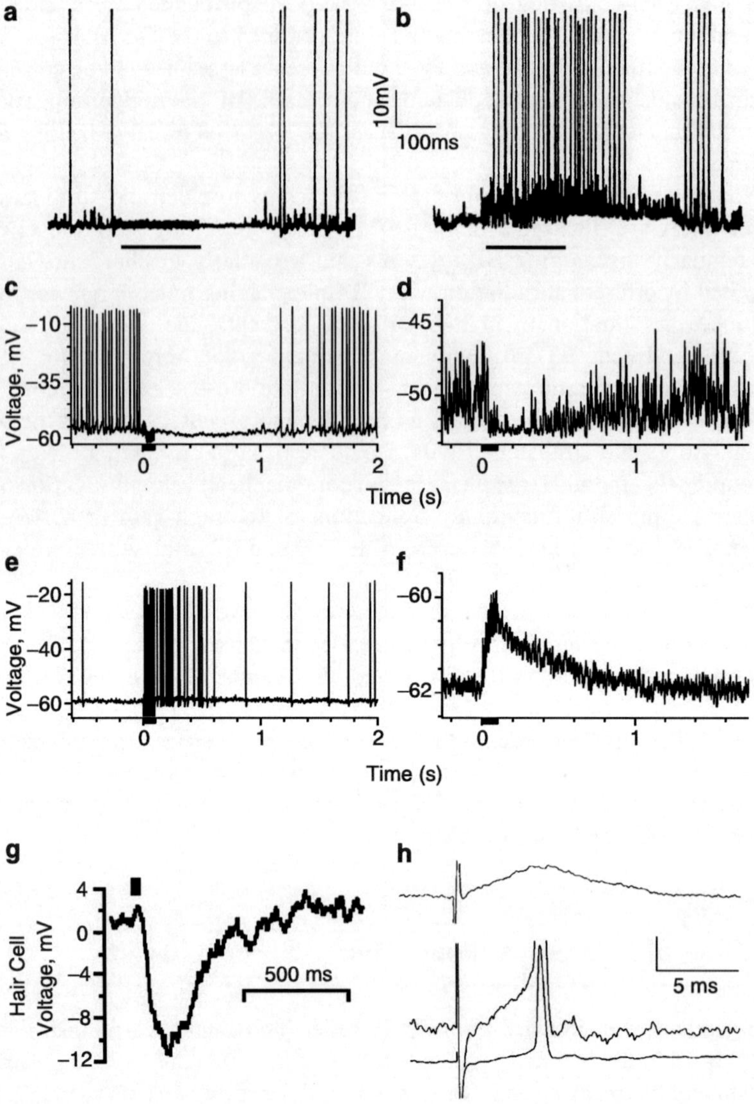

Fig. 6.12 (**a, b**) Efferent synaptic actions from posterior canal afferents in the frog, *Rana esculenta*. Inhibition (**a**) and excitation (**b**) of afferent discharge in two separate units are elicited by brief shock trains (200/s for 200 ms) to the anterior-horizontal canal nerves. The mEPSPs, which represent quantal transmitter release from hair cells, are seen as small depolarizations along the base of action potentials. They are eliminated in (**a**) and increased in (**b**). The increase in mEPSP traffic is associated with a depolarizing shift. (Modified with permission from Rossi et al. 1980. Copyright 1980, Elsevier Science.) (**c, d**) The effect of efferent actions on postsynaptic recordings from posterior crista afferents of the red-eared turtle. In bouton afferents located near the torus (BT), activation of efferent fibers results in a cessation of discharge (**c**) correlated with a reduction of quantal activity in the same unit (**d**). Efferent activation in calyx-bearing (CD, calyx or dimorphic) afferents increases discharge. (**e**) An underlying efferent-mediated EPSP from another CD

Efferent fibers in the red-eared turtle innervate both hair cells and afferent terminals, including both calyces and boutons (Lysakowski 1996; Holt et al. 2006). The efferent inhibition seen in BT and BM fibers is the result of a marked reduction in mEPSP rate and, hence, is mediated by a hair-cell action (Fig. 6.12c, d). As mEPSP rates fall, the afferent hyperpolarizes due to the loss of depolarizing quantal activity. In contrast, the excitation observed in CD units is associated with a direct efferent-mediated EPSP (Fig. 6.12e, f) consistent with efferent terminals synapsing on calyx endings. Hair cell–mediated efferent inhibition in BT/BM fibers typically masks an afferent excitation attributed to the efferent innervation of afferent boutons. This direct excitation can be revealed, however, using several pharmacological blockers of hair cell inhibition (Holt et al. 2006). Currently, intracellular recordings of efferent actions in BP units are lacking.

In the oyster toadfish, there is an efferent innervation of both hair cells (Sans and Highstein 1984; Holstein et al. 2004a) and afferents (Sans and Highstein 1984). In response to single or multiple EVS shocks, long-latency IPSPs have been recorded from hair cells (Boyle et al. 2009) (Fig. 6.12g) and monosynaptic EPSPs from afferents (Highstein and Baker 1985) (Fig. 6.12h). The hair-cell IPSPs, although delayed, lead to such large conductance changes that they might be expected to reduce spike discharge. Yet, efferent stimulation almost always increases discharge (Boyle and Highstein 1990b). How hair-cell inhibition might interact with afferent excitation to produce this result remains to be determined.

6.8 Pharmacology of Efferent Neurotransmission

The predominant efferent neurotransmitter is acetylcholine (ACh). To account for the diversity of efferent responses in vestibular organs based solely on the actions of ACh would require differences in cholinergic receptors and/or subsequent intracellular signaling. Pharmacological studies, conducted largely in frogs and turtles, have delineated the receptor and signaling mechanisms underlying the three principal effects of efferent activation: hair-cell inhibition, hair-cell excitation, and afferent excitation. These are fast efferent activations. Evidence to be considered in this section indicates that they are mediated by nAChR mechanisms.

Fig. 6.12 (continued) unit is shown (**f**). Shock train, 20 shocks, 200 Hz. (Adapted with permission from Holt et al. 2006. Copyright 2006, Society for Neuroscience.) (**g, h**) Intracellular recordings from a hair cell (**g**) and afferents (**h**) in the horizontal canal of the oyster toadfish during efferent stimulation. (**g**) Large IPSP generated in a canal hair cell during the delivery of an efferent shock train (100/s, *gray bar*). (Modified with permission from Boyle et al. 2009. Copyright 2009, The American Physiological Society.) (**h**) Depolarization of a canal afferent elicited by a single efferent shock (*top*). The single-shock, efferent-mediated EPSP can elicit action potentials (*bottom*). Voltage *scale bar* in *bottom* applies to both: *top*, 4 mV; *bottom*, 8 mV (Modified with permission from Highstein and Baker 1985. Copyright 1985, The American Physiological Society.)

In addition, there is a slow excitation, which involves mAChR-mediated and/or possibly non-cholinergic actions.

6.8.1 Hair-Cell Inhibition

When it was first reported that efferent stimulation exerts an exclusively excitatory action on mammalian vestibular afferents (Goldberg and Fernández 1980), it seemed rather surprising since efferent actions in other hair-cell organs were known to be inhibitory (Fex 1962; Russell 1968; Ashmore and Russell 1982). Even in frogs, almost half of vestibular afferents were excited by efferent stimulation (Rossi et al. 1980). At that time, it was also known that: (1) inhibitory efferent actions in the cochlea are the result of cholinergic nicotinic neurotransmission (Bobbin and Konishi 1971, 1974); and (2) activation of nicotinic receptors typically results in excitation (Cooper et al. 2002). The initial challenge, it seems, was to explain how an efferent action based on nicotinic receptors could give rise to inhibition, rather than excitation.

A possible clue was provided by hair-cell recordings in the red-eared turtle basilar papilla, which showed that efferent inhibition was mediated by a hyperpolarization that was preceded by a brief depolarization; the suggestion was made that the early depolarization was the primary synaptic event that triggered the slower hyperpolarization (Art et al. 1984). Consistent with this idea, central recordings indicated that nicotinic transmission led to inhibition by triggering a calcium-dependent increase in a potassium conductance (Wong and Gallagher 1991). Within a few years, a similar mechanism was shown to be responsible for efferent inhibition in auditory hair cells in the chick basilar papilla (Fuchs and Murrow 1992a, b). Later work showed that inhibition is the result of the efferent-mediated activation of α9/α10-nicotinic ACh receptors (α9/10nAChRs) (Elgoyhen et al. 1994, 2001), whose opening allows the entry of Ca^{2+} ions (Weisstaub et al. 2002) that activate small-conductance, calcium-dependent potassium (SK) channels (Yuhas and Fuchs 1999; Oliver et al. 2000). Outward K^+ currents through SK channels hyperpolarize the hair cell, inhibit neurotransmitter release, and reduce afferent discharge.

As noted previously, inhibition of afferent discharge during efferent stimulation is seen in recordings from the vestibular organs of anurans (Rossi et al. 1980; Bernard et al. 1985; Sugai et al. 1991) and red-eared turtles (Brichta and Goldberg 2000b). As with auditory hair cells, efferent inhibition in the turtle vestibular labyrinth also involves the linked activation of α9/10nAChRs and SK channels. The situation is illustrated by recordings from a BT afferent innervating the posterior crista (Fig. 6.13a, b). The reduction of quantal activity and the ensuing afferent hyperpolarization resulting from efferent stimulation are completely blocked by the α9/10nAChR antagonist tropisetron (ICS-205390 or ICS) (Fig. 6.13a). In line with the sequential activation of α9/10nAChRs and SK channels, single efferent shocks generate a biphasic voltage response in the afferent comprised of an initial

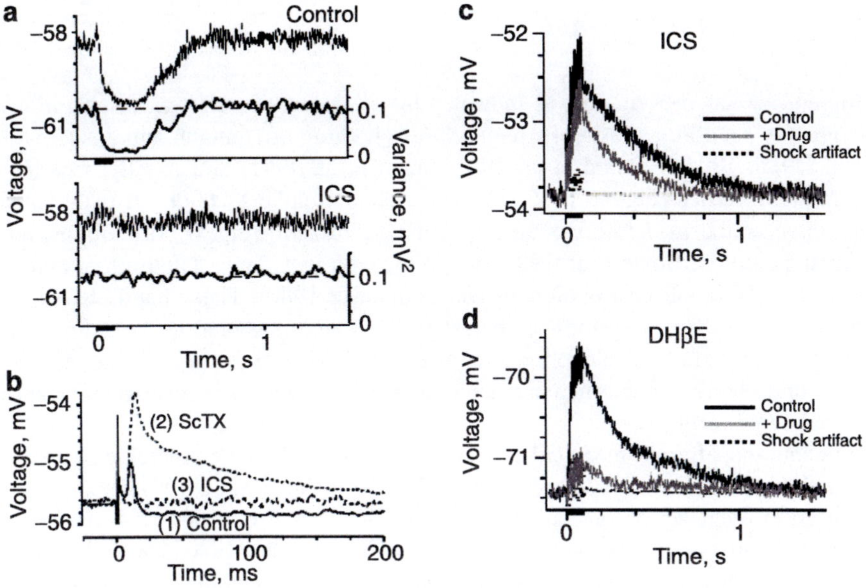

Fig. 6.13 Pharmacological dissection of efferent responses in the posterior crista of the red-eared turtle. (**a**) Tropisetron (ICS), a pharmacological blocker of α9/α10 nicotinic receptors (α9/10nAChRs), eliminates the response of a BT afferent to efferent shock trains (20 shocks at 200/s, *black bar*). Average responses are compared before and during the application of 10 μM ICS. In each panel, the ensemble mean voltage (*top trace*) and variance (*bottom trace*) are shown. Variance should be proportional to quantal rate. *Dashed line* indicates the average prestimulus variance. ICS blocks the effects of efferent stimulation on both voltage and variance. (**b**) Scyllatoxin (ScTX), an SK blocker, converts afferent responses to single efferent shocks from a biphasic excitation–inhibition to a monophasic excitation. In the red eared turtle, the biphasic voltage response (*1*, Control) to single efferent shocks, seen here in a BT unit, is a signature for the sequential activation of α9/10nAChRs and SK where a brief depolarization is followed by a prolonged hyperpolarization. The biphasic response is converted to an exclusively depolarizing response by 1 μM ScTX (*2*), which is then completely blocked by 10 μM ICS (*3*). (**c**, **d**) Pharmacology of postsynaptic depolarizing responses in CD afferents can be distinguished from presynaptic responses in BT/BM afferents. The effects of ICS and the nicotinic antagonist, dihydro-β-erythroidine (DHβE), on the response of CD afferents to efferent shocks (20 shocks, 200/s) are shown. (**c**) In a CD afferent, 10 μM ICS has only a small effect on the response. (**d**) In another CD unit, 300 nM DHβE blocks most of the response (All panels reproduced with permission from Holt et al. 2006. Copyright 2006, Society for Neuroscience.)

brief depolarization followed by a more prolonged hyperpolarization (Fig. 6.13b, control). Application of scyllatoxin (ScTX), an SK antagonist, blocks the hyperpolarization and unmasks a substantial afferent depolarization (Fig. 6.13b, ScTX). That the depolarization is mediated by α9/10nAChRs is confirmed by its subsequent blockade with ICS (Fig. 6.13b, ICS). Other blockers of α9/10nAChR (e.g., strychnine) and SK channels (e.g., apamin) confirm the linked participation of the nAChR and SK (Holt et al 2006). Similar mechanisms have been identified in vibratory and lateral line organs (Yoshida et al. 1994; Holt et al. 2001; Dawkins et al. 2005).

6.8.2 Hair-Cell Excitation

Efferent-mediated excitation, as indicated by an increase in afferent discharge, is commonly seen in a subset of irregularly discharging afferents in anurans (Rossi et al. 1980, 1994; Bernard et al. 1985; Sugai et al. 1991) and in calyx-bearing afferents in turtles (Figs. 6.10–6.12) (Brichta and Goldberg 2000b). It is the predominant action in the horizontal crista of *Opsanus tau* (Fig. 6.11a) (Boyle and Highstein 1990b; Boyle et al. 1991) and in the entire vestibular labyrinth of mammals (Figs. 6.8 and 6.9) (Goldberg and Fernández 1980). The synaptic bases of excitation in different species are heterogeneous and attributed to the actions of efferents synapsing on hair cells and/or afferents. This section discusses the mechanisms underlying efferent-mediated excitation of hair cells as seen in anurans and the red-eared turtle.

In anurans, efferent-mediated excitation is associated with an increase in quantal activity (Rossi et al. 1980; Bernard et al. 1985; Sugai et al. 1991), consistent with an efferent action on hair cells. A slow excitation of frog canal afferents, generated with cholinergic agonists, is attributed to mAChRs also on hair cells; however, such slow excitation has not been demonstrated with genuine efferent stimulation (Guth et al. 1986; Holt et al. 2003; Derbenev et al. 2005). The observation in red-eared turtles that efferent inhibition can be converted to excitation when SK channels are blocked suggests that hair cells expressing α9/10nAChRs uncoupled from SK could account for such hair-cell excitation (Holt et al. 2006). However, several pharmacological observations suggest, at least in frogs, that the receptor underlying hair-cell excitation is nicotinic but not α9/10nAChR: (1) application of the nicotinic agonist 1,1-dimethyl-4-phenylpiperazinium (DMPP) or carbachol mimics efferent stimulation in increasing the rate of both mEPSPs (Fig. 6.14a) and action potentials (Fig. 6.14b–d), and can do so in the same afferents previously excited by efferent stimulation (Bernard et al. 1985; Holt et al. 2003); (2) efferent-mediated hair-cell excitation and DMPP responses are blocked by *d*-tubocurarine (dTC) (Fig. 6.14c) but not by strychnine (Fig. 6.14d), the latter a potent α9/10nAChR antagonist; (3) finally, the nAChRs underlying hair-cell excitation are more sensitive to ACh and DMPP than are α9/10nAChRs (Fig. 6.14e, f)

Fig. 6.14 (continued) rate of mEPSPs. (**b**) Single unit spike histograms illustrating the actions of the cholinergic agonists, DMPP and carbachol (in mM), upon the firing rates of frog vestibular afferents. NS, normal saline. DMPP elicits a fast excitation (*top*) whereas carbachol produces both fast and slow excitation. (Modified with permission from Bernard et al. 1985. Copyright 1985, Elsevier Science.) (**c, d**) Effects of the nicotinic antagonists *d*-tubocurarine (dTC) and strychnine on the response of multiunit afferent firing to DMPP in posterior semicircular canal of *Rana pipiens*. Similar to single-unit observations in (**b**), the application of 10 μM DMPP (*small bars*) results in a rapid increase in background discharge of canal afferents (**c, d**, *left*). This DMPP-mediated excitation was mostly antagonized by 1 μM dTC (**c**, *right*) but not by 10 μM strychnine (**d**, *right*). (**e, f**) Current clamp recordings demonstrate that frog semicircular-canal hair cells are strongly depolarized by low concentrations of both ACh (1 μM) and DMPP (0.1 μM). AP, artificial perilymph (Modified with permission from Holt et al. 2003. Copyright 2003, The American Physiological Society.)

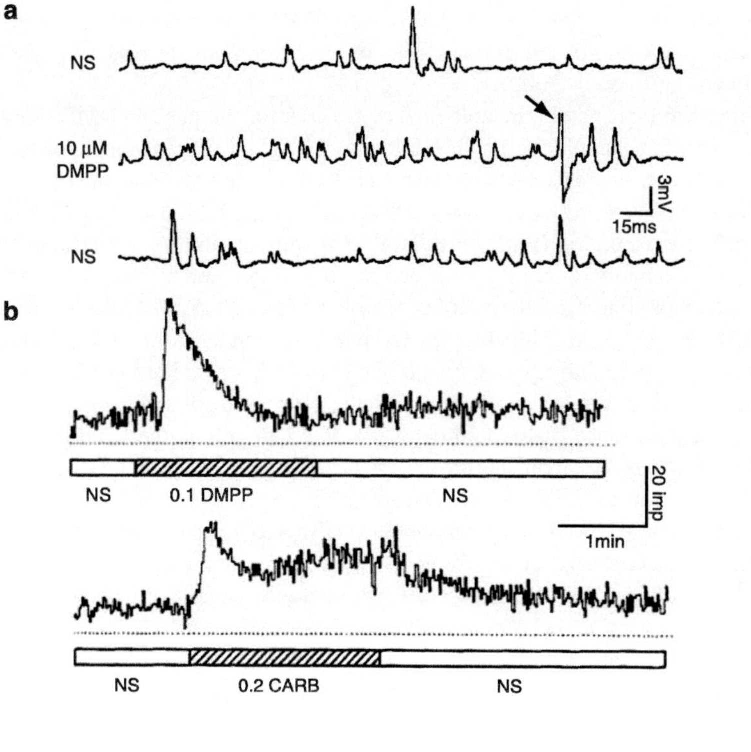

Fig. 6.14 Pharmacological dissection of excitatory efferent responses in frogs, *Rana temporaria* and *Rana pipiens*. (**a**) Effect of the nicotinic agonist 1,1-dimethyl-4-phenyl-piperazinium (DMPP) on spontaneous synaptic potentials recorded intracellularly from vestibular afferents. Spontaneous mEPSPs were recorded before (*top*), during (*middle*), and after (*bottom*) the application of 10 μM DMPP. A single action potential has been truncated (*arrow*). DMPP significantly increases the

(Elgoyhen et al. 2001; Holt et al. 2003). This DMPP-sensitive excitatory nAChR may contain α4 and/or α6 subunits but its exact composition has not yet been determined (Guth et al. 2002).

Although there is no pharmacological evidence for a comparable DMPP-sensitive nAChR in turtle vestibular hair cells (Holt et al. 2006), an efferent-mediated hair-cell excitation can be seen in recordings from a subset of afferents innervating the turtle posterior crista. With few exceptions, efferent-mediated inhibition in BM afferents is followed by excitation (Fig. 6.11b, BM). This post-inhibitory excitation (PIE) is associated with an increase in quantal activity that is initiated by the preceding inhibition (Brichta and Goldberg 2000b; Holt et al. 2006). Blockade of either α9/10nAChRs or SK also blocks PIE. Despite its dependence on an α9/10nAChR/SK-mediated hyperpolarization of type II hair cells, PIE is peculiar to BM, but not BT, afferents. These observations suggest that the efferent-mediated hyperpolarization activates a conductance that would depolarize the hair cell as the hyperpolarization terminates. Two likely candidates are a T-type calcium current or an I_h (HCN) current. PIE is also seen in vestibular and lateral-line afferents of anurans (Rossi and Martini 1991; Dawkins et al. 2005). Here, similar mechanisms could be involved. Alternatively, there could be a convergence of efferent-mediated inhibition (e.g. α9/10nAChR/SK) and excitation (e.g. DMPP-sensitive nAChRs) onto the same hair cells, or on different hair cells innervating the same afferent. PIE could result were the inhibition to outweigh the excitation during the evoking shock train, but to decay faster in the post-train period.

6.8.3 Fast Afferent Excitation

While efferent terminals in the frog are confined to hair cells, those in most other species also contact afferent terminals (Sect. 6.4.4). In species containing both type I and type II hair cells, calyx endings and afferent boutons receive an efferent innervation. As far as we know, this postsynaptic efferent innervation is excitatory in all species where it is present. In addition, there is circumstantial evidence that excitation may also occur in the mammalian cochlea at efferent synapses onto afferent dendrites underneath inner hair cells (Walsh et al. 1998; Zheng et al. 1999).

Pharmacological experiments in the red-eared turtle have further demonstrated that the direct efferent-mediated excitation of both kinds of afferent terminals involves nicotinic ACh receptors distinct from α9/10nAChRs (Holt et al. 2006, 2010). In particular, the receptors underlying efferent-mediated afferent EPSPs can be distinguished from α9/10nAChRs in the potency of their blockade by various agents. Strychnine and ICS are more potent blockers of α9/10nAChRs (cf. Fig. 6.13a, c), whereas dihydro-β-erythroidine (DHβE) is a qualitatively more effective blocker of the efferent-mediated EPSPs (Fig. 6.13d). Molecular biological and pharmacological data suggest that α4β2-containing nAChRs may underlie efferent-mediated excitation of vestibular afferents (Wackym et al. 1995; Anderson et al. 1997; Holt et al. 2006, 2008, 2010).

How do the efferent actions on hair cells and afferent terminals interact? Most afferents in mammals and turtles receive synaptic inputs from type II hair cells. This connectivity is obviously the case for afferents with bouton endings, but even the calyx terminals innervating type I hair cells can be contacted on their outer faces by ribbon synapses from type II hair cells (Lysakowski and Goldberg 1997, 2008; Holt et al. 2007). Type II hair cells also receive a conspicuous efferent innervation; yet inhibition is not seen in mammalian afferents, even those receiving type II inputs. This observation as well as recent work in the oyster toadfish (Boyle et al. 2009) and the red-eared turtle (Holt et al. 2006) suggests one of two possibilities: (1) there is a presynaptic efferent inhibition of hair cells that is outweighed by the postsynaptic excitation of calyx endings and other afferent processes; or (2) the presynaptic action is also excitatory. Concerning the latter possibility, a presynaptic excitatory action might be mediated by a novel receptor, as is suggested by work in the frog (Bernard et al. 1985; Rossi et al. 1994; Holt et al. 2003), or it could result from an α9/10nAChR-mediated excitation that is not completely checked by an activation of SK channels. The matter is currently unresolved.

6.8.4 Slow Afferent Excitation

The fast excitatory component seen in mammals has kinetics of 10–100 ms, while the slow excitatory component can take several seconds to increase and decrease (Goldberg and Fernández 1980; McCue and Guinan 1994; Marlinski et al. 2004). Slow responses have also been recorded in the oyster toadfish (Boyle and Highstein 1990b; Boyle et al. 1991) and in turtles (Brichta and Goldberg 2000b) by the electrical activation of the EVS as well as in frogs by the application of cholinergic agonists (Bernard et al. 1985; Holt et al. 2003). Because of the slow kinetics involved, it is natural to suspect a G-protein–coupled receptor such as a mAChR. At the same time, it should be noted that fast and slow efferent effects in the mammalian cochlea can involve one and the same nicotinic receptor, likely α9/10nAChR (Sridhar et al. 1995, 1997).

There is pharmacological evidence for the participation of mAChRs in the etiology of slow responses. Such receptors are present in vestibular neuroepithelia, nerve fibers, and ganglia (Wackym et al. 1996; Drescher et al. 1999; Li et al. 2007). Application of mAChR agonists in the crista of frogs and turtles results in a slow excitation (Bernard et al. 1985; Holt et al. 2003; Jordan et al. 2010). Furthermore, in red-eared turtles, mAChR antagonists block slow responses in CD afferents evoked by prolonged electrical stimulation of efferent fibers (Jordan et al. 2010). How might mAChR activation lead to slow responses? One possibility is the activation of the G protein (G_q) that inhibits so-called M currents through depletion of intramembranous phosphatidylinositol 4,5-bisphosphate and/or the elevation of intracellular calcium (Hernandez et al. 2008; Brown and Passmore 2009). M currents are outwardly rectifying K^+ channels whose suppression by activation of mAChRs or other metabotropic receptors results in a slow depolarization coupled to an increase in impedance (Fukuda et al. 1988; Brown 1988; Delmas and Brown 2005). KCNQ4, one of several channels that can give rise to M currents (Selyanko et al. 2000; Brown and Passmore 2009), has been immunolocalized to type I hair cells and calyx endings (Kharkovets et al. 2000; Hurley et al. 2006; Sousa et al. 2009). Calyx endings are also immunoreactive for KCNQ5 (Hurley et al. 2006). M-currents have been recorded from chick and rat vestibular ganglion cells (Yamaguchi and Ohmori 1993; Pérez et al. 2009a, b).

It is also possible that mAChRs tap into the nitric oxide (NO) pathway by activating nNOS in the calyx or in efferent terminals to produce NO that then diffuses to block $I_{K,L}$, an M-like current in type I hair cells, and thereby to depolarize the hair cell and enhance neurotransmitter release (Chen and Eatock 2000; Lysakowski and Singer 2000; Hurley et al. 2006). Whether the slow response arises from hair cells or afferents or whether KCNQ channels and/or NO are involved is currently unresolved.

Alternatively, vestibular efferent neurons also contain CGRP (Tanaka et al. 1989; Wackym et al. 1991; Ishiyama et al. 1994) and possibly other neuroactive peptides (Sects. 6.5.3 and 6.5.4; review: Goldberg et al. 2000). In lateral lines, CGRP causes a slow excitation (Sewell and Starr 1991; Bailey and Sewell 2000a, b). In mammals, postsynaptic actions could be mediated by CGRP-containing efferent axons, which have been observed to contact calyces and other afferent processes (Tanaka et al. 1989; Wackym et al. 1991; Ishiyama et al. 1994).

6.9 Efferent Modulation of Afferent Responses to Natural Stimulation

Most studies of the physiological actions of the EVS have looked at the consequences of efferent activation on the background activity of vestibular afferents. However, it is equally important to understand how different efferent actions impact the afferent's response to natural stimulation. In posterior-canal afferents of frog and turtle, efferent inhibition can completely abolish responses to cupular deflections produced either by rotations (Rossi et al. 1980) or indentations of the canal duct (Holt 2008). More modest changes in rotational sensitivity are seen with efferent-mediated fast excitation. In the squirrel monkey (*Saimiri sciureus*), when fast efferent excitation is paired with rotation, there can be a small decrease in the rotational gain of irregularly discharging units (Goldberg and Fernández 1980). Similar gain reductions have been observed during efferent-mediated fast excitation in the oyster toadfish (Boyle and Highstein 1990b). A somewhat different result was obtained in the cat under conditions that may have favored slow responses; here there was an enhanced sensitivity of saccular afferents to intense air-borne sounds (McCue and Guinan 1994).

A conventional parallel-conductance model suggested that fast and slow efferent excitation could have opposite effects on afferent gain (Fig. 6.15a). Fast responses, by increasing conductance, should decrease gain; slow responses, were they to decrease conductance, for example, by inhibiting an M current, should have the opposite effect. The prediction has been confirmed in the posterior canal of the red-eared turtle; pairing indentation of the canal duct with an efferent-mediated fast response is associated with a modest gain decrease (Fig. 6.15b), whereas pairing with a slow response is associated with a more substantial gain increase (Fig. 6.15c) (Holt 2008; Shah et al. 2010). In both situations, there is an increase in the background discharge and in the average discharge during sinusoidal stimulation.

6.10 Functional Studies of the EVS

The afferent responses to electrical stimulation of the EVS can provide clues as to function. But to go beyond speculation, an understanding is needed of how efferent neurons are influenced by natural stimulation and how they, in turn, can modify afferent discharge under physiological conditions. Studies of efferent discharge characteristics have almost exclusively been done in fish and anurans.

6.10.1 Response of EVS Neurons to Natural Stimulation

Efferent neurons respond to vestibular stimulation, including activity arising from semicircular canals (Schmidt 1963; Gleisner and Henriksson 1964; Precht et al. 1971; Blanks and Precht 1976; Hartmann and Klinke 1980) and otolith organs (Klinke and Schmidt 1968). These studies provide evidence that efferent neurons receive a convergent input from several vestibular organs in both ears. Possibly reflecting such a bilateral convergence, efferent neurons respond in a type III manner, increasing their discharge for angular rotations in either direction (Precht et al. 1971; Blanks and Precht 1976). A type III response may be contrasted with the invariable type I responses of afferents, where discharge is increased (excited) by rotations in one direction and reduced (inhibited) by oppositely directed rotations. Only inconsistent rotational responses were recorded from efferent neurons in the oyster toadfish (Highstein 1991), even though these neurons receive a monosynaptic excitatory input from the vestibular nerve (Highstein and Baker 1985).

Efferent neurons also respond to nonvestibular stimulation, including pressure applied to the skin (Schmidt 1963; Precht et al. 1971), passive movement of limbs (Schmidt 1963; Precht et al. 1971), and visual stimulation (Klinke and Schmidt 1970). As has also been observed in lateral-line efferents (Russell 1971; Roberts and Russell 1972), vestibular efferents also respond in anticipation of active body movements (Schmidt 1963; Gleisner and Henriksson 1964; Precht et al. 1971). In the oyster toadfish, efferents are excited by a large variety of sensory stimuli (Highstein and Baker 1985; Highstein 1991). Here, the responses have a long latency (≈150 ms) and may outlast the stimulus by 500 ms or more. Efferent activation in the oyster toadfish is most likely associated with arousal, a stereotyped

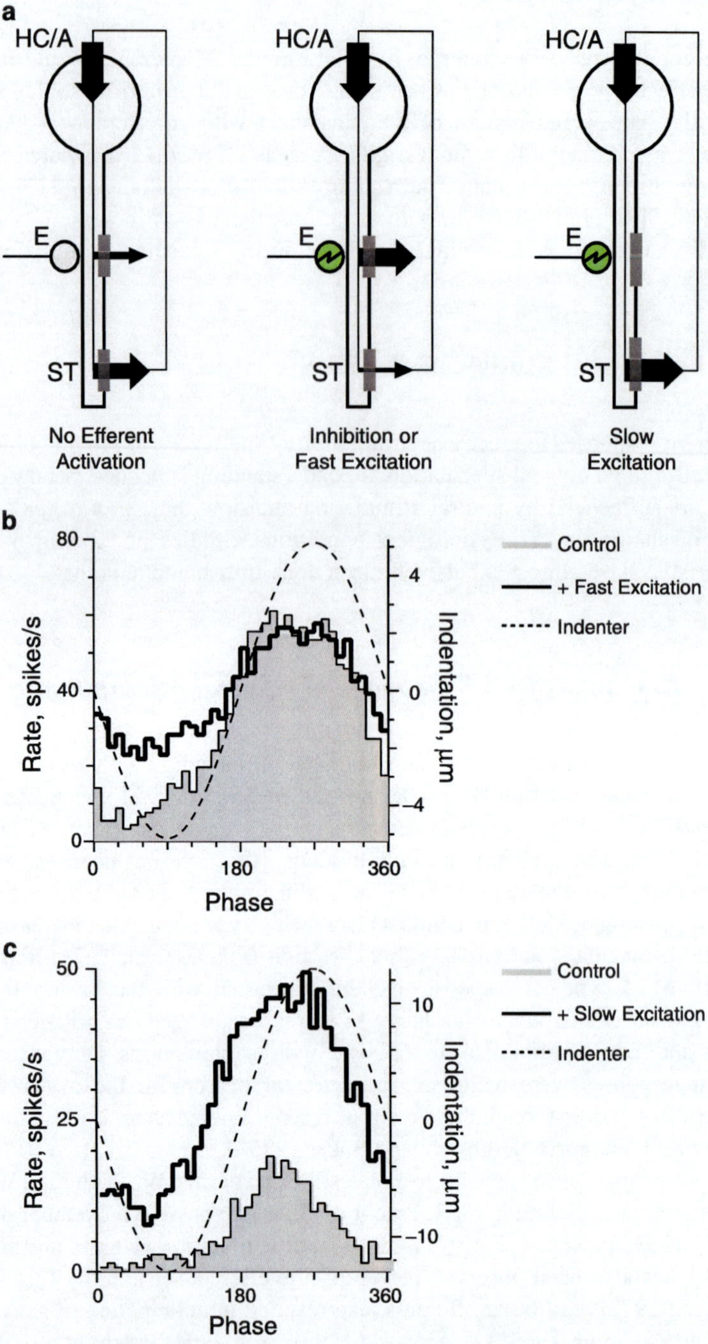

Fig. 6.15 Effects of efferent activation on the indenter response of posterior canal afferents of the red-eared turtle. (**a**) Parallel conductance model indicating how efferent activation (E) can attenuate or augment synaptic currents arising from a hair cell or afferent (HC/A) and their subsequent interaction with synaptic transmission or the spike-trigger site (ST), respectively. Upon efferent-mediated

behavior that can be evoked by sensory stimulation and can also occur spontaneously (Highstein and Baker 1985). The behavior can be a prelude to movement. With one possible exception (Marlinsky 1995), discharge properties of efferent neurons have not been studied in mammals.

6.10.2 *Efferent-Mediated Modulation of Afferent Discharge*

A potential function of efferents is to modify afferent discharge on a short time scale. An example is provided by the arousal response in the oyster toadfish, which is associated with an excitation of afferents, as well as of efferents (Highstein and Baker 1985; Boyle and Highstein 1990b). To study efferent-mediated vestibular responses, a mechanical indenter was used in the decerebrate pigeon to stimulate the horizontal canal contralateral to the afferents being recorded (Dickman and Correia 1993). Some afferents were excited by contralateral stimulation, others were inhibited, and still others showed mixed responses. The diversity of afferent responses presumably reflects a similar diversity in the peripheral actions evoked by electrical stimulation of efferent pathways in birds.

Efferent-mediated rotational responses were obtained in the decerebrate chinchilla (Plotnik et al. 2002) from otolith afferents, which do not otherwise respond to head rotations (Goldberg and Fernández 1975), and from canal afferents after positioning the head so that conventional rotational responses of each fiber were nulled by placing the innervated canal nearly orthogonal to the plane of motion. High-intensity (320°/s) rotations led to type III responses (Fig. 6.16a–c), which resembled those obtained by electrical stimulation of efferent pathways in several ways. The responses were always excitatory. They were considerably larger in irregular, as compared to regular, afferents (Fig. 6.16d–f). In irregular units, both fast and slow responses were seen, whereas the responses in regular units were predominantly slow. Canal-plugging and labyrinthine galvanic polarization were used to show that type III responses could be

Fig. 6.15 (continued) inhibition and/or fast excitation, opening of ion channels (e.g., nAChRs or SK) in the hair cell and/or afferent will shunt current away from ST effectively reducing the sensitivity of the afferent to the same vestibular stimulus. Closure of ion channels, as is thought to occur with efferent-mediated slow excitation, should have the opposite effect. (**b**) Phase histograms illustrating the effect of fast excitation on a CD afferent's response to sinusoidal canal-duct indentation. For this particular unit, there was no discernible efferent-mediated slow excitation. Comparing the afferent's response to the indenter alone to that when the efferent stimulus (+ Fast Excitation) and the same indenter stimulus are given simultaneously demonstrates that fast excitation reduces the peak-to-peak modulation. (**c**) Phase histograms of another CD afferent illustrating the effect of slow excitation on the indenter response. Slow excitation was generated using an efferent shock protocol similar to that shown in Fig. 6.11c. The afferent's response (control, 18 cycles, 0.3 Hz) to indentation alone is significantly enhanced when the same indenter stimulus is applied during efferent-mediated slow excitation

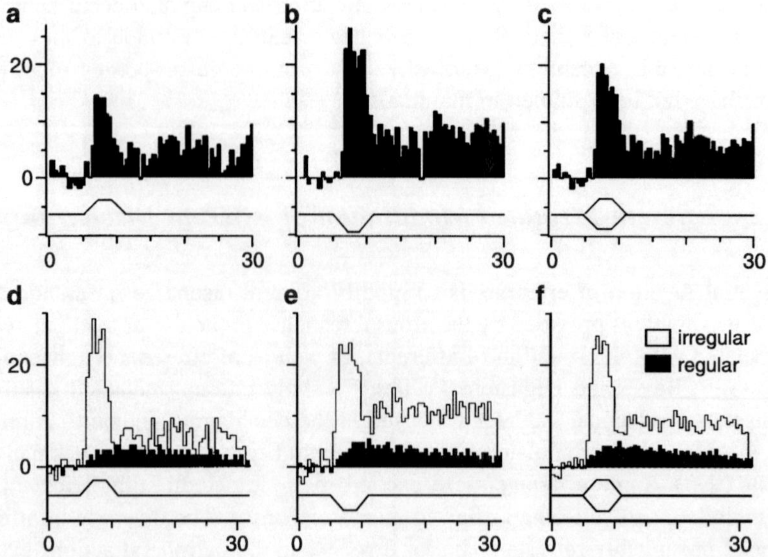

Fig. 6.16 (**a–c**) Responses (discharge rate minus background rate) in response to (**a**) counterclockwise (CCW) and (**b**) clockwise (CW) rotations, 320°/s in the horizontal plane, for an irregular otolith afferent ($cv^* = 0.32$) from a decerebrate preparation. Velocity profiles are below each trace and directions are as viewed from above. Both rotations gave excitatory responses, which are averaged in (**c**). (Adapted from Plotnik et al. 2005, with permission from the Association for Research in Otolaryngology.) (**d–f**) Comparison of responses of two otolith units, one regular and the other irregular, to horizontal angular head rotations. Other details as above in (**a–c**) (Modified with permission from Plotnik et al. 2002. Copyright 2002, The American Physiological Society.)

obtained from stimulation of either the ipsilateral or contralateral labyrinths. Remarkably, after unilateral canal plugging, efferent-mediated excitatory responses could be produced by rotations in either direction, including those resulting in the excitation or inhibition of afferents on the unplugged side. Some of the potential mediating pathways are depicted in Fig. 6.17a. There is evidence that group *e* receives direct inputs from the ipsilateral vestibular nerve (White 1985; Li et al. 2005) and bilaterally from the vestibular nuclei (Chi et al. 2007). Because of the bilateral projections of the ipsilateral and contralateral efferent cell groups, it is easy to see how an afferent could be excited by excitatory rotations of either ear. How inhibitory rotations are converted into excitation is less clear. In Fig. 6.17a, the conversion is accomplished by disinhibition, that is, the inhibition of crossing inhibitory fibers.

Efferent-mediated rotational responses, which have also been seen in alert monkeys (Sadeghi et al. 2009), are small, typically less than 20 spikes/s even in irregular afferents. These small responses suggest that the efferent system has only weak actions when not stimulated with high-frequency shock trains. That the EVS can exert powerful effects on afferent discharge is shown by large fluctuations in background discharge in the decerebrate chinchilla (Plotnik et al. 2005). A particularly striking example is shown in Fig. 6.17b1. Even in the absence of stimulation, there are nearly periodic fluctuations in the background discharge, which ranges from less than 50 to almost 300 spikes/s. Fluctuations are unusually large in this case. More

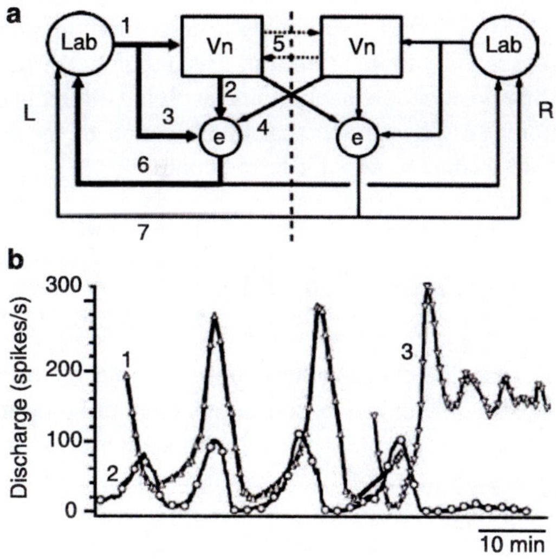

Fig. 6.17 (**a**) A bilaterally symmetric model of central efferent vestibular pathways. *Bold lines* indicate a positive-feedback loop. Labyrinth (*Lab*) sends direct projections (*3*) to ipsilateral efferent group (*e*) and indirect projections via the ipsilateral (*1*, *2*) and contralateral (*4*) vestibular nuclei (*Vn*). The Vn on the two sides are interconnected by inhibitory fibers (*5*). All other connections are excitatory. Efferent neurons to each labyrinth come from the ipsilateral (*6*) and contralateral (*7*) efferent group. (Adapted from Plotnik et al. 2005, with permission from the Association for Research in Otolaryngology.) (**b**) Large fluctuations in the background discharge of three irregular units (otolith afferents (*1* and *2*); horizontal-canal afferent (*3*) with animal stationary in the horizontal plane (Adapted from Plotnik et al. 2005, with permission from the Association for Research in Otolaryngology.)

typically, the oscillations have peak-to-peak amplitudes of 50–100 spikes/s (Fig. 6.17b2) and may be damped (Fig. 6.17b3). In all cases the fluctuations have a period lasting several minutes. Such fluctuations are confined to irregular units in decerebrates and are not seen in either regular or irregular units in anesthetized or alert preparations. Several lines of evidence suggest that the fluctuations are efferent mediated, of which the two most salient are the fact that (1) they are abolished when the vestibular nerve is cut central to the recording site and (2) the presence of a positive correlation between fluctuation amplitude and the size of type III, efferent-mediated rotation responses. A possible explanation for the fluctuations is provided by the positive feedback loops between efferents and afferents (Fig. 6.17a, thick lines): afferents and efferents are mutually excitatory. A theoretical model indicates that such a positive feedback loop could give rise to the observed periodic fluctuations (Plotnik et al. 2005). The excitation provided by efferents could also explain another finding, viz., the increase in background discharge of decerebrate, as compared to anesthetized, animals (Perachio and Correia 1983; Plotnik et al. 2005). Once again, the effect is targeted to irregular afferents, even those not showing large fluctuations in background discharge.

There can be no question that the fluctuations are an artifact of decerebration since they are not seen in alert, behaving animals (Keller 1976; Louie and Kimm

1976; Sadeghi et al. 2007, 2009). Presumably the decerebration releases the efferent cell groups from a descending tonic inhibition. Even though the fluctuations are an artifact, they show that positive feedback loops involving efferents can have a powerful impact on afferent discharge and that modification of these loops by other systems could provide a novel mode of efferent control.

6.10.3 Possible Functions of the EVS

The functions of efferent modulation have not been clearly established in any vertebrate, including mammals. One reason for a lack of progress may relate to the properties of the peripheral efferent synapse. Large responses of mammalian afferents to electrical stimulation of the EVS require several closely spaced shocks. Since single shocks are relatively ineffective, the implication is that large responses require potentiation at the peripheral efferent synapse, likely attributed to the presynaptic facilitation of neurotransmitter release and/or the amplification of postsynaptic effects. Regardless of the mechanisms involved, the efferent synapse acts as a filter, maximizing the effects of high-frequency bursts of activity in central efferent neurons and minimizing the influence of lower, tonic discharge rates. As a result of such filtering, a sensory stimulus may be quite effective in exciting efferents, yet have only a small or no influence on afferent discharge. This property suggests that recordings from efferent neurons might be more revealing than recordings from afferents. Yet, although there have been several recordings from afferent fibers in alert, behaving mammals (Keller 1976; Louie and Kimm 1976; Lisberger and Pavelko 1986; Cullen and Minor 2002; Sadeghi et al. 2009), the only study possibly recording from efferents in mammals was done in decerebrate, decerebellate animals (Marlinsky 1995).

Large efferent-mediated afferent responses require a high rate of efferent discharge, which is likely to occur in bursts associated with active head movements. As a specific hypothesis, it has been supposed that the resulting excitation would serve among other things to prevent the silencing of afferent discharge during rapid head movements in the inhibitory direction (Goldberg and Fernández 1980; Highstein 1991). Afferent recordings in monkeys free to move their heads have been used to test this hypothesis (Cullen and Minor 2002). Contrary to expectations, there were no differences between active and passive head movements. The conclusion can be related to recordings from head-restrained animals (Keller 1976; Louie and Kimm 1976). In the latter studies, it was found that vestibular-nerve discharge was insensitive to eye saccades. When an animal makes a gaze saccade, there is usually a coordinated head and eye movement (Sadeghi et al. 2007). When the head is restrained, a head torque can be measured and indicates that the motor strategy still includes a head movement. So the head-restrained studies imply that efferents are not sufficiently excited by efference-copy signals to affect afferent discharge.

These last results imply that active head movements do not lead to an efferent modulation of afferent discharge in mammals over and above the vestibular stimulation they produce. It remains possible that active head movements play a role in other vertebrates, specifically in red-eared turtles. Recall that BT/BM units are characterized by much higher rotational gains than BP and CD units (Sect. 6.6.4) (Brichta and Goldberg 2000b). The former units are inhibited by EVS stimulation, whereas the latter units are excited. One implication of the difference in rotational gains is that BT/BM afferents have limited dynamic ranges, responding well to small head movements (<10°/s), but approaching saturation during modest head movements of 20–50°/s. In addition, responses become highly distorted for even modest head velocities. Excited (BP and CD) units, in contrast, have dynamic ranges in excess of 100°/s. Given their responses to head rotations, BT/BM units would seem suited to monitor the small head movements characterizing postural control during quiet standing. BP and CD units, on the other hand, could provide useful information during rapid, voluntary head movements, for example, during prey striking. If one assumes that efferents in turtles fire in anticipation of rapid head movements, the crista could be switched from a postural to a volitional mode. It must be emphasized that the suggestion is speculative, as there have been no observations of either afferent or efferent discharge in behaving turtles. Nor would the scheme work in the oyster toadfish because the high-gain afferents, which might function in postural control, are excited, rather than inhibited by efferents (Sect. 6.6.2) (Boyle and Highstein 1990b; Boyle et al. 1991).

Most speculations about efferent function have focused on the rapid modification of afferent discharge. One theme that emerges from studies in mammals and in other vertebrates is that any such actions are likely to be nonspecific, affecting afferent discharge in several vestibular organs similarly. This theme is exemplified in lower vertebrates by the anatomical organization of both central and peripheral efferent pathways. A similar conclusion in mammals is suggested by the fact that efferent-mediated afferent responses from all organs are similar for head rotations in either direction and in several canal planes (Plotnik et al. 2002).

In addition to rapid modifications of afferent discharge, it is quite possible that the EVS could also serve a slower, modulatory role, shaping the function of the peripheral organs during development and/or during adult life. An indication of this possibility is the presence of slow responses, which suggests that some efferent actions are mediated by metabotropic, rather than by ionotropic, receptors (see Sect. 6.8.4). Of particular interest in this regard are the efferent responses of regularly discharging afferents. These are quite slow and quite small. Were the only function of the efferents to modify spike discharge on a relatively fast time scale, it would be difficult to understand why the peripheral/extrastriolar zones of the cristae and maculae, where regular fibers reside (Baird et al. 1988; Goldberg et al. 1990a), receive as rich or richer an efferent innervation as central/striolar zones (Nomura et al. 1965; Lysakowski and Goldberg 1997; Purcell and Perachio 1997).

A possible target of efferent modulation relates to the balance between the vestibular organs on the two sides. Such a balance is important functionally. When the activity of one labyrinth is depressed, for example, by some pathological process, it is necessary to redress or compensate for the imbalance. Much of compensation involves central mechanisms that rebalance the activity of the two vestibular nuclei (Curthoys and Halmagyi 1995). But because of the bilateral organization of the efferent pathways (Figs. 6.6 and 6.17a), the activity of one labyrinth can influence that of the other labyrinth. From this relationship, it is conceivable that some of the compensation could involve an efferent modification of afferent activity. This hypothesis has been investigated in behaving monkeys by recording from one vestibular nerve after destruction of the contralateral labyrinth. Remarkably, no obvious change in overall activity was detected (Sadeghi et al. 2007).

6.11 Summary

This chapter summarizes our current knowledge of the EVS, including its neuroanatomical organization, the responses of afferents to electrical stimulation of the EVS, the extent to which these responses are the result of actions on hair cells or afferent terminals, the neurotransmitters and neurotransmitter receptors involved, and the efferent modulation of afferent responses to natural stimulation. While studies of mammals were emphasized, nonmammalian vertebrates were also considered. When compared to other hair-cell systems, the peripheral actions of the EVS are unusually diverse and involve excitation and inhibition targeted to hair cells and afferent terminals. Both ionotropic and metabotropic receptors participate.

Several questions remain, of which two may be mentioned here. (1) ACh is a neurotransmitter at efferent synapses. Other transmitters, including CGRP, ATP, and nitric oxide, may be colocalized with ACh in the same efferent terminals. The physiological effects of these agents remain to be determined, as does the role of various cholinergic receptors. (2) To explore possible functions of the EVS, the responses of efferents and the efferent modification of afferent discharge have been studied. Despite this effort, the functions of the EVS have remained elusive. It is suggested that a promising approach to a study of function would be to record from efferent neurons under a variety of behavioral circumstances.

Acknowledgments Research in the authors' laboratories is supported by Grants NIDCD DC08981 (JCH), NIDCD DC02521 (AL), and NIDCD DC02508 (JMG). The authors also thank Erika Bruss for providing the animal illustrations used in several of the figures.

References

Abbracchio MP, Burnstock G, Verkhratsky A, Zimmermann H (2009) Purinergic signalling in the nervous system: an overview. Trends Neurosci 32:19–29

Albuquerque EX, Pereira EF, Alkondon M, Rogers SW (2009) Mammalian nicotinic acetylcholine receptors: from structure to function. Physiol Rev 89:73–120

Almanza A, Navarrete F, Vega R, Soto E (2007) Modulation of voltage-gated Ca^{2+} current in vestibular hair cells by nitric oxide. J Neurophysiol 97:1188–1195

Anderson AD, Troyanovskaya M, Wackym PA (1997) Differential expression of alpha2–7, alpha9 and beta2–4 nicotinic acetylcholine receptor subunit mRNA in the vestibular end organs and Scarpa's ganglia of the rat. Brain Res 778:409–413

Andrianov GN, Ryzhova IV (1999) Opioid peptides as possible neuromodulators of the afferent synaptic transmission in the frog semicircular canal. Neuroscience 93:801–806

Annoni JM, Cochran SL, Precht W (1984) Pharmacology of the vestibular hair cell-afferent fiber synapse in the frog. J Neurosci 4:2106–2116

Art JJ, Fettiplace R, Fuchs PA (1984) Synaptic hyperpolarization and inhibition of turtle cochlear hair cells. J Physiol 356:525–550

Ashmore JF, Russell IJ (1982) Effects of efferent nerve stimulation on hair cells of the frog sacculus. J Physiol 329:25P–26P

Aubert A, Norris CH, Guth PS (1994) Influence of ATP and ATP agonists on the physiology of the isolated semicircular canal of the frog (*Rana pipiens*). Neuroscience 62:963–974

Aubert A, Norris CH, Guth PS (1995) Indirect evidence for the presence and physiological role of endogenous extracellular ATP in the semicircular canal. Neuroscience 64:1153–1160

Bailey GP, Sewell WF (2000a) Calcitonin gene-related peptide suppresses hair cell responses to mechanical stimulation in the *Xenopus* lateral line organ. J Neurosci 20:5163–5169

Bailey GP, Sewell WF (2000b) Pharmacological characterization of the CGRP receptor in the lateral line organ of *Xenopus laevis*. J Assoc Res Otolaryngol 1:82–88

Baird RA, Desmadryl G, Fernández C, Goldberg JM (1988) The vestibular nerve of the chinchilla. II. Relation between afferent response properties and peripheral innervation patterns in the semicircular canals. J Neurophysiol 60:182–203

Barbas-Henry HA, Lohman AHM (1988) Primary projections and efferent cells of the VIIIth cranial nerve in the monitor lizard, *Varanus exanthematicus*. J Comp Neurol 277:234–249

Bean BP (2007) The action potential in mammalian central neurons. Nat Rev Neurosci 8:451–465

Behrend O, Schwark C, Kunihiro T, Strupp M (1997) Cyclic GMP inhibits and shifts the activation curve of the delayed-rectifier (I[K1]) of type I mammalian vestibular hair cells. Neuroreport 8:2687–2690

Bernard C, Cochran SL, Precht W (1985) Presynaptic actions of cholinergic agents upon the hair cell-afferent fiber synapses in the vestibular labyrinth of the frog. Brain Res 338:225–236

Blanks RHI, Precht W (1976) Functional characterization of primary vestibular afferents in the frog. Exp Brain Res 25:369–390

Bobbin RP, Konishi T (1971) Acetylcholine mimics crossed olivocochlear bundle stimulation. Nature 231:222–223

Bobbin RP, Konishi T (1974) Action of cholinergic and anticholinergic drugs at the crossed olivochlear bundle-hair cell junction. Acta Otolaryngol (Stockh) 77:55–65

Bowery NG, Bettler B, Froestl W, Gallagher JP, Marshall F, Raiteri M, Bonner TI, Enna SJ (2002) International Union of Pharmacology. XXXIII. Mammalian gamma-aminobutyric acid (B) receptors: structure and function. Pharmacol Rev 54:247–264

Boyle R, Highstein SM (1990a) Resting discharge and response dynamics of horizontal semicircular canal afferents in the toadfish, *Opsanus tau*. J Neurosci 10:1557–1569

Boyle R, Highstein SM (1990b) Efferent vestibular system in the toadfish: action upon horizontal semicircular canal afferents. J Neurosci 10:1570–1582

Boyle R, Carey JP, Highstein SM (1991) Morphological correlates of response dynamics and efferent stimulation in horizontal semicircular canal afferents of the toadfish, *Opsanus tau*. J Neurophysiol 66:1504–1521

Boyle RD, Rabbitt RD, Highstein SM (2009) Efferent control of hair cell and afferent responses in the semicircular canals. J Neurophysiol 102:1513–1525

Brichta AM, Goldberg JM (1996) Afferent and efferent responses from morphological fiber classes in the turtle posterior crista. Ann N Y Acad Sci 781:183–195

Brichta AM, Goldberg JM (2000a) Morphological identification of physiologically characterized afferents innervating the turtle posterior crista. J Neurophysiol 83:1202–1223

Brichta AM, Goldberg JM (2000b) Responses to efferent activation and excitatory response-intensity relations of turtle posterior-crista afferents. J Neurophysiol 83:1224–1242

Brichta AM, Peterson EH (1994) Functional architecture of vestibular primary afferents from the posterior semicircular canal of a turtle, *Pseudemys (Trachemys) scripta elegans*. J Comp Neurol 344:481–507

Brontë-Stewart HM, Lisberger SG (1994) Physiological properties of vestibular primary afferents that mediate motor learning and normal performance of the vestibulo-ocular reflex in monkeys. J Neurosci 14:1290–1308

Brown DA (1988) M currents: an update. Trends Neurosci 11:294–299

Brown DA, Passmore GM (2009) Neural KCNQ (Kv7) channels. Br J Pharmacol 156: 1185–1195

Carpenter MB, Chang L, Pereira AB, Hersch LB, Bruce G, Wu J-Y (1987) Vestibular and cochlear efferent neurons in the monkey identified by immunocytochemical methods. Brain Res 408:275–280

Caulfield MP, Birdsall NJ (1998) International Union of Pharmacology. XVII. Classification of muscarinic acetylcholine receptors. Pharmacol Rev 50:279–290

Chebib M, Johnston GAB (1999) The "ABC" of GABA receptors: a brief review. Clin Exp Pharmacol Physiol 26:937–940

Chen JW, Eatock RA (2000) Major potassium conductance in type I hair cells from rat semicircular canals: characterization and modulation by nitric oxide. J Neurophysiol 84:139–151

Chi FL, Jiao Y, Liu HJ, Wang J, Shi Y, Barr JJ (2007) Retrograde neuron tracing with microspheres reveals projection of CGRP-immunolabeled vestibular afferent neurons to the vestibular efferent nucleus in the brainstem of rats. Neuroendocrinology 85:131–138

Churchill JA, Schuknecht HF, Doran R (1956) Acetylcholinesterase activity in the cochlea. Laryngoscope 66:1–15

Claas B, Fritzsch B, Münz H (1981) Common efferents to lateral line and labyrinthine hair cells in aquatic vertebrates. Neurosci Lett 27:231–235

Connor JA (1978) Slow repetitive activity from fast conductance changes in neurons. Fed Proc 37:2139–2145

Cooper JR, Bloom FE, Roth RH (2002) The biochemical basis of neuropharmacology. Raven Press, New York

Cullen KE, Minor LB (2002) Semicircular canal afferents similarly encode active and passive head-on-body rotations: implications for the role of vestibular efference. J Neurosci 22(RC226):1–7

Curthoys IS (1982) The response of primary horizontal semicircular canal neurons in the rat and guinea pig to angular acceleration. Exp Brain Res 47:286–294

Curthoys IS, Halmagyi GM (1995) Vestibular compensation: a review of oculomotor, neural, and oculomotor consequences of unilateral vestibular loss. J Vestib Res 5:67–107

Dawkins R, Keller SL, Sewell WF (2005) Pharmacology of acetylcholine-mediated cell signaling in the lateral line organ following efferent stimulation. J Neurophysiol 93:2541–2551

de Robertis ED, Bennett HS (1955) Some features of the submicroscopic morphology of synapses in frog and earthworm. J Biophys Biochem Cytol 1:47–58

Dechesne C, Raymond J, Sans A (1984) The efferent vestibular system in the cat: a horseradish peroxidase and fluorescent retrograde tracer study. Neuroscience 11:893–901

Delmas P, Brown DA (2005) Pathways modulating neural KCNQ/M (Kv7) potassium channels. Nat Rev Neurosci 6:850–862

Derbenev AV, Linn CL, Guth PS (2005) Muscarinic ACh receptor activation causes transmitter release from isolated frog vestibular hair cells. J Neurophysiol 94:3134–3142

Desai SS, Dhaliwal J, Lysakowski A (2004) NOS immunochemical staining in calyces in chinchilla vestibular endorgans. Association for Research in Otolaryngology Midwinter Meeting Abstracts 27:853

Desai SS, Ali H, Lysakowski A (2005a) Comparative morphology of rodent vestibular periphery. II. Cristae ampullares. J Neurophysiol 93:267–280

Desai SS, Zeh C, Lysakowski A (2005b) Comparative morphology of rodent vestibular periphery. I. Saccular and utricular maculae. J Neurophysiol 93:251–266

Desmadryl G, Dechesne CJ (1992) Calretinin immunoreactivity in chinchilla and guinea pig vestibular end organs characterizes the calyx unit subpopulation. Exp Brain Res 89:105–108

Desmedt JE, Monaco P (1961) Mode of action of the efferent olivo-cochlear bundle on the inner ear. Nature 192:1263–1265

Dickman JD, Correia MG (1993) Bilateral communication between vestibular labyrinths in pigeons. Neuroscience 57:1097–1108

Didier A, Dupont J, Cazals Y (1990) GABA immunoreactivity of calyceal nerve endings in the vestibular system of the guinea pig. Cell Tissue Res 260:415–419

Dohlman G, Farkashidy J, Salonna F (1958) Centrifugal nerve fibers to the sensory epithelium of the vestibular labyrinth. J Laryngol 72:984–991

Dowdall MJ, Boyne AF, Whittaker VP (1974) Adenosine triphosphate. A constituent of cholinergic synaptic vesicles. Biochem J 140:1–12

Drescher DG, Kerr TP, Drescher MJ (1999) Autoradiographic demonstration of quinuclidinyl benzilate binding sites in the vestibular organs of the gerbil. Brain Res 845:199–207
Drescher DG, Ramakrishnan NA, Drescher MJ, Chun W, Wang X, Myers SF, Green GE, Sadrazodi K, Karadaghy AA, Poopat N, Karpenko AN, Khan KM, Hatfield JS (2004) Cloning and characterization of alpha9 subunits of the nicotinic acetylcholine receptor expressed by saccular hair cells of the rainbow trout (*Oncorhynchus mykiss*). Neuroscience 127:737–752
Eglen RM (2005) Muscarinic receptor subtype pharmacology and physiology. Prog Med Chem 43:105–136
Elgoyhen AB, Johnson DS, Boutler J, Vetter DE, Heinemann S (1994) α9: an acetylcholine receptor with novel pharmacological properties expressed in rat chochlear hair cells. Cell 79:705–715
Elgoyhen AB, Vetter DE, Katz E, Rothlin CV, Heinemann SF, Boulter J (2001) Alpha10, a determinant of nicotinic cholinergic receptor function in mammalian vestibular and cochlear mechanosensory hair cells. Proc Natl Acad Sci U S A 98:3501–3506
Engström H (1958) On the double innervation of the sensory epithelia of the inner ear. Acta Otolaryngol (Stockh) 49:109–118
Engström H, Wersäll J (1958) The ultrastructural organization of the organ of Corti and of the vestibular sensory epithelia. Exp Cell Res 5:460–492
Eybalin M (1993) Neurotransmitters and neuromodulators of the mammalian cochlea. Physiol Rev 73:309–373
Ezure K, Cohen MS, Wilson VJ (1983) Response of cat semicircular canal afferents to sinusoidal polarizing currents: implications for input-output properties of second-order neurons. J Neurophysiol 49:639–648
Favre D, Sans A (1979) Morphological changes in afferent vestibular hair cell synapses during the postnatal development of the cat. J Neurocytol 8:765–775
Fernández C, Goldberg JM (1976) Physiology of peripheral neurons innervating otolith organs of the squirrel monkey. III. Response dynamics. J Neurophysiol 39:996–1008
Fernández C, Baird RA, Goldberg JM (1988) The vestibular nerve of the chinchilla. I. Peripheral innervation patterns in the horizontal and superior semicircular canals. J Neurophysiol 60:167–181
Fernández C, Goldberg JM, Baird RA (1990) The vestibular nerve of the chinchilla. III. Peripheral innervation patterns in the utricular macula. J Neurophysiol 63:767–780
Fernández C, Lysakowski A, Goldberg JM (1995) Hair cell counts and afferent innervation patterns in the cristae ampullares of the squirrel monkey with a comparison to the chinchilla. J Neurophysiol 73:1253–1269
Fex J (1959) Augmentation of cochlear microphonic by stimulation of efferent fibres to the cochlea; preliminary report. Acta Otolaryngol 50:540–541
Fex J (1962) Auditory activity in centrifugal and centripetal cochlear fibres in cat. Acta Physiol Scand Suppl 189:1–68
Flock Å, Lam DM (1974) Neurotransmitter synthesis in inner ear and lateral line sense organs. Nature 249:142–144
Flock Å, Russell IJ (1973) The post-synaptic action of efferent fibres in the lateral line organ of the burbot *Lota lota*. J Physiol 235:591–605
Flock Å, Russell IJ (1976) Inhibition by efferent nerve fibres: action on hair cells and afferent synaptic transmission in the lateral line canal organ of the burbot *Lota lota*. J Physiol 257:45–62
Flores A, Soto E, Vega R (2001) Nitric oxide in the afferent synaptic transmission of the axolotl vestibular system. Neuroscience 103:457–464
Fontaine B, Klarsfeld A, Hökfelt T, Changeux JP (1986) Calcitonin gene-related peptide, a peptide present in spinal cord motoneurons, increases the number of acetylcholine receptors in primary cultures of chick embryo myotubes. Neurosci Lett 71:59–65
Fritzsch B (1981) Efferent neurons to the labyrinth of *Salamandra salamandra* as revealed by retrograde transport of horseradish peroxidase. Neurosci Lett 26:191–196
Fritzsch B, Crapon De Caprona D (1984) The origin of centrifugal inner ear fibers of *Gymnophions* (amphibia). A horseradish peroxidase study. Neurosci Lett 45:131–136
Fritzsch B, Dubuc R, Ohta Y, Grillner S (1989) Efferents to the labyrinth of the river lamprey (*Lampetra fluviatilis*) as revealed with retrograde tracing techniques. Neurosci Lett 96:241–246

Fuchs PA, Murrow BW (1992a) Cholinergic inhibition of short (outer) hair cells of the chick's cochlea. J Neurosci 12:800–809
Fuchs PA, Murrow BW (1992b) A novel cholinergic receptor mediates inhibition of chick cochlear hair cells. Proc R Soc Lond B Biol Sci 248:35–40
Fukuda K, Higashida H, Kubo T, Maeda A, Akiba I, Bujo H, Mishina M, Numa S (1988) Selective coupling with K^+ currents of muscarinic acetylcholine receptor subtypes in NG108-15 cells. Nature 335:355–358
Gacek RR (1960) Efferent component of the vestibular nerve. In: Rasmussen GL, Windle WF (eds) Neural mechanisms of the auditory and vestibular systems. Charles C. Thomas, Springfield, pp 276–284
Gacek RR, Lyon M (1974) The localization of vestibular efferent neurons in the kitten with horseradish peroxidase. Acta Otolaryngol (Stockh) 77:92–101
Gacek RR, Rasmussen GL (1961) Fiber analysis of statoacoustic nerve of guinea pig, cat and monkey. Anat Rec 139:455–463
Galambos R (1956) Suppression of the auditory nerve activity by stimulation of efferent fibers to the cochlea. J Neurophysiol 19:424–437
Garthwaite J (2008) Concepts of neural nitric oxide-mediated transmission. Eur J Neurosci 27:2783–2802
Gleisner L, Henriksson NG (1964) Efferent and afferent activity pattern in the vestibular nerve of the frog. Acta Otolaryngol Suppl (Stockh) 192:90–103
Goldberg JM (2000) Afferent diversity and the organization of central vestibular pathways. Exp Brain Res 130:277–297
Goldberg JM, Chatlani S (2009) Repetitive discharge in vestibular nerve afferents. Association for Research in Otolaryngology Midwinter Meeting Abstracts 32:1106
Goldberg JM, Fernández C (1971) Physiology of peripheral neurons innervating semicircular canals of the squirrel monkey. I. Resting discharge and response to constant angular accelerations. J Neurophysiol 34:635–660
Goldberg JM, Fernández C (1975) Responses of peripheral vestibular neurons to angular and linear accelerations in the squirrel monkey. Acta Otolaryngol (Stockh) 80:101–110
Goldberg JM, Fernández C (1977) Conduction times and background discharge of vestibular afferents. Brain Res 122:545–550
Goldberg JM, Fernández C (1980) Efferent vestibular system in the squirrel monkey: anatomical location and influence on afferent activity. J Neurophysiol 43:986–1025
Goldberg JM, Fernández C, Smith CE (1982) Responses of vestibular-nerve afferents in the squirrel monkey to externally applied galvanic currents. Brain Res 252:156–160
Goldberg JM, Smith CE, Fernández C (1984) Relation between discharge regularity and responses to externally applied galvanic currents in vestibular nerve afferents of the squirrel monkey. J Neurophysiol 51:1236–1256
Goldberg JM, Desmadryl G, Baird RA, Fernández C (1990a) The vestibular nerve of the chinchilla. V. Relation between afferent discharge properties and peripheral innervation patterns in the utricular macula. J Neurophysiol 63:791–804
Goldberg JM, Lysakowski A, Fernández C (1990b) Morphophysiological and ultrastructural studies in the mammalian cristae ampullares. Hear Res 49:89–102
Goldberg JM, Brichta AM, Wackym PW (2000) Efferent vestibular system: anatomy, physiology and neurochemistry. In: Anderson JH, Beitz AJ (eds) Neurochemistry of the vestibular system. CRC Press, Boca Raton, pp 61–94
Guth PS, Norris CH, Guth SL, Quine DB, Williams WH (1986) Cholinomimetics mimic efferent effects on semicircular canal afferent activity in the frog. Acta Otolaryngol 102:194–203
Guth PS, Perin P, Norris CH, Valli P (1998) The vestibular hair cells: post-transductional signal processing. Prog Neurobiol 54:193–247
Guth PS, Holt JC, Lioudyno M, McIntosh JM, Hendricson AW, Athas GB, Shipon S (2002) The pharmacology of the non–$\alpha9/\alpha10$ nicotinic receptor of hair cells: clues as to subunit composition. Association for Research in Otolaryngology Midwinter Meeting Abstracts 25:480
Haque A, Angelaki DE, Dickman JD (2004) Spatial tuning and dynamics of vestibular semicircular canal afferents in rhesus monkeys. Exp Brain Res 155:81–90

Hartmann R, Klinke R (1980) Efferent activity in the goldfish vestibular nerve and its influence on afferent activity. Pflügers Arch 388:123–128

Hernandez CC, Zaika O, Tolstykh GP, Shapiro MS (2008) Regulation of neural KCNQ channels: signalling pathways, structural motifs and functional implications. J Physiol 586:1811–1821

Hiel H, Elgoyhen AB, Drescher DG, Morley BJ (1996) Expression of nicotinic acetylcholine receptor mRNA in the adult rat peripheral vestibular system. Brain Res 738:347–352

Highstein SM (1991) The central nervous system efferent control of the organs of balance and equilibrium. Neurosci Res 12:13–30

Highstein SM, Baker R (1985) Action of the efferent vestibular system on primary afferents in the toadfish, *Opsanus tau*. J Neurophysiol 54:370–384

Highstein SM, Baker R (1986) Organization of the efferent vestibular nuclei and nerves in the toadfish, *Opsanus tau*. J Comp Neurol 243:309–325

Hilding D, Wersäll J (1962) Cholinesterase and its relation to the nerve endings in the inner ear. Acta Otolaryngol (Stockh) 55:205–217

Hillman DE (1969) Light and electron microscopical study of the relationships between the cerebellum and the vestibular organ of the frog. Exp Brain Res 9:1–15

Hoffman LF, Honrubia V (2002) Fiber diameter distributions in the chinchilla's ampullary nerves. Hear Res 172:37–52

Holstein GR, Martinelli GP, Boyle R, Rabbitt RD, Highstein SM (2004a) Ultrastructural observations of efferent terminals in the crista ampullaris of the toadfish, *Opsanus tau*. Exp Brain Res 157:128–136

Holstein GR, Martinelli GP, Henderson SC, Friedrich VL Jr, Rabbitt RD, Highstein SM (2004b) Gamma-aminobutyric acid is present in a spatially discrete subpopulation of hair cells in the crista ampullaris of the toadfish, *Opsanus tau*. J Comp Neurol 471:1–10

Holstein GR, Rabbitt RD, Martinelli GP, Friedrich VL Jr, Boyle RD, Highstein SM (2004c) Convergence of excitatory and inhibitory hair cell transmitters shapes vestibular afferent responses. Proc Natl Acad Sci U S A 101:15766–15771

Holt JC (2008) The effect of efferent stimulation on an afferent's response to sinusoidal indentation in the turtle posterior crista. Association for Research in Otolaryngology Midwinter Meeting Abstracts 31:1249

Holt JC, Lioudyno M, Athas G, Garcia MM, Perin P, Guth PS (2001) The effect of proteolytic enzymes on the alpha9-nicotinic receptor-mediated response in isolated frog vestibular hair cells. Hear Res 152:25–42

Holt JC, Lioudyno M, Guth PS (2003) A pharmacologically distinct nicotinic ACh receptor is found in a subset of frog semicircular canal hair cells. J Neurophysiol 90:525–553

Holt JC, Lysakowski A, Goldberg JM (2006) Mechanisms of efferent-mediated responses in the turtle posterior crista. J Neurosci 26:13180–13193

Holt JC, Chatlani S, Lysakowski A, Goldberg JM (2007) Quantal and nonquantal transmission in calyx-bearing fibers of the turtle posterior crista. J Neurophysiol 98:1083–1101

Holt JC, Klapczynski M, Price SD, McIntosh JM, Goldberg JM, Lysakowski A (2008) Efferent cholinergic receptors in the turtle posterior crista. Association for Research in Otolaryngology Midwinter Meeting Abstracts 31:1108

Holt JC, Jordan P, Shah A, Barsz K (2010) Efferent-mediated excitation of turtle calyx-bearing afferents does not involve α9/10nAChRs. Association for Research in Otolaryngology Midwinter Meeting Abstracts 33:543

Hullar TE, Della Santina CC, Hirvonen T, Lasker DM, Carey JP, Minor LB (2005) Responses of irregularly discharging chinchilla semicircular canal vestibular-nerve afferents during high-frequency head rotations. J Neurophysiol 93:2777–2786

Hurley KM, Gaboyard S, Zhong M, Price SD, Wooltorton JR, Lysakowski A, Eatock RA (2006) M-like K^+ currents in type I hair cells and calyx afferent endings of the developing rat utricle. J Neurosci 26:10253–10269

Ireland PE, Farakashidy J (1961) Studies on the efferent innervation of the vestibular end-organs. Trans Am Otol Soc 49:20–30

Ishiyama A, López I, Wackym PA (1994) Subcellular innervation patterns of the calcitonin gene related peptidergic efferent terminals in the chinchilla vestibular periphery. Otolaryngol Head Neck Surg 111:385–395

Iurato S, Luciano L, Pannese E, Reale E (1972) Efferent vestibular fibers in mammals: morphological and histochemical aspects. Prog Brain Res 37:429–443

Jordan P, Shah A, Barsz K, Holt JC (2010) Activation of muscarinic ACh receptors underlies efferent-mediated slow excitation in calyx-bearing afferents of the turtle posterior semicircular canal. Association for Research in Otolaryngology Midwinter Meeting Abstracts 33:555

Keller EL (1976) Behavior of horizontal semicircular canal afferents in alert monkey during vestibular and optokinetic stimulation. Exp Brain Res 24:459–471

Kharkovets T, Hardelin J-P, Safieddine S, Schweizer M, El-Amraoui A, Petit C, Jentsch TJ (2000) KCNQ4, a K^+ channel mutated in a form of dominant deafness, is expressed in the inner ear and the central auditory pathway. Proc Nat Acad Sci U S A 97:4333–4338

Klinke R, Schmidt CL (1968) Efferente Impulse in Nervus vestibularis bei Reizung des kontralateralen Otolithenorgans. Pflügers Arch 304:183–188

Klinke R, Schmidt C (1970) Efferent influence on the vestibular organ during active movement of the body. Pflügers Arch 318:325–332

Kong WJ, Egg G, Hussl B, Spoendlin H, Schrott-Fischer A (1994) Localization of CHaT-like immunoreactivity in the vestibular endorgans of the rat. Hear Res 75:191–200

Kong WJ, Hussl B, Thumfart WF, Schrott-Fischer A (1998) Ultrastructural localization of GABA-like immunoreactivity in the vestibular periphery of the rat. Acta Otolaryngol 118:90–95

Lasker DM, Han GC, Park HJ, Minor LB (2008) Rotational responses of vestibular-nerve afferents innervating the semicircular canals in the C57BL/6 mouse. J Assoc Res Otolaryngol 9:334–348

Leonard RB, Kevetter GA (2002) Molecular probes of the vestibular nerve. I. Peripheral termination patterns of calretinin, calbindin and peripherin containing fibers. Brain Res 928:8–17

Li C, Zhang YK, Guan ZL, Shum DK, Chan YS (2005) Vestibular afferent innervation in the vestibular efferent nucleus of rats. Neurosci Lett 385:36–40

Li GQ, Kevetter GA, Leonard RB, Prusak DJ, Wood TG, Correia MJ (2007) Muscarinic acetylcholine receptor subtype expression in avian vestibular hair cells, nerve terminals and ganglion cells. Neuroscience 146:384–402

Lincoln J, Hoyle CHV, Burnstock G (1997) Nitric oxide in health and disease. Cambridge University Press, Cambridge

Lindeman HH (1969) Studies on the morphology of the sensory regions of the vestibular apparatus. Ergeb Anat Entwicklungsgesch 42:1–113

Lioudyno MI, Verbitsky M, Glowatzki E, Holt JC, Boulter J, Zadina JE, Elgoyhen AB, Guth PS (2002) The alpha9/alpha10-containing nicotinic ACh receptor is directly modulated by opioid peptides, endomorphin-1, and dynorphin B, proposed efferent cotransmitters in the inner ear. Mol Cell Neurosci 20:695–711

Lioudyno M, Hiel H, Kong JH, Katz E, Waldman E, Parameshwaran-Iyer S, Glowatzki E, Fuchs PA (2004) A "synaptoplasmic cistern" mediates rapid inhibition of cochlear hair cells. J Neurosci 24:11160–11164

Lisberger SG, Pavelko TA (1986) Vestibular signals carried by pathways subserving plasticity of the vestibulo-ocular reflex in monkeys. J Neurosci 6:346–354

Llinás R, Precht W (1969) The inhibitory vestibular efferent system and its relation to the cerebellum in the frog. Exp Brain Res 9:16–29

López I, Juiz JM, Altschuler RA, Meza G (1990) Distribution of GABA-like immunoreactivity in guinea pig vestibular cristae ampullaris. Brain Res 530:170–175

López I, Wu JY, Meza G (1992) Immunocytochemical evidence for an afferent GABAergic neurotransmission in the guinea pig vestibular system. Brain Res 589:341–348

Louie AW, Kimm J (1976) The response of 8th nerve fibers to horizontal sinusoidal oscillation in the alert monkey. Exp Brain Res 24:447–457

Luebke AE, Maroni PD, Guth SM, Lysakowski A (2005) Alpha-9 nicotinic acetylcholine receptor immunoreactivity in the rodent vestibular labyrinth. J Comp Neurol 492:323–333

Lustig LR, Hiel H, Fuchs PA (1999) Vestibular hair cells of the chick express the nicotinic acetylcholine receptor subunit alpha 9. J Vestib Res 9:359–367

Lysakowski A (1996) Synaptic organization of the crista ampullaris in vertebrates. Ann N Y Acad Sci 781:164–182

Lysakowski A, Goldberg JM (1997) A regional ultrastructural analysis of the cellular and synaptic architecture in the chinchilla cristae ampullares. J Comp Neurol 389:419–443

Lysakowski A, Goldberg JM (2004) Morphophysiology of the vestibular periphery. In: Highstein SM, Popper A, Fay RR (eds) The vestibular system. Springer, New York, pp 57–152

Lysakowski A, Goldberg JM (2008) Ultrastructural analysis of the cristae ampullares in the squirrel monkey (*Saimiri sciureus*). J Comp Neurol 511:47–64

Lysakowski A, Singer M (2000) Nitric oxide synthase localized in a subpopulation of vestibular efferents with NADPH diaphorase histochemistry and nitric oxide synthase immunohistochemistry. J Comp Neurol 427:508–521

Lysakowski A, Minor LB, Fernández C, Goldberg JM (1995) Physiological identification of morphologically distinct afferent classes innervating the cristae ampullares of the squirrel monkey. J Neurophysiol 73:1270–1281

Lysakowski A, Alonto A, Jacobson L (1999) Peripherin immunoreactivity labels small diameter vestibular "bouton" afferents in rodents. Hear Res 133:149–154

Marco J, Lee W, Suarez C, Hoffman L, Honrubia V (1993) Morphologic and quantitative study of the efferent vestibular system in the chinchilla: 3-D reconstruction. Acta Otolaryngol (Stockh) 113:229–234

Marlinski V, Plotnik M, Goldberg JM (2004) Efferent actions in the chinchilla vestibular labyrinth. J Assoc Res Otolaryngol 5:126–143

Marlinsky VV (1995) The effect of somatosensory stimulation on second-order and efferent vestibular neurons in the decerebrate decerebellate guinea-pig. Neuroscience 69:661–669

Matsubara A, Usami S, Fujita S, Shinkawa H (1995) Expression of substance P, CGRP, and GABA in the vestibular periphery, with special reference to species differences. Acta Otolaryngol Suppl 519:248–252

Matsuda Y (1996) Localization of choline acetyltransferase and calcitonin gene-related peptide immunoreactivities in the vestibular end-organs of the guinea pig. Osaka City Med J 42:61–76

McCue MP, Guinan JJ Jr (1994) Influence of efferent stimulation on acoustically responsive vestibular afferents in the cat. J Neurosci 14:6071–6083

Meredith GE (1988) Comparative view of the central organization of afferent and efferent circuitry for the inner ear. Acta Biol Hung 39:229–249

Meredith GE, Roberts BL (1987) Distribution and morphological characteristics of efferent neurons innervating end organs in the ear and lateral line of the european eel. J Comp Neurol 265:494–506

Millar NS, Gotti C (2009) Diversity of vertebrate nicotinic acetylcholine receptors. Neuropharmacology 56:237–246

Moncada S, Palmer RM, Higgs EA (1991) Nitric oxide: physiology, pathophysiology, and pharmacology. Pharmacol Rev 43:109–142

Narins PM, Lewis ER (1984) The vertebrate ear as an exquisite seismic sensor. J Acoust Soc Am 76:1384–1387

New HV, Mudge AW (1986) Calcitonin gene-related peptide regulates muscle acetylcholine receptor synthesis. Nature 323:809–811

Nomura Y, Gacek RR, Balogh KJ (1965) Efferent innervation of vestibular labyrinth. Arch Otolaryngol 81:335–339

Ohno K, Takeda N, Yamano M, Matsunaga T, Tohyama M (1991) Coexistence of acetylcholine and calcitonin gene-related peptide in the vestibular efferent neurons in the rat. Brain Res 566:103–107

Ohno K, Takeda N, Kiyama H, Kato H, Fujita S, Matsunaga T, Tohyama M (1993) Synaptic contact between vestibular afferent nerve and cholinergic efferent terminal: its putative mediation by nicotinic receptors. Brain Res Mol Brain Res 18:343–346

Oliver D, Klocker N, Schuck J, Baukrowitz T, Ruppersberg JP, Fakler B (2000) Gating of Ca^{2+}-activated K^+ channels controls fast inhibitory synaptic transmission at auditory outer hair cells. Neuron 26:595–601

Palay SL, Palade GE (1955) The fine structure of neurons. J Biophys Biochem Cytol 1:69–88

Pellergrini M, Ceccotti F, Magherini P (1985) The efferent vestibular neurons in the toad (*Bufo bufo L.*): their location and morphology. A horseradish peroxidase study. Brain Res 344:1–8
Perachio AA, Correia MJ (1983) Responses of semicircular canal and otolith afferents to small angle static head tilts in the gerbil. Brain Res 280:287–298
Perachio AA, Kevetter GA (1989) Identification of vestibular efferent neurons in the gerbil: histochemical and retrograde labelling. Exp Brain Res 78:315–332
Pérez C, Limón A, Vega R, Soto E (2009a) The muscarinic inhibition of the potassium M-current modulates the action-potential discharge in the vestibular primary-afferent neurons of the rat. Neuroscience 158:1662–1674
Pérez C, Vega R, Soto E (2009b) Phospholipase C-mediated inhibition of the M-potassium current by muscarinic-receptor activation in the vestibular primary-afferent neurons of the rat. Neurosci Lett 468:238–242
Plotnik M, Marlinski V, Goldberg JM (2002) Reflections of efferent activity in rotational responses of chinchilla vestibular afferents. J Neurophysiol 88:1234–1244
Plotnik M, Marlinski V, Goldberg JM (2005) Efferent-mediated fluctuations in vestibular nerve discharge: a novel, positive-feedback mechanism of efferent control. J Assoc Res Otolaryngol 6:311–323
Popper P, Wackym PA (2001) Opioid peptides and receptors in the vestibular epithelia in rats. In: 31st SFN meeting, abstract 298.28
Popper P, Cristobal R, Wackym PA (2004) Expression and distribution of mu opioid receptors in the inner ear of the rat. Neuroscience 129:225–233
Precht W, Llinás R, Clarke M (1971) Physiological responses of frog vestibular fibers to horizontal angular rotation. Exp Brain Res 13:378–407
Prigioni I, Valli P, Casella C (1983) Peripheral organization of the vestibular efferent system in the frog: an electrophysiological study. Brain Res 269:83–90
Purcell IM, Perachio AA (1997) Three-dimensional analysis of vestibular efferent neurons innervating semicircular canals of the gerbil. J Neurophysiol 78:3234–3248
Rasmussen GL (1946) The olivary peduncle and other fiber connections of the superior olivary complex. J Comp Neurol 84:141–219
Rasmussen GL (1953) Further observations of the efferent cochlear bundle. J Comp Neurol 99:61–74
Rasmussen GL, Gacek RR (1958) Concerning the question of the efferent fiber component of the vestibular nerve of the cat. Anat Rec 130:361–362
Rennie KJ (2002) Modulation of the resting potassium current in type I vestibular hair cells by cGMP. In: Berlin CI, Hood LJ, Ricci AJ (eds) Hair cell micromechanics and otoacoustic emissions. Singular Press, Clifton Park, pp 79–89
Rennie KJ, Ashmore JF (1993) Effects of extracellular ATP on hair cells isolated from the guinea-pig semicircular canals. Neurosci Lett 160:185–189
Roberts BL, Russell IJ (1972) The activity of lateral line efferent neurones in stationary and swimming dogfish. J Exp Biol 57:433–448
Rosenfeld MG, Mermod JJ, Amara SG, Swanson LW, Sawchenko PE, Rivier J, Vale WW, Evans RM (1983) Production of a novel neuropeptide encoded by the calcitonin gene via tissue-specific RNA processing. Nature 304:129–135
Rossi G, Cortesina G (1965) The "efferent cochlear and vestibular system" in *Lepus cuniculus L.* Acta Anat 60:362–381
Rossi ML, Martini M (1991) Efferent control of posterior canal afferent receptor discharge in the frog labyrinth. Brain Res 555:123–134
Rossi ML, Prigioni I, Valli P, Casella C (1980) Activation of the efferent system in the isolated frog labyrinth: effects on the afferent EPSPs and spike discharge recorded from single fibres of the posterior nerve. Brain Res 185:125–137
Rossi ML, Martini M, Pelucchi B, Fesce R (1994) Quantal nature of synaptic transmission at the cytoneural junction in the frog labyrinth. J Physiol (Lond) 478:17–35
Russell IJ (1968) Influence of efferent fibres on a receptor. Nature 219:177–178
Russell IJ (1971) The role of the lateral-line efferent system in *Xenopus laevis*. J Exp Biol 54:621–641
Ryan AF, Simmons DM, Watts AG, Swanson LW (1991) Enkephalin mRNA production by cochlear and vestibular efferent neurons in the gerbil brainstem. Exp Brain Res 87:259–267

Sadeghi SG, Minor LB, Cullen KE (2007) Response of vestibular-nerve afferents to active and passive rotations under normal conditions and after unilateral labyrinthectomy. J Neurophysiol 97:1503–1514

Sadeghi SG, Goldberg JM, Minor LB, Cullen KE (2009) Efferent-mediated responses in vestibular nerve afferents of the alert macaque. J Neurophysiol 101:988–1001

Sala O (1965) The efferent vestibular system. Electrophysiological research. Acta Otolaryngol Suppl 197:1–34

Sans A, Highstein SM (1984) New ultrastructural features in the vestibular labyrinth of the toadfish, *Opsanus tau*. Brain Res 306:191–195

Schmidt RS (1963) Frog labyrinthine efferent impulses. Acta Otolaryngol (Stockh) 56:51–64

Schneider LW, Anderson DJ (1976) Transfer characteristics of first and second order lateral canal vestibular neurons in gerbil. Brain Res 112:61–76

Schuknecht HF, Churchill JA, Doran R (1959) The localization of acetylcholine-esterase in the cochlea. AMA Arch Otolaryngol 69:549–559

Schwarz DWF, Satoh K, Schwarz IE, Hu K, Fibiger HC (1986) Cholinergic innervation of the rat's labrinth. Exp Brain Res 64:19–26

Schweitzer E (1987) Coordinated release of ATP and ACh from cholinergic synaptosomes and its inhibition by calmodulin antagonists. J Neurosci 7:2948–2956

Selyanko AA, Hadley JK, Wood IC, Abogadie FC, Jentsch TJ, Brown DA (2000) Inhibition of KCNQ1-4 potassium channels expressed in mammalian cells via M1 muscarinic acetylcholine receptors. J Physiol (Lond) 522(Pt 3):349–355

Sewell WF, Starr PA (1991) Effects of calcitonin gene-related peptide and efferent nerve stimulation on afferent transmission in the lateral line organ. J Neurophysiol 65:1158–1169

Shah A, Barsz K, Jordan P, Parker D, Holt JC (2010) Efferent actions differentially affect afferent sensitivity to sinusoidal indentation in the turtle posterior semicircular canal. Association for Research in Otolaryngology Midwinter Meeting Abstracts 33:516

Shinder ME, Purcell IM, Kaufman GD, Perachio AA (2001) Vestibular efferent neurons project to the flocculus. Brain Res 889:288–294

Silinsky EM (1975) On the association between transmitter secretion and the release of adenine nucleotides from mammalian motor nerve terminals. J Physiol 247:145–162

Smith CE, Goldberg JM (1986) A stochastic afterhyperpolarization model of repetitive activity in vestibular afferents. Biol Cybern 54:41–51

Smith CA, Rasmussen GL (1968) Nerve ending in the maculae and cristae of the chinchilla vestibule, with a special reference to the efferents. In: Graybiel A (ed) The third symposium on the role of the vestibular organs in space exploration. U.S. Government Printing Office, Washington, pp 183–201

Sousa AD, Andrade LR, Salles FT, Pillai AM, Buttermore ED, Bhat MA, Kachar B (2009) The septate junction protein caspr is required for structural support and retention of KCNQ4 at calyceal synapses of vestibular hair cells. J Neurosci 29:3103–3108

Sridhar TS, Liberman MC, Brown MC, Sewell WF (1995) A novel cholinergic "slow effect" of efferent stimulation on cochlear potentials in the guinea pig. J Neurosci 15:3667–3678

Sridhar TS, Brown MC, Sewell WF (1997) Unique postsynaptic signaling at the hair cell efferent synapse permits calcium to evoke changes on two time scales. J Neurosci 17:428–437

Stocker M (2004) Ca^{2+}-activated K^{+} channels: molecular determinants and function of the SK family. Nat Rev Neurosci 5:758–770

Strutz J (1981) The origin of centrifugal fibers to the inner ear in *Caiman crocodilus*. A HRP study. Neurosci Lett 27:65–100

Strutz J (1982a) The origin of efferent fibers to the inner ear in a turtle (*Terrapene ornata*). A horseradish peroxidase study. Brain Res 244:165–168

Strutz J (1982b) The origin of efferent vestibular fibers in the guinea pig. Acta Otolaryngol (Stockh) 94:299–305

Strutz J, Schmidt CL (1982) Acoustic and vestibular efferent neurons in the chicken (*Gallus domesticus*). Acta Otolaryngol (Stockh) 94:45–51

Sugai T, Sugitani M, Ooyama H (1991) Effects of activation of the divergent efferent fibers on the spontaneous activity of vestibular afferent fibers in the toad. Jpn J Physiol 41:217–232

Syeda SN, Lysakowski A (2001) P2X2 purinergic receptor localized in the inner ear. Association for Research in Otolaryngology Midwinter Meeting Abstracts 24:19

Takumida M, Anniko M (2002) Nitric oxide in the inner ear. Curr Opin Neurol 15:11–15

Taly A, Corringer PJ, Guedin D, Lestage P, Changeux JP (2009) Nicotinic receptors: allosteric transitions and therapeutic targets in the nervous system. Nat Rev Drug Discov 8:733–750

Tanaka M, Takeda N, Senba E, Tohyama M, Kubo T, Matsunaga T (1989) Localization, origin and fine structure of calcitonin gene related peptide-containing fibers in the vestibular end-organs of the rat. Brain Res 504:31–35

Tomko DL, Peterka RJ, Schor RH, O'Leary DP (1981) Response dynamics of horizontal canal afferents in barbiturate-anesthetized cats. J Neurophysiol 45:376–396

Troyanovskaya M, Wackym PA (1998) Evidence for three additional P2X2 purinoceptor isoforms produced by alternative splicing in the adult rat vestibular end-organs. Hear Res 126:201–209

Usami S-I, Igarashi M, Thompson G (1987) GABA-like immunoreactivity in the squirrel monkey vestibular endorgans. Brain Res 417:367–370

Usami S, Hozawa J, Tazawa M, Igarashi M, Thompson GC, Wu JY, Wenthold RJ (1989) Immunocytochemical study of the GABA system in chicken vestibular endorgans and the vestibular ganglion. Brain Res 503:214–218

Vega R, Soto E (2003) Opioid receptors mediate a postsynaptic facilitation and a presynaptic inhibition at the afferent synapse of axolotl vestibular hair cells. Neuroscience 118:75–85

Wackym PA, Popper P, Ward PH, Micevych PE (1991) Cell and molecular anatomy of nicotinic acetylcholine receptor subunits and calcitonin gene-related peptide in the rat vestibular system. Otolaryngol Head Neck Surg 105:493–510

Wackym PA, Popper P, López I, Ishiyama A, Micevych PE (1995) Expression of alpha 4 and beta 2 nicotinic acetylcholine receptor subunit mRNA and localization of alpha-bungarotoxin binding proteins in the rat vestibular periphery. Cell Biol Int 19:291–300

Wackym PA, Chen CT, Ishiyama A, Pettis RM, López IA, Hoffman L (1996) Muscarinic acetylcholine receptor subtype mRNAs in the human and rat vestibular periphery. Cell Biol Int 20:187–192

Walsh EJ, McGee J, McFadden SL, Liberman MC (1998) Long-term effects of sectioning the olivocochlear bundle in neonatal cats. J Neurosci 18:3859–3869

Warr WB (1975) Olivocochlear and vestibular efferent neurons of the feline brain stem: their location, morphology and number determined by retrograde axonal transport and acetylcholinesterase histochemistry. J Comp Neurol 161:159–182

Weisstaub N, Vetter DE, Elgoyhen AB, Katz E (2002) The alpha9alpha10 nicotinic acetylcholine receptor is permeable to and is modulated by divalent cations. Hear Res 167:122–135

Wersäll J, Bagger-Sjöbäck D (1974) Morphology of the vestibular sense organ. In: Kornhuber HH (ed) Handbook of sensory physiology, vol. VI. Vestibular system. Part 1. Basic mechanisms. Springer, Berlin, pp 123–170

White JS (1985) Fine structure of vestibular efferent neurons in the albino rat. Soc Neurosci Abstr 11:322

Whitehead MC, Morest DK (1981) Dual populations of efferent and afferent cochlear axons in the chicken. Neuroscience 6:2351–2365

Wilson VJ, Melvill Jones G (1979) Mammalian vestibular physiology. Plenum Press, New York

Wong LA, Gallagher JP (1991) Pharmacology of nicotinic receptor-mediated inhibition in rat dorsolateral septal neurones. J Physiol 436:325–346

Yagi T, Simpson NE, Markham CH (1977) The relationship of conduction velocity to other physiological properties of the cat's horizontal canal neurons. Exp Brain Res 30:587–600

Yamaguchi K, Ohmori H (1993) Suppression of the slow K^+ current by cholinergic agonists in cultured chick cochlear ganglion neurones. J Physiol (Lond) 464:213–228

Yoshida N, Shigemoto T, Sugai T, Ohmori H (1994) The role of inositol trisphosphate on ACh-induced outward currents in bullfrog saccular hair cells. Brain Res 644:90–100

Yuhas WA, Fuchs PA (1999) Apamin-sensitive, small-conductance, calcium-activated potassium channels mediate cholinergic inhibition of chick auditory hair cells. J Comp Physiol A 185:455–462

Zheng XY, Henderson D, McFadden SL, Ding DL, Salvi RJ (1999) Auditory nerve fiber responses following chronic cochlear de-efferentation. J Comp Neurol 406:72–86

Chapter 7
Development of the Inner Ear Efferent System

Dwayne Simmons, Jeremy Duncan, Dominique Crapon de Caprona, and Bernd Fritzsch

7.1 Introduction

Roberts and Meredith (1992) wrote: "For more than forty years, the efferent supply to the mammalian ear provided by the olivocochlea bundle has been an enigma," and this is still true today, in particular for the development of efferents. The inner ear efferents are so unique in their physiology, axonal course, and distribution that this adds to the mystery of their role in hearing and balance (Christopher Kirk and Smith 2003). However, analyzing the development of the vestibulocochlear efferent system may not only give us new insight into the development of this system but may also help to understand how the distribution and neurochemcial properties of the adult vestibulocochlear efferent system all come about.

To begin with, we will need to understand where the efferents originate during development because they are clearly related to the brain's only other efferent system, the motor neurons. Known similarities are their transmitter (acetylcholine [ACh]) and the fact that only motoneurons and efferents leave the brain with their axon to innervate a target outside the brain. In the hindbrain, motor neurons have been historically divided into three functional columns: the somatic, branchial, and visceral motor neurons. All hindbrain motor neurons become postmitotic near the floor plate, no matter what functional group they belong to. However, they later migrate away from their "birthplace," thereby confusing the issue of their adjacent and partially overlapping origin. Even though these three classes of motor neurons vary in distribution, morphology, and target, they are all cholinergic and project out of the hindbrain to innervate their peripheral targets, muscle fibers or neural crest derived visceral ganglia (Fritzsch and Northcutt 1993).

B. Fritzsch (✉)
Department of Biology, University of Iowa, Iowa City, IA 52242, USA
e-mail: bernd-fritzsch@uiowa.edu

D.K. Ryugo et al. (eds.), *Auditory and Vestibular Efferents*,
Springer Handbook of Auditory Research 38, DOI 10.1007/978-1-4419-7070-1_7,

The vestibulocochlear efferent population, however, does not fit into any of these three motor categories. These efferent cells project only to placodally derived tissue, such as hair cell receptors or sensory neuron dendrites. Apart from most motoneurons and shared only by some oculomotor neurons, many efferents project contralaterally, or even may project bilaterally (Fritzsch 1999). In addition, some vestibulocochlear efferents change their neurotransmitter from acetylcholine (typical for motoneurons) to one of a variety of other neurotransmitters (Simmons 2002).

This chapter provides an overview of pertinent findings and specifies uncertainties surrounding the development of the inner ear efferents. Table 7.1 provides a timeline to the major findings related to inner ear development in rodents. The developmental origin of the cholinergic inner ear efferents from branchial motoneurons, while originally met with understandable skepticism, is now accepted. The motoneuron nature of inner ear effernts suggests that these cells belong to the earliest forming neurons in the brain and are not the late developers they were thought to be. However, major issues of the molecular basis of cell migration, axonal pathfinding, and onset and dynamics of transmitter and receptor regulations remain at a descriptive level with little molecular basis for a mechanistic explanation of the somewhat disputed events.

The most profound gains in our understanding of developmental processes are in the molecular basis of segregation of facial branchial motoneurons from inner ear efferents, but even here we see primarily an increasing aggregation of data while a causal link to the actual migration and axonal navigation processes of efferents remains unclear. Some insights into the molecular basis of efferent axon navigation along inner ear afferents are beginning to emerge, but the basis of neurotrophic efferent support, which reduces efferents dramatically if they do not reach their proper target, remains elusive.

The maturation of efferent terminals on hair cells is even more contentious in part because of the differing sensitivities of the techniques employed by various investigators. Nevertheless, a theme of a developing motor neuron synapse is emerging, with much more work to be done to consolidate this current impression. In particular, more studies on the appearance of pre- and postsynaptic contact proteins and studies of knockout mouse models of these proteins are needed to go beyond the present status of the description of maturation. On the positive side, virtually all molecules that have been identified in the development of motoneurons exist now as simple or conditional knockouts, and studies using ear specific deletions of these genes should be able to resolve some of the pertinent issues around inner ear efferent development in the near future.

The goal of this chapter is to review what is currently known about the ontogeny of the vestibulocochlear neurons and their molecular specification in the hindbrain, the development of their projections to and within the ear, and their neurochemical maturation including development of the acetylcholine receptor. Additional information not presented here can be found in other recent reviews summarizing previous work (Fritzsch 1999; Puel et al. 2002; Simmons 2002).

Table 7.1 Timetable of mouse vestibulocochlear efferent development

E8–8.5	*Hoxb1* is initially expressed in all cells of rhombomere 4
E9?	Vestibulocochlear efferents become postmitotic in rhombomere 4
E9.5	Earliest known *GATA3* expression near the floor plate of rhombomere 4
E11.5	Vestibulocochlear efferent axons reach the vestibular and spiral ganglion of the otocyst
E12	Vestibulocochlear efferents become bilaterally distributed. Vestibular fibers arrive at vestibular epithelia, commissure of von Oort is formed segregating cochlear efferent from vestibular efferent axons in the ear
E13.5	Vestibulocochlear efferents complete segregation from facial branchial motor neurons through lateral migration of efferents and caudal migration of facial branchial motor neurons
E14.5	Vestibular efferents are segregated from cochlear efferents through dorsal and ventral migration near the point of segregation of facial branchial motor from efferent axons. Intraganglionic spiral bundles begin to form, the first efferent fibers grow along radial fibers toward the organ of Corti
E16.5	Medial and lateral olivocochlear efferents begin to become segregated. GABAergic expression occurs throughout the superior olivary complex
	Medial olivocochlear efferents express cholinergic properties
	Efferents reach hair cells in the vestibular sensory epithelia and inner hair cells of the cochlea
	Hair cells express *Chrna9* subunits
E18.5	Lateral olivocochlear efferents express cholinergic properties. Efferents expand through the tunnel of Corti to outer hair cells
	Segregation of medial and lateral olivocochlear efferents is nearly completed
	Hair cells begin *Chrna10* subunit expression
P0	Efferents below inner hair cells show cholinergic properties including AChE. Inner hair cell surface expression of nicotinic receptors
	Efferents below vestibular hair cells show cholinergic properties
P2	Presynaptic terminals on inner hair cells and vestibular hair cell. Efferents to inner hair cells show cholinergic properties including VAChT and ChT1 expression
	Efferents below outer hair cells show cholinergic properties, including AChE.
	Outer hair cell surface expression of nicotinic receptors
	Efferents below vestibular hair cells express CGRP
P4	Efferents to outer hair cells show mature cholinergic properties, including VAChT and ChT1 expression. Inner hair cells respond to acetylcholine
	Efferent terminals on vestibular afferent calyxes

(continued)

Table 7.1 (continued)

P7	Mature efferent synapses on outer hair cells. Lateral olivocochlear efferents express CGRP
P14	Lateral olivocochlear efferents express mature GABAergic properties. ChT1 expression reduced below inner hair cells *Chrna9* expression reduced below inner hair cells Onset of hearing

7.2 Central Development

The hindbrain is divided into rostrocaudal compartments called rhombomeres. The patterning of each rhombomere is due to variable expression of *Hox* genes. The mammalian hindbrain has four different classes of *Hox* genes. Rhombomere 4 is unique in that it is the only rhombomere where the gene *Hoxb1* is expressed. The vestibulocochlear and facial branchial motor neuron precursor cells are "born" near the floorplate of this rhombomere and will be exposed to Hoxb1 (Fig. 7.1). Their birthdates have not been determined in mice and the suggestions in rats are tentative (Altman and Bayer 1982).

Vestibulocochlear efferents remain in rhombomere 4 and migrate laterally due to the influence of sonic hedgehog (Shh) and Netrin, both produced in the floor plate, as well as other unknown factor(s) (Fig. 7.1). The separation of mammalian facial branchial motor and vestibulocochlear efferents begins early in embryonic development and is in mice completed 6 days before birth (Fritzsch et al. 1993; Bruce et al. 1997). The vestibular efferent cell bodies are fully segregated from the cochlear efferents also long before birth (Bruce et al. 1997). The vestibulocochlear efferents migrate laterally, with the vestibular efferents taking a more dorsolateral course and the cochlear efferents migrating in a ventrolateral direction, possibly under the guidance of the bHLH transcription factor Mash-1 (Tiveron et al. 2003). Because of the common origin in rhombomere 4, vestibulocochlear efferents and facial branchial motor neurons have axons that course together through the facial genu and then laterally for a short distance. In mammals, the vestibulocochlear efferent axons diverge from the facial branchial motor neurons after the facial genu. Vestibulocochlear axons exit the brain stem with the VIIIth cranial nerve (Fig. 7.1).

The mammalian vestibulocochlear efferents that project contralaterally do so most likely by extending their axon during development across the midline floor plate (Fritzsch et al. 1993; Bruce et al. 1997). This process contrasts to that of the contralateral efferents in the chicken, where cell bodies migrate across the midline leaving behind their axonal projections (Fritzsch et al. 1993; Simon and Lumsden 1993). Axonal fibers that cross to the contralateral side appear to depend on certain molecules that are present in the floor plate; for example, the ephrin receptor, EphB2, is expressed in rhombomere 4. EphB2 null mutant mice show a delayed and reduced crossing of axons in the floor plate (Cowan et al. 2000). In addition, EphB2 null mice have axons that extend for several rhombomeres caudally along the floor

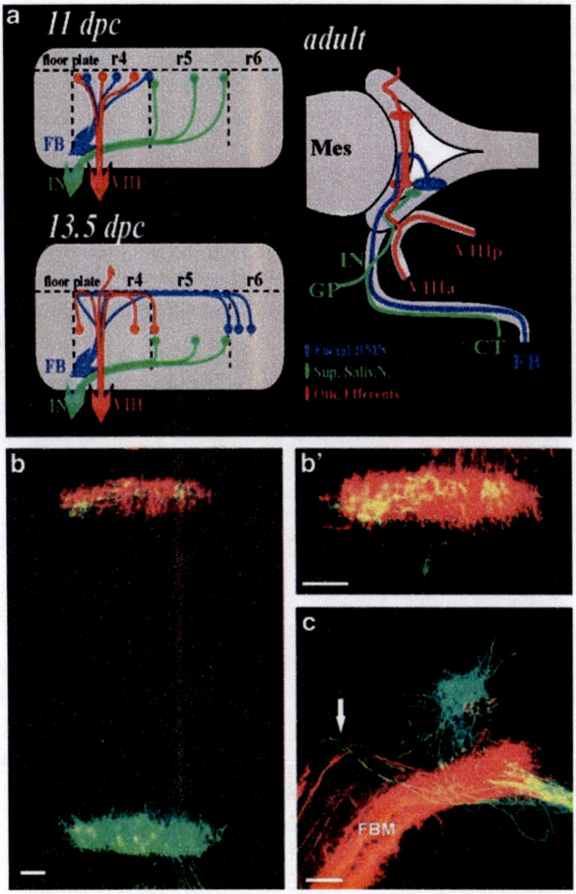

Fig. 7.1 Diagram and backfilling data showing the distribution of facial branchial motor neurons, facial visceral motor neurons, and vestibulocochlear efferents. **(a)** At E11 all motor neurons of rhombomeres 4 and 5 are adjacent to the floor plate and have extended their axons into the appropriate peripheral nerves (*FB* facial branchial motor nerve; *IN* intermediate nerve; *VIII* eighth nerve). At E13.5 facial branchial motor neurons have migrated along the floor plate, through rhombomere 5 and into rhombomere 6. Their trailing axons form part of the internal facial genu. Visceral motor neurons have migrated laterally, and do not contribute to the internal facial genu. In contrast, otic efferents, which have also migrated laterally, send their axons into the internal facial genu and develop axonal collaterals that cross the floor plate in rhombomere 4. As a consequence of their migration patterns, facial branchial motor neurons and otic efferent neurons form an almost continuous column of cells medial to the visceral efferents of the superior salivatory nucleus. The facial branchial motor fibers extend only through the facial nerve to the periphery. In contrast, visceral motor neurons, which join facial branchial motor axons via the intermediate nerve, diverge to different targets through the greater petrosal (GP), and chorda tympani nerves (CT). Otic efferent neurons are the only population that develops contralaterally projecting fibers, all of which cross in rhombomere 4. Peripherally, otic efferents contribute to the anterior (VIIIa) and posterior (VIIIp) rami of the VIIIth nerve (modified after Bruce et al. 1997). **(b)** Embryonic day (E) 14.5 ventral view of olivocochlear (OC) cell population in rhombomere 4. The olivocochlear fibers were backfilled with different colored Neurovue dye from each cochlea. Red or green cells are ipsilateral to the labeled ear. *Yellow* results from overlap in these collapsed stack. **(b′)** Higher magnification of the cell population shown in (**b**, *top*). *Red* is ipsilateral and contralateral olivocochlear cells are labeled in *green*. Contralateral olivocochlear cells (*green*) are starting to segregate anterior compared to the ipsilateral olivocochlear cells (*red*). **(c)** Vestibular efferents are labeled in green in this coronal section. They are located just rostral to the facial genu (*red*) carrying the axons of the facial branchial motor neurons (FBM) and efferent fibers continue toward the ear (*arrow*). Rostral is to the left. Scale bars = 100 μm

plate midline, as if they are unable to cross. Similar axons running along the midline have been described in Semaphorin3a mutant mice. Semaphorin 3a apparently provides a stop signal that prevents peripheral growth of axons beyond the sensory epithelia (Gu et al. 2003). These data show that some genes can now be associated with axonal crossing of efferent neurons, but the overall molecular guidance of axon crossing is essentially unclear. Contralateral projections are one of the most puzzling features of comparative development of the inner ear efferent system in vertebrates, with profound differences even in closely related taxa (Fritzsch 1999).

It is possible that vestibular and cochlear efferent neurons are already molecularly distinct at the time they become postmitotic. However, they become distinct cell populations only after they have segregated through lateral migration away from the caudally migrating facial branchial motor population (Muller et al. 2003). It is during lateral migration that vestibulocochlear efferents express the transcription factor *GATA3* (Karis et al. 2001). As efferents migrate laterally, they reach an area where efferent axons diverge from facial branchial motor axons (Fig. 7.1). This area is identified by the convergence of dorsal acoustic stria axons (Gurung and Fritzsch 2004). After the initial lateral migration from the basal plate toward the alar plate that separates vestibulocochlear from facial branchial motor neurons, vestibular and cochlear neurons migrate differently: Cochlear efferents migrate ventrally toward the meninges near the facial nerve root, whereas vestibular neurons migrate dorsally toward the IVth ventricle (Bruce et al. 1997). Vestibular efferents form a single nucleus (Fig. 7.1), as revealed by backfilling from different end organs (Maklad and Fritzsch 2003). In contrast, olivocochlear efferents have long been recognized as two distinct populations. One population is a medial group near the ventral nucleus of the trapezoid body (the medial olivocochlear system, which is distributed bilaterally). The other population is in a lateral position near the lateral superior olive (the lateral olivocochlear system, which is nearly exclusively ipsilateral) (Wilson et al. 1991; Vetter and Mugnaini 1992). Labeling from both ears shows that initially the two populations of olivocochlear efferents are nearly completely overlapping. However, as early as embryonic day 14.5 (E14.5), some segregation can be seen, with ipsilateral projecting cells being more lateral and contralateral projecting cells more medial. In addition, these cells show a rostrocaudal progression of segregation, with the more rostral cells being nearly completely segregated and the more caudal cells being more overlapping (Fig. 7.1). The segregation into the lateral and medial olivocochlear system is completed before birth.

7.3 Defects of Efferent Development Revealed Through Targeted Mutations

Attempts have been made to molecularly dissect the migration of facial branchial motoneurons. Because both facial branchial motoneurons and efferents share a very close origin, many of these mutations also affected efferent development.

For example, in *Hoxb1* null mice facial branchial motor neurons fail to migrate caudally and the vestibulocochlear efferents in these mutant mice show loss of contralateral projections (Studer et al. 1996). The improper migration and projection of motor neurons in rhombomere 4 suggests that *Hoxb1* plays a role in defining proper cell identity. *Hoxb1* null mice later lose facial branchial motor neurons (Gaufo et al. 2000; Studer 2001) due to programmed cell death. This cell death may relate to the absence of expression of the transcription factor *Phox2b*, and failure to down-regulate the bHLH gene *Mash1* (Gaufo et al. 2000). *Phox2b* is needed in the formation of visceral and branchial motor neurons (Pattyn et al. 2000; Tiveron et al. 2003) while *Mash1* is a marker for neuronal progenitor cells (Ma et al. 1997; Ohsawa et al. 2005). Unfortunately, little is known about the onset of expression and development of efferents relative to *Hoxb1*, and more work is needed.

Recently data have emerged about the role of genes involved in planar cell polarity (PCP) pathways during neuronal development. Planar cell polarity genes have been implicated in neuronal migration as well as in axonal/dendritic outgrowth and guidance. In rhombomere 4, most of the studies have focused on the pathways influencing the caudal and subsequent tangential migration of the facial branchial motor perikarya, while very few considered the effects of planar cell polarity genes on the vestibulocochlear efferents. Interestingly, *Hoxb1a*, the zebrafish homolog of mammalian *Hoxb1*, regulates the expression of *prickle1b* (a gene involved in PCP). In mice, *prickle1b*-deficient facial branchial motor neurons fail to initiate caudal migration (Rohrschneider et al. 2007). Other studies in mice on another transcription factor have shown that T-Box 20 (*Tbx20*) regulates the PCP pathways in facial branchial motor neurons and vestibulocochlear efferents. As with zebrafish, in the absence of this PCP protein, facial branchial motor neurons fail to migrate caudally. In addition, the vestibulocochlear neurons have no contralateral projections (Song et al. 2006). These mutants also show that in the absence of *Tbx20*, PCP proteins are reduced in facial branchial motoneurons. Although this new evidence shows the necessity of PCP pathways in facial branchial motoneurons, additional work is needed to move forward in this area, and to show the effect of PCP pathways in the vestibulocochlear efferents.

Because of the relatedness between facial branchial motor neurons and vestibulocochlear efferents, many of the same genetic makers identify both neuronal populations such as *Islet 1* and *2* (Briscoe and Ericson 2001). However, there is one additional gene, *Gata3*, that is not expressed by the facial branchial motor neurons but is expressed by the developing vestibulocochlear neurons. *Gata3* (a zinc finger transcription factor) is expressed in vestibulocochlear neurons as well as the otic placode and is an early marker which identifies vestibular and olivocochlear efferents prior to the onset of migration (Karis et al. 2001). In *Gata3* knockout mice, vestibulocochlear axons project with the facial branchial motor neurons and do not innervate the inner ear (Karis et al. 2001). Thus, these null mutations showing aberrant axonal fiber growth reveal that this gene may play a role in axonal guidance. However, the vestibulocochlear efferent cell bodies still segregate from the facial branchial motor neurons as well as from each other. These data indicate that *Gata3* does not play a role in cell fate determination. Owing to the complexity of *Gata3* action (e.g., defects in the ear and the possible direct effect on olivocochlear efferents), an understanding of *Gata3* function requires more sophisticated analysis knocking out *Gata3* conditionally only in the hindbrain or the ear.

7.4 Neurochemical Development of Auditory Efferents

Although inner ear efferent neurons in mammals are neurochemically heterogeneous, the relation between the neurotransmitters in peripheral efferent terminals and in brain stem cell bodies remains enigmatic. Studies of the development of these neurotransmitter systems in vestibulocochlear efferents have provided some additional clarity, but the mechanisms of neurotransmitter regulation are unknown. Track-tracing studies have established the location, projections, and terminations of lateral and medial olivocochlear neurons across mammalian species (Smith 1961; Spoendlin 1966; Iurato et al. 1978; Brown 1987), and an account of their developmental segregation from facial branchial motor neurons is summarized above. As previously mentioned, lateral olivocochlear neurons are found in and/or around the lateral superior olive; they project mainly to the ipsilateral cochlea and terminate mostly on dendritic fibers below inner hair cells (IHCs). In contrast, medial olivocochlear neurons are found mostly in rostral and ventral periolivary regions (Figs. 7.1 and 7.2); they project mainly to the contralateral cochlea, and terminate directly on outer hair cells (OHCs). The development of the complex distribution of cells and their axons has been described in Sect. 7.2.

Immunocytochemical studies agree in general that efferent terminals in the cochlea contain a variety of small molecular-weight neurotransmitters, including acetylcholine (ACh), γ-aminobutyric acid (GABA), dopamine, and several peptidergic transmitters such as calcitonin gene-related peptide (CGRP), enkephalins, and neuronal nitric oxide synthase (Eybalin 1993; Reuss et al. 2009). However, there is a lack of clarity as to whether all transmitters are expressed in the IHC and OHC regions, whether cytochemical subgroups exist within the lateral and medial olivocochlear populations, and, if so, which transmitters are colocalized and which are not (Vetter et al. 1991; Maison et al. 2003a, b). Because olivocochlear efferents originate from regions of the embryonic brain stem giving rise to motor neurons (Fig. 7.1), it is not surprising that ACh serves as a primary neurotransmitter found in olivocochlear efferent cell bodies and nerve terminals. Despite its lack of complete specificity, the histochemical demonstration of acetylcholinesterase (AChE), the degradative enzyme for acetylcholine, provided much of the early ideas on the morphology and development of the efferent system (Gacek et al. 1965; Osen et al. 1984).

7.4.1 Cholinergic Development

The antibody to choline acetyltransferase (ChAT), the synthesizing enzyme for acetylcholine, allows a more accurate identification of cholinergic neurons in the brain stem and efferent terminals in the cochlea (Altschuler et al. 1984; Raji-Kubba et al. 2002; Zidanic 2002). In the superior olivary complex of adult animals, ChAT immunoreactivity is found in periolivary regions (usually ventral and medial

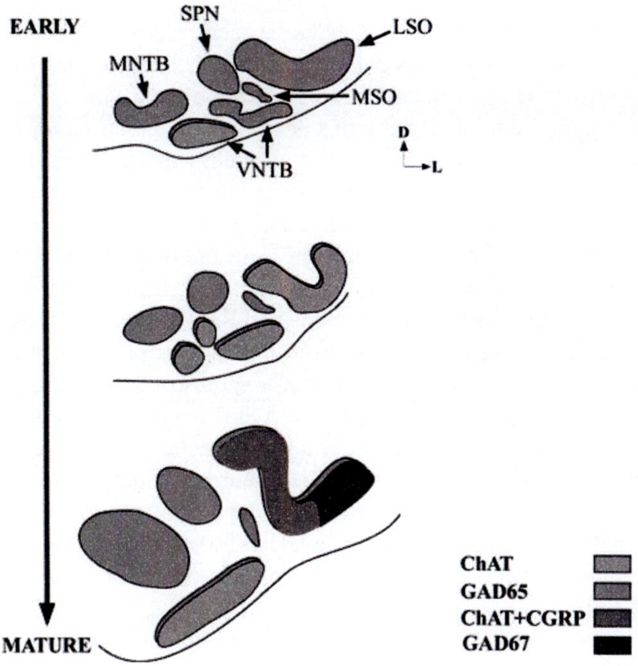

Fig. 7.2 Development of efferent neurotransmitter-like characteristics in superior olive. This schematic picture illustrates the neurochemical development of cholinergic and GABAergic neurons in olivocochlear regions of the superior olive. The figure shows coronal sections through the superior olivary complex of a hamster at three developmental ages. The nuclear boundaries are based on previous studies (Simmons and Raji-Kubba 1993). Shown are the medial nucleus of the trapezoid body (MNTB), ventral nucleus of the trapezoid body (VNTB), the lateral superior olive (LSO), the medial superior olive (MSO), and the superior paraolivary nucleus (SPN). Medial olivocochlear neurons are located in VNTB regions and lateral olivocochlear neurons are located typically in the LSO or the shell surrounding the LSO from previous adult studies. Choline acetyltransferease (ChAT) expression occurs early and is restricted to regions of the VNTB. Also, glutamic acid decarboxylase isoform 65 (GAD65) expression occurs early throughout most superior olivary nuclei. The neurons in the LSO shell express ChAT before intrinsic neurons in the LSO. Intrinsic LSO neurons express ChAT after LSO shell neurons, and are the only olivocochlear group that coexpresses CGRP. Intrinsic LSO neurons express ChAT several days prior to the onset of CGRP. Mature GAD67 expression occurs developmentally on or after the onset of hearing (based on studies from Raji-Kubba et al. 2002; Bergeron et al. 2005; and Jenkins and Simmons 2006)

locations) and in or around the lateral superior olive (Vetter et al. 1991; Moore et al. 1999). Brain stem studies of cholinergic olivocochlear development show that these neurons are chemically mature well before the onset of auditory function. As schematized in Fig. 7.2, a temporal and spatial developmental pattern of cholinergic expression is found in the superior olivary complex, where the distribution and number of cholinergic neurons change prior to and immediately after the onset of hearing (Simmons and Raji-Kubba 1993; Raji-Kubba et al. 2002). ChAT expression occurs first in ventral and medial periolivary populations and then in neurons associated with the lateral superior olive. Ventral and medial periolivary populations express AChE after ChAT. Except for the absence of calcitonin gene related protein (CGRP) expression, the embryonic pattern of ChAT expression in ventral and medial regions is similar to cranial motor nuclei (Raji-Kubba et al. 2002). However, the onset and pattern of ChAT expression in the lateral superior olive differs qualitatively and quantitatively from periolivary regions. In most studies of brain stem superior olivary regions in rodents, the intrinsic neurons of the lateral superior olive are the only olivocochlear neurons that coexpress CGRP. In the hamster's lateral superior olive, CGRP and AChE are expressed after ChAT expression (Simmons and Raji-Kubba 1993). Shell neurons surrounding the lateral superior olive also differ in their pattern of ChAT expression from the intrinsic neurons. Shell neurons demonstrate an intermediate pattern of cholinergic expression where ChAT and AChE expression occur after medial periolivary neurons but before intrinsic neurons of the lateral superior olive (Vetter and Mugnaini 1992). In addition, these shell neurons in the hamster and rat lack CGRP expression. The human fetal brain stem demonstrates a temporal and spatial sequence similar to hamsters (Moore et al. 1999). Also, medial olivocochlear efferents remain CGRP negative while lateral olivocochlear neurons coexpress CGRP (Fig. 7.2).

As mentioned previously, vestibulocochlear efferents have several unique characteristics when compared to hindbrain motor neurons, one of which is that they express more than one neurotransmitter phenotype. GABA, dopamine, and enkephalins are found in some efferent cell bodies and terminals. Although GABA is more prominent than either dopamine or enkephalins within olivocochlear neurons, the role of GABA in efferent function remains uncertain (Eybalin and Pujol 1984; Vetter et al. 1991). Antibodies made directly against either GABA or its synthesizing enzyme, glutamate decarboxylase (GAD), show immunoreactivity in cell bodies located in the superior olivary complex and terminals located below hair cells in the cochlea. There is a well documented descending projection from the lateral olivocochlear system of GABAergic neurons to the cochlea (Felix and Ehrenberger 1992; Jenkins and Simmons 2006). Studies of GABAergic development in the hamster's superior olivary complex are consistent with numerous other studies, suggesting that GABAergic neurons in the lateral limb of the lateral superior olive project to the cochlea and are thus a part of the lateral olivocochlear system (Adams 1983). However, few studies distinguish between the two dominant GAD isoforms: GAD65 and GAD67. At least in the hamster, GAD67 distinguishes a separate, descending inhibitory pathway which uses GABA as its primary neurotransmitter (Jenkins and Simmons 2006). This suggestion contradicts the interpretation of results in other studies about efferent neurotransmitters in the cochlea (Maison et al. 2003a, b). At least in the hamster, an early GAD65 expression appears extensively throughout the superior olive and may reflect a more generic function found in olivary neurons across different neurotransmitter types (Jenkins and Simmons 2006). Overall, these data on the morphological and neurochemical development are consistent: much of the central development is more or less completed before birth in most rodents. These data challenge the traditional view driven by the apparent postnatal arrival of efferents in the ear but are in line with the embryonic peripheral growth by motor neuron axons. This topic is discussed next.

7.5 Peripheral Development

The axonal projection of each neuron is mediated by molecular cues that are used by the extending growth cone to navigate. This includes both attractant and repellent molecular cues. The axonal trajectory of the vestibulocochlear efferent fibers is initially combined with the facial branchial motor neurons, but only efferent axon branches cross the floor plate. As efferent axons reach the alar plate of rhombomere 4, vestibulocochlear efferent, and facial branchial motor efferents diverge: vestibulocochlear efferents extend to and project via the VIIIth cranial nerve to their end organs. In contrast, facial branchial motor neurons project ventrally to exit through the facial motor nerve root (Fig. 7.1). It is only at the periphery that the vestibular and cochlear efferents segregate from one another, creating the commissure of von Oort (Figs. 7.3 and 7.4). Thus, all vestibulocochlear efferents have a common pathway selection to reach the vestibular ganglion. It can only be speculated here that these molecular cues may be related to the different path finding properties of vestibular and cochlear afferents which enables them to reach the vestibular and cochlear nerve, respectively. It is noteworthy that the longest common peripheral trajectory is between cochlear and saccular efferents, indicating perhaps that the mammalian cochlea evolved from the saccule (Fritzsch 1992; Nichols et al. 2008).

The peripheral development of vestibulocochlear efferent fibers is mediated in part by growing along afferent fibers (Fritzsch et al. 1999; Ma et al. 2000). In the absence of afferents, the vestibulocochlear efferent axons are unable to reach the ear and many fibers redirect axons and project with the facial nerve. Those "misguided" efferents indicate some degree of flexibility in efferents to interpret cues along their trajectory as if they are facial branchial motor neurons. This is in line with the interpretation that efferents are both evolutionary and developmentally derived from facial branchial motor neurons.

Efferents grow along the spiral ganglion axons toward the cochlea. Within the spiral ganglion, efferents form the intraganglionic spiral bundle (Fig. 7.3). In mutants with an altered migration of the spiral ganglion into the modiolus, the intraganglionic spiral bundle does not form (Morris et al. 2006). Similar defects occur when spiral neurons are partially or completely lost (Kim et al. 2001). These data suggest that the formation of this bundle depends on the presence of spiral ganglion neurons or their processes but does not reflect an intrinsic property of efferents. At least one of the genes involved in this process (*Neurod1*) is currently being investigated. Obviously, this association of efferents to grow along afferent fiber tracks cannot apply to fiber growth to OHCs, which are known to be accessed via different fiber crossing pathways in the tunnel of Corti (Maison et al. 2003a, b). Indeed, using double labeling techniques made recently available (Fritzsch et al. 2005), we are able to demonstrate that even the initial crossing of efferents and type

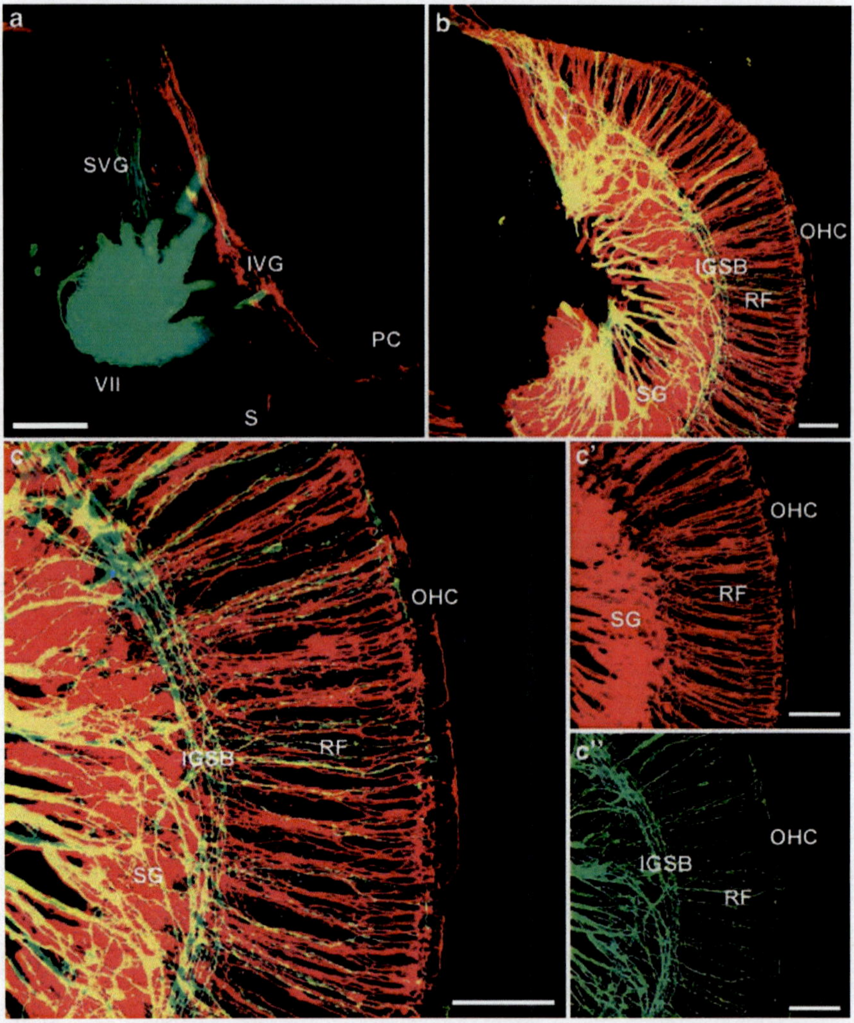

Fig. 7.3 Efferent outgrowth at embryonic day 11 and at birth in a mouse. (**a**) Neurovue dye application made into the cerebellum of an embryonic day 11 mouse to label afferent fibers (*red*) and brain stem to label efferents and facial branchial motor neurons (*green*). Facial nerve (VII) can be seen as it courses around the ear near the superior vestibular ganglion (SVG) and inferior vestibular ganglion (IVG). The *green* fibers in these ganglia are labeled efferents to the ear. Individual efferent axons can be seen at this level as they are adjacent to afferent vestibular fibers. These data show that efferent outgrowth is as early as afferent growth. (**b**) Low-power image of cochlear basal turn showing efferents in *green* and afferents in *red* fibers labeled from similar injections as in (**a**). Note the formation of the intraganglionic spiral bundle (IGSB) at the lateral edge of the spiral ganglion (SG). Radial fibers (RF) are mixed efferents and afferents. (**c**) Higher magnification of the basal turn with efferent fibers (*green*) projecting to the inner hair cells, forming partial inner spiral bundles along inner hair cells and projecting occasionally to outer hair cells. Notice the course of efferent axons near the cochlear margin of the spiral ganglion. The described differences in fiber trajectory and absence of spiral ganglion cell labeling are more obvious if both channels are separated (**c′**, **c**). Note the afferent growth cones extending toward the base below outer hair cells. *VII* facial nerve; *IGSB* intraganglionic spiral bundle; *IVG* inferior vestibular ganglia; *OHC* outer hair cell; *PC* posterior canal; *RF* radial fibers; *S* saccule; *SG* spiral ganglion; *SVG* superior vestibular ganglia. Scale bar = 100 μm

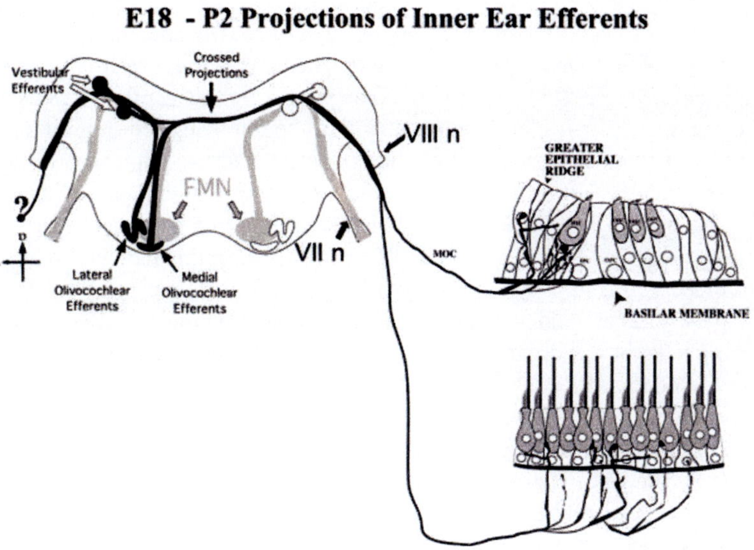

Fig. 7.4 Projections of inner ear efferents during late embryonic and early postnatal periods in rodents. During early periods of vestibulocochlear innervation to the inner ear, anterograde labeling and immunocytochemical studies in rodents show efferent axons are intimately associated with the facial branchial motor axons prior to their entry into the VIIIth cranial nerve (VIII n). In mammals, vestibular and auditory efferent neurons are completely separate. Auditory efferents lay rostral to the facial branchial motor nucleus (FMN) whereas vestibular efferents are associated with the abducens motor nucleus (not shown). Within the inner ear, the sensory neuroepithelium of cochlear and vestibular end organs have efferent terminals at the base of hair cells. Efferent terminals make axosomatic synapses on to hair cells and axodendritic synapses are found on dendritic axons. Auditory efferents are divided into two populations: medial and lateral olivocochlear neurons. Medial olivocochlear neurons are located in the ventral nucleus of the trapezoid body, which is medial to the lateral superior olive. Vestibular efferents are also divided into subgroups: a large group located dorsolateral to the facial genu and a smaller group located ventral to the genu. In this schematic drawing, medial olivocochlear projections occur from the contralateral side of the brainstem whereas lateral olivocochlear projections occur from the ipsilateral side. In the *early* stage of efferent axonal projections, medial olivocochlear (MOC) axons project to the region of the greater epithelial ridge and beneath the inner hair cells (IHCs). The projections of the mostly ipsilateral lateral olivocochlear axons are unknown. In the immature organ of Corti, the inner and outer pillar cells (IPC and OPC) have yet to form the tunnel of Corti. Projections from the contralateral vestibular efferent nuclei reach the immature sensory cristae or maculae and contact both type I and type II hair cells

II afferents are distinct (Fig. 7.3). Such dye tracings confirm and extend previous work using AChE (Sobkowicz and Emmerling 1989).

Inner ear efferents project their axon along medial–lateral gradients emanating from the floor plate and the alar plate (Fritzsch et al. 2006). They either project toward the highest floor plate gradient to cross to the contralateral side or away from it toward the alar plate. Near the alar plate, efferents take different trajectories from facial branchial motor neuron axons, which may relate to an attraction provided by the VIIIth nerve root. After having reached the vestibular part of the VIIIth cranial nerve root, efferents diverge along various afferent fiber paths. Vestibular efferents will project to vestibular organs innervated by the inferior vestibular ganglion, to vestibular organs innervated by the superior vestibular ganglion (Maklad and Fritzsch 1999), and reach the spiral ganglion via the commissure of von Oort (Fig. 7.4). It is clear that formation of the intraganglionic spiral bundle is critically dependent on normal distribution of spiral neurons, but it remains unclear which molecular features are actually used to form that bundle. After having entered the organ of Corti via the radial fibers, efferents sort much earlier than previously suggested between inner and OHCs (Fig. 7.3); this indicates that similar cues to those segregating type I and type II spiral afferents may play a role. What exactly those cues are and how they relate to the emerging molecular complexity of supporting cells of the organ of Corti (Puligilla et al. 2007; Dabdoub et al. 2008) remains unknown.

In the mouse, efferent olivocochlear axons arrive in the primordial cochlear epithelium as early as embryonic day 12 (Fritzsch and Nichols 1993) and are below hair cells before birth (Sobkowicz and Emmerling 1989). Early investigations of the development of efferent innervation of the mammalian cochlea assumed that the initial efferent projections to IHCs originated from lateral olivocochlear neurons. During cochlear development, there are numerous efferent axosomatic synapses on IHCs that mostly disappear in the mature cochlea with the possible exception of some GABAergic terminals (Liberman et al. 1990; Nitecka and Sobkowicz 1996). Studies in hamster and rat seem to support the idea of an initial transient innervation of IHCs by medial olivocochlear terminals (Simmons et al. 1990, 1998). As schematized in Fig. 7.4, anterograde labeling of the crossed olivocochlear projections in neonatal animals results in labeled efferent axons terminating in the greater epithelial ridge region of the organ of Corti, as well as beneath IHCs. In contrast, anterograde labeling of crossed olivocochlear projections in adult animals results in labeled efferent axons terminating below OHCs (Cole and Robertson 1992; Simmons et al. 1998). In line with a medial olivocochlear origin, tracer injections into the crossed olivocochlear bundles labels retrogradely cell bodies in medial and ventral periolivary regions (Fig. 7.1). This period during which medial olivocochlear axons accumulate below IHCs is reminiscent of a developmental waiting period. Such "waiting periods" have been studied extensively during the development of thalamocortical projections (Rakic 1977; Shatz et al. 1990).

In order to reach hair cell targets, efferent axons may be guided by molecular cues along afferent fiber tracks. It is now well documented that brain-derived neurotrophic factor (BDNF) and neurotrophin-3 (NT-3), and their specific tyrosine kinase receptors, trkB and trkC, are essential for the survival and guidance of inner ear afferent neurons (Pirvola et al. 1992; Fritzsch et al. 1999). Once efferent axons reach the vicinity of the hair cells, they may respond to molecular cues other than afferents to find their appropriate synaptic target. Extracellular matrix glycoproteins, such as tenascin-C or laminins, could direct local afferent and efferent synaptogenesis (Whitlon et al. 1999). Interactions between neurotransmitters released by efferents and their receptors in hair cells and afferent dendrites could also play a role in synapse formation (Simmons and Morley 1998; Zuo et al. 1999).

7.6 Onset of Neurotransmitter-Related Expression Within Cochlea

Efferent neurochemistry in the cochlea show AChE-stained fibers visible in the developing inner spiral bundle before birth in mice (Sobkowicz and Emmerling 1989). An association between growing, immature axons with efferent fibers has been made using a monoclonal antiserum against growth-associated protein 43 (GAP43) (Merchan-Perez et al. 1993; Simmons et al. 1996a, b). The GAP43-labeled axons may contact IHCs before birth, consistent with AChE studies. In rats, weak ChAT immunoreactivity is observed below IHCs at birth (Merchan-Perez et al. 1994). While AChE and ChAT activities can be measured at birth in the mouse cochlea, these levels increase sharply during the first 10 days of postnatal development (Sobkowicz and Emmerling 1989), supporting the view that mature levels of cholinergic protein expression follow the arrival of efferent fibers in the cochlea.

In addition to AChE and ChAT, the temporal expression patterns of other cholinergic proteins such as vesicular acetylcholine transporter (VAChT), and the high-affinity choline transporter (ChT1) have also been investigated in the rodent cochlea (Bergeron et al. 2005). Immunoreactivity to VAChT is present as early as 2 days after birth (P2) in the mouse and at birth in the rat. At P2, VAChT immunoreactivity is visible in a subset of GAP43-labeled efferent fibers that are below IHCs. In the mouse, ChT1 is detected after P2 and in the rat after birth. In both cases, ChT1-positive terminals are not visible until after synapsin immunoreactivity also occurs. Significant ChT1 is not visible until P4 in the mouse, well after efferent axons arrive below IHCs. The temporal expression patterns of the ChT1 and VAChT suggest that the induction of these two cholinergic markers may occur after the arrival of efferent axon terminals in the cochlea.

The temporal separation between the arrival of efferent axon growth cones and the appearance of either ChT1 or VAChT in the mouse and rat cochlea also coincides with a decrease in GAP-43 immunoreactivity and an increase in synapsin immunoreactivity. This temporal correspondence suggests that direct or indirect contact between efferent olivocochlear neuron growth cones and hair cells may be important for the induction and/or regulation of a mature cholinergic phenotype. Alternatively, cholinergic efferent axons may differentiate via a target-independent mechanism that does not require contact with hair cells, and thus, the timing among VAChT, ChT1, and synapsin could be coincidental. Irrespective of the trigger, it appears that VAChT precedes ChT1 in cholinergic differentiation (Bergeron et al. 2005). At least in the mouse, the pattern of ChT1 development is also consistent with the developmental innervation patterns of medial olivocochlear axons as shown by anterograde and immunocytochemical labeling studies (Simmons et al. 1996a, b). Figure 7.5 schematizes the proposed relationship among efferent axon extension (growth), synapse formation, and, for IHCs, synapse retraction based on studies of GAP43, synapsin, synaptophysin, and cholinergic marker expressions.

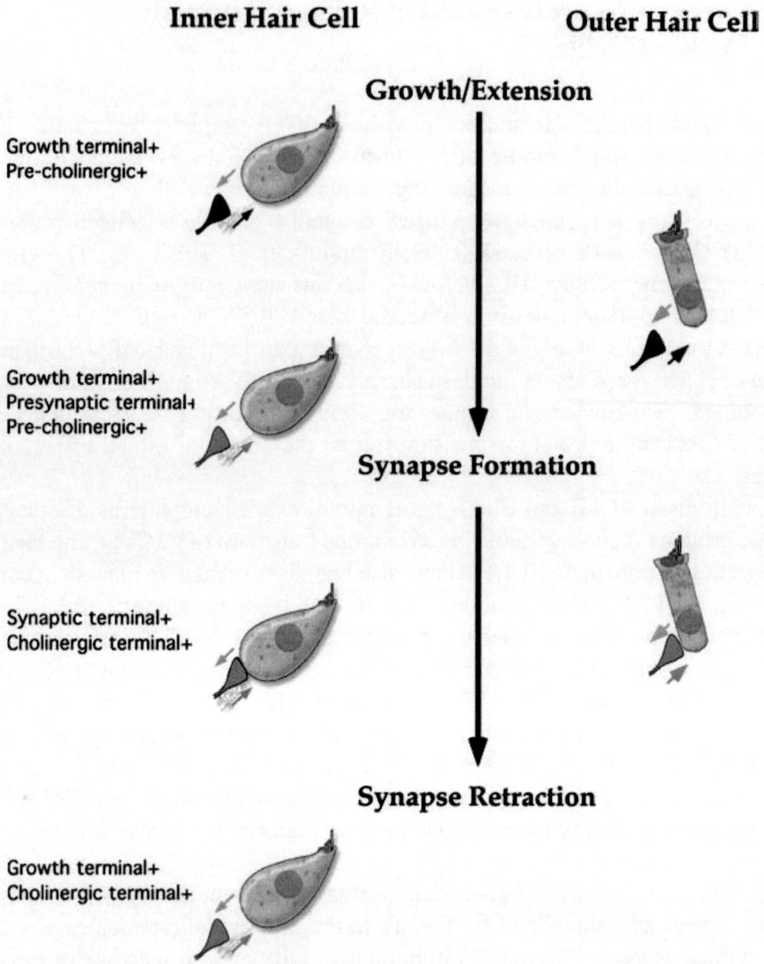

Fig. 7.5 Cholinergic efferent synaptogenesis on inner and outer hair cells. Different periods of efferent synaptogenesis are depicted that coincide with the growth and immunocytochemical differentiation of cholinergic markers (see text). As ChAT-positive efferent axons invade the organ of Corti, they express growth-associated proteins. These efferent axons make transient contacts with inner hair cells. While these efferent axons undergo synapse formation with inner hair cells, they express synaptic proteins (e.g., synapsin) and cholinergic proteins (e.g., vesicular acetylcholine transferase and choline transporter). These transient efferent innervations undergo synapse retraction. On the outer hair cells, efferent terminals extend from the inner hair cell area past the pillar cells. During extension, they probably express early cholinergic markers. As efferent axons form final synapses with outer hair cells, they express synaptic proteins and the choline transporter

In addition to ACh, efferents carry other peptides such as CGRP. The developmental pattern of CGRP expression in the cochlea is in line with the idea that lateral olivocochlear efferents have a delayed development of the neurotransmitter phenotype. CGRP is typically identified with cholinergic lateral olivocochlear neurons in the brainstem and inner ear (Vetter et al. 1991; Safieddine and Eybalin 1992). In the hamster, CGRP is not expressed in either efferent cell bodies or axons until after medial olivocochlear axons have accumulated below IHCs (Raji-Kubba et al. 2002). In the mouse cochlea, cholinergic fibers contact IHCs before birth, whereas CGRP immunoreactivity is not present until around birth (Sobkowicz 1992). Cholinergic fibers appear below OHCs in the mouse first, and CGRP immunoreactivity appears later (Sobkowicz 1992). In the hamster cochlea, ChAT-positive fibers and CGRP-positive fibers show a comparable spatiotemporal pattern (Simmons et al. 1998). If the presence of CGRP is not changing between cell types, then these findings support that the cholinergic lateral efferents mature later than medial efferents (Simmons and Raji-Kubba 1993; Raji-Kubba et al. 2002). Of course, the absence of CGRP expression in the cochlea does not necessarily mean that cholinergic lateral olivocochlear terminals are absent.

Another complex transmitter development in efferents is GABA. In mouse and rat, GABAergic terminals appear in the IHC area at birth and on OHCs much later (Whitlon and Sobkowicz 1989; Maison et al. 2006). In the developing IHC area, these terminals contact both IHCs and afferent terminals. In the adult, GABAergic innervation of the IHCs extends throughout the cochlea, whereas the extent of GABAergic OHCs innervation varies with species: in the gerbil, guinea pig, and the rat, it is restricted to the apical half of the cochlea (Vetter et al. 1991), whereas so far only in the mouse, it is reported from base to apex, and GABAergic and cholinergic markers colocalize in efferent terminals in both IHC and OHC areas (Maison et al. 2006).

In contrast to recent studies of GAD65 and GAD67 immunoreactivity in the superior olive (Jenkins and Simmons 2006), most studies in the inner ear have used only a pan-GAD antibody that recognizes both GAD65 and GAD67. Several immunocytochemical studies suggest that pan-GAD positive fibers appear before CGRP-positive fibers during cochlear development (Sobkowicz 1992; Sobkowicz 2001). In neonatal rats and mice, immunoreactivity to pan GAD is present at birth, and CGRP immunoreactivity is present by the end of the first postnatal week. Immunoreactivity to markers for presynaptic terminals such as synaptophysin and synapsin generally appear below IHCs just before efferent terminals arriving below OHCs (Bergeron et al. 2005). Taken together, these studies suggest that part of the early efferent innervation to the cochlea comes from a pan GABAergic population of olivocochlear efferents. However, it is not clear whether these efferents have GAD65 or GAD67. Adult brainstem studies strongly suggest that GAD67 is responsible for any mature GABAergic innervation coming from lateral olivocochlear neurons since GABAergic lateral olivocochlear neurons contain mostly GAD67 and not GAD65. If the data from hamster can be generalized to other species, it is possible that the early GABAergic labeling seen in the inner ear reflects a ubiquitous type of immature labeling due mostly to the GAD65 isoform, which is expressed early in most neurons of the superior olive. This early GABAergic innervation would not express GAD67 and could represent both medial and lateral olivocochlear neurons. The mature GABAergic innervation would express GAD67 and come mostly from lateral olivocochlear neurons. As studies using a pan-GAD antibody in the adult inner ear show clearly a type of GAD-positive labeling, we propose that this is due to the GAD65 isoform, which may be expressed residually in non-GABAergic neurons in which GABA is not the primary neurotransmitter. This scenario may explain why ChAT and pan-GAD immunoreactivities show extensive colocalization in medial olivocochlear terminals.

7.7 Acetylcholine Receptors on Hair Cells

IHCs, at least transiently, express both nicotinic ACh receptor (Chrna) 9 and Chrna10 subunits and respond to ACh. This circumstance is in a way similar to that for OHCs, corroborating the idea of a transient cholinergic innervation of IHCs during synaptogenesis and supporting the notion that efferents have a motor neuron embryonic origin (Glowatzki and Fuchs 2000; Katz et al. 2004). Studies in rats and mice show a transient developmental expression of α9 and α10 transcripts in IHCs (Luo et al. 1998; Katz et al. 2004). In situ hybridization studies in rats (Luo et al. 1998; Simmons and Morley 1998), as well as transgenic studies in mice (Zuo et al. 1999), show a progressive up-regulation and subsequent down-regulation of α9 expression in IHCs in basal cochlear regions that is followed by a radial expansion of expression in the OHCs and a longitudinal expansion into the apical regions. There is also a subsequent down-regulation of α9 expression in basal turn IHCs, and it has been proposed that α9 expression is hair-cell autonomous (i.e., occurs in the absence of efferent innervation) during development (Jagger et al. 2000; He et al. 2001).

The available data suggest that most α9 *Chrna* expression in hair cells occurs before efferent synaptogenesis, whereas α10 *Chrna* expression occurs concomitant with the presence of efferent fibers. Studies of the α10 nicotinic subunit indicate that it is expressed developmentally in both IHCs and OHCs, but, unlike α9, is detected only in OHCs and not in IHCs in the adult rat (Elgoyhen et al. 2001; Morley and Simmons 2002). α10 subunit expression peaks in newborn mice for IHCs and around P10 for OHCs (Morley and Simmons 2002). Although α9 expression occurs prior to birth in the cochlea, α10 expression cannot be detected before birth. By P21 in the rat, α10 expression virtually disappears from the IHCs, but remains robust in the OHCs, which is in contrast to the expression patterns found with α9 in rats. Consistent with in situ hybridization studies which show that α10 mRNA disappears after the onset of hearing in IHCs, immunocytochemical studies also suggest an absence of α10 and the calcium-dependent potassium channels (SK2), but not α9, after the onset of hearing (Katz et al. 2004). These data suggest that α9 and α10 subunits as well as SK2 channels may be regulated differentially by synaptic activity.

7.8 Nicotinic Synapse Formation and Maturation of ACh Receptors

Studies that show early hair cell responses to ACh suggest that IHCs and OHCs may assemble and cluster nicotinic receptors during efferent synaptogenesis in the inner ear. However, the molecules associated with receptor assembly and clustering in hair cells are only now being investigated. The cholinergic mechanisms that induce synapse formation and regulate the differentiation of synaptic specializations are best known at the neuromuscular junction (NMJ). The earliest Chrna clusters in muscle apparently require no nerve-derived signals (Willmann and Fuhrer 2002). As development proceeds, muscle specific receptor tyrosine kinase (MuSK) and rapsyn (receptor-associated protein of the synapse) are essential for clustering and localization of Chrnas at the developing NMJ (Apel et al. 1997; Lin et al. 2001). Also at the NMJ, RIC-3, a transmembrane protein located within the endoplasmic reticulum, may chaperone nicotinic and serotonergic receptors (Cheng et al. 2007). Recent studies (Osman et al. 2008) suggest that MuSK, rapsyn and RIC-3 are not only expressed in rodent hair cells but may regulate Chrna clusters during development (Fig. 7.6).

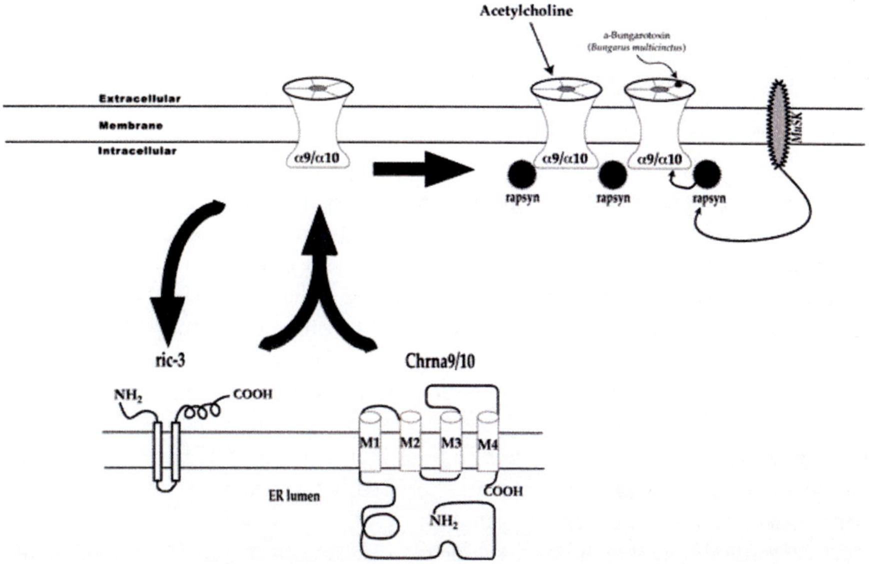

Fig. 7.6 Putative role of rapsyn, MuSK, and RIC-3 at the nicotinic hair cell synapse. Based on heterologous cell expression studies, rapsyn, MuSK, and Ric-3 are believed necessary for cell surface expression of nAChRs in mammalian cells. Both rapsyn and Ric-3 interact with the *Chrna9* subunit. The proposed scheme follows known roles of rapsyn, MuSK, and Ric-3 at the neuromuscular junction. Ric-3 facilitates receptor folding, assembly, and/or cell surface targeting, while rapsyn enhances clustering and/or functioning of receptors at the cell surface that is mediated via MuSK

Unlike at the NMJ, the use of genetic experimental approaches for the study of the development of cholinergic synapses in the inner ear are of limited use because nicotinic synapses form after E18 beyond which mice, for example, with null mutations of rapsyn do not survive. Conditional knockouts may be informative for future molecular dissection of interactions leading to clustering of receptors and synaptogenesis of efferents on hair cells.

α-Bungarotoxin (Bgtx) labeling of hair cell cholinergic receptors has been difficult to demonstrate reliably on sensory hair cells (Canlon et al. 1989; Wackym et al. 1995). The first Chrnas labeling with α-Bgtx during development was performed on freshly isolated cochlear hair cells from newborn rats (Osman et al. 2008) and demonstrated distinct plaque-like structures restricted almost exclusively to hair cells. As previously described for efferent innervation (Simmons 2002), α-Bgtx labeling follows a similar developmental progression: it is first seen mostly on IHCs at P1 and then becomes progressively localized to OHCs by P10. Although α-Bgtx labeling exhibited the same spatial development as efferent innervation at P1, double labeling with a cholinergic terminal marker (e.g., VAChT) revealed no obvious association between the early cholinergic innervation and α-Bgtx-labeled puncta. These results suggest that the initial appearance of α-Bgtx labeling may be independent of efferent innervation.

7.9 Maturation of Efferent Connections and Efferent-Induced Hair Cell Responses

In line with a transient cholinergic innervation, functional nicotinic cholinergic receptors on immature rat IHCs have been reported with α9 and α10 receptor pharmacology (Glowatzki and Fuchs 2000; Gomez-Casati et al. 2005). Application of ACh to IHCs in apical turns of the rat results in the activation of small-conductance calcium-dependent potassium (SK2) channels. The pharmacological and biophysical characteristics of this cholinergic receptor on IHCs closely resemble those of the recombinant α9/α10 receptors, reinforcing the hypothesis that the functional nicotinic receptors at the olivocochlear efferent synapse are composed of both the α9 and α10 subunits. These studies suggest that neonatal IHCs are inhibited by cholinergic synaptic input before the onset of hearing, consistent with the anatomical tracing data. The cholinergic inhibition of IHCs could be important for the maturation of IHC and auditory nerve responses. Studies conducted in gerbils and rats suggest that the onset of ACh-induced responses in OHCs begins on or after P6 and becomes functionally mature by P12 (He and Dallos 1999). It is during this period that the number of nicotinic receptors and the number of calcium-activated potassium channels dramatically increases (Dulon and Lenoir 1996; Klocker et al. 2001). Further, in a series of studies in the gerbil cochlea, the development of ACh-induced responses in isolated OHCs seems to coincide with the time period when OHCs develop motility but before the onset of auditory function (He et al. 1994).

Evidence that efferent neurons are capable of providing activity to the ear during early postnatal periods is suggested by brain slice studies of efferent neurons in the rat (Fujino et al. 1997) as well as by neonatal studies of deefferentation (Walsh et al. 1998). The idea that the early presence of cochlear efferents plays a role in the normal maturation of afferent responses is supported by observations following either the ablation or stimulation of efferent projections during the early postnatal period. Unlike deefferentation in adults, neonatally deefferented animals show a number of changes in auditory fiber responses including elevated thresholds, decreased sharpness of tuning, as well as lower spontaneous discharge rates (Walsh et al. 1998). In these findings (Walsh et al. 1998) it appeared as if the OHC contribution to frequency tuning and sensitivity had been compromised, but the majority of efferent connections are to the IHCs at the time of deefferentation (Ginzberg and Morest 1984). Taken together, these findings provide additional evidence that early cholinergic inhibition of IHCs could be important for the maturation of IHC and auditory nerve responses.

The developmental contribution of olivocochlear efferent neurons to cochlear maturation processes requires more than the presence of either the α9 and α10 subunits or SK2 channels. Work on either α9 or combined α10 and SK2 double null mouse mutants indicates that the absence of these nicotinic receptor subunits does not result in the same type of deficits seen in the neonatal deefferentation studies (Walsh et al. 1998; Murthy et al. 2009). In studies of α9 knockout adult mice, OHCs show altered efferent innervation and also lack suppression of cochlear responses during efferent fiber activation (Vetter et al. 1999). Double-null mice have been created which lack both the α10 *Chrna* gene, the loss of which results in hypertrophied olivocochlear terminals in adult animals, and the SK2 gene, the loss of which results in an altered synaptic phenotype. This loss also leads in these double null mutants to down-regulation of α9 *Chrna* gene expression (Murthy et al. 2009). Based on such studies, the expression of SK2 may establish an environment upon which *Chrna* subunit gene expression can exert further effects to facilitate olivocochlear synapse maintenance. That these mutant mouse models lack a significant developmental effect similar to the neonatal deefferentation does not negate the influence of efferent terminals on functional maturation in the cochlea. It should, however, be noticed that hair cell development is normal in mice lacking all afferent and efferent innervations at least until birth (Ma et al. 2000) and that tissue culture shows normal differentiation of hair cells without any innervations. Given the likely presence of medial efferent axons below IHCs and the likely absence of efferent connections to the OHCs, these studies in neonatal cats (Pujol et al. 1978; Walsh et al. 1998) suggest at the very least that efferent fibers are capable of transmitting information directly to the IHCs and afferent fibers.

7.10 Efferent Connections to Vestibular hair Cells

In mammals, efferent vestibular fibers innervating the crista ampullaris or macula originate from a population of neurons lateral to the abducens motor nuclei (Fig. 7.1) and medial to the vestibular nuclei (Warr 1975; Chi et al. 2007). The majority, but not all, vestibular efferent neurons contain cholinergic enzymes, ChAT and AChE, and CGRP (Perachio and Kevetter 1989; Purcell and Perachio 1997). Vestibular efferents may also contain encephalin mRNA (Ryan et al. 1991).

Vestibular efferent neurons send projections to the crista or macula (see Chap. 6). On reaching the neuroepithelium of each crista or macula, they innervate hair cells and afferent terminals with extensive axonal branching (Goldberg et al. 1992; Lysakowski and Goldberg 2008). In mammals, two types of hair cells, type I and type II, are found in the crista and macular vestibular end-organs. Type I hair cells are flask-shaped and are surrounded by a nerve calyx from one of the terminal branches of the vestibular nerve. Type II hair cells are shaped like cylinders and contacted by multiple afferent synapses. Aside from the calyx, one of the main differences between type I and type II hair cells is that the former typically lack direct contact (i.e., axosomatic synapses) with the efferent terminals, whereas type II hair cells have axosomatic synapses with efferent boutons.

Relatively little is known about the spatial and temporal development of vestibular efferent axons. Early investigations suggested that efferent nerve endings develop directly on both type I and type II hair cells (Favre and Sans 1977; Mbiene et al. 1988). Prior to the formation of the type I afferent calyx, efferent fibers are in contact either with hair cells or afferent fibers. These studies conclude that afferent nerve terminals develop prior to efferent nerve endings and that maturation at least in the rodents occurs mostly during postnatal periods after hair cells have reached maturity. However, more contemporary studies challenge the assumption that efferent axons arrive well after afferent axons. In contrast to almost all previous reports, Fritzsch and co-workers indicate an embryonic arrival of efferent fibers (Fritzsch et al. 1993; Bruce et al. 1997). In mice, efferent fibers begin to invade the future vestibular neurosensory epithelia by embryonic day 12, 1 week before birth. The early neuroanatomic arrival of efferents at the future sensory epithelia demonstrates an as yet unexplored interaction of efferent fibers with the differentiating vestibular hair cells. This sequence of efferent development in vestibular end organs occurs much earlier than efferent development in the cochlea (Simmons 2002). Figure 7.4 shows a proposed scheme for efferent innervation of vestibular hair cells compared to efferent innervation of cochlear hair cells in the mouse.

The dynamic changes of *Chrna9* expression in the utricle do not correlate with the pattern of efferent innervation or the distribution of type I and II hair cells. Recent data suggest that both types of hair cells are present in the striola and extrastriola, and that there are more type I hair cells in the extrastriola than in the striola of neonatal and adult mice (Rusch et al. 1998). The transgenic studies by Zuo et al. (1999) describe an embryonic up-regulation and subsequent postnatal down-regulation of α9 expression. In these studies, GFP is expressed by both type I and II hair cells in vestibular sensory epithelia. GFP expression is first detected in these vestibular sensory epithelia at E16, up-regulated in striolar regions at P2, down-regulated in the striolar region, and up-regulated in the extrastriolar region at P13 that persists into adulthood. In contrast, the gradient of GFP expression throughout the semicircular sensory epithelia remains constant during development. In developing utricles, hair cells in striolar regions appear to differentiate earlier than those in extrastriola regions (Rusch et al. 1998). It is thus likely that expression of α9 AChR is an intrinsic characteristic of hair cells, and independent of efferent fiber regulation (Simmons 2002).

The identity of efferent neurotransmitters involved in the vestibular periphery is similar to the cochlea: ACh and GABA have been postulated as likely to act as an efferent neurotransmitter in vestibular organs (Lopez and Meza 1990). Using AChE histochemistry and ChAT immunoreactivity to visualize efferent terminals in adult animals, cholinergic efferents were shown to form a complicated plexus in all vestibular end organs and to make axodendritic synapses onto afferent calyx endings and axosomatic synapses onto type II hair cells (Ohno et al. 1993; Schrott-Fischer et al. 2007). Most studies of cholinergic vestibular efferents also suggest that they coexpress CGRP. Like ChAT axons and terminals, CGRP axons and terminals ramify throughout the neurosensory epithelium of the maculae and cristae and make contact with both type II hair cells and afferent calyceal endings on type I hair cells (Wackym 1993; Popper et al. 2002). The suggestion that GABA may also be an efferent neurotransmitter has been more problematic. In fact, GABA has been postulated as both an afferent and efferent neurotransmitter based on biochemical, immunocytochemical, in situ hybridization, and molecular biological techniques (Didier et al. 1990; Meza 2008). Consistent with its role as an efferent neurotransmitter, GABA is found in vesiculated fibers and bouton-type terminals (Usami et al. 1987; Schrott-Fischer et al. 2007). However, GABA has yet to be identified in vestibular efferent neurons within the brain stem.

Neurotransmitter data regarding the early efferent innervation of the vestibular periphery is mostly limited to studies using tracers in mice (see Chap. 5) or CGRP immunocytochemistry in rats (Dememes et al. 2001). Studies of CGRP immunoreactivity show a progressive early postnatal innervation. At birth, a few CGRP-positive fibers are found beneath the sensory epithelium, but no CGRP-positive terminals contact hair cells. These then massively invade the neuroepithelium between P2 and P4. At P8, CGRP-positive fibers in the utricle follow transient courses before contacting hair cells. At subsequent stages, a lower density of CGRP-positive fibers and synapses is found in the utricle. It is unclear how these observations reconcile either with previous ultrastructural findings (Nordemar 1983; Mbiene et al. 1988) or with DiI tracer studies (Fritzsch and Nichols 1993; Bruce et al. 1997). The lack of either vesiculated endings or CGRP-immunoreactive fibers in contact with hair cells at birth in the rat does not signify a lack of efferent axons. It is possible that efferent fibers at these early stages do not express CGRP or even cholinergic enzymes at high enough levels to allow detection. In fact, ultrastructural characteristics of immature efferent axons are virtually indistinguishable from immature afferent axons contacting cochlear IHCs (Bruce et al. 2000), another source of possible uncertainty underlying earlier interpretations.

7.11 Conclusion and Outlook

This chapter provides a snapshot of the current level of understanding of efferent development from the early specification of efferent cell bodies near the floor plate through their migration, embryonic development of the axonal trajectories across the floor plate and into the ear, neurochemical maturation at various levels, and effects of specific mutation on pathfinding to and within the ear. A picture of a progressive definition of efferents away from a facial branchial motor neuron development begins to emerge at many levels, but details remain obscure. The motor neuron nature of efferents is also obvious with the presence of ACh and CGRP as transmitters and nAChR receptors on hair cells. Likewise, data on synapse formation indicate that the efferent synapse on hair cells resembles a modified muscle synapse. Future research would need to uncover how much divergence and similarity exists between motor neuron development and efferents. The late appearance of various peptide transmitters may indicate such divergence. It is important to realize that only mammals have evolved a novel contractile mechanism, prestin, which is somehow controlled by efferents, an apparent flashback of their motor neuron nature.

The progress over the last 20 years has been remarkable, but it has also raised even more questions that require more advanced techniques to find answers. It is the authors' hope that the brief outline of current data and the indications of open questions or controversies throughout the text will stimulate and accelerate more investigations in this underexplored field of ear development.

Acknowledgments This work was supported by several NIH/NIDCD funds to some of the authors. We are indebted to the editors for help in stimulating this review.

References

Adams JC (1983) Cytology of periolivary cells and the organization of their projections in the cat. J Comp Neurol 215:275–289

Altman J, Bayer SA (1982) Development of the cranial nerve ganglia and related nuclei in the rat. Adv Anat Embryol Cell Biol 74:1–90

Altschuler RA, Fex J, Parakkal MH, Eckenstein F (1984) Co-localization of enkephalin-like and choline acetyltransferase-like immunoreactivities in olivocochlear neurons of the guinea pig. J Histochem Cytochem 32:839–843

Apel ED, Glass DJ, Moscoso LM, Yancopoulos GD, Sanes JR (1997) Rapsyn is required for MuSK signaling and recruits synaptic components to a MuSK-containing scaffold. Neuron 18:623–635

Bergeron AL, Schrader A, Yang D, Osman AA, Simmons DD (2005) The final stage of cholinergic differentiation occurs below inner hair cells during development of the rodent cochlea. J Assoc Res Otolaryngol 6:401–415

Briscoe J, Ericson J (2001) Specification of neuronal fates in the ventral neural tube. Curr Opin Neurobiol 11:43–49

Brown MC (1987) Morphology of labeled efferent fibers in the guinea pig cochlea. J Comp Neurol 260:605–618

Bruce LL, Kingsley J, Nichols DH, Fritzsch B (1997) The development of vestibulocochlear efferents and cochlear afferents in mice. Int J Dev Neurosci 15:671–692

Bruce LL, Christensen MA, Warr WB (2000) Postnatal development of efferent synapses in the rat cochlea. J Comp Neurol 423:532–548

Canlon B, Cartaud J, Changeux JP (1989) Localization of alpha-bungarotoxin binding sites on outer hair cells from the guinea-pig cochlea. Acta Physiol Scand 137:549–550

Cheng A, Bollan KA, Greenwood SM, Irving AJ, Connolly CN (2007) Differential subcellular localization of RIC-3 isoforms and their role in determining 5-HT3 receptor composition. J Biol Chem 282:26158–26166

Chi FL, Jiao Y, Liu HJ, Wang J, Shi Y, Barr JJ (2007) Retrograde neuron tracing with microspheres reveals projection of CGRP-immunolabeled vestibular afferent neurons to the vestibular efferent nucleus in the brainstem of rats. Neuroendocrinology 85:131–138

Christopher Kirk E, Smith DW (2003) Protection from acoustic trauma is not a primary function of the medial olivocochlear efferent system. J Assoc Res Otolaryngol 4:445–465

Cole KS, Robertson D (1992) Early efferent innervation of the developing rat cochlea studied with a carbocyanine dye. Brain Res 575:223–230

Cowan CA, Yokoyama N, Bianchi LM, Henkemeyer M, Fritzsch B (2000) EphB2 guides axons at the midline and is necessary for normal vestibular function. Neuron 26:417–430

Dabdoub A, Puligilla C, Jones JM, Fritzsch B, Cheah KS, Pevny LH, Kelley MW (2008) Sox2 signaling in prosensory domain specification and subsequent hair cell differentiation in the developing cochlea. Proc Natl Acad Sci USA 105:18396–18401

Dememes D, Dechesne CJ, Venteo S, Gaven F, Raymond J (2001) Development of the rat efferent vestibular system on the ground and in microgravity. Brain Res Dev Brain Res 128:35–44

Didier A, Dupont J, Cazals Y (1990) GABA immunoreactivity of calyceal nerve endings in the vestibular system of the guinea pig. Cell Tissue Res 260:415–419

Dulon D, Lenoir M (1996) Cholinergic responses in developing outer hair cells of the rat cochlea. Eur J Neurosci 8:1945–1952

Elgoyhen AB, Vetter DE, Katz E, Rothlin CV, Heinemann SF, Boulter J (2001) Alpha10: a determinant of nicotinic cholinergic receptor function in mammalian vestibular and cochlear mechanosensory hair cells. Proc Natl Acad Sci USA 98:3501–3506

Eybalin M (1993) Neurotransmitters and neuromodulators of the mammalian cochlea. Physiol Rev 73:309–373

Eybalin M, Pujol R (1984) Immunofluorescence with Met-enkephalin and Leu-enkephalin antibodies in the guinea pig cochlea. Hear Res 13:135–140

Favre D, Sans A (1977) Synaptogenesis of the efferent vestibular nerve endings of the cat: ultrastructural study. Arch Otorhinolaryngol 215:183–186

Felix D, Ehrenberger K (1992) The efferent modulation of mammalian inner hair cell afferents. Hear Res 64:1–5

Fritzsch B (1992) The water-to-land transition: evolution of the tetrapod basilar papilla, middle ear and auditory nuclei. In: Webster DB, Fay RR, Popper AN (eds) The evolutionary biology of hearing. Springer, New York, pp 351–375

Fritzsch B (1999) Ontogenetic and evolutionary evidence for the motoneuron nature of vestibular and cochlear efferents. In: Berlin C (ed) The efferent auditory system: basic science and clinical applications. Singular Publishing, San Diego, p 31

Fritzsch B, Nichols DH (1993) DiI reveals a prenatal arrival of efferents at the differentiating otocyst of mice. Hear Res 65:51–60

Fritzsch B, Northcutt RG (1993) Origin and migration of trochlear, oculomotor and abducent motor neurons in *Petromyzon marinus* L. Brain Res Dev Brain Res 74:122–126

Fritzsch B, Christensen MA, Nichols DH (1993) Fiber pathways and positional changes in efferent perikarya of 2.5- to 7-day chick embryos as revealed with DiI and dextran amines. J Neurobiol 24:1481–1499

Fritzsch B, Pirvola U, Ylikoski J (1999) Making and breaking the innervation of the ear: neurotrophic support during ear development and its clinical implications. Cell Tissue Res 295:369–382

Fritzsch B, Muirhead KA, Feng F, Gray BD, Ohlsson-Wilhelm BM (2005) Diffusion and imaging properties of three new lipophilic tracers, NeuroVue Maroon, NeuroVue Red and NeuroVue Green and their use for double and triple labeling of neuronal profile. Brain Res Bull 66:249–258

Fritzsch B, Pauley S, Beisel KW (2006) Cells, molecules and morphogenesis: the making of the vertebrate ear. Brain Res 1091:151–171

Fujino K, Koyano K, Ohmori H (1997) Lateral and medial olivocochlear neurons have distinct electrophysiological properties in the rat brain slice. J Neurophysiol 77:2788–2804

Gacek RR, Nomura Y, Balogh K (1965) Acetylcholinesterase activity in the efferent fibers of the stato-acoustic nerve. Acto Otolaryngol 59:541–553

Gaufo GO, Flodby P, Capecchi MR (2000) Hoxb1 controls effectors of sonic hedgehog and *Mash1* signaling pathways. Development (Cambridge, UK) 127:5343–5354

Ginzberg RD, Morest DK (1984) Fine structure of cochlear innervation in the cat. Hear Res 14:109–127

Glowatzki E, Fuchs PA (2000) Cholinergic synaptic inhibition of inner hair cells in the neonatal mammalian cochlea. Science 288:2366–2368

Goldberg JM, Lysakowski A, Fernandez C (1992) Structure and function of vestibular nerve fibers in the chinchilla and squirrel monkey. Ann NY Acad Sci 656:92–107

Gomez-Casati ME, Fuchs PA, Elgoyhen AB, Katz E (2005) Biophysical and pharmacological characterization of nicotinic cholinergic receptors in rat cochlear inner hair cells. J Physiol 566:103–118

Gu C, Rodriguez ER, Reimert DV, Shu T, Fritzsch B, Richards LJ, Kolodkin AL, Ginty DD (2003) Neuropilin-1 conveys semaphorin and VEGF signaling during neural and cardiovascular development. Dev Cell 5:45–57

Gurung B, Fritzsch B (2004) Time course of embryonic midbrain thalamic and thalamic auditory connection development in mice as revealed by carbocyannine dye injection. J Comp Neurol 479:309–327

He DZ, Dallos P (1999) Development of acetylcholine-induced responses in neonatal gerbil outer hair cells. J Neurophysiol 81:1162–1170

He DZ, Evans BN, Dallos P (1994) First appearance and development of electromotility in neonatal gerbil outer hair cells. Hear Res 78:77–90

He DZ, Zheng J, Dallos P (2001) Development of acetylcholine receptors in cultured outer hair cells. Hear Res 162:113–125

Iurato S, Smith CA, Eldredge DH, Henderson D, Carr C, Ueno Y, Cameron S, Richter R (1978) Distribution of the crossed olivocochlear bundle in the chinchilla's cochlea. J Comp Neurol 182:57–76

Jagger DJ, Griesinger CB, Rivolta MN, Holley MC, Ashmore JF (2000) Calcium signalling mediated by the alpa 9 acetylcholine receptor in a cochlear cell line from the Immortomouse. J Physiol 527:49–54

Jenkins SA, Simmons DD (2006) GABAergic neurons in the lateral superior olive of the hamster are distinguished by differential expression of GAD isoforms during development. Brain Res 1111:12–25

Karis A, Pata I, van Doorninck JH, Grosveld F, de Zeeuw CI, de Caprona D, Fritzsch B (2001) Transcription factor GATA-3 alters pathway selection of olivocochlear neurons and affects morphogenesis of the ear. J Comp Neurol 429:615–630

Katz E, Elgoyhen AB, Gomez-Casati ME, Knipper M, Vetter DE, Fuchs PA, Glowatzki E (2004) Developmental regulation of nicotinic synapses on cochlear inner hair cells. J Neurosci 24:7814–7820

Kim WY, Fritzsch B, Serls A, Bakel LA, Huang EJ, Reichardt LF, Barth DS, Lee JE (2001) NeuroD-null mice are deaf due to a severe loss of the inner ear sensory neurons during development. Development 128:417–426

Klocker N, Oliver D, Ruppersberg JP, Knaus HG, Fakler B (2001) Developmental expression of the small-conductance $Ca^{(2+)}$-activated potassium channel SK2 in the rat retina. Mol Cell Neurosci 17:514–520

Liberman MC, Dodds LW, Pierce S (1990) Afferent and efferent innervation of the cat cochlea: quantitative analysis with light and electron microscopy. J Comp Neurol 301:443–460
Lin W, Burgess RW, Dominguez B, Pfaff SL, Sanes JR, Lee KF (2001) Distinct roles of nerve and muscle in postsynaptic differentiation of the neuromuscular synapse. Nature 410:1057–1064
Lopez I, Meza G (1990) Comparative studies on glutamate decarboxylase and choline acetyltransferase activities in the vertebrate vestibule. Comp Biochem Physiol B 95:375–379
Luo L, Bennett T, Jung HH, Ryan AF (1998) Developmental expression of alpha 9 acetylcholine receptor mRNA in the rat cochlea and vestibular inner ear. J Comp Neurol 393:320–331
Lysakowski A, Goldberg JM (2008) Ultrastructural analysis of the cristae ampullares in the squirrel monkey (*Saimiri sciureus*). J Comp Neurol 511:47–64
Ma Q, Sommer L, Cserjesi P, Anderson DJ (1997) Mash1 and neurogenin1 expression patterns define complementary domains of neuroepithelium in the developing CNS and are correlated with regions expressing notch ligands. J Neurosci 17:3644–3652
Ma Q, Anderson DJ, Fritzsch B (2000) Neurogenin 1 null mutant ears develop fewer, morphologically normal hair cells in smaller sensory epithelia devoid of innervation. J Assoc Res Otolaryngol JARO 1:129–143
Maison SF, Adams JC, Liberman MC (2003a) Olivocochlear innervation in the mouse: immunocytochemical maps, crossed versus uncrossed contributions, and transmitter colocalization. J Comp Neurol 455:406–416
Maison SF, Emeson RB, Adams JC, Luebke AE, Liberman MC (2003b) Loss of alpha CGRP reduces sound-evoked activity in the cochlear nerve. J Neurophysiol 90:2941–2949
Maison SF, Rosahl TW, Homanics GE, Liberman MC (2006) Functional role of GABAergic innervation of the cochlea: phenotypic analysis of mice lacking GABA(A) receptor subunits alpha 1, alpha 2, alpha 5, alpha 6, beta 2, beta 3, or delta. J Neurosci 26:10315–10326
Maklad A, Fritzsch B (1999) Incomplete segregation of endorgan-specific vestibular ganglion cells in mice and rats. J Vestib Res 9:387–399
Maklad A, Fritzsch B (2003) Development of vestibular afferent projections into the hindbrain and their central targets. Brain Res Bull 60:497–510
Mbiene JP, Favre D, Sans A (1988) Early innervation and differentiation of hair cells in the vestibular epithelia of mouse embryos: SEM and TEM study. Anat Embryol (Berl) 177:331–340
Merchan-Perez A, Bartolome MV, Ibanez MA, Gil-Loyzaga P (1993) Expression of GAP-43 in growing efferent fibers during cochlear development. ORL J Otorhinolaryngol Relat Spec 55:208–210
Merchan-Perez A, Gil-Loyzaga P, Eybalin M, Fernandez Mateos P, Bartolome MV (1994) Choline-acetyltransferase-like immunoreactivity in the organ of Corti of the rat during postnatal development. Brain Res Dev Brain Res 82:29–34
Meza G (2008) Modalities of GABA and glutamate neurotransmission in the vertebrate inner ear vestibule. Neurochem Res 33:1634–1642
Moore JK, Simmons DD, Guan Y (1999) The human olivocochlear system: organization and development. Audiol Neurootol 4:311–325
Morley BJ, Simmons DD (2002) Developmental mRNA expression of the alpha10 nicotinic acetylcholine receptor subunit in the rat cochlea. Brain Res Dev Brain Res 139:87–96
Morris JK, Maklad A, Hansen LA, Feng F, Sorensen C, Lee KF, Macklin WB, Fritzsch B (2006) A disorganized innervation of the inner ear persists in the absence of ErbB2. Brain Res 1091:186–199
Muller M, Jabs N, Lorke DE, Fritzsch B, Sander M (2003) Nkx6.1 controls migration and axon pathfinding of cranial branchio-motoneurons. Development (Cambridge, UK) 130:5815–5826
Murthy V, Maison SF, Taranda J, Haque N, Bond CT, Elgoyhen AB, Adelman JP, Liberman MC, Vetter DE (2009) SK2 channels are required for function and long-term survival of efferent synapses on mammalian outer hair cells. Mol Cell Neurosci 40:39–49
Nichols DH, Pauley S, Jahan I, Beisel KW, Millen KJ, Fritzsch B (2008) Lmx1a is required for segregation of sensory epithelia and normal ear histogenesis and morphogenesis. Cell Tissue Res 334:339–358

Nitecka LM, Sobkowicz HM (1996) The GABA/GAD innervation within the inner spiral bundle in the mouse cochlea. Hear Res 99:91–105

Nordemar H (1983) Embryogenesis of the inner ear. II. The late differentiation of the mammalian crista ampullaris in vivo and in vitro. Acta Otolaryngol 96:1–8

Ohno K, Takeda N, Kiyama H, Kato H, Fujita S, Matsunaga T, Tohyama M (1993) Synaptic contact between vestibular afferent nerve and cholinergic efferent terminal: its putative mediation by nicotinic receptors. Brain Res Mol Brain Res 18:343–346

Ohsawa R, Ohtsuka T, Kageyama R (2005) Mash1 and Math3 are required for development of branchiomotor neurons and maintenance of neural progenitors. J Neurosci 25:5857–5865

Osen KK, Mugnaini E, Dahl AL, Christiansen AH (1984) Histochemical localization of acetylcholinesterase in the cochlear and superior olivary nuclei. A reappraisal with emphasis on the cochlear granule cell system. Arch Ital Biol 122:169–212

Osman AA, Schrader AD, Hawkes AJ, Akil O, Bergeron A, Lustig LR, Simmons DD (2008) Muscle-like nicotinic receptor accessory molecules in sensory hair cells of the inner ear. Mol Cell Neurosci 38:153–169

Pattyn A, Hirsch M, Goridis C, Brunet JF (2000) Control of hindbrain motor neuron differentiation by the homeobox gene *Phox2b*. Development (Cambridge, UK) 127:1349–1358

Perachio AA, Kevetter GA (1989) Identification of vestibular efferent neurons in the gerbil: histochemical and retrograde labelling. Exp Brain Res 78:315–326

Pirvola U, Ylikoski J, Palgi J, Lehtonen E, Arumae U, Saarma M (1992) Brain-derived neurotrophic factor and neurotrophin 3 mRNAs in the peripheral target fields of developing inner ear ganglia. Proc Natl Acad Sci USA 89:9915–9999

Popper P, Ishiyama A, Lopez I, Wackym PA (2002) Calcitonin gene-related Peptide and choline acetyltransferase colocalization in the human vestibular periphery. Audiol Neurootol 7:298–302

Puel JL, Ruel J, Guitton M, Wang J, Pujol R (2002) The inner hair cell synaptic complex: physiology, pharmacology and new therapeutic strategies. Audiol Neurootol 7:49–54

Pujol R, Carlier E, Devigne C (1978) Different patterns of cochlear innervation during the development of the kitten. J Comp Neurol 177:529–536

Puligilla C, Feng F, Ishikawa K, Bertuzzi S, Dabdoub A, Griffith AJ, Fritzsch B, Kelley MW (2007) Disruption of fibroblast growth factor receptor 3 signaling results in defects in cellular differentiation, neuronal patterning, and hearing impairment. Dev Dyn 236:1905–1917

Purcell IM, Perachio AA (1997) Three-dimensional analysis of vestibular efferent neurons innervating semicircular canals of the gerbil. J Neurophysiol 78:3234–3248

Raji-Kubba J, Micevych PE, Simmons DD (2002) Superior olivary complex of the hampster has multiple periods of cholinergic neuron development. J Chem Neuroanat 24:75–93

Rakic P (1977) Prenatal development of the visual system in rhesus monkey. Philos Trans R Soc Lond B Biol Sci 278:245–260

Reuss S, Disque-Kaiser U, Antoniou-Lipfert P, Gholi MN, Riemann E, Riemann R (2009) Neurochemistry of olivocochlear neurons in the hamster. Anat Rec 292:461–471

Roberts BL, Meredith GE (1992) The efferent innervation of the ear: variations on an enigma. In: Webster DB, Fay RR, Popper AN, editors. The Evolutionary Biology of Hearing. New York: Springer Verlag. pp. 185–210

Rohrschneider MR, Elsen GE, Prince VE (2007) Zebrafish *Hoxb1a* regulates multiple downstream genes including prickle1b. Dev Biol 309:358–372

Rusch A, Lysakowski A, Eatock RA (1998) Postnatal development of type I and type II hair cells in the mouse utricle: acquisition of voltage-gated conductances and differentiated morphology. J Neurosci 18:7487–7501

Ryan AF, Simmons DM, Watts AG, Swanson LW (1991) Enkephalin mRNA production by cochlear and vestibular efferent neurons in the gerbil brainstem. Exp Brain Res 87:259–267

Safieddine S, Eybalin M (1992) Triple immunofluorescence evidence for the coexistence of acetylcholine, enkephalis, and calcitonin gene-related peptide within efferent neurons of rats and guinea pigs. Eur J Neurosci 4:981–992

Schrott-Fischer A, Kammen-Jolly K, Scholtz A, Rask-Andersen H, Glueckert R, Eybalin M (2007) Efferent neurotransmitters in the human cochlea and vestibule. Acta Otolaryngol 127: 13–19
Shatz CJ, Ghosh A, McConnell SK, Allendoerfer KL, Friauf E, Antonini A (1990) Pioneer neurons and target selection in cerebral cortical development. Cold Spring Harb Symp Quant Biol 55:469–480
Simmons DD (2002) Development of the inner ear efferent system across vertebrate species. J Neurobiol 53:228–250
Simmons DD, Morley BJ (1998) Differential expression of the alpha 9 nicotinic acetylcholine receptor subunit in neonatal and adult cochlear hair cells. Brain Res Mol Brain Res 56:287–292
Simmons DD, Raji-Kubba J (1993) Postnatal calcitonin gene-related peptide in the superior olivary complex. J Chem Neuroanat 6:407–418
Simmons DD, Manson-Gieseke L, Hendrix TW, McCarter S (1990) Reconstructions of efferent fibers in the postnatal hamster cochlea. Hear Res 49:127–139
Simmons DD, Mansdorf NB, Kim JH (1996a) Olivocochlear innervation of inner and outer hair cells during postnatal maturation: evidence for a waiting period. J Comp Neurol 370:551–562
Simmons DD, Moulding HD, Zee D (1996b) Olivocochlear innervation of inner and outer hair cells during postnatal maturation: an immunocytochemical study. Brain Res Dev Brain Res 95:213–226
Simmons DD, Bertolotto C, Kim J, Raji-Kubba J, Mansdorf N (1998) Choline acetyltransferase expression during a putative developmental waiting period. J Comp Neurol 397:281–295
Simon H, Lumsden A (1993) Rhombomere-specific origin of the contralateral vestibulo-acoustic efferent neurons and their migration across the embryonic midline. Neuron 11:209–220
Smith CA (1961) Innervation pattern of the cochlea. The internal hair cell. Ann Otol Rhinol Laryngol 70:381–394
Sobkowicz HM (1992) The development of innervation in the organ of Corti. In: Romand R (ed) Development of auditory and vestibular systems 2. Elsevier, Amsterdam, pp 59–100
Sobkowicz HM (2001) The development of the GABAergic innervation in the organ of corti of the mouse. CRC, Boca Raton, pp 117–136
Sobkowicz HM, Emmerling MR (1989) Development of acetylcholinesterase-positive neuronal pathways in the cochlea of the mouse. J Neurocytol 18:209–224
Song MR, Shirasaki R, Cai CL, Ruiz EC, Evans SM, Lee SK, Pfaff SL (2006) T-Box transcription factor Tbx20 regulates a genetic program for cranial motor neuron cell body migration. Development 133:4945–4955
Spoendlin H (1966) The organization of the cochlear receptor. Fortschr Hals Nasen Ohrenheilkd 13:1–227
Studer M (2001) Initiation of facial motoneurone migration is dependent on rhombomeres 5 and 6. Development (Cambridge, UK) 128:3707–3716
Studer M, Lumsden A, Ariza-McNaughton L, Bradley A, Krumlauf R (1996) Altered segmental identity and abnormal migration of motor neurons in mice lacking Hoxb-1. Nature 384: 630–634
Tiveron MC, Pattyn A, Hirsch MR, Brunet JF (2003) Role of Phox2b and Mash1 in the generation of the vestibular efferent nucleus. Dev Biol 260:46–57
Usami S, Igarashi M, Thompson GC (1987) GABA-like immunoreactivity in the squirrel monkey vestibular endorgans. Brain Res 417:367–370
Vetter DE, Mugnaini E (1992) Distribution and dendritic features of three groups of rat olivocochlear neurons. A study with two retrograde cholera toxin tracers. Anat Embryol (Berl) 185:1–16
Vetter DE, Adams JC, Mugnaini E (1991) Chemically distinct rat olivocochlear neurons. Synapse 7:21–43
Vetter DE, Liberman MC, Mann J, Barhanin J, Boulter J, Brown MC, Saffiote-Kolman J, Heinemann SF, Elgoyhen AB (1999) Role of alpha9 nicotinic ACh receptor subunits in the development and function of cochlear efferent innervation. Neuron 23:93–103

Wackym PA (1993) Ultrastructural organization of calcitonin gene-related peptide immunoreactive efferent axons and terminals in the vestibular periphery. Am J Otol 14:41–50

Wackym PA, Popper P, Lopez I, Ishiyama A, Micevych PE (1995) Expression of alpha 4 and beta 2 nicotinic acetylcholine receptor subunit mRNA and localization of alpha-bungarotoxin binding proteins in the rat vestibular periphery. Cell Biol Int 19:291–300

Walsh EJ, McGee J, McFadden SL, Liberman MC (1998) Long-term effects of sectioning the olivocochlear bundle in neonatal cats. J Neurosci 18:3859–3869

Warr WB (1975) Olivocochlear and vestibular efferent neurons of the feline brain stem: their location, morphology and number determined by retrograde axonal transport and acetylcholinesterase histochemistry. J Comp Neurol 161:159–181

Whitlon DS, Sobkowicz HM (1989) GABA-like immunoreactivity in the cochlea of the developing mouse. J Neurocytol 18:505–518

Whitlon DS, Zhang X, Pecelunas K, Greiner MA (1999) A temporospatial map of adhesive molecules in the organ of Corti of the mouse cochlea. J Neurocytol 28:955–968

Willmann R, Fuhrer C (2002) Neuromuscular synaptogenesis: clustering of acetylcholine receptors revisited. Cell Mol Life Sci 59:1296–1316

Wilson JL, Henson MM, Henson OW Jr (1991) Course and distribution of efferent fibers in the cochlea of the mouse. Hear Res 55:98–108

Zidanic M (2002) Cholinergic innervation of the chick basilar papilla. J Comp Neurol 445:159–175

Zuo J, Treadaway J, Buckner TW, Fritzsch B (1999) Visualization of alpha9 acetylcholine receptor expression in hair cells of transgenic mice containing a modified bacterial artificial chromosome. Proc Natl Acad Sci USA 96:14100–14105

Chapter 8
Evolution of the Octavolateral Efferent System

Christine Köppl

8.1 Introduction

Efferent, or downstream, modulation of sensory processing is a common attribute of all sensory systems. However, feedback from the brain directly onto the sensory cells themselves, that is, at the transduction stage, is rare (Robertson 2009). The mechanosensory hair cells of vertebrates are arguably the most sophisticated example of this relationship, and much remains to be learned about efferent modulation of their function under natural conditions. In trying to understand the role of the efferent system, it is instructive to consider its evolution. This chapter discusses the various manifestations of efferent innervation to the hair-cell organs across vertebrates and what we know about its function. From these data, major steps in the evolution of the efferent system can be inferred that in turn allow us to deduce some general rules about efferent function and to define interesting questions for future investigation. The emphasis will be on the auditory system, as Goldberg, Lysakowski and Holt, Chap. 6, already discussed the vestibular efferent system across vertebrates. For orientation on the phylogenetic relationships of the various animal groups discussed here, please refer to the recent review in Manley and Clack (2004).

8.2 Anatomical Layout and Neurochemistry of the Efferent System

This section reviews the location and number of efferent neurons in the brain stem of different vertebrates, their connections to hair cells in the different end organs, and the transmitters and neuroactive substances found.

C. Köppl (✉)
Cochlear and Auditory Brainstem Physiology, IBU, Carl von Ossietzky University Oldenburg, 26111 Oldenburg, Germany
e-mail: christine.koeppl@uni-oldenburg.de

D.K. Ryugo et al. (eds.), *Auditory and Vestibular Efferents*,
Springer Handbook of Auditory Research 38, DOI 10.1007/978-1-4419-7070-1_8,

8.2.1 The Origin of Octavolateral Efferents

Efferent innervation is likely to be as old as hair cells, although at present many questions still shroud the origin of hair cells themselves and the early evolution of the inner ear and lateral line, collectively termed the octavolateralis system. There is, however, no doubt that hair cells go back at least as far as the earliest vertebrates (reviews: Coffin et al. 2004; Manley and Ladher 2008). Sensory cells that share crucial similarities with hair cells are found in the coronal organ of sea squirts (tunicates, living close relatives of vertebrates), and have been suggested to be homologous to hair cells (Burighel et al. 2003; Manni et al. 2004, 2006; Caicci et al. 2007). Interestingly, coronal cells typically receive both afferent and efferent synapses, although the neurons providing the efferent terminals have not been identified. The suggestion of homology to hair cells is currently very tentative (see also Northcutt 2005). Several crucial features of coronal cells, for example, their transduction mechanism, are as yet unknown.

Among vertebrates, the interpretation of the early evolution of the inner ear and lateral line is still ambiguous. Much of the current uncertainty is owed to the unresolved phylogenetic status of hagfishes (recent review in Janvier 2008). Hagfishes, together with lampreys, are the only surviving jawless vertebrates. They represent early offshoots from the main vertebrate line, but it remains contentious whether they themselves belong to a monophyletic group or not. In addition, hagfishes and lampreys have long independent histories during which they adapted to rather specialized lifestyles and now differ from each other in many respects. Thus, the common assumption that they represent a plesiomorphic (ancestral) stage of vertebrate evolution clearly does not hold for all of their features. With respect to the hair-cell systems, lampreys may be a reasonable match to the plesiomorphic state if both inner ear and lateral line are considered. They possess an inner ear with two semicircular canal ampullae and a macula communis (Lowenstein et al. 1968), as well as a lateral line consisting of mechanoreceptive neuromasts (Yamada 1973) and ampullary-like electroreceptors (Bodznick and Northcutt 1981; Ronan and Bodznick 1986). Only inner-ear hair cells, but not those of the lateral line, receive efferent innervation in lampreys (Lowenstein et al. 1968; Yamada 1973; Hoshino 1975; Popper and Hoxter 1987; Ronan and Northcutt 1987; Koyama et al. 1990). Tracing studies have located the inner-ear efferents in the brain stem, in the vicinity of facial motorneurons and on the ipsilateral side only (Fritzsch et al. 1989; Koyama et al. 1989). They are immunopositive for choline acetyltransferase, that is, they appear to be cholinergic (Pombal et al. 2001). Hagfishes have a similar but somewhat simpler inner ear (Lowenstein and Thornhill 1970). Their unique "lateral line" is not composed of neuromasts or electroreceptors and varies considerably between species (review: Braun and Northcutt 1997). Efferent innervation appears to be absent altogether in hagfishes, both from the inner ear and the "lateral line." Neither efferent terminals on the sensory cells nor retrogradely labeled neurons after tracer application could be found (Lowenstein and Thornhill 1970; Amemiya et al. 1985; Kishida et al. 1987). Based on developmental studies and strong evidence for a lateral-line system in many fossil jawless vertebrates, Wicht and Northcutt (1995) concluded that hagfishes most likely display a secondary, degenerate state of the lateral line. Assuming this interpretation is correct and, by extension, also applies to the hagfish inner ear, this could mean that the wholesale absence of efferent innervation represents a regressive loss. Following from that, the most parsimonious explanation would then place the origin of efferents at least as far back as the last common ancestor of hagfishes and lampreys and the lack of lateral-line efferents in lampreys would also represent a regressive loss. Alternatively, the condition in lampreys might indicate a later acquisition of efferent innervation, first to the inner ear, then to the lateral line (where it is universally present in later vertebrates; see later). None of these scenarios is more plausible than the other, so at present the issue remains unresolved.

During development, efferents to inner-ear hair cells in birds and mammals derive from the same population of neurons that give rise to the facial motorneurons. If the inner ear, which is their normal target, is ablated early, the prospective efferents partly revert to the facial-motorneuron phenotype (reviews: Fritzsch 1999; Simmons et al., Chap. 7). These findings strongly suggest that octavolateral efferents are, in fact, motorneurons that have re-routed to the inner ear. Facial motorneurons innervate, in the plesiomorphic condition, branchiomeric striated muscles attached to the pharyngeal arches. The evolution of this musculature appears to have been driven by a functional switch in early vertebrates from filter feeding to active respiration through gills and may predate the appearance of the inner ear and lateral line (Northcutt 2005). Thus, it is plausible that octavolateral efferents are as old as vertebrate hair cells themselves and have coevolved with them. However, it is doubtful whether the same neuron population could be the source of efferent innervation in earlier sensory cells discussed as hair-cell homologues, such as the ascidian coronal cells mentioned previously.

8.2.2 *The Plesiomorphic Condition as Seen in Fish*

There is a rich literature on the anatomical layout of the octavolateral efferent system in various species of fish. Although none of the species studied are members of the Sarcopterygii, the line of direct ancestors to land vertebrates, the data are consistent enough in salient aspects to allow the identification of a typical pattern that probably approximates the plesiomorphic and unspecialized case for vertebrates (summarized schematically in Fig. 8.1).

8.2.2.1 A Small Number of Efferent Neurons Innervates a Large Number of Both Lateral-Line and Inner-Ear Hair Cells

In fish, all inner-ear end organs and lateral-line neuromasts receive efferent innervation. Electron microscopic studies have identified efferent terminals in all end

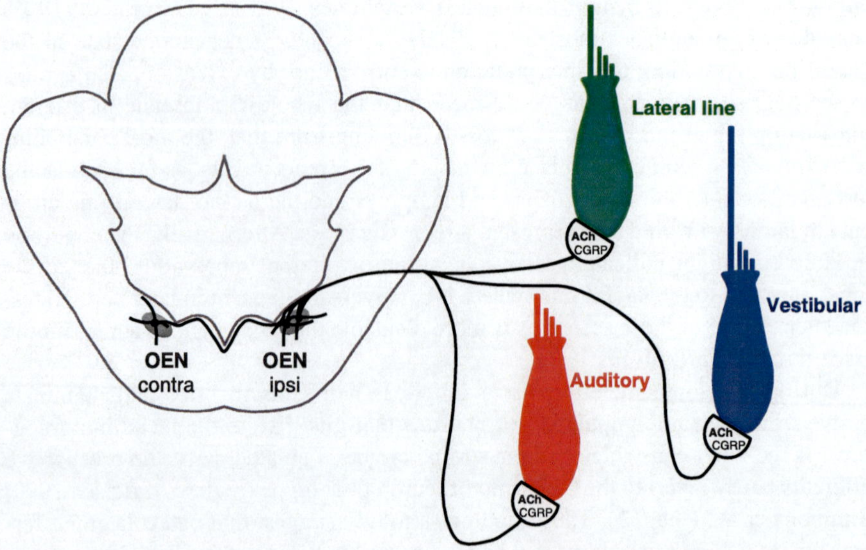

Fig. 8.1 Schematic summary of the plesiomorphic pattern of the octavolateral efferent system, as seen in basal fish and amphibia. The *gray* outline on the *left* shows a schematic cross-section of the brain stem, with a single octavolateral efferent nucleus (OEN, *filled gray*) on each side. Neurons in the OEN (drawn in *black*) have extensive dendritic trees which remain, however, confined to one side of the brain stem. About two-thirds of OEN neurons project their axon to the ipsilateral side, about one third to the contralateral side. An individual axon may innervate hair cells in lateral-line, vestibular and auditory end organs (symbolized by different colors). The principal transmitter is ACh, with CGRP colocalized

organs investigated (Flock 1965; Hama 1965, 1969; Yamada and Hama 1972; Nakajima and Wang 1974; Corwin 1977; Popper and Saidel 1990; Chang et al. 1992) and tracing experiments (see below) found backfilled efferent neurons from every nerve branch tested. Efferent terminals typically synapse directly onto the hair cells (Hama 1965, 1969; Yamada and Hama 1972; Corwin 1977; Popper and Saidel 1990; Chang et al. 1992), whereas axodendritic synapses onto afferent fibers appear to be rare exceptions (Flock 1965; Nakajima and Wang 1974). This does not necessarily mean that every single hair cell is efferently innervated. The available, limited evidence suggests that there may be considerable variation in the prevalence of efferent terminals on hair cells of different end organs and even within one end organ (Yamada and Hama 1972; Wegner 1982; Popper and Saidel 1990; Chang et al. 1992; Edds-Walton et al. 1999).

The source of this efferent innervation is typically a single hindbrain region termed the octavolateral efferent nucleus (OEN, by Meredith and Roberts 1986). The OEN is located close to the medial longitudinal fasciculus of the reticular formation, and at a similar rostrocaudal position to the facial motorneurons, the Mauthner cell, and the VIIIth nerve entry. It typically consists of only a few tens of cells on each side (Claas and Münz 1980; Bell 1981; Zottoli and van Horne 1983; Meredith and Roberts 1986, 1987; Bleckmann et al. 1991; Schellart et al. 1992;

New and Singh 1994; Wagner and Schwartz 1996; Tomchik and Lu 2006a). Careful tracing experiments, restricted to different end-organ branches, tended to yield similar numbers of backfilled neurons to bulk labeling of several branches (Claas and Münz 1980; Bell 1981; Meredith and Roberts 1986, 1987). Extensive branching of individual efferent neurons to many peripheral targets was thus inferred early on. More recently, double-labeling with differently colored fluorescent tracers confirmed that a substantial proportion of efferent neurons project axon collaterals into more than one major nerve branch and thus appears to innervate many end organs (Bleckmann et al. 1991; Tomchik and Lu 2005). Intriguingly, about 30% of efferents were seen to innervate both inner-ear and lateral-line targets (Bleckmann et al. 1991), consistent with direct observations of collateral branching into both VIIIth nerve and lateral-line nerve branches in a range of different species (Claas et al. 1981; Metcalfe et al. 1985; Meredith and Roberts 1986). Although no cases of single efferents labeled in their entirety have yet been reported, individual collaterals showed large terminal fields within the sacculus, much larger than that of any afferent fiber (Edds-Walton et al. 1999). Similarly, in the lateral line, individual efferent axons were seen to innervate several consecutive neuromasts (Sapede et al. 2005). In summary, these data suggest that the plesiomorphic condition for octavolateral efferent neurons is an extensive peripheral target area, both in terms of numbers of hair cells contacted and in terms of number and variety of target end organs.

8.2.2.2 Are There Any Subpopulations of Efferents?

The fact that a substantial proportion of efferents branches profusely does not preclude, of course, the co-existence of more restricted subpopulations. However, the scope for that appears limited in the face of the typically low total number of efferents. Furthermore, several studies that have probed for regional differentiation within the OEN, according to peripheral targets, have found none (Claas and Münz 1980; Bell 1981; Meredith and Roberts 1986; Bleckmann et al. 1991). In cases were there is some evidence for differential innervation it appears to be mostly between different parts of the lateral line (New and Northcutt 1984; Meredith and Roberts 1987; Song and Northcutt 1991; Wagner and Schwartz 1996; Bricaud et al. 2001). However, less effort has been exerted exploring any possible segregation of efferents to the different end organs of the inner ear. A well-known exception is the oyster toadfish, *Opsanus tau* (see also Holt et al., Chap. 6), a member of the advanced group of neoteleost fishes and a vocalizing species. The toadfish has an unusually high number (~400) of octavolateral efferent neurons and shows evidence of selective innervation to the semicircular canals (Highstein and Baker 1986).

Several researchers have observed a division of the OEN into two distinct cell groups, located rostral and caudal relative to one another (Claas and Münz 1980; Zottoli and van Horne 1983; Metcalfe et al. 1985; Puzdrowski 1989; Song and Northcutt 1991; New and Singh 1994; Wagner and Schwartz 1996). The significance of this division remains unclear, and salient characteristics appear to differ between species. For example, in some cases the two cell groups have different peripheral projection patterns along the lateral-line system (Metcalfe et al. 1985; Song and Northcutt 1991), whereas in others no differentiation was found (Claas and Münz 1980; Puzdrowski 1989; Wagner and Schwartz 1996). Furthermore, there is no obvious phylogenetic pattern to the presence or absence of such subpopulations. They have been reported for the Florida gar *Lepisosteus platyrhincus*, a basal ray-finned fish (Song and Northcutt 1991) and for several species from more advanced teleost groups. Conversely, in closely related species, such as two species of catfish, one may show it (New and Singh 1994) but another one not (Bleckmann et al. 1991). Taken together, these observations suggest that subdivisions of the OEN may have arisen several times independently in fish and for different reasons. In the zebrafish, *Danio rerio*, cell migration patterns during early development indicated an independent origin of its caudal OEN subdivision, possibly from the glossopharyngeal motorneuron population (Sapede et al. 2005; see also Simmons et al., Chap. 7).

Another specialized development is an additional efferent cell group in the diencephalon (Zottoli and van Horne 1983; Metcalfe et al. 1985; Puzdrowski 1989). Interestingly, this specialization has only been found in some species of otophysan fish, an advanced group of teleosts renowned for its sensitive, high-frequency hearing (Popper and Fay 1999). However, it appears unlikely that the additional efferents are directly related to the hearing specialization. Diencephalic efferents are few in number and the rare cases that have been, at least partly, traced, indicate a profuse axon branching pattern that not only includes both major lateral-line nerves and the VIIIth nerve but also collaterals within the brain and a projection into the spinal cord (Metcalfe et al. 1985; Bricaud et al. 2001).

8.2.2.3 Bilateral Distribution of Efferent Somata and Dendrites

To the author's knowledge, there is no evidence in fish for octavolateral efferents sending output (i.e., axon collaterals) to both ipsi- and contralateral sides. With respect to possible sources of input to the efferents, the question whether efferent cell bodies typically reside ipsilateral to the innervated body side or distribute bilaterally has caused some confusion. In the small-spotted catshark, *Scyliorhinus canicula*, a cartilaginous fish, and in the Florida gar, a basal bony fish, the OENs of both sides are clearly separate nuclei and about one-third of the neurons projects to end organs on the contralateral side (Meredith and Roberts 1986; Song and Northcutt 1991). A similar pattern can be inferred for the shovelnose sturgeon, *Scuphirhynchus platorynchus*, another basal bony fish (New and Northcutt 1984). Furthermore, although not specifically stated, description and illustrations for the small-spotted catshark suggest that the dendrites of efferent neurons remain unilateral to the soma side (Meredith and Roberts 1986). A reasonable guess for the plesiomorphic condition would thus be that efferent axons innervating end organs on one side of the body originate bilaterally in the hindbrain, but both soma and dendrites of an individual neuron remain confined to one side, either ipsi- or contralateral to the innervated ear.

The more advanced, teleost group of bony fishes shows a different configuration that is likely to be derived but may not be that different functionally. Here, the OENs tend to lie more medially and form a continuous midline-spanning region. Only very minor proportions, if any, of contralaterally projecting neurons have been found in teleost species (Claas and Münz 1980; Strutz et al. 1980; Bell 1981; Claas et al. 1981; Zottoli and van Horne 1983; Highstein and Baker 1986; Meredith and Roberts 1987; Puzdrowski 1989; Bleckmann et al. 1991; New and Singh 1994; Wagner and Schwartz 1996; Tomchik and Lu 2005); the only exception is the rainbow trout *Oncorhynchus mykiss* (Schellart et al. 1992). However, several authors emphasized that at least some efferent neurons' dendrites are bilateral, clearly extending over the midline (Claas and Münz 1980; Metcalfe et al. 1985; Highstein and Baker 1986; Meredith and Roberts 1987; Tomchik and Lu 2005). Again, the rainbow trout makes for an interesting exception, with only unilateral dendritic arborizations (Schellart et al. 1992). These observations imply that the efferent output to one body side in fish comes either (1) from bilaterally located neurons with individually unilateral dendrites or (2) from predominantly ipsilateral neurons, some of which have bilateral dendritic trees.

Efferent neurons with bilateral dendritic trees in particular can show far-reaching arborizations (Metcalfe et al. 1985; Highstein and Baker 1986; Tomchik and Lu 2006a; Fig. 8.2). However, not many individual dendritic trees of efferents have been reconstructed in fish, so it is unclear how typical these are.

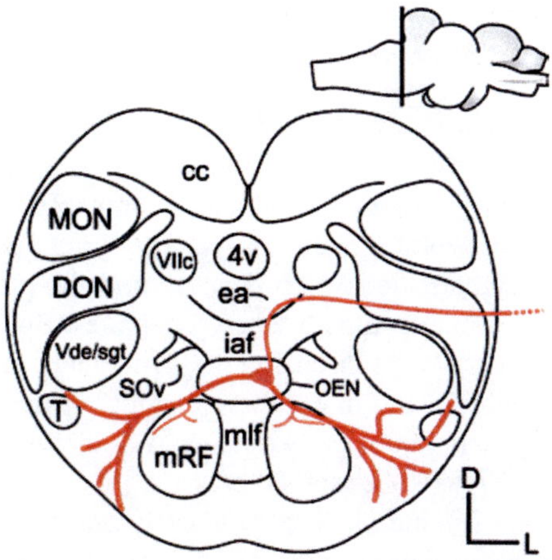

Fig. 8.2 Extensive, bilateral dendritic tree of an individual OEN neuron in the sleeper goby. Line drawing of a cross-section of the brain stem, at the level indicated in the *inset*, and several landmark brain stem regions within it. Note that the OEN forms a single midline-spanning region. A schematic efferent neuron is shown in *red*, based on dye fills of a number of such neurons. *4v* fourth ventricle; *cc* cerebellar crest; *ea* efferent axon(s); *DON* descending octaval nucleus; *iaf* internal arcuate fibers; *mRF* medial reticular formation; *mlf* medial longitudinal fasciculus; *MON* medial octavolateral nucleus; *OEN* octavolateral efferent nucleus; *sgt* secondary gustatory tract; *SOv* ventral secondary octaval population; *T* tangential octaval nucleus; *Vde* descending trigeminal tract; *VIIc* central tract of the facial nerve (reprinted with kind permission from Springer Science+Business Media and Tomchik: Tomchik and Lu 2006a, their Fig. 1a)

8.2.2.4 Efferent Transmitters and Neuropeptides

There is no doubt that efferents in fish are typically cholinergic. Efferent cell bodies and synaptic terminals consistently label for choline acetyl transferase (ChAT; Brantley and Bass 1988; Danielson et al. 1988; Roberts et al. 1994; Anadon et al. 2000; Clemente et al. 2004; Mueller et al. 2004) or acetyl cholinesterase (AchE; Meredith and Roberts 1987; Khan et al. 1991; Sugihara 2001; Clemente et al. 2004). Whether 100% of efferents are cholinergic is, of course, more difficult to show and very few studies have probed for any other neurotransmitters. Most efferents are also immunopositive for calcitonin gene-related peptide (CGRP; Molist et al. 1995) and sequential double-labeling suggested extensive colocalization with ACh (Roberts et al. 1994). As this is also typical of cranial motorneurons (Takami et al. 1985; Martínez-García et al. 2002), it appears that octavolateral efferents have retained that heritage.

Interestingly, the group of diencephalic efferents found in some otophysan fish has a different neurochemical phenotype, providing additional evidence that this is an independently derived specialization. Diencephalic efferents are not cholinergic (Danielson et al. 1988; Mueller et al. 2004) but label positively for tyrosine hydroxylase (Bricaud et al. 2001), suggesting the presence of catecholaminergic transmitters (dopamine, adrenaline, noradrenaline). Hindbrain efferents appear to be negative for tyrosine hydroxylase (Bricaud et al. 2001).

8.2.3 The Most Derived Case: Separate Subsystems of Vestibular and Auditory Efferents of High Complexity in Mammals

The octavolateral efferent system of mammals is arguably the most thoroughly investigated among all vertebrates. Here, only a very brief summary will be provided of the salient characteristics which deviate from the plesiomorphic pattern and are crucial for the comparison with other vertebrates. Both the vestibular and auditory efferents of mammals are discussed in considerably more depth in several other chapters of this volume and in excellent previous reviews (e.g., Warr 1992; Guinan 1996; Goldberg et al. 2000; Lysakowski and Goldberg 2004; Le Prell 2007; Russell and Lukashkin 2008; Robertson 2009).

Of course, mammals (including the secondarily aquatic mammals) have no lateral-line system, so all of the octavolateral efferent innervation goes to the hair cells of the inner ear. Efferents innervating the vestibular end organs and the auditory cochlea, respectively, are completely separate populations. Both populations have greatly increased in number relative to the plesiomorphic condition for vertebrates and, in adult mammals, show no more spatial association with facial motorneurons. Further, both vestibular and auditory efferents form anatomically and functionally distinct subpopulations and have diversified their neurochemical makeup.

The auditory efferents are located in close proximity to the olivary and periolivary nuclei of the afferent auditory pathway. They comprise two major groups, the medial (MOC) and lateral (LOC) olivocochlear efferents, with clearly segregated peripheral targets. MOC efferents target the outer hair cells (OHCs), where they form large terminals on the hair cells' basolateral pole. MOC neurons to one cochlea typically originate in a 1:2 ratio in the ipsi- and contralateral brain stem (uncrossed and crossed MOC efferents). A small proportion (typically <5%) sends an axon to both cochleae. MOC efferents are all cholinergic and may coexpress GABA, but appear to have abandoned CGRP. LOC efferents do not innervate hair cells but instead synapse onto the afferent terminals on inner hair cells (IHCs). They originate predominantly from the brain stem half ipsilateral to the innervated cochlea (uncrossed). LOC efferents express an entended range of neuroactive substances in addition to ACh and CGRP, often in the same individual neuron. Interestingly, both MOC and LOC efferent fibers also send collaterals into the cochlear nucleus and to brain stem vestibular nuclei. No such connections have been observed in any nonmammal.

Unfortunately, there is no information on the olivocochlear efferent system of monotreme or of marsupial mammals. Monotremes in particular would be interesting, as their differentiation of IHCs and OHCs differs in several respects from that in eutherian mammals. It should be enlightening to determine what form and degree of differentiation their olivocochlear efferent system has taken.

8.2.4 *An Intriguing Case with Many Similarities to Mammals: The Archosaurs (Birds and Crocodilians)*

The basilar papilla of archosaurs presents a most intriguing and interesting case of convergent evolution to the mammalian cochlea (reviewed in Manley et al. 1989; Manley and Köppl 1998; Manley and Clack 2004). Although the range of sensitive hearing remains restricted to below 10 kHz in archosaurs, they and mammals have independently developed a differentiation of auditory hair cells into one type taking the classical role of sensory cell (archosaur tall hair cells and mammalian IHCs) and another type specializing in stimulus amplification (archosaur short hair cells and mammalian OHCs). Although there is presently no direct evidence for the assumed amplificatory role of short hair cells, their innervation pattern rules out any primary sensory role. Short hair cells receive large efferent terminals but are devoid of afferent innervation. This may be the only example of a sensory cell losing its primary function entirely (review: Fischer 1994a) and is even more extreme than the mammalian OHCs, which retain a sparse afferent innervation. How much have the octaval efferents specialized in archosaurs? Figure 8.3 schematically summarizes the following discussion of archosaurs.

8.2.4.1 Separation of Auditory and Vestibular Efferents

The question whether vestibular and auditory efferents are separate populations in archosaurs has been a difficult one to resolve. The main practical problem is that

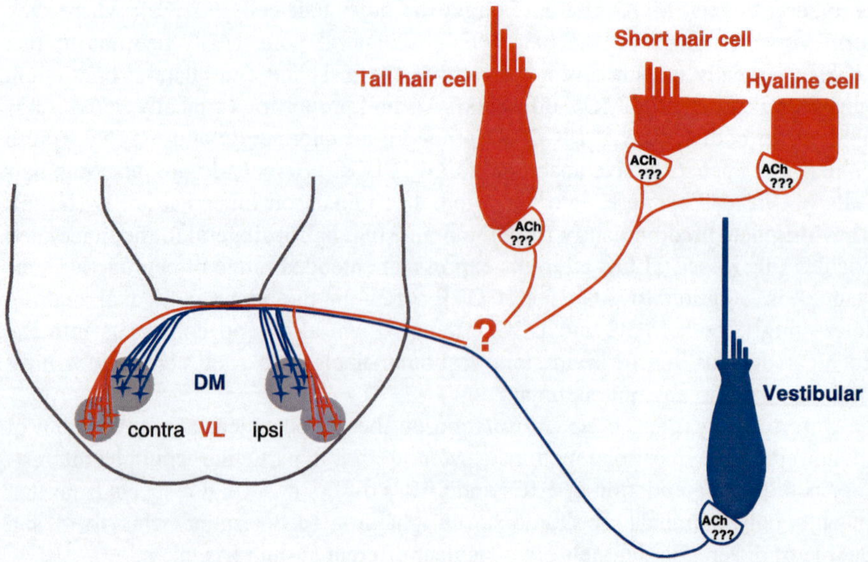

Fig. 8.3 Schematic summary of the pattern of the octavolateral efferent system in archosaurs (birds and crocodilians). The *gray* outline on the *left* shows a schematic cross-section of the brain stem. There are two clusters of efferent neurons on each side of the brain stem, located dorsomedial (DM) and ventrolateral (VL). DM neurons innervate vestibular hair cells and they project ipsi- and contralaterally in about equal numbers; this pathway is shown in *blue*. VL neurons innervate auditory hair cells in the basilar papilla, as well as the nonsensory hyaline cells; this pathway is shown in *red*. There are probably two separate populations of auditory efferents, going to the tall hair cells and short hair cells/hyaline cells, respectively, but this remains to be definitely shown, as indicated by the *question mark*. More auditory VL neurons project contralaterally than ipsilaterally. The principal transmitter is ACh for both auditory and vestibular efferens. No other neurochemicals have been identified

a vestibular end organ, the lagenar macula, is situated at the apical end of the cochlear duct and is supplied by the same branch of the VIIIth nerve as the basilar papilla. Inner-ear efferent neurons are located in the hindbrain, in the general vicinity of facial motorneurons and the superior olive, and form two loosely spaced, partly overlapping clusters, termed dorsomedial and ventrolateral according to their relative positions (reviewed in Code 1997). Retrograde tracing from the cochlear duct consistently labeled neurons in both clusters (Strutz 1981; Whitehead and Morest 1981; Strutz and Schmidt 1982; Cole and Gummer 1990; Schwarz et al. 1992; Kaiser 1993; Code and Carr 1994; Raabe and Köppl 2003a), whereas tracing from vestibular parts of the labyrinth revealed neurons in the dorsomedial cluster only (Schwarz et al. 1981; Strutz 1981; Strutz and Schmidt 1982). Further, neurons selectively backfilled from the lagenar macula were confined to the dorsomedial cell group (Code 1995; Kaiser and Manley 1996). Finally, independently derived counts of efferent axons supplying the basilar papilla and lagenar macula, respectively, agree well with the maximal numbers of neurons labeled in the ventrolateral and dorsomedial cell groups after tracer injections into the cochlear duct. This correlation was shown for both the chicken, *Gallus gallus*, and the barn owl,

Table 8.1 Separation of auditory and vestibular efferents in birds

	Backfilled cell bodies		Efferent axon counts	
	Dorsomedial	Ventrolateral	Lagena	Basilar papilla
Chicken	260	224	220	120
Barn owl	215	1,119	130	1,200

Comparison of independently derived counts of efferent cell bodies backfilled after tracer injections into the cochlea, and efferent axons in the auditory nerve. Maximal numbers of backfilled neurons are given from samples of six chicken and 14 barn owl ears, respectively (Raabe and Köppl, unpublished). Efferent axons were identified via AChE label in ultrathin sections at different levels along the auditory nerve in one specimen each (Köppl 2001). In both species, the number of neurons in the ventrolateral cell cluster corresponds well to the number of efferent axons supplying the basilar papilla, whereas the number of neurons in the dorsomedial cluster approximates the number of axons supplying the lagenar macula. This comparison is one piece of evidence that the ventrolateral cell cluster is exclusively auditory and the dorsomedial cell cluster is vestibular

Tyto alba, two species with very different numbers of basilar-papilla efferents (Köppl 2001; Raabe and Köppl 2003b; Table 8.1). In summary, efferents to the vestibular system thus appear to originate only from the dorsomedial group, whereas the ventrolateral group is exclusively auditory. A small ambiguity remains regarding neurons in the region where both clusters overlap.

Importantly, there appear to be no efferent neurons that innervate both the basilar papilla and any vestibular end organs. In a series of careful tracing experiments using selective application of DiI to the lagenar nerve branchlets in fixed specimens, Kaiser and Manley (1996) labeled large proportions of both afferent and efferent fibers of the lagenar macula. They did not report any labeled collaterals to the basilar papilla, although they observed such collaterals to other vestibular end organs. This suggests strongly that while the lagenar macula may share efferents with other vestibular organs, the supply to the auditory basilar papilla is distinct.

The total number of efferent neurons is not reliably known for any bird species, but an educated guess can be made of typically about 200 vestibular efferents and upwards of 150 auditory efferents. Axon counts of the supply to the basilar papilla and lagenar macula numbered 130–1,070 and 120–220, respectively, in different species (Köppl 2001). These numbers agree well with labeled cell counts in the ventrolateral and dorsomedial clusters from the most successful cochlear-duct tracing experiments (Schwarz et al. 1992; Kaiser 1993; Raabe 2004). Individual vestibular efferents appear to send axon collaterals into several end organs (Kaiser and Manley 1996), such that bulk tracer injections and more selective labeling from individual end organs tend to yield similar numbers of backfilled neurons (Schwarz et al. 1981). Thus, numbers of lagenar efferents may be assumed to approximate the total number of vestibular efferents. Taken together, these data suggest that vestibular efferent numbers vary only moderately between species and are typically around 200. Numbers of auditory efferents, in contrast, vary considerably and show some correlation with auditory specializations. The highest number was observed in the barn owl and is clearly related to other well-known papillar specializations in this species (see Sect. 8.2.4.4). Interestingly, the chicken, a bird with unspecialized hearing and a popular species in auditory research, appears to have an unusually low number of auditory efferents (Köppl 2001).

8.2.4.2 Bilateral Distribution in Archosaurs (Crossed and Uncrossed Efferents)

Efferents to one inner ear are bilaterally distributed in the brain stem. Thus, there is both an uncrossed, ipsilateral, and a crossed, contralateral component to its innervation (Strutz 1981; Whitehead and Morest 1981; Cole and Gummer 1990; Schwarz et al. 1992; Kaiser 1993; Code and Carr 1994; Raabe and Köppl 2003a). The proportions are nearly equal for backfills from the cochlear duct as a whole. However, auditory efferents may be more contralaterally biased than vestibular ones, as neurons in the dorsomedial (vestibular) cluster consistently show a higher ipsi:contra ratio than neurons in the ventromedial (auditory) cluster (Strutz 1981; Whitehead and Morest 1981; Kaiser 1993; Raabe 2004).

Efferent neurons sending axons to both ipsi- and contralateral sides do exist but are a very small population. Their detection is thus partly a function of overall tracing success, which explains why early experiments reported conflicting findings (Cole and Gummer 1990; Schwarz et al. 1992). More recently, a comparison of chicken and barn owl suggested that efferents projecting bilaterally to the cochlear ducts predominantly innervate the auditory basilar papilla as opposed to the lagenar macula (Raabe and Köppl 2003a). In both species, bilaterally projecting efferents comprised no more than 2% of all neurons labeled.

8.2.4.3 Evidence for Subpopulations of Auditory Efferents

Both tall and short hair cells receive efferent innervation, as do the hyaline cells, a nonsensory cell type adjacent to the abneural fringe of the papillar epithelium (von Düring et al. 1974; Oesterle et al. 1992). There is evidence that different populations of neurons may be the sources of efferent innervation to the tall hair cells on the one hand and short hair cells/hyaline cells on the other. However, this issue is not conclusively resolved. Part of the problem is that avian tall and short hair cells themselves are not sharply distinct groups like mammalian inner and OHCs. With respect to most parameters studied, from the name-giving soma shape (review: Gleich and Manley 2000) to the complement of ion channels (review: Fuchs et al. 1998), tall hair cells gradually change into short hair cells as one moves across the basilar papilla from the inner, neural side to the outer, abneural side. Afferent innervation also appears to gradually thin out and is entirely absent in varying proportions of abneurally situated hair cells, depending on tonotopic location and species (Chandler 1984; Fischer 1992, 1994b, 1998; Fischer et al. 1992). Thus, while tall and short hair cells are well defined and clearly different in their extremes, the border between them is somewhat variable and depends on the criterion used.

Like most other parameters, the efferent innervation of tall and short hair cells is clearly different for the extremes. Tall hair cells receive small, bouton-type terminals, in contrast to large, cup-like terminals on typical short hair cells (Takasaka and Smith 1971; Hirokawa 1978; Tanaka and Smith 1978; Firbas and Müller 1983; Chandler 1984; Fischer 1992, 1994b; Fischer et al. 1992). Furthermore, several investigators have observed efferent axons to take two distinct courses, often correlated with different axon diameters. Thicker axons tend to form calyx-like terminals on short hair cells and continue into the hyaline cell area, where they form a network of small terminals. Thinner axons, in contrast, ramify only among the tall hair cells on the inner, neural side of the basilar papilla where they form bouton contacts (Takasaka and Smith 1971; Whitehead and Morest 1981; Keppler et al. 1994; Zidanic 2002; Fig. 8.4). Cross sections of the efferent nerve bundle

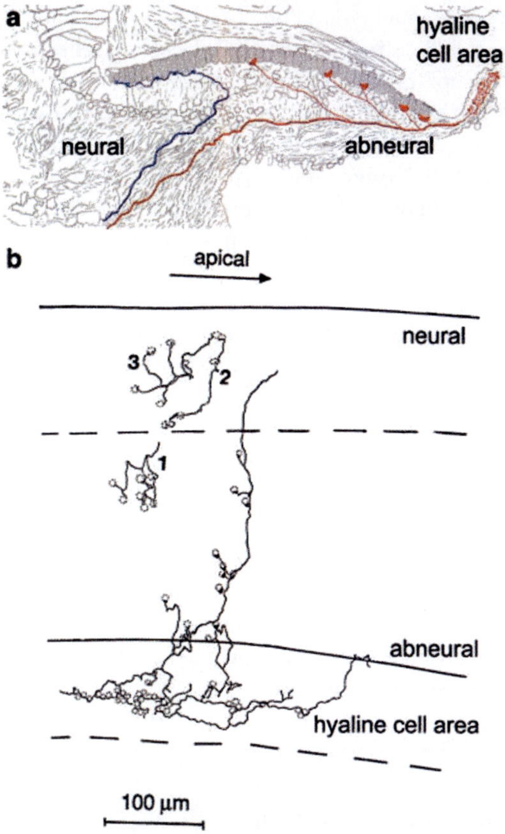

Fig. 8.4 Two populations of auditory efferents in birds, as distinguished by their peripheral branching patterns. (**a**) Schematic drawing of a cross-section of the basilar papilla, with the hair cells shown as *filled gray* shapes. Nerve fibers enter from the neural (inner, superior) side on the *left*. Two representative, efferent axons are shown schematically. The axon in *blue* remains confined to the neural side where is makes small bouton-like contacts on a number of tall hair cells. The axon in *red* runs straight toward abneural, giving off branches that lead to cup-like terminals on short hair cells, and continues to form a network of small terminals on hyaline cells (modified with kind permission from John Wiley & Sons: Zidanic 2002, his Fig. 12a). (**b**) Drawings of individual, dye-filled terminal fields of efferent axons; three examples of axons ramifying in the neural area of tall hair cells are shown (numbered 1–3) and one example of an axon coursing abneurally. *Dotted circles* indicate the nuclei of presumed target hair cells or hyaline cells (modified and reprinted from Keppler et al. 1994, with kind permission from Elsevier and Keppler)

shortly after its exit from the brain stem show myelinated and unmyelinated populations of axons which differ in average diameter (Fischer et al. 1994; Köppl 2001). Thus, the shape of terminals, the diameter and myelination of axons, and the course of axons within the basilar papilla imply that at least two different populations of efferents are the sources of innervation. However, this remains to be definitely shown by tracing individual efferents from the cell body to the peripheral terminals. Even if confirmed, it remains an interesting question how sharply separated those populations are. With respect to terminal sizes, some authors emphasized a gradual enlargement from small terminals on tall hair cells to large ones on short hair cells (Fischer 1992, 1994b; Fischer et al. 1992), whereas others reported an abrupt transition (Zidanic and Fuchs 1996; Zidanic 2002). Axon diameters, while significantly different on average, heavily overlap in their distributions and some of the initially myelinated fibers may lose their myelination at the transition from central oligodendrocyte to peripheral Schwann cell myelin, blurring the distinction between types (Köppl 2001). Finally, at least some of the "larger axons" observed in papillar whole mounts may in fact be aggregations of small axons (Ofsie and Cotanche 1996).

Studies of the emu, *Dromaius novaehollandiae*, a member of the flightless birds and thus a basal sister group of all other modern birds (e.g., Hackett et al. 2008), have suggested a possibly quite recent origin of auditory efferent subpopulations in birds. The emu differs strikingly from all other avian species in that it has a very low proportion of the larger myelinated axons in its efferent nerve bundle (Köppl 2001). Further, the tall and short hair cell populations are not as fully differentiated as in more derived birds (Fischer 1998; Köppl et al. 1998). This condition suggests that the evolution of the specialized short hair cell population may have occurred concurrently with the specialization of a dedicated octaval efferent population innervating them, possibly linked to an extension of the high-frequency hearing range where short hair cells are known to predominate. It should be interesting to explore the state of efferent differentiation in crocodilians, the archosaur sister group of birds, where a surprisingly abrupt transition in shape from tall to short hair cells has been consistently observed (review: Gleich and Manley 2000). There are, however, currently no data relevant to the question of auditory efferent subgroups in crocodilians.

The suggested subgroups of basilar-papilla efferents in birds bear many similarities to the MOC and LOC groups of mammalian cochlear efferents. There is one important difference, however: in birds, all avian efferents form axosomatic contacts directly with the hair cells. Axodendritic synapses on afferent processes were only rarely observed (Rebillard and Pujol 1983; Whitehead and Morest 1985; Fischer 1992) and it is unclear whether these are a regular feature of the mature basilar papilla. The only study on a crocodilian, the caiman, implied that axodendritic contacts are moderately common (von Düring et al. 1974).

Data on efferent soma and dendritic morphology currently show only tentative evidence for any subtypes that might correlate with the different populations suggested to innervate tall hair cells and short hair cells/hyaline cells, respectively. Although two different morphologies, spindle-shaped and radiate, were identified

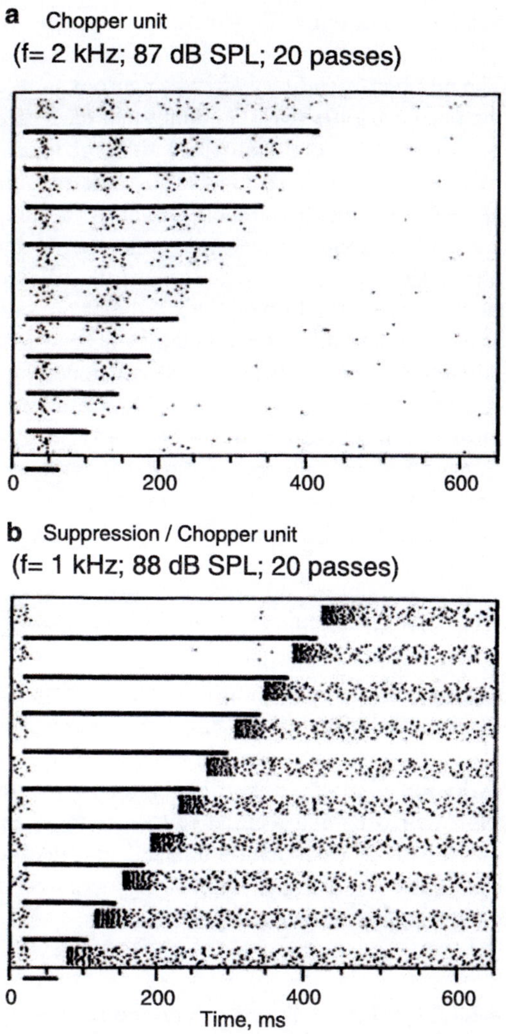

Fig. 8.5 Physiological response types of presumed auditory efferents in the chicken. *Dot* display of tone-evoked activity with variable duration of the test tone (50–400 ms, indicated as *bars*). Each *dot* represents an action potential; responses to 20 repetitions of each stimulus are shown. (**a**) A chopper-type excitatory response. (**b**) A suppressive response, followed by an excitatory off-discharge (used with permission from the American Physiological Society and Kaiser: Kaiser and Manley 1994, their Fig. 8)

early in the chicken (Whitehead and Morest 1981), a classification into distinct groups was not supported by quantitative analyses of soma parameters (Schwarz et al. 1992; Kaiser 1993). Preliminary data from the barn owl suggest that an analysis based on dendritic morphometrics may be more powerful (Raabe 2004). Code and Carr (1994) subjectively classified five neuron types in the chicken, which were homogeneously distributed within the efferent cell clusters.

Perhaps the most tantalizing evidence for subgroups of auditory efferents in birds comes from physiology. In the only study to date of single-unit responses in the efferent cell clusters, Kaiser and Manley (1994) found three distinct response types to acoustic stimulation, including one type that suppressed (Fig. 8.5). No correlations with morphology or peripheral innervation pattern have yet been shown.

8.2.4.4 Tonotopic Distribution Along the Avian Basilar Papilla

The number of auditory efferent neurons in relation to the number of hair cells in the basilar papilla leave no doubt that each efferent axon must supply many cells and there are numerous observations of local branching of efferents among the hair cells (see Sect. 8.2.4.3). Nevertheless, efferents do not distribute homogeneously along the tonotopic gradient. As a rule, fewer efferent fibers enter the very apical, low-frequency regions, compared to more basal locations (Takasaka and Smith 1971; Köppl 2001). Because the basilar papilla progressively widens and carries progressively more hair cells for the same unit distance from base to apex, efferent innervation is predicted to be considerably less dense per hair cell and individual axons should branch considerably more in the apex. Indeed, in the chicken, even a conspicuous void of efferent terminals was observed at apical locations in whole mounts, where only strips of hair cells along the neural and abneural edges, and the hyaline cells, appeared to be innervated (Zidanic and Fuchs 1996; Zidanic 2002). Similarly, in both the starling and the barn owl, Fischer (1992, 1994b) found no efferent synapses on several hair cells of the most apical transects evaluated.

In the only attempt to date to evaluate the differential distribution of possible efferent subpopulations, myelinated and unmyelinated efferent axons were counted repeatedly along the basilar papilla of the barn owl, assuming that any decrease represented axons leaving to innervate hair cells at the corresponding locations (Köppl 2001). Both populations showed distributions skewed towards the specialized high-frequency papillar base, termed the "auditory fovea" of the barn owl (Köppl et al. 1993). The distribution of myelinated axons which may innervate short hair cells and hyaline cells, was most extreme (Fig. 8.6), with about three quarters of that population directed towards locations corresponding to 3–8 kHz. This extreme example implies an increased demand for differentiated efferent control at high frequencies.

8.2.4.5 Efferent Transmitters and Neuropeptides

Acetylcholine remains the main transmitter of the octaval efferents in birds as well. Although there are currently no data on crocodilians, it appears safe to assume that acetylcholine is used by all archosaurs. The somata of efferent neurons, their axons and their terminals have all been shown to label for either AChE (Takasaka and Smith 1971; Firbas and Müller 1983; Köppl 2001) or ChAT (Code and Carr 1994; Medina and Reiner 1994; Ofsie et al. 1997; Zidanic 2002). Both types of axon profiles label, as do all types of efferent terminals in the basilar papilla, on hair cells and hyaline cells. These observations suggest that ACh is present in all efferent subtypes, although not necessarily in 100% of neurons.

Based on double-labeling experiments with ChAT and a generic synaptic marker, Zidanic (2002) concluded that the vast majority, if not all, of the efferent

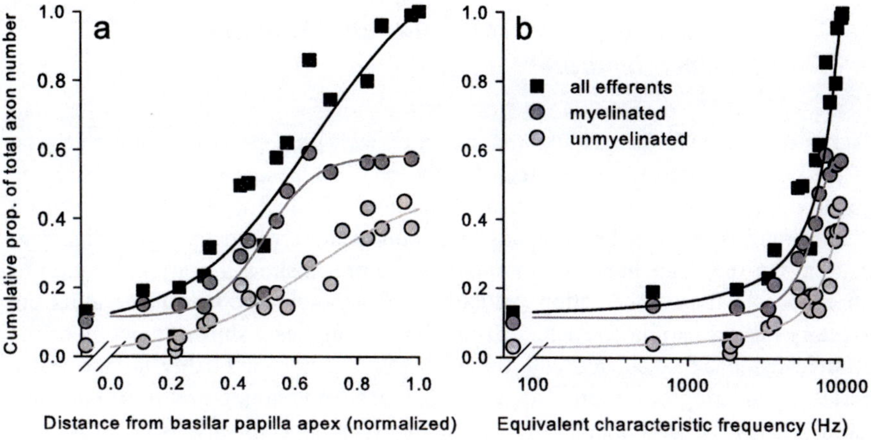

Fig. 8.6 Distribution of efferent fibers along the tonotopic gradient in the basilar papilla of the barn owl. Normalized, cumulative counts for three ears are shown, as a function of papillar location (**a**) or equivalent characteristic frequency (**b**). Data are shown separately for total efferent fiber number, myelinated and unmyelinated axons. As a visual aid, logistic functions (drawn as *lines*) were fitted. Data points to the left of the *x*-axis break represent axon counts in the lagenar fiber supply continuing beyond the papillar apex. Note that the steepest parts of the curves represent the most densely innervated papillar regions; for a full description of the method, see Köppl (2001)

terminals in the basilar papilla are cholinergic. This finding is at odds with Code and Carr's (1994) combined tracing and immunolabeling study which found only about 70% of efferent neurons backfilled from the cochlear duct to be ChAT-positive. As they had a high proportion of backfilled cells in the ventrolateral, auditory cluster, the discrepancy is unlikely to be due to differences between auditory and vestibular efferents. Further, expression of the α9-subunit of hair-cell ACh receptor was absent or at least below detection threshold in most tall hair cells although short hair cells, hyaline cells and vestibular hair cells showed a clear signal (Lustig et al. 1999; Hiel et al. 2000). Similarly, tall hair cells did not respond to ACh applied in vitro whereas short hair cells did (Fuchs and Murrow 1992). These findings are at odds with the consistent observations of cholinergic efferent terminals on both hair-cell types. Finally, Shigemoto and Ohmori (1990) found that both CGRP and glycine potentiated the ACh-evoked response of chicken basilar-papilla hair cells under some conditions in vitro. All considered, the question remains open whether all of the efferent innervation to the avian basilar papilla is cholinergic or whether there is a role for noncholinergic transmission as well.

Intriguingly, all attempts to immunohistochemically label avian efferents for noncholinergic transmitters or neuropeptides have failed so far. Assuming that the presence of both ACh and CGRP represents the plesiomorphic phenotype, birds appear to have abandoned the CGRP (Code et al. 1996). There is also no evidence for GABA, enkephalin or dynorphin (Code and Carr 1995; Code 1996; Zidanic 2002).

8.2.5 When and Why Did Vestibular and Auditory Efferents Separate?

The discussion so far has highlighted two extremes with regard to the differentiation of the octavolateral efferent system. On the one hand, there is the plesiomorphic condition as seen in fish, where a small number of efferent neurons innervates a diverse range of hair-cell end organs in an almost indiscriminate, highly divergent pattern. On the other hand, archosaurs and mammals show a sharp separation of efferents into those innervating vestibular end organs and others innervating the auditory basilar papilla or cochlea. This division implies a shift in function of the efferent feedback associated with the advent of a new submodality in land vertebrates, the hearing of airborne sound. A look at the remaining groups of land vertebrates may be enlightening for deducing more specifically what drove the separation of auditory and vestibular efferents. All major groups of land vertebrates evolved the sensitive hearing of airborne sound independently of each other (review: Manley and Clack 2004). Did they all specialize the efferent system according to the same principles?

8.2.5.1 Amphibians

In amphibians, the octavolateral efferent system appears to have undergone very little, if any, change compared to the plesiomorphic pattern. Tracing studies unanimously report small numbers of efferent brain stem neurons with a highly divergent peripheral innervation pattern. Typically, 10–20 neurons comprise the entire efferent supply to all hair-cell organs on one body side (Fritzsch 1981; Strutz et al. 1982; Will 1982; Fritzsch and Crapon de Caprona 1984; Pellegrini et al. 1985; Gonzalez et al. 1993; Hellmann and Fritzsch 1996; Birinyi et al. 2001). Individual efferent neurons have been observed to send axon collaterals to several end organs (Gonzalez et al. 1993; Birinyi et al. 2001). In larval amphibians and adults that retain a lateral line, there is, as in fish, a subpopulation of efferents supplying both the lateral line and the inner ear (Claas et al. 1981; Will 1982; Fritzsch and Wahnschaffe 1987; Hellmann and Fritzsch 1996). Extensive branching has also been suggested by physiological experiments where electrical stimulation at an isolated VIIIth-nerve branchlet tends to elicit similar evoked potentials or effects on afferent activity in all other VIIIth-nerve branches (Schmidt 1963, 1965; Rossi et al. 1980; Prigioni et al. 1983).

There is little evidence for any regional specialization of efferents with respect to the end organs innervated but not many experiments have addressed the question directly. In a careful tracing study that labeled most VIIIth-nerve branches individually, Birinyi et al. (2001) found tentative evidence for a partial segregation of the efferent supply to the vestibular ampullae vs. the maculae. Unfortunately, they did not include the auditory end organs in their analysis. Amphibians evolved two hair-cell organs specialized for pressure reception of airborne sound, the amphibian papilla and the basilar papilla. All three major groups of amphibians – Gymnophiona (caecilians), Anura (frogs and toads), and Caudata (salamanders and newts) – possess the amphibian papilla whereas the basilar papilla appears to have been lost in most Caudata and some Gymnophiona (reviewed in Lewis and Narins 1999; Smotherman and Narins 2004). The basilar papilla generally codes for the upper frequencies of a species' hearing range and the amphibian papilla tonotopically represents the lower range (Smotherman and Narins 2004). The amphibian papilla was consistently found to be innervated by efferent fibers (Flock and Flock 1966; Robbins et al. 1967; Fritzsch and Wahnschaffe 1987; Simmons et al. 1995; Hellmann and Fritzsch 1996), whereas reports for the basilar papilla vary. Early work had suggested an efferent innervation to the basilar papilla in salamanders (White 1986) but a complete absence thereof in anurans (Robbins et al. 1967; Frishkopf and Flock 1974). However, anurans are clearly heterogeneous and efferents to their basilar papilla have since been observed in the African clawed frog, *Xenopus laevis* (Schucker 1972; Hellmann and Fritzsch 1996) and the cane toad, *Bufo marinus* (unpublished). It is currently unclear whether the efferent innervation to the anuran basilar papilla is so sparse that it may conceivably be missed or whether there has been a true loss in some species. Either way, it appears that efferent innervation has been reduced, which raises the interesting question which conditions might favor that. There is evidence for regional differences in efferent innervation within the frog amphibian papilla as well, such that the caudal part which represents the relatively higher frequencies is less densely or not at all innervated (Flock and Flock 1966; Simmons et al. 1995). Thus, in anurans at least, there may be a correlation between higher-frequency auditory processing and a reduction or loss of efferent feedback.

Another yet unexplained finding is that the efferent cell bodies are found strictly ipsilateral to the innervated hair-cell organs in the Anura and Gymnophiona, whereas they distribute bilaterally in the Caudata. The Caudata are thought to represent the plesiomorphic condition among amphibians (reviews: Fritzsch 1997; Smotherman and Narins 2004). This situation is reminiscent of that in fish where the efferents innervating one body side originate bilaterally in the brain stem in the plesiomorphic condition but remain confined ipsilaterally in the more advanced teleosts (see Sect. 8.2.2.3). Indeed, there is evidence that the dendrites of individual efferent neurons may extend into the contralateral brain stem in gymnophionans and frogs (Strutz et al. 1982; Will and Fritzsch 1988), similar to the derived condition in fish.

Acetylcholine is the main efferent transmitter in amphibians also. This is supported by both histochemical (Lopez and Meza 1988; Gonzalez et al. 1993; Hellmann and Fritzsch 1996) and pharmacological studies (Russell 1971a; Rossi et al. 1980; Caston and Roussel 1984; Sugai et al. 1992). To the author's knowledge, the presence of other transmitters or neuropeptides has not been explored at all.

8.2.5.2 Turtles

Turtles are commonly regarded as something akin to living fossils among modern reptiles. Although their exact classification remains the subject of debate, turtles undoubtedly have preserved a range of ancient features. Their basilar papilla is also regarded as a fairly unspecialized hearing organ, representing an early stage in the hearing of airborne sound (Manley and Köppl 1998). Turtles' hearing is restricted to low frequencies (below about 1 kHz) and only moderately sensitive (Patterson 1966). Efferent innervation is present throughout the turtle basilar papilla (Baird 1974; Sneary 1988). However, only one species, the red-eared turtle, *Chrysemys scripta*, has been examined in any detail (Sneary 1988). Here, there appears to be an interesting shift from about equal numbers of axosomatic and axodendritic efferent synapses in the lowest-frequency, apical regions to axosomatic innervation exclusively in the basal regions. In parallel, the overall efferent innervation density was relatively increased in the basal regions.

In the only tracing study, on the ornate box turtle, *Terrapene ornata*, Strutz (1982) backfilled a surprisingly high number of up to 78 neurons. However, tracer injections into different parts of the inner ear revealed no evidence for any separation between efferent neurons to the basilar papilla and the vestibular end organs, respectively. Individual axons' branching patterns have not yet been followed. Efferent neurons to one inner ear were found on both sides of the brain stem, with the majority originating ipsilaterally. There is the predictable evidence for acetylcholine being the principal transmitter (Art et al. 1984; Powers and Reiner 1993; Dailey et al. 2000) but no study appears to have probed for other neurochemicals.

8.2.5.3 The Lepidosauromorphs (Tuataras, Lizards, Snakes, and Amphisbaenids)

Lizards in general have evolved good hearing, including a significant extension of the hearing range towards higher frequencies. In contrast, snakes and amphisbaenids are sensitive only to low-frequency sounds, below 1 kHz (reviews: Wever 1978; Manley 1990). This condition may well be a regressive development, associated with their burrowing lifestyles or feeding habits, as snakes and amphisbaenids are each closely related to different families of lizard (e.g., Vidal and Hedges 2009). The only lepidosauromorphs that are probably primitively restricted to low-frequency hearing are the rare tuataras (*Sphenodon*) of New Zealand (Gans and Wever 1976; Wever 1978). Although the *Sphenodon* basilar papilla shows many similarities to that of turtles, there are currently no data on innervation.

The basilar papilla of lizards is characteristically subdivided into a high-frequency segment (responding best to frequencies above 1 kHz) and a low-frequency segment (<1 kHz; reviews: Köppl and Manley 1992; Manley 2004). Characteristic anatomical differences between the two regions, such as the hair-bundle orientation pattern or the type of tectorial structure covering them, vary with lizard family but the basic division is always present. The low-frequency segment is typically smaller than the high-frequency segment. For a long time, it was believed that lizards, as a rule, lacked any efferent supply to the high-frequency segment of their basilar papilla, as electron-microscopic studies on a range of lizard families had suggested (Baird 1974; Bagger-Sjöbäck 1976; Mulroy and Oblak 1985; Teresi 1985; Mulroy 1986; Miller and Beck 1988, 1990). However, a consistent cholinergic innervation

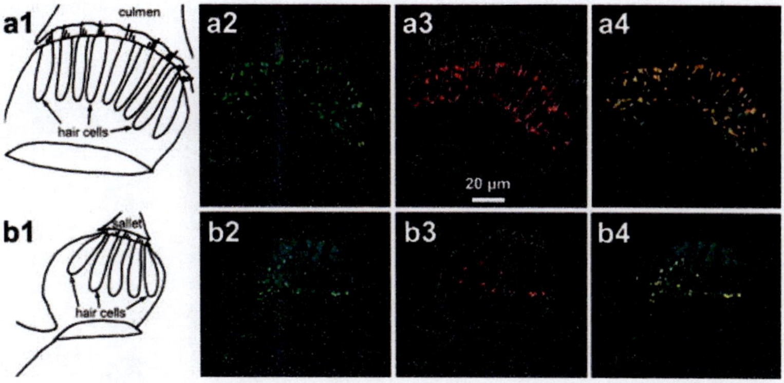

Fig. 8.7 Immunolabeling showing cholinergic terminals in both low- and high-frequency segments of the basilar papilla of a lizard, the three-toed skink, *Saiphos equalis*. Cross sections of the basilar papilla were double-labeled with anti-ChAT (shown *green*) and anti-synaptic vesicle protein 2 (SV2, shown *magenta*). (**a1–4**) A section from the apical, low-frequency segment. (**b1–4**) A section from the basal, high-frequency segment. (**a1**, **b1**) Schematic drawings of the sections for general orientation; for added clarity, the same drawings are overlaid in a3 and b3, respectively. Note the high degree of colocalization between ChAT- and SV2-immunoreactivity, evident as yellow color in the combined images in the third column (modified with kind permission from John Wiley & Sons: Wibowo et al. 2009, their Fig. 7)

to all parts of the basilar papilla was recently demonstrated in scincid lizards, although it was less dense in the high-frequency regions (Wibowo et al. 2009; Fig. 8.7). It is currently unclear why this innervation was not seen earlier, so the issue of efferent innervation to the basilar papilla of lizards needs further clarification. The situation is somewhat reminiscent of that in frogs (see Sect. 8.2.5.1). Here too, sparser efferent innervation to the higher-frequency regions may have led to contradictory results about the presence or absence of efferent terminals. If confirmed, this pattern would represent amazing parallel evolutionary trends in frogs and lizards, toward a reduction or even abolition of efferent feedback to the more recently acquired parts of the hearing organs processing higher-frequency airborne sound above about 1 kHz.

Tracing experiments in one species of monitor lizard (*Varanus exanthematicus*) yielded about 60 backfilled neurons, the majority of which (>85%) originated ipsilateral to the innervated ear (Barbas-Henry and Lohman 1988). Intriguingly, those data indicated the presence of two efferent cell groups in the brain stem, reminiscent of the dorsomedial and ventrolateral groups in archosaurs. However, attempts to label auditory and vestibular efferents separately have not been successful, so this remains speculative and needs further investigation. Evidence for the presence of acetylcholine (Medina et al. 1993; Wibowo et al. 2009) is consistent with all other vertebrates, but no other neurochemicals have been specifically looked for. A general study on the distribution of CGRP-immunoreactive structures in the brain of the Iberian wall lizard, *Podarcis hispanica*, did not identify any prospective octaval efferents (Martínez-García et al. 2002). Figure 8.8 summarizes schematically what is known about octavolateral efferents in lizards.

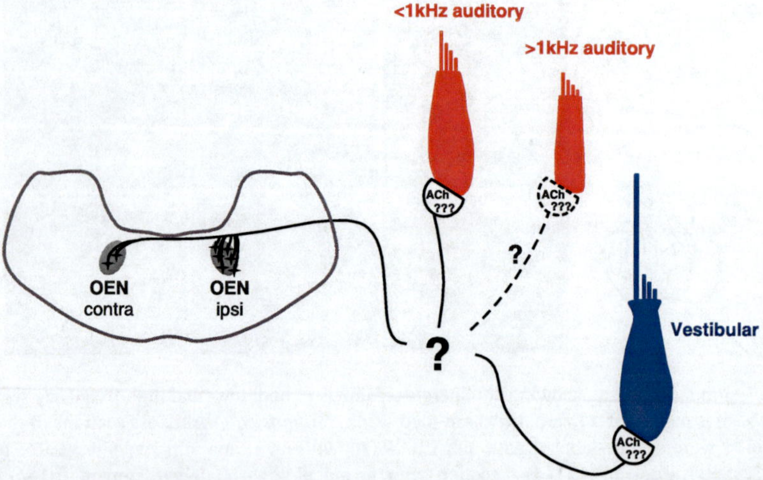

Fig. 8.8 Schematic summary of the pattern of the octavolateral efferent system in lizards. The *gray* outline on the *left* shows a schematic cross-section of the brain stem, with a single octavolateral efferent nucleus (OEN, *filled gray*) on each side. Most OEN neurons project to the ipsilateral side, only a minority appears to send their axon contralaterally. The projection pattern of individual neurons is unknown; they may contact both auditory and vestibular hair cells or have more restricted targets. It is also unclear whether all auditory hair cells receive efferent innervation. The innervation of the high-frequency segment of the basilar papilla may vary among lizard families, as indicated by the *dashed line*. The principal transmitter is ACh, however, other neurochemicals have not been explored

8.3 Function of Efferent Innervation to Hair Cells

8.3.1 Transferring the Efferents' Neurochemical Heritage to Hair Cells

When discussing the function of an efferent innervation to the hair-cell end organs, the origins of the system, i.e. cranial motorneurons, need to be considered. What are their basic characteristics that octavolateral efferents would have inherited originally and how have they changed?

8.3.1.1 Cholinergic Inhibition

Perhaps the most consistent feature that octavolateral efferents have carried over from their motorneuron origin is the use of acetylcholine as the principal transmitter. However, the receptor that mediates its action on hair cells is of a unique nicotinic acetylcholine receptor (nAChR) subtype not expressed in muscle cells and the effect on hair cells is a predominantly inhibitory one. At least two variants of the nAChR $\alpha 9$ appear to be expressed in fish hair cells and may represent the plesiomorphic, heteromeric, native receptor (Drescher et al. 2004). In mammals, the native receptor in both vestibular and auditory hair cells is a nAChR α9/10 heteromer, with the α10 subunit possibly being a uniquely mammalian variant (Drescher et al. 2004; Franchini and Elgoyhen 2006). Ligand binding leads to calcium influx, which in turn opens Ca^{2+}-dependent K^{+}-channels that eventually hyperpolarize the hair cell and reduce its electrical resistance. This basic, fast-acting cholinergic action is conserved in lateral-line, vestibular and auditory hair cells across a wide variety of vertebrates and is discussed in more detail in Katz, Elgoyhen, and Fuchs, Chap. 5.

8.3.1.2 CGRP

Many motorneurons, including the great majority of facial motorneurons across many vertebrates, immunostain for both acetylcholine and CGRP (Takami et al. 1985; Martínez-García et al. 2002). The functions that have been suggested for CGRP include a role in inducing and maintaining postsynaptic nAChR expression, a trophic influence on motorneuron survival, and regeneration and enhancement of the ACh-induced muscle contraction (reviewed in Changeaux et al. 1992; Micevych and Kruger 1992; Arvidsson et al. 1993; Daniels 1997). Although CGRP can clearly be released upon electrical stimulation, it is not entirely clear if and how its release is linked to the release of ACh. Interestingly, there is evidence that the type and/or activity of target muscles has considerable feedback influence on the expression level of CGRP in the motorneuron (Arvidsson et al. 1993; McWilliam et al. 1995).

Many octavolateral efferents retained the co-localization of CGRP, however, there may be a trend toward a reduction and specialization of its expression in more advanced efferent systems. In the only fish species studied (European eel, *Anguilla anguilla*), most efferent neurons were, like motorneurons, double-positive for ChAT and CGRP (Roberts et al. 1994). In contrast, the proportions of CGRP-positive efferents appear to be reduced in both vestibular and auditory efferents in mammals (Ohno et al. 1991; Vetter et al. 1991). Further, CGRP is then only found in defined subpopulations of efferents, which make a predominance of axodendritic contacts onto afferent boutons or calyces (Vetter et al. 1991; Goldberg et al. 2000). As discussed above (Sect. 8.2.4.5), birds may have abandoned CGRP in the efferent system altogether, although this awaits further investigation, as CGRP can be difficult to demonstrate immunohistochemically (Arvidsson et al. 1993) and there is evidence for an effect of CGRP on chicken auditory hair cells in vitro (Shigemoto and Ohmori 1990).

The function of CGRP in the octavolateral efferent system is poorly known. Developmental or trophic-like effects, similar to those suggested for CGRP in motorneurons, have not been explored so far. The most extensive studies on the potential role of CGRP have been carried out on the lateral line of the frog *Xenopus laevis* (see also Sewell, Chap. 4). Here, there is evidence for a slow-acting (over minutes) reduction of afferent sensitivity, over and above that induced by ACh (Sewell and Starr 1991; Bailey and Sewell 2000). In contrast to the action of ACh, however, this effect is largely brought about by an increase in the spontaneous rate of afferent fibers (Fig. 8.9a). Given that axodendritic efferent terminals are very rare

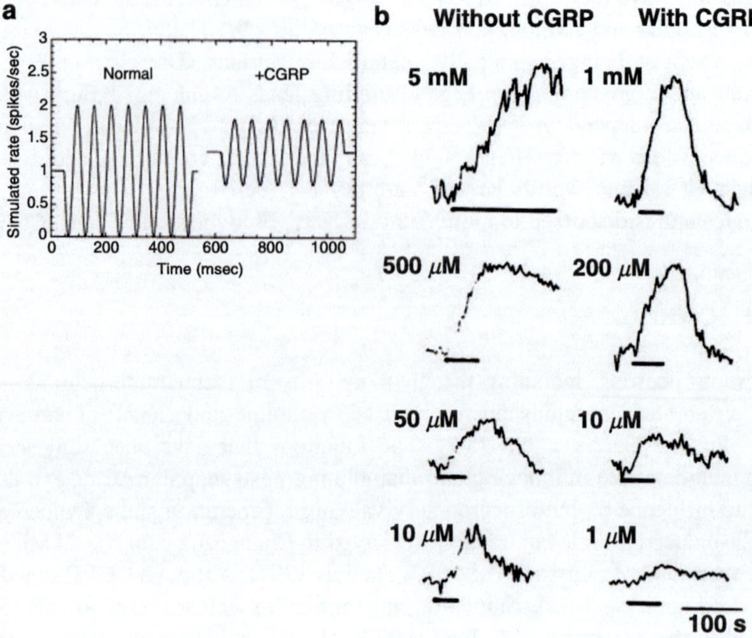

Fig. 8.9 Effects of CGRP in nonmammalian preparations. (**a**) Conceptual summary of the effects of CGRP application to an in vitro preparation of the fish lateral line. Simulations of afferent discharge are shown, which follow a sinusoidal stimulus in a cycle-by-cycle fashion. In the presence of CGRP, spontaneous rate is increased and the depth of modulation is reduced, i.e., the response is desensitized (reproduced with permission of Society for Neuroscience via Copyright Clearance Center, and Sewell: Bailey and Sewell 2000, their Fig. 8). (**b**) Effects of CGRP pre-incubation on the responses of isolated chicken basilar-papilla hair cells to short puffs of ACh (indicated by the *bars*). Without CGRP (*left column*), increasing concentrations of ACh triggered increasingly large, transient rises in intracellular [Ca^{2+}]. After preincubation with CGRP, significantly lower concentrations of ACh (as indicated next to each graph) produced the same responses (modified with kind permission from Wiley-Blackwell, UK, and Shigemoto and Ohmori (1990), their Fig. 4)

in lateral-line neuromasts (Flock 1965; Jande 1966), the source of this appears to be axosomatic efferent terminals causing an increased spontaneous transmitter release by the hair cells (Bailey and Sewell 2000). Net excitatory effects of efferent stimulation are commonly observed in vestibular afferents of many vertebrates (review: Goldberg et al. 2000), which has led to the suggestion that perhaps these are also mediated by CGRP (e.g., Bailey and Sewell 2000) but this idea has not been experimentally addressed so far. In isolated avian basilar-papilla hair cells, CGRP did not exert any effects on its own, but, after long-term exposure, potentiated the initial Ca^{2+}-influx in response to ACh application (Shigemoto and Ohmori 1990; Fig. 8.9b). A chronic excitatory effect of CGRP was confirmed for the mammalian cochlea (Maison et al. 2003; Le Prell et al. 2007). However, the precise mechanisms are likely to be different to lateral-line and avian hair cells, as CGRP in the cochlea mainly acts postsynaptically on afferent terminals instead of directly on the hair cells. In summary, CGRP appears to act on a much slower time scale than ACh and is thus predestined for a modulatory role. Its main effect appears to be excitatory on the afferent output. However, the underlying mechanisms and the physiological roles may differ across lateral-line, vestibular and auditory systems.

8.3.2 *Adding New Levels of Sophistication to the Auditory Efferents*

8.3.2.1 Specializing Together with the Hair Cells: Modulating the Cochlear Amplifier

Specialization of an auditory hair-cell subpopulation for mechanical amplification has occurred at least twice independently, in archosaurs and mammals (see Sect. 8.2.4). In both cases, there is a concomitant specialization of the efferent innervation. In mammals, the OHCs receive large efferent terminals from neurons of the medial olivocochlear subpopulation (MOC). In archosaurs, the short hair cells are similarly surrounded by large cup-like terminals that likely originate from only a subpopulation of auditory efferents. For both cases, the classic net hyperpolarizing effects of ACh have been well documented in vitro (e.g., Housley and Ashmore 1991; Fuchs and Murrow 1992). The physiological role of this hyperpolarization has been most extensively studied by stimulating the MOC system in the mammalian cochlea and observing its desensitizing effect on various physiological responses, predominantly at low to medium sound levels and around the characteristic frequency of the respective response (reviewed in Guinan, Chap. 3). From these studies, the concept of efferents adjusting the gain of mechanical amplification by the OHCs has emerged and has become firmly associated with cholinergic inhibition. There appear to be at least two mechanisms by which the efferent hyperpolarization affects outer OHC electromotility: a fast effect as a direct consequence of the membrane hyperpolarization and associated reduction of the receptor-potential driving force, and a slow effect mediated by Ca-triggered second messenger signaling and protein phosphorylation resulting in stiffness changes (reviews: Russell and Lukashkin 2008; Guinan, Chap. 3).

Mechanical amplification intimately associated with the hair cells' transduction apparatus (commonly called hair-bundle motility) appears to be an ancient feature of all hair cells (e.g., Manley 2001; Martin 2008). Somatic electromotility, however, is a novel mechanism, unique to mammalian OHCs (e.g., Okoruwa et al. 2008). It is therefore an obvious but largely unexplored question whether modulation of hair-cell amplification is a plesiomorphic function of the efferent system or a later evolutionary novelty. Birds are not immediately enlightening here, as their hair-cell and efferent subpopulations are also highly derived and specialized. Birds do imply one important aspect, however: that efferent hyperpolarization is able to modulate the plesiomorphic hair-bundle amplifier. This is not an entirely trivial point, as none of the mechanisms that are currently discussed to underlie hair-bundle motility is directly voltage-dependent. However, Ca^{2+}-entry through the transduction channels

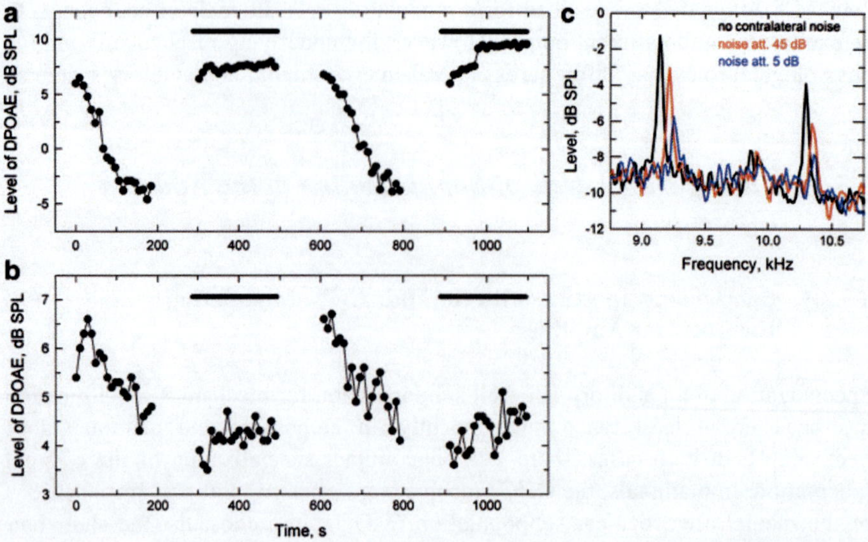

Fig. 8.10 Influence of presumed crossed auditory efferent activity on otoacoustic emissions in the barn owl. (**a, b**) Repeated measurements over time of the amplitude of a distortion product otoacoustic emission in response to a standard primary-tone pair. The *bars* indicate periods of contralateral noise stimulation to activate crossed efferents. Examples from two different experiments, on different owls, are shown (**a, b**). Emission amplitudes were not consistently enhanced or suppressed during periods of presumed efferent stimulation, but appeared to stabilize. (**c**) Effect on spontaneous otoacoustic emissions. Power spectra of the sound field in an owl's ear canal showing two prominent peaks corresponding to spontaneous emissions in quiet (*black line*). Contralateral noise stimulation induced an upward shift of emission frequencies and a decrease in their amplitudes which was graded with noise amplitude, i.e., with presumed efferent activity level (*colored lines*; note that level is given in dB attenuation, such that 45 dB att. indicates the lower-level noise) (modified and reprinted from Manley et al. (1999), their Figs. 3 and 7, with kind permission from Elsevier and Manley)

is a crucial modulator of hair-bundle motility and would be affected by any change in membrane potential (Martin 2008). Whether this is indeed a plausible scenario remains to be investigated. The strongest, albeit indirect, experimental evidence for an efferent modulation of cochlear amplification in birds are the clear effects of contralateral sound stimulation, presumed to activate crossed auditory efferents, on spontaneous and distortion-product otoacoustic emissions in the barn owl (Manley et al. 1999; Fig. 8.10).

In the plesiomorphic condition, hair-bundle motility is likely to be inextricably linked with electrical resonance of the hair cell's basolateral membrane and both may be concomitantly affected by cholinergic efferent activity. In turtle basilar-papilla hair cells, which display both hair-bundle motility and electrical resonance, efferent activity causes desensitization and a profound loss of frequency tuning (Art et al. 1985). The exact mechanisms and the relative contributions of efferent effects on hair-bundle motility and electrical resonance to these outcomes are as yet unknown. Turtle hair cells are confined to characteristic frequencies below about

1 kHz. Chiappe et al. (2007) have argued that at higher frequencies, a separation of transducing and amplifying hair-cell groups should be favored as the optimal operating points for these different tasks become increasingly divergent. They also suggested that this separation has happened at least three times independently, in mammals, archosaurs, and geckos. Intriguingly, the presumed amplifying hair-cell population in geckos (called neural, preaxial or tectorial) appears to receive no innervation at all, neither afferent nor efferent (Chiappe et al. 2007; Wibowo et al. 2008; but see Miller and Beck 1988).

8.3.2.2 Modulating Afferents Instead of Hair Cells: A Mammalian Speciality?

In the auditory system, the lateral olivocochlear (LOC) efferent subsystem of mammals appears to be unique. First, the LOC system synapses almost exclusively onto the afferent terminals, which in turn synapse on IHCs. Such axodendritic efferent contacts are rare exceptions in any nonmammalian auditory end organ (see Sect. 8.2). Second, the LOC system correlates with an unusual abundance of typically 15–20 afferent terminals on a single IHC (reviewed, e.g., in Slepecky 1996). Again, such dense afferent innervation is restricted to very rare exceptions in nonmammals, such as the barn owl (Fischer 1994b). Thirdly, an unusual range of neurotransmitters and neuroactive peptides, in addition to ACh and CGRP, is found in LOC terminals, typically extensively co-localized (e.g., Robertson and Mulders 2000; Le Prell 2007; Ruel et al. 2007; Sewell, Chap. 4). One caveat is that the absence of such a rich complement in nonmammals is currently more an absence of evidence than real evidence of absence (see Sect. 8.2).

The dense afferent innervation of mammalian IHCs comprises two or three overlapping physiological subclasses that are distinguished by spontaneous rate, threshold sensitivity, and dynamic range of discharge (review: Taberner and Liberman 2005). Among the physiological roles discussed for the LOC efferents is the adjustment of afferent spontaneous rate and dynamic response range and they probably do so by exerting both inhibitory and excitatory effects (e.g., Le Prell 2007; Robertson 2009; Guinan, Chap. 3). Physiological subclasses of afferents are also known for birds. However, these appear to emanate from different individual hair cells and thus reflect different inputs (reviewed in Köppl and Yates 1999). The LOC efferents of mammals may be uniquely specialized to establish and fine-adjust physiological subclasses of afferents which signal different aspects of their common hair-cell input to the brain.

8.3.2.3 Branching Out to Nonsensory Cell Types

Some efferent neurons have a third target, besides hair cells and their afferent terminals. In the mammalian cochlea, synapse-like structures have been observed on Deiters and Hensen cells, predominantly in apical regions. These terminals appear to be partly efferent, although their precise origin within the cochlear efferent subgroups is still unresolved (reviewed in Burgess et al. 1997). Deiters cells carry muscarinic receptors (Khan et al. 2002) and in vitro application of ACh caused a slow rise of intracellular calcium (Matsunobu et al. 2001), which in turn triggered a mechanical response (Dulon et al. 1994). This raises the intriguing possibility that supporting cells are involved in homeostatic mechanical adjustments which are under efferent control.

Perhaps the most striking example of this – as yet speculative – function are the hyaline cells of the archosaur basilar papilla. This nonsensory cell type adjoins the hair-cell field and populates the outer-most part of the basilar membrane before it anchors to the abneural cartilagenous limbus. Further, hyaline cells contain the contractile elements typical of smooth muscle cells, and actin filaments form a strictly organized pattern in the transverse plane (Drenckhahn et al. 1991; Cotanche et al. 1992; Oesterle et al. 1992). Hyaline cells are thus theoretically in a prime position to modulate local, transverse basilar-membrane tension. As discussed in Sect. 8.2.4.3, hyaline cells are robustly innervated by the same efferent neurons that also contact the short hair cells. Similar to Deiters cells, they respond to ACh with a rise in intracellular calcium, mediated by receptors with muscarinic pharmacology (Lippe et al. 2002). However, it is still unknown whether hyaline cells are actually capable of a mechanical response, so their functional role remains speculative. It has also been suggested that they may have a secretory function (Oesterle et al. 1992).

To the author's knowledge, there are no other examples of nonsensory cell types that receive efferent innervation in auditory or vestibular end organs. Whatever the precise function, it thus appears that this is a relatively late evolutionary novelty, perhaps associated with larger hearing organs and more spatially differentiated macromechanics.

8.3.3 Still an Enigma: Natural Conditions of Efferent Activity

One of the most persistent difficulties in studying the function of the octavolateral efferents at the systems level is that they appear to have no mandatory role in the basic operation of mature hair cells and whole end organs. Elimination of efferent input in a mature system generally causes no changes in basic response parameters in quiet (e.g., Hartmann and Klinke 1980; Robertson 2009). It is thus a matter of elucidating the specific conditions where efferents are active and where their activity makes a difference. There are two conditions which appear to be ancient, plesiomorphic functions of octavolateral efferents: Protection from overstimulation and improvement of signal detection in a noisy environment.

8.3.3.1 Protection from Predictable Damage

Early groundbreaking work on the fish and frog lateral line showed that efferents were consistently activated in the context of voluntary movements (Fig. 8.11a) and that

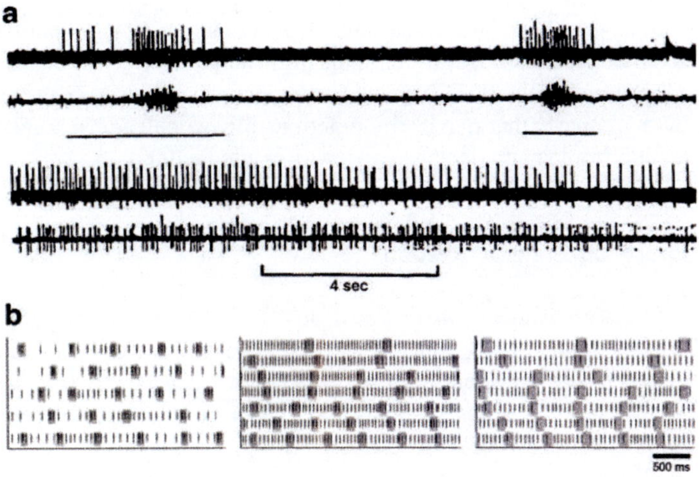

Fig. 8.11 Natural activation of efferents. (**a**) Efferent activity accompanying voluntary movements in the small-spotted catshark. Two pairs of records are shown; the *top trace* in each case represents a recording from efferent fibers running in the posterior lateral-line nerve, the *bottom trace* an electromyogram. In the first example, efferent and motor activity occurred in response to tactile stimulation to the body (stimulus duration marked by *bars*). In the second example, muscle fibers contracted tonically during voluntary movements and this was accompanied by a sustained efferent discharge (adapted with permission from the Company of Biologists and Russell: Roberts and Russell 1972, their Fig. 4). (**b**) Activity of octavolateral efferents during vocalizations in the midshipman fish. Each panel represents the activity of a different neuron, shown as a raster diagram of action potentials (*black*). Efferent discharge changed during fictive vocalizations (*gray boxes*). Note that different patterns occurred, including both excitation and suppression during vocalization (reproduced with permission of Society for Neuroscience via Copyright Clearance Center, and Bass: Weeg et al. 2005, their Fig. 3)

afferent sensitivity was substantially reduced during that time (Russell 1971b; Roberts and Russell 1972; Russell and Roberts 1974). It was therefore suggested that efferent activity may serve to protect the peripheral receptors from overstimulation or fatigue during vigorous self-induced movements (Roberts and Russell 1972). Interestingly, it proved nearly impossible to mimic the natural activation by electrical stimulation of the efferents (Russell and Roberts 1972). Further, while the efferents responded to motor control center activity and a whole range of stimulus modalities, they did not respond to lateral-line stimulation, that is, they did not appear to be part of a closed feedback loop control system (Russell 1971b; Roberts and Russell 1972). A protective effect from self-induced overstimulation in a very similar manner was also suggested for the auditory system of the plainfin midshipman fish, *Porichthys notatus*. In this highly vocal species, activity of the vocal motor system elicited a robust activation of efferents (Weeg et al. 2005; Fig. 8.11b). Again, this was not an acoustically driven closed feedback loop, and the effect was only demonstrable in awake fish.

Protection from overstimulation and even damage has also repeatedly been suggested as a function of the olivocochlear system in mammals (e.g., Liberman and Kujawa 1999; Rajan 2000). Here, however, the effect was shown to be mediated through acoustically activated reflex loops. It has been pointed out that external sounds of damaging level are essentially a phenomenon of our industrialized societies and that protection from such unnatural sounds is likely to be an epiphenomenon (Kirk and Smith 2003). This and other inconsistencies have recently cast doubt on the hypothesis that one of the functions olivocochlear efferents evolved for is protection from sound damage (review: Robertson 2009).

8.3.3.2 Improving Signal Detection

Efferents to auditory end organs, where tested, do respond to sound stimulation (Kaiser and Manley 1994; Guinan 1996; Tomchik and Lu 2006a; Guinan, Chap. 3). In fish and amphibians, where an individual efferent may target many end organs, they also respond to other types of sensory stimulation and higher-order motor control and arousal signals (Schmidt 1963, 1965; Russell 1976; Hartmann and Klinke 1980; Caston and Bricout-Berthout 1982; Tricas and Highstein 1991; Tomchik and Lu 2006a). Whether such multi-modal activation still persists in auditory efferents of birds and mammals has barely been explored (Robertson 2009; Guinan, Chap. 3; Schofield, Chap. 9). Regardless, some sort of acoustically-driven feedback appears to be a basic feature of octavolateral efferent function. One physiological role that has been suggested for both fish and mammals is an antimasking effect of efferent feedback inhibition, improving the signal-to-noise ratio for salient auditory stimuli above environmental noise (Tomchik and Lu 2006b; Robertson 2009; Guinan, Chap. 3; Fig. 8.12).

8.3.3.3 A Role for Efferents in Auditory Development?

Finally, efferents may have an important role in early development which is quite distinct from their influence in the mature system. Relevant data are currently scarce and only available for the mammalian cochlea. Here, two possible developmental roles have emerged for efferents.

First, chronic elimination of crossed MOC efferents early in development led to severe abnormalities in adult cochlear responses suggesting a trophic role for the normal differentiation of OHC function (Walsh et al. 1998). Interestingly, similar long-term, trophic roles are discussed for CGRP in motorneuron development (see Sect. 8.3.1.2). Although CGRP appears not to be expressed in MOC efferents themselves (Raji-Kubba et al. 2002), a whole cascade of events could conceivably lead to the abnormal mature responses observed. It may be worth investigating whether CGRP is involved and whether this may be a common mechanism across vertebrates.

Second, MOC efferents form transient synapses with IHCs, but these are retracted again around the onset of hearing (review: Pujol et al. 1998). Immature IHCs are responsive to ACh at this stage, suggesting the synapses are functional (Glowatzki, Chap. 8). Furthermore, in both birds and mammals, during a limited pre-hearing developmental phase, afferent auditory nerve fibers fire spontaneously in a rhythmic pattern (Jones et al. 2001, 2007). This structured activity is believed to

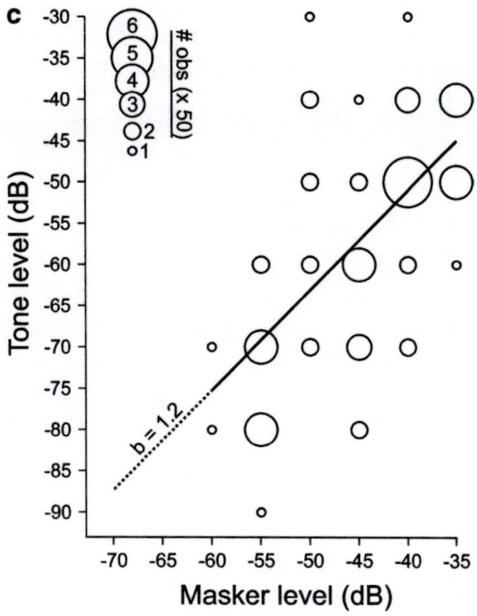

Fig. 8.12 Improvement in the signal to noise ratio of saccular afferent responses during efferent stimulation in the sleeper goby, a fish. A range of different levels of a 200-Hz tone, masked by different levels of broadband noise, was tested. The size of each *circle* indicates the number of times that an increase of signal to noise ratio was observed at the given combination, in a sample of ten fish. The regression line emphasizes the clear correlation of tone and masker levels. Thus, efferent stimulation had the greatest effect at a certain relative level of tone and masker (used with permission from the American Physiological Society and Tomchik: Tomchik and Lu 2006b, their Fig. 10b)

be crucial for the activity-dependent refinement of central circuits, analogous to spontaneous retinal activity waves in the visual system (review: Köppl 2007). If efferent input is eliminated, the rhythmicity of the immature discharge disappears. Thus, the efferents may rhythmically drive or modulate the activity of small groups of hair cells, and generate synchronized patterns of activity in the associated afferent fibers (Walsh and McGee 1997). Interestingly, an alternative mechanism – spontaneous release of ATP by specialized supporting cells – has recently been shown to excite IHCs and elicit similar rhythmic activity in vitro (Tritsch et al. 2007). The relative roles of efferents and supporting cells in this crucial phase of pre-hearing auditory development need to be critically explored.

8.4 Conclusions and Outlook

The landmark comparative review by Roberts and Meredith (1992) was entitled "The efferent innervation of the ear: Variations on an enigma." Nearly 20 years on, the enigma is being solved. While the functions of efferent feedback for normal hearing and vestibular sensation in humans and other mammals are still intensely debated, our understanding of the origins of the efferent system and major trends during its evolution has improved and should help to pose instructive, new experimental questions.

8.4.1 A Plausible Story of Efferent Evolution

The neurons efferent to hair cells are derived from the facial motorneuron population. Why they re-routed from their traditional targets to form the octavolateral efferents innervating hair cells is an interesting and unresolved question (see Fritzsch and Simmons, Chap. 7) and it is likely to have occurred very early in the evolution of hair cells. Initially, there were probably very few octavolateral efferents, servicing a vastly larger number of hair cells in diverse end organs. This plesiomorphic configuration persists in unspecialized modern fish and amphibians. Functional studies in these species suggest that the efferent system was initially a global "shut-off" system to protect peripheral sensors from desensitization and possibly damage during voluntary movements, vocalizations and similar situations of self-induced sensory overload. The same efferent neurons, driven by environmental stimuli, probably also produced an anti-masking effect at more moderate stimulation levels, improving signal detection.

Circumstances changed gradually but dramatically with life on land and the advent of the hearing of airborne sound via a middle ear. The entire system of lateral-line hair cells was lost and with it the immediate, strong coupling of voluntary movements and the need for hair-cell protection in that context. The modes of locomotion changed with land vertebrates gradually raising their bodies above ground and isolating the head from the body via flexible necks. This change would have further reduced the coupling of whole-body movements to the vestibular system and created new demands to maintain balance and head stability. As Weeg et al. (2005) pointed out, in land vertebrates, protection from overstimulation during loud, self-generated vocalizations became the classic role of the middle-ear reflex (which is unrelated to the octavolateral efferents). All of these developments reduced the importance of, perhaps even made redundant, the global protective function of the efferent system. This may have accelerated the evolution of ever more sophisticated, specialized roles for the efferents in signal detection and processing, with much more restricted peripheral targets, resulting also in a complete separation of vestibular and auditory efferents. Salient parallels between the hearing organs in modern archosaurs and mammals suggest that the evolution of hair cells specializing in mechanical amplification was tied to the co-evolution of an efferent population that modulates this amplifier – although geckos might yet prove to be an interesting exception. The extension of the sensitive hearing range to higher frequencies in the kHz range, which happened independently in several vertebrate lines, may have compounded the pressure to specialize. There are some puzzling parallel trends towards reduction or even loss of the efferent innervation in the higher-frequency regions of the hearing organs of frogs and lizards. Assuming that these animals still have no or incomplete separation of vestibular and auditory efferents (which is currently unknown), this condition may indicate an evolutionary watershed of “abandon or separate.” In other words, there may be something about the function of efferents common to vestibular and auditory end organs that is detrimental to higher-frequency auditory processing.

8.4.2 Interesting Open Questions

The enigma is far from solved. Part of the evolutionary story outlined above is still very speculative and rests on assumptions that need to be experimentally tested. There is an especially large gap in knowledge regarding nonmammalian land vertebrates. The question whether any separation of vestibular and auditory efferents has taken place in amphibians and reptiles is a crucial, unanswered one. The LOC efferent system of mammals currently stands out as using an unusually rich cocktail of transmitters and neuroactive substances but hardly any of those substances have been tested for in nonmammalian systems. Archosaurs deserve a closer look. They are tantalizingly similar to mammals in many aspects of the concomitant specialization of auditory hair cells and efferents into subgroups and might reveal important common principles of this division of labor. Also, archosaurs arguably represent the most extreme case of hair cells using bundle motility for mechanical amplification at high frequencies and we know very little about how this type of amplifier is influenced by efferent activity.

These are just some of the questions that offer themselves when looking at the evolution of the octavolateral efferent system. From humble beginnings, efferent neurons have greatly multiplied and diversified in function. The complexity of the mammalian efferent system in particular can be overwhelming in the face of virtually total ignorance as to when the different parts are normally active. It might be more instructive to examine a simpler system. We should not ignore the great potential of studying the natural variations of octavolateral efferents in the context of different species’ neuroethology, in order to solve the enigma of efferent systems.

References

Amemiya F, Kishida R, Goris RC, Onishi H, Kusunoki T (1985) Primary vestibular projections in the hagfish *Eptatretus burgeri*. Brain Res 337:73–80

Anadon R, Molist P, Rodriguez-Moldes I, Lopez JM, Quintela I, Cervino MC, Barja P, Gonzalez A (2000) Distribution of choline acetyltransferase immunoreactivity in the brain of an elasmobranch, the lesser spotted dogfish (*Scyliorhinus canicula*). J Comp Neurol 420:139–170

Art JJ, Fettiplace R, Fuchs PA (1984) Synaptic hyperpolarization and inhibition of turtle cochlear hair cells. J Physiol 356:525–550

Art JJ, Crawford AC, Fettiplace R, Fuchs PA (1985) Efferent modulation of hair cell tuning in the cochlea of the turtle. J Physiol 360:397–421

Arvidsson U, Piehl F, Johnson H, Ulfhake B, Cullheim S, Hökfelt T (1993) The peptidergic motoneurone. Neuroreport 4:849–856
Bagger-Sjöbäck D (1976) The cellular organization and nervous supply of the basilar papilla in the lizard, *Calotes versicolor*. Cell Tissue Res 165:141–156
Bailey GP, Sewell WF (2000) Calcitonin gene-related peptide suppresses hair cell responses to mechanical stimulation in the *Xenopus* lateral line organ. J Neurosci 20:5163–5169
Baird IL (1974) Some aspects of the comparative anatomy and evolution of the inner ear in submammalian vertebrates. Brain Behav Evol 10:11–36
Barbas-Henry HA, Lohman AHM (1988) Primary projections and efferent cells of the VIIIth cranial nerve in the monitor lizard, *Varanus exanthematicus*. J Comp Neurol 277:234–249
Bell C (1981) Central distribution of octavolateral afferents and efferents in a teleost (Mormyridae). J Comp Neurol 195:391–414
Birinyi A, Straka H, Matesz C, Dieringer N (2001) Location of dye-coupled second order and of efferent vestibular neurons labeled from individual semicircular canal or otolith organs in the frog. Brain Res 921:44–59
Bleckmann H, Niemann U, Fritzsch B (1991) Peripheral and central aspects of the acoustic and lateral line system of a bottom dwelling catfish, *Ancistrus* sp. J Comp Neurol 314:452–466
Bodznick D, Northcutt RG (1981) Electro reception in lampreys *Lampetra tridentata*: evidence that the earliest vertebrates were electro receptive. Science 212:465–467
Brantley RK, Bass AH (1988) Cholinergic neurons in the brain of a teleost fish (*Porichthys notatus*) located with a monoclonal antibody to choline acetyltransferase. J Comp Neurol 275:87–105
Braun CB, Northcutt RG (1997) The lateral line system of hagfishes (Craniata: Myxinoidea). Acta Zool 78:247–268
Bricaud O, Chaar V, Dambly-Chaudiere C, Ghysen A (2001) Early efferent innervation of the zebrafish lateral line. J Comp Neurol 434:253–261
Burgess BJ, Adams JC, Nadol JB Jr (1997) Morphologic evidence for innervation of Deiters' and Hensen's cells in the guinea pig. Hear Res 108:74–82
Burighel P, Lane NJ, Fabio G, Stefano T, Zaniolo G, Daniela M, Carnevali C, Manni L (2003) Novel, secondary sensory cell organ in ascidians: in search of the ancestor of the vertebrate lateral line. J Comp Neurol 461:236–249
Caicci F, Burighel P, Manni L (2007) Hair cells in an ascidian (Tunicata) and their evolution in chordates. Hear Res 231:63–72
Caston J, Bricout-Berthout A (1982) Responses of afferent and efferent neurons to visual inputs in the vestibular nerve of the frog. Brain Res 240:141–145
Caston J, Roussel H (1984) Curare and the efferent vestibular system. Acta Otolaryngol 97:19–26
Chandler JP (1984) Light and electron microscopic studies of the basilar papilla in the duck. *Anas platyrhynchos*. I. The hatchling. J Comp Neurol 222:506–522
Chang JSY, Popper AN, Saidel WM (1992) Heterogeneity of sensory hair cells in a fish ear. J Comp Neurol 324:621–640
Changeaux J-P, Duclert A, Sekine S (1992) Calcitonin gene-related peptides and neuromuscular interactions. Ann N Y Acad Sci 657:361–378
Chiappe ME, Kozlov AS, Hudspeth AJ (2007) The structural and functional differentiation of hair cells in a lizard's basilar papilla suggests an operational principle of amniote cochleas. J Neurosci 27:11978–11985
Claas B, Münz H (1980) Bonyfish lateral line efferent neurons identified by retrograde axonal transport of horseradish peroxidase (HRP). Brain Res 193:249–253
Claas B, Fritzsch B, Münz H (1981) Common efferents to lateral line and labyrinthine hair cells in aquatic vertebrates. Neurosci Lett 27:231–235
Clemente D, Porteros A, Weuaga E, Alonso JR, Arenzana FJ, Aijón J, Arévalo R (2004) Cholinergic elements in the zebrafish central nervous system: histochemical and immunohistochemical analysis. J Comp Neurol 474:75–107
Code RA (1995) Efferent neurons to the macula lagena in the embryonic chick. Hear Res 82:26–30

Code RA (1996) Chick auditory terminals contain dynorphin-like immunoreactivity. Neuroreport 7:2917–2920

Code RA (1997) The avian cochlear efferent system. Poultry Avian Biol Rev 8:1–8

Code RA, Carr CE (1994) Choline acetyltransferase-immunoreactive cochlear efferent neurons in the chick auditory brainstem. J Comp Neurol 340:161–173

Code RA, Carr CE (1995) Enkephalin-like immunoreactivity in the chick brainstem: possible relation to the cochlear efferent system. Hear Res 87:69–83

Code RA, Darr MS, Carr CE (1996) Chick cochlear efferent neurons are not immunoreactive for calcitonin gene-related peptide. Hear Res 97:127–135

Coffin AB, Kelley M, Manley GA, Popper AN (2004) Evolution of sensory hair cells. In: Manley GA, Popper A, Fay RR (eds) Evolution of the vertebrate auditory system. Springer, New York, pp 55–94

Cole KS, Gummer AW (1990) A double-label study of efferent projections to the cochlea of the chicken, *Gallus domesticus*. Exp Brain Res 82:585–588

Corwin JT (1977) Morphology of the macula neglecta in sharks of the genus *Carcharhinus*. J Morphol 152:341–362

Cotanche DA, Henson MM, Henson OW Jr (1992) Contractile proteins in the hyaline cells of the chicken cochlea. J Comp Neurol 324:353–364

Dailey SH, Wackym PA, Brichta AM, Gannon PJ, Popper P (2000) Topographic distribution of nicotinic acetylcholine receptors in the cristae of a turtle. Hear Res 141:51–56

Daniels MP (1997) Intercellular communication that mediates formation of the neuromuscular junction. Mol Neurobiol 14:143–170

Danielson PD, Zottoli SJ, Corrodi JG, Rhodes KJ, Mufson EJ (1988) Localization of choline acetyltransferase to somata of posterior lateral line efferents in the goldfish. Brain Res 448:158–161

Drenckhahn D, Merte C, von Düring M, Smolders J, Klinke R (1991) Actin, myosin and alpha-actinin containing filament bundles in hyaline cells of the caiman cochlea. Hear Res 54:29–38

Drescher DG, Ramakrishnan NA, Drescher MJ, Chun W, Wang X, Myers SF, Green GE, Sadrazodi K et al (2004) Cloning and characterization of alpha 9 subunits of the nicotinic acetylcholine receptor expressed by saccular hair cells of the rainbow trout (*Oncorhynchus mykiss*). Neuroscience 127:737–752

Dulon D, Blanchet C, Laffon E (1994) Photo-released intracellular Ca^{2+} evolkes reversible mechanical responses in supporting cells of the guinea-pig organ of Corti. Biochem Biophys Res Commun 201:1263–1269

Edds-Walton PL, Fay RR, Highstein SM (1999) Dendritic arbors and central projections of physiologically characterized auditory fibers from the saccule of the toadfish, *Opsanus tau*. J Comp Neurol 411:212–238

Firbas W, Müller G (1983) The efferent innervation of the avian cochlea. Hear Res 10:109–116

Fischer FP (1992) Quantitative analysis of the innervation of the chicken basilar papilla. Hear Res 61:167–178

Fischer FP (1994a) General pattern and morphological specializations of the avian cochlea. Scanning Microsc 8:351–364

Fischer FP (1994b) Quantitative TEM analysis of the barn owl basilar papilla. Hear Res 73:1–15

Fischer FP (1998) Hair-cell morphology and innervation in the basilar papilla of the emu (*Dromaius novaehollandiae*). Hear Res 121:112–124

Fischer FP, Miltz C, Singer I, Manley GA (1992) Morphological gradients in the starling basilar papilla. J Morphol 213:225–240

Fischer FP, Eisensamer B, Manley GA (1994) Cochlear and lagenar ganglia of the chicken. J Morphol 220:71–83

Flock A (1965) The ultrastructure and microphonic potential of the lateral line canal organ. Acta Otolaryngol 59(suppl 199):7–90

Flock A, Flock B (1966) Ultrastructure of the amphibian papilla in the bullfrog. J Acoust Soc Am 40:1262

Franchini LF, Elgoyhen AB (2006) Adaptive evolution in mammalian proteins involved in cochlear outer hair cell electromotility. Mol Phylogenet Evol 41:622–635
Frishkopf LS, Flock A (1974) Ultrastructure of the basilar papilla, an auditory organ in the bullfrog. Acta Otolaryngol (Stockh) 77:176–184
Fritzsch B (1981) Efferent neurons to the labyrinth of *Salamandra salamandra* as revealed by retrograde transport of horseradish peroxydase. Neurosci Lett 26:191–196
Fritzsch B (1997) On the role played by ontogenetic remodeling and functional transformation in the evolution of terrestrial hearing. Brain Behav Evol 50:38–49
Fritzsch B (1999) Ontogenetic and evolutionary evidence for the motoneuron nature of vestibular and cochlear efferents. In: Berlin CI (ed) The efferent auditory system: basic science and clinical applications. Singular Publishing, San Diego, pp 31–59
Fritzsch B, Crapon de Caprona D (1984) The origin of centrifugal inner ear fibers of gymnophions (amphibia). A horseradish peroxidase study. Neurosci Lett 46:131–136
Fritzsch B, Wahnschaffe U (1987) Electron microscopical evidence for common inner ear and lateral line efferents in urodeles. Neurosci Lett 81:48–52
Fritzsch B, Dubuc R, Otha Y, Grillner S (1989) Efferents to the labyrinth of the river lamprey (*Lampetra fluviatilis*) as revealed with retrograde tracing techniques. Neurosci Lett 96:241–246
Fuchs PA, Murrow BW (1992) Cholinergic inhibition of short (outer) hair cells of the chick's cochlea. J Neurosci 12:800–809
Fuchs P, Zidanic M, Michaels R, Yuhas W, Jiang GJ (1998) Ion channels and synaptic function in chick cochlear hair cells. In: Palmer AR, Rees A, Summerfield AQ, Meddis R (eds) Psychophysical and psychological advances in hearing. Whurr Publishers, London, pp 97–104
Gans C, Wever EG (1976) Ear and hearing in *Sphenodon punctatus*. Proc Natl Acad Sci U S A 73:4244–4246
Gleich O, Manley GA (2000) The hearing organ of birds and crocodilia. In: Dooling RJ, Fay RR, Popper AN (eds) Comparative hearing: birds and reptiles. Springer, New York, pp 70–138
Goldberg JM, Brichta AM, Wackym PA (2000) Efferent vestibular system: anatomy, physiology, and neurochemistry. In: Beitz AJ, Anderson JH (eds) Neurochemistry of the vestibular system. CRC Press, Boca Raton, pp 61–94
Gonzalez A, Meredith GE, Roberts BL (1993) Choline acetyltransferase immunoreactive neurons innervating labyrinthine and lateral line sense organs in amphibians. J Comp Neurol 332:258–268
Guinan JJJ (1996) Physiology of olivocochlear efferents. In: Dallos P, Popper AN, Fay RR (eds) The cochlea. Springer, New York, pp 435–502
Hackett SJ, Kimball RT, Reddy S, Bowie RCK, Braun EL, Braun MJ, Chojnowski JL, Cox A et al (2008) A phylogenomic study of birds reveals their evolutionary history. Science 320:1763–1768
Hama K (1965) Some observations on the fine structure of the lateral line organ of the Japanese sea eel *Lyncozymba nystromi*. J Cell Biol 24:193–210
Hama K (1969) A study on the fine structure of the saccular macula of the gold fish. Z Zellforsch 94:155–171
Hartmann R, Klinke R (1980) Efferent activity in the goldfish vestibular nerve and its influence on afferent activity. Pflügers Arch 388:123–128
Hellmann B, Fritzsch B (1996) Neuroanatomical and histochemical evidence for the presence of common lateral line and inner ear efferents and of efferents to the basilar papilla in a frog, *Xenopus laevis*. Brain Behav Evol 47:185–194
Hiel H, Luebke AE, Fuchs PA (2000) Cloning and expression of the alpha 9 nicotinic acetylcholine receptor subunit in cochlear hair cells of the chick. Brain Res 858:215–225
Highstein SM, Baker R (1986) Organsization of the efferent vestibular nuclei and nerves of the toadfish, *Opsanus tau*. J Comp Neurol 243:309–325
Hirokawa N (1978) The ultrastructure of the basilar papilla of the chick. J Comp Neurol 181:361–374
Hoshino T (1975) An electron microscopic study of the otolithic maculae of the lamprey (*Entosphenus japonicus*). Acta Otolaryngol 80:43–53

Housley GD, Ashmore JF (1991) Direct measurement of the action of acetylcholine on isolated outer hair cells of the guinea pig cochlea. Proc R Soc Lond B 244:161–167

Jande SS (1966) Fine structure of lateral-line organs of frog tadpoles. J Ultrastruct Res 15:496–509

Janvier P (2008) Early jawless vertebrates and cyclostome origins. Zool Sci 25:1045–1056

Jones TA, Jones SM, Paggett KC (2001) Primordial rhythmic bursting in embryonic cochlear ganglion cells. J Neurosci 21:8129–8135

Jones TA, Leake PA, Snyder RL, Stakhovskaya O, Bonham B (2007) Spontaneous discharge patterns in cochlear spiral ganglion cells before the onset of hearing in cats. J Neurophysiol 98:1898–1908

Kaiser A (1993) Das efferente Bündel des Hörorgans beim Huhn – Ursprung, Projektionen und Physiologie [Dr.rer.nat.]. Garching, Technische Universität München, p 167

Kaiser A, Manley GA (1994) Physiology of single putative cochlear efferents in the chicken. J Neurophysiol 72:2966–2979

Kaiser A, Manley GA (1996) Brainstem connections of the macula lagenae in the chicken. J Comp Neurol 374:108–117

Keppler C, Schermuly L, Klinke R (1994) The course and morphology of efferent nerve fibres in the papilla basilaris of the pigeon (*Columba livia*). Hear Res 74:259–264

Khan KM, Hatfield JS, Drescher MJ, Drescher DG (1991) The histochemical localization of acetylcholinesterase in the rainbow trout saccular macula by electron microscopy. Neurosci Lett 131:109–112

Khan KM, Drescher MJ, Hatfield JS, Khan A-M, Drescher DG (2002) Muscarinic receptor subtypes are differentially distributed in the rat cochlea. Neuroscience 111:291–302

Kirk EC, Smith DW (2003) Protection from acoustic trauma is not a primary function of the medial olivocochlear efferent system. J Assoc Res Otolaryngol 4:445–465

Kishida R, Goris RC, Nishizawa H, Koyama H, Kadota T, Amemiya F (1987) Primary neurons of the lateral line nerves and their central projections in hagfishes. J Comp Neurol 264:303–310

Köppl C (2001) Efferent axons in the avian auditory nerve. Eur J Neurosci 13:1889–1901

Köppl C (2007) Spontaneous generation in early sensory development. Invited Focus on: Spontaneous discharge patterns in cochlear spiral ganglion cells prior to the onset of hearing in cats. J Neurophysiol 98:1843–1844

Köppl C, Manley GA (1992) Functional consequences of morphological trends in the evolution of lizard hearing organs. In: Webster DB, Fay RR, Popper AN (eds) The evolutionary biology of hearing. Springer, New York, pp 489–510

Köppl C, Yates GK (1999) Coding of sound pressure level in the barn owl's auditory nerve. J Neurosci 19:9674–9686

Köppl C, Gleich O, Manley GA (1993) An auditory fovea in the barn owl cochlea. J Comp Physiol A 171:695–704

Köppl C, Gleich O, Schwabedissen G, Siegl E, Manley GA (1998) Fine structure of the basilar papilla of the emu: implications for the evolution of avian hair-cell types. Hear Res 126:99–112

Koyama H, Kishida R, Goris RC, Kusunoki TT (1989) Afferent and efferent projections of the VIIIth cranial nerve in the lamprey, *Lampetra japonica*. J Comp Neurol 280:663–671

Koyama H, Kishida R, Goris RC, Kusunoki T (1990) Organisation of the primary projections of the lateral line in nerves in the lamprey *Lampetra japonica*. J Comp Neurol 295:277–289

Le Prell CG (2007) Role for the lateral olivocochlear neurons in auditory function. Focus on: Selective removal of lateral olivocochlear efferents increases vulnerability to acute acoustic injury. J Neurophysiol 97:963–965

Le Prell CG, Dolan DF, Halsey K, Hughes L (2007) Chronic intra-cochlear delivery of CGRP, a lateral olivocochlear transmitter, enhances auditory nerve activity. Assoc Res Otolaryngol Abstr 30:120

Lewis ER, Narins PM (1999) The acoustic periphery of amphibians: anatomy and physiology. In: Fay RR, Popper AN (eds) Comparative hearing: fish and amphibians. Springer, New York, pp 101–154

Liberman MC, Kujawa SG (1999) The olivocochlear system and protection from acoustic injury: Acute and chronic effects. In: Berlin CI (ed) The efferent auditory system: basic science and clinical applications. Singular Publishing, San Diego, pp 1–29

Lippe WR, Zirpel L, Stone JS (2002) Muscarinic receptors modulate intracellular Ca^{2+} concentration in hyaline cells of the chicken basilar papilla. J Comp Physiol A 188:381–395

Lopez I, Meza G (1988) Neurochemical evidence for afferent GABAergic and efferent cholinergic neurotransmission in the frog vestibule. Neuroscience 25:13–18

Lowenstein O, Thornhill RA (1970) The labyrinth of *Myxine*: anatomy, ultrastructure and electrophysiology. Proc R Soc B Biol Sci 176:21–42

Lowenstein O, Osborne MP, Thornhill RA (1968) The anatomy and ultrastructure of the labyrinth of the lamprey (*Lampetra fluviatilis* L.). Proc R Soc B Biol Sci 170:113–134

Lustig LR, Hiel H, Fuchs PA (1999) Vestibular hair cells of the chick express the nicotinic acetylcholine receptor subunit α9. J Vestib Res 9:359–367

Lysakowski A, Goldberg JM (2004) Morphophysiology of the vestibular periphery. In: Highstein SM, Fay RR, Popper AN (eds) The vestibular system. Springer, New York, pp 57–152

Maison SF, Emeson RB, Adams JC, Luebke AE, Liberman MC (2003) Loss of alpha CGRP reduces sound-evoked activity in the cochlear nerve. J Neurophysiol 90:2941–2949

Manley GA (1990) Peripheral hearing mechanisms in reptiles and birds. Springer, Berlin, p 288

Manley GA (2001) Evidence for an active process and a cochlear amplifier in nonmammals. J Neurophysiol 86:541–549

Manley GA (2004) The lizard basilar papilla and its evolution. In: Manley GA, Popper A, Fay RR (eds) Evolution of the vertebrate auditory system. Springer, New York, pp 200–223

Manley GA, Clack JA (2004) An outline of the evolution of vertebrate hearing organs. In: Manley GA, Popper A, Fay RR (eds) Evolution of the vertebrate auditory system. Springer, New York, pp 1–26

Manley GA, Köppl C (1998) Phylogenetic development of the cochlea and its innervation. Curr Opin Neurobiol 8:468–474

Manley GA, Ladher R (2008) Phylogeny and evolution of ciliated mechanoreceptor cells. In: Dallos P, Oertel D (eds) Audition. Academic Press, San Diego, pp 1–34

Manley GA, Gleich O, Kaiser A, Brix J (1989) Functional differentiation of sensory cells in the avian auditory periphery. J Comp Physiol A 164:289–296

Manley GA, Taschenberger G, Oeckinghaus H (1999) Influence of contralateral acoustic stimulation on distortion-product and spontaneous otoacoustic emissions in the barn owl. Hear Res 138:1–12

Manni L, Caicci F, Gasparini F, Zaniolo G, Burighel P (2004) Hair cells in ascidians and the evolution of lateral line placodes. Evol Dev 6:379–381

Manni L, Mackie GO, Caicci F, Zaniolo G, Burighel P (2006) Coronal organ of ascidians and the evolutionary significance of secondary sensory cells in chordates. J Comp Neurol 495: 363–373

Martin P (2008) Active hair-bundle motility of the hair cells of vestibular and auditory organs. In: Manley GA, Fay RR, Popper AN (eds) Active processes and otoacoustic emissions in hearing. Springer, New York, pp 93–144

Martínez-García F, Novejarque A, Landete JM, Moncho-Bogani J, Lanuza E (2002) Distribution of calcitonin gene-related peptide-like immunoreactivity in the brain of the lizard *Podarcis hispanica*. J Comp Neurol 447:99–113

Matsunobu T, Chung JW, Schacht J (2001) Acetylcholine-evoked calcium increases in Deiters' cells of the guinea pig cochlea suggest alpha 9-like receptors. J Neurosci Res 63:252–256

McWilliam PN, Maqbool A, Batten TFC, Kaye JC (1995) Influence of peripheral targets on the expression of calcitonin gene-related peptide immunoreactivity in rat cranial motoneurones. J Neurobiol 28:506–514

Medina L, Reiner A (1994) Distribution of choline acetyltransferase immunoreactivity in the pigeon brain. J Comp Neurol 342:497–537

Medina L, Smeets WJAJ, Hoogland PV, Puelles L (1993) Distribution of choline acetyltransferase immunoreactivity in the brain of the lizard *Gallotia galloti*. J Comp Neurol 331:261–285

Meredith GE, Roberts BL (1986) Central organization of the efferent supply to the labyrinthine and lateral line receptors of the dogfish. Neuroscience 17:225–233
Meredith GE, Roberts BL (1987) Distribution and morphological characteristics of efferent neurons innervating end organs in the ear and lateral line of the European eel. J Comp Neurol 265:494–506
Metcalfe WK, Kimmel CB, Schabtach E (1985) Anatomy of the posterior lateral line system in young larvae of the zebrafish. J Comp Neurol 233:377–389
Micevych PE, Kruger L (1992) The status of calcitonin gene-related peptide as an effector peptide. Ann N Y Acad Sci 657:379–396
Miller MR, Beck J (1988) Auditory hair cell innervational patterns in lizards. J Comp Neurol 271:604–628
Miller MR, Beck J (1990) Further serial transmission electron microscopy studies of auditory hair cell innervation in lizards and in a snake. Am J Anat 188:175–184
Molist P, Rodriguez-Moldes I, Batten TFC, Anadon R (1995) Distribution of calcitonin gene-related peptide-like immunoreactivity in the brain of the small-spotted dogfish, *Scyliorhinus canicula* L. J Comp Neurol 352:335–350
Mueller T, Vernier P, Wullimann MF (2004) The adult central nervous cholinergic system of a neurogenetic model animal, the zebrafish *Danio rerio*. Brain Res 1011:156–169
Mulroy MJ (1986) Patterns of afferent synaptic contacts in the alligator lizard's cochlea. J Comp Neurol 248:263–271
Mulroy MJ, Oblak TG (1985) Cochlear nerve of the alligator lizard. J Comp Neurol 233:463–472
Nakajima Y, Wang DW (1974) Morphology of afferent and efferent synapses in hearing organ of the goldfish. J Comp Neurol 156:403–416
New JG, Northcutt RG (1984) Central projections of the lateral line nerves in the shovelnose sturgeon. J Comp Neurol 225:129–140
New JG, Singh S (1994) Central topography of anterior lateral line nerve projections in the channel catfish, *Ictalurus punctatus*. Brain Behav Evol 43:34–50
Northcutt RG (2005) The new head hypothesis revisited. J Exp Zool B 304B:274–297
Oesterle EC, Cunningham DE, Rubel EW (1992) Ultrastructure of hyaline, border, and vacuole cells in chick inner ear. J Comp Neurol 318:64–82
Ofsie MS, Cotanche DA (1996) Distribution of nerve fibers in the basilar papilla of normal and sound-damaged chick cochleae. J Comp Neurol 370:281–294
Ofsie MS, Hennig AK, Messana EP, Cotanche DA (1997) Sound damage and gentamicin treatment produce different patterns of damage to the efferent innervation of the chick cochlea. Hear Res 113:207–223
Ohno K, Takeda N, Yamano M, Matsunaga T, Tohyama M (1991) Coexistence of acetylcholine and calcitonin gene-related peptide in the vestibular efferent neurons in the rat. Brain Res 566:103–107
Okoruwa OE, Weston MD, Sanjeevi DC, Millemon AR, Fritzsch B, Hallworth R, Beisel KW (2008) Evolutionary insights into the unique electromotility motor of mammalian outer hair cells. Evol Dev 10:300–315
Patterson WC (1966) Hearing in the turtle. J Audit Res 6:453–464
Pellegrini M, Ceccotti F, Magherini P (1985) The efferent vestibular neurons in the toad (*Bufo bufo* L.): their location and morphology. A horseradish peroxidase study. Brain Res 344:1–8
Pombal MA, Marin O, Gonzalez A (2001) Distribution of choline acetyltransferase-immunoreactive structures in the lamprey brain. J Comp Neurol 431:105–126
Popper AN, Fay RR (1999) The auditory periphery in fishes. In: Fay RR, Popper AN (eds) Comparative hearing: fish and amphibians. Springer, New York, pp 43–100
Popper AN, Hoxter B (1987) Sensory and nonsensory ciliated cells in the ear of the sea lamprey, *Petromyzon marinus*. Brain Behav Evol 30:43–61
Popper AN, Saidel WM (1990) Variations in receptor cell innervation in the saccule of a teleost fish ear. Hear Res 46:211–228
Powers SA, Reiner A (1993) The distribution of cholinergic neurons in the central nervous system of turtles. Brain Behav Evol 41:326–345

Prigioni I, Valli P, Casella C (1983) Peripheral organization of the vestibular efferent system in the frog: an electrophysiological study. Brain Res 269:83–90

Pujol R, Lavigne-Rebillard M, Lenoir M (1998) Development of sensory and neural structures in the mammalian cochlea. In: Rubel EW, Popper AN, Fay RR (eds) Development of the auditory system. Springer, New York, pp 146–192

Puzdrowski RL (1989) Peripheral distribution and central projections of the lateral-line nerves in goldfish, *Carassius auratus*. Brain Behav Evol 34:110–131

Raabe T (2004) Ursprung der cochleären Efferenzen im Hirnstamm der Schleiereule (*Tyto alba*) [Dr.rer.nat.]. Freising-Weihenstephan, Technische Universität München, p 101

Raabe T, Köppl C (2003a) Bilaterally-projecting efferent neurones to the basilar papilla in the barn owl and the chicken. Brain Res 986:124–131

Raabe T, Köppl C (2003b) Cochlear efferents in different avian species – a way to distinguish between auditory and vestibular? In: Abstracts 26th midwinter research meeting ARO:92

Rajan R (2000) Centrifugal pathways protect hearing sensitivity at the cochlea in noisy environments that exacerbate the damage induced by loud sound. J Neurosci 20:6684–6693

Raji-Kubba J, Micevych PE, Simmons DD (2002) The superior olivary complex of the hamster has multiple periods of cholinergic neuron development. J Chem Neuroanat 24:75–93

Rebillard M, Pujol R (1983) Innervation of the chicken basilar papilla during its development. Acta Otolaryngol (Stockh) 96:379–388

Robbins RG, Bauknight MS, Honrubia V (1967) Anatomical distribution of efferent fibers in the VIIIth cranial nerve of the bullfrog (*Rana catesbeiana*). Acta Otolaryngol 64:436–448

Roberts BL, Meredith GE (1992) The efferent innervation of the ear: variations on an enigma. In: Webster DB, Fay RR, Popper AN (eds) The evolutionary biology of hearing. Springer, New York, pp 185–210

Roberts BL, Russell IJ (1972) The activity of lateral-line efferent neurones in stationary and swimming dog-fish. J Exp Biol 57:435–448

Roberts BL, Maslam S, Los I, Van DJB (1994) Coexistence of calcitonin gene-related peptide and choline acetyltransferase in EEL efferent neurons. Hear Res 74:231–237

Robertson D (2009) Centrifugal control in mammalian hearing. Clin Exp Pharmacol Physiol 36:603–611

Robertson D, Mulders WHAM (2000) Distribution and possible functional roles of some neuroactive peptides in the mammalian superior olivary complex. Microsc Res Tech 51:307–317

Ronan MC, Bodznick D (1986) End buds non-ampullary electroreceptors in adult lampreys. J Comp Physiol A 158:9–16

Ronan M, Northcutt RG (1987) Primary projections of the lateral line nerves in adult lampreys. Brain Behav Evol 30:62–81

Rossi ML, Prigioni I, Valli P, Casella C (1980) Activation of the efferent system in the isolated frog labyrinth: effects on the afferent EPSPs and spike discharge recorded from single fibres of the posterior nerve. Brain Res 185:125–137

Ruel J, Wang J, Rebillard G, Eybalin M, Lloyd R, Pujol R, Puel JL (2007) Physiology, pharmacology and plasticity at the inner hair cell synaptic complex. Hear Res 227:19–27

Russell IJ (1971a) The pharmacology of efferent synapses in the lateral-line system of *Xenopus laevis*. J Exp Biol 54:643–658

Russell IJ (1971b) The role of the lateral-line efferent system in *Xenopus laevis*. J Exp Biol 54:621–641

Russell IJ (1976) Central inhibition of lateral line input in the medulla of the goldfish by neurons which control active body movements. J Comp Physiol 111:335–358

Russell IJ, Lukashkin AN (2008) Cellular and molecular mechanisms in the efferent control of cochlear nonlinearities. In: Manley GA, Fay RR, Popper AN (eds) Active processes and otoacoustic emissions in hearing. Springer, New York, pp 343–380

Russell IJ, Roberts BL (1972) Inhibition of spontaneous lateral-line activity by efferent nerve stimulation. J Exp Biol 57:77–82

Russell IJ, Roberts BL (1974) Active reduction of lateral-line sensitivity in swimming dogfish. J Comp Physiol 94:7–15
Sapede D, Rossel M, Dambly-Chaudiere C, Ghysen A (2005) Role of SDF1 chemokine in the development of lateral line efferent and facial motor neurons. Proc Natl Acad Sci USA 102:1714–1718
Schellart NAM, Prins M, Kroese ABA (1992) The pattern of trunk lateral line afferents and efferents in the rainbow trout (*Salmo gairdneri*). Brain Behav Evol 39:371–380
Schmidt RS (1963) Frog labyrinthine efferent impulses. Acta Otolaryngol 56:51–64
Schmidt R (1965) Amphibian acoustico-lateralis efferents. J Cell Comp Physiol 65:155–162
Schucker FA (1972) A preliminary report on light and electron microscopic studies of the basilar papilla in the toad, *Xenopus laevis*. Anat Rec 172:400
Schwarz IE, Schwarz DWF, Frederickson JM, Landolt JP (1981) Efferent vestibular neurons: A study employing retrograde tracer methods in the pigeon (*Columba livia*). J Comp Neurol 196:1–12
Schwarz DWF, Schwarz IE, Dezsoe A (1992) Cochlear efferent neurons projecting to both ears in the chicken, *Gallus domesticus*. Hear Res 60:110–114
Sewell WF, Starr PA (1991) Effects of calcitonin gene-related peptide and efferent nerve stimulation on afferent transmission in the lateral line organ. J Neurophysiol 65:1158–1169
Shigemoto T, Ohmori H (1990) Muscarinic agonists and ATP increase the intracellular Ca^{2+} concentration in chick cochlear hair cells. J Physiol 420:127–148
Simmons DD, Bertolotto C, Leong M (1995) Synaptic ultrastructure within the amphibian papilla of *Rana pipiens pipiens*: rostrocaudal differences. Audit Neurosci 1:183–193
Slepecky NB (1996) Structure of the mammalian cochlea. In: Dallos P, Popper AN, Fay RR (eds) The cochlea. Springer, New York, pp 44–129
Smotherman M, Narins P (2004) Evolution of the amphibian ear. In: Manley GA, Popper A, Fay RR (eds) Evolution of the vertebrate auditory system. Springer, New York, pp 164–199
Sneary MG (1988) Auditory receptor of the red-eared turtle: II. Afferent and efferent synapses and innervation patterns. J Comp Neurol 276:588–606
Song J, Northcutt RG (1991) The primary projections of the lateral-line nerves of the Florida gar, *Lepisosteus platyrhincus*. Brain Behav Evol 37:38–63
Strutz J (1981) The origin of centrifugal fibers to the inner ear in *Caiman crocodilus*. A horseradish peroxidase study. Neurosci Lett 27:95–100
Strutz J (1982) The origin of efferent fibers to the inner ear in a turtle (*Terrapene ornata*). A horseradish peroxidase study. Brain Res 244:165–168
Strutz J, Schmidt CL (1982) Acoustic and vestibular efferent neurons in the chicken (*Gallus domesticus*). Acta Otolaryngol (Stockh) 94:45–51
Strutz J, Schmidt CL, Stürmer C (1980) Origin of efferent fibers of the vestibular apparatus in goldfish. A horseradish peroxidase study. Neurosci Lett 18:5–9
Strutz J, Bielenberg K, Spatz WB (1982) Location of efferent neurons to the labyrinth of the green tree frog (*Hyla cinerea*). Arch Otorhinolaryngol 234:245–251
Sugai T, Yano J, Sugitani M, Ooyama H (1992) Actions of cholinergic agonists and antagonists on the efferent synapse in the frog sacculus. Hear Res 61:56–64
Sugihara I (2001) Efferent innervation in the goldfish saccule examined by acetylcholinesterase histochemistry. Hear Res 153:91–99
Taberner AM, Liberman MC (2005) Response properties of single auditory nerve fibers in the mouse. J Neurophysiol 93:557–569
Takami K, Kawai Y, Shiosaka S, Lee Y, Girgis S, Hillyard CJ, MacIntyre I, Emson PC et al (1985) Immunohistochemicai evidence for the coexistence of calcitonin gene-related peptide- and choline acetyltransferase-like immunoreactivity in neurons of the rat hypoglossal, facial and ambiguus nuclei. Brain Res 328:386–389
Takasaka T, Smith CA (1971) The structure and innervation of the pigeon's basilar papilla. J Ultrastruct Res 35:20–65

Tanaka K, Smith CA (1978) Structure of the chicken's inner ear: SEM and TEM study. Am J Anat 153:251–272

Teresi PC (1985) Hair cell innervation patterns in the papilla basilaris of the fence lizard *Sceloporus occidentalis*. Dissertation, University of California, San Francisco

Tomchik SM, Lu ZM (2005) Octavolateral projections and organization in the medulla of a teleost fish, the sleeper goby (*Dormitator latifrons*). J Comp Neurol 481:96–117

Tomchik SM, Lu ZM (2006a) Auditory physiology and anatomy of octavolateral efferent neurons in a teleost fish. J Comp Physiol A 192:51–67

Tomchik SM, Lu ZM (2006b) Modulation of auditory signal-to-noise ratios by efferent stimulation. J Neurophysiol 95:3562–3570

Tricas TC, Highstein SM (1991) Action of the octavolateralis efferent system upon the lateral line of free-swimming toadfish, *Opsanus tau*. J Comp Physiol A 169:25–37

Tritsch NX, Yi EY, Gale JE, Glowatzki E, Bergles DE (2007) The origin of spontaneous activity in the developing auditory system. Nature 450:50–56

Vetter DE, Adams JC, Mugnaini E (1991) Chemically distinct rat olivocochlear neurons. Synapse 7:21–43

Vidal N, Hedges SB (2009) The molecular evolutionary tree of lizards, snakes, and amphisbaenians. C R Biol 332:129–139

von Düring M, Karduck A, Richter H-G (1974) The fine structure of the inner ear in *Caiman crocodilus*. Z Anat Entwickl Gesch 145:41–65

Wagner T, Schwartz E (1996) Efferent neurons of the lateral line system and their innervation of lateral line branches in a euteleost and an osteoglossomorph. Anat Embryol 194:271–278

Walsh EJ, McGee J (1997) Does activity in the olivochochlear bundle affect development of the auditory periphery? In: Lewis ER, Long G, Lyon RF, Narins PM, Steele CR, Hecht-Poinar E (eds) Diversity in auditory mechanics. World Scientific, Singapore, pp 376–385

Walsh EJ, McGee J, McFadden SL, Liberman MC (1998) Long-term effects of sectioning the olivocochlear bundle in neonatal cats. J Neurosci 18:3859–3869

Warr WB (1992) Organization of olicocochlear efferent systems in mammals. In: Webster DB, Popper AN, Fay RR (eds) The mammalian auditory pathway: neuroanatomy. Springer, New York, pp 410–448

Weeg MS, Land BR, Bass AH (2005) Vocal pathways modulate efferent neurons to the inner ear and lateral line. J Neurosci 25:5967–5974

Wegner N (1982) A qualitative and quantitative study of a sensory epithelium in the inner ear of a fish (*Colisa labiosa*; Anabnatidae). Acta Zool 63:133–146

Wever EG (1978) The reptile ear. Princeton University Press, Princeton

White J (1986) Morphological and fine structural features of the basilar papilla in ambystomatid salamanders (Amphibia; Caudata). J Morphol 187:181–199

Whitehead MC, Morest DK (1981) Dual populations of efferent and afferent cochlear axons in the chicken. Neuroscience 6:2351–2365

Whitehead MC, Morest DK (1985) The growth of cochlear fibers and the formation of their synaptic endings in the avian inner ear: a study with the electron microscope. Neuroscience 14:277–300

Wibowo E, Brockhausen J, Köppl C (2008) Efferent innervation is present to both low- and high-frequency regions of the auditory basilar papilla of lizards. In: 38th annual meeting of the society for neuroscience. Society for Neuroscience, Washington, DC, Program No. 259.251

Wibowo E, Brockhausen J, Köppl C (2009) Efferent innervation to the auditory basilar papilla of scincid lizards. J Comp Neurol 516:74–85

Wicht H, Northcutt RG (1995) Ontogeny of the head of the Pacific hagfish (*Eptatretus stouti*, Myxinoidea): Development of the lateral line system. Philos Trans R Soc Lond B 349:119–134

Will U (1982) Efferent neurons of the lateral-line system and the VIII cranial nerve in the brainstem of anurans. Cell Tissue Res 225:673–685

Will U, Fritzsch B (1988) The eighth nerve of amphibians. Peripheral and central distribution. In: Fritzsch B, Ryan MJ, Wilczynski W, Hetherington TE, Walkowiak W (eds) The evolution of the amphibian auditory system. Wiley, New York, pp 159–183

Yamada Y (1973) Fine structure of the ordinary lateral line organ. I. The neuromast of lamprey *Entosphenus japonicus*. J Ultrastruct Res 43:1–17

Yamada Y, Hama K (1972) Fine structure of the lateral-line organ of the common eel, *Anguilla japonica*. Z Zellforsch 124:454–464

Zidanic M (2002) Cholinergic innervation of the chick basilar papilla. J Comp Neurol 445:159–175

Zidanic M, Fuchs PA (1996) Synapsin-like immunoreactivity in the chick cochlea: specific labeling of efferent terminals. Audit Neurosci 2:347–362

Zottoli SJ, van Horne C (1983) Posterior lateral line afferent and efferent pathways withing the central nervous system the goldfish with special reference to the Mauthner cell. J Comp Neurol 219:100–111

Chapter 9
Central Descending Auditory Pathways

Brett R. Schofield

Abbreviations

A1, AII	Primary auditory cortex, area II of AC
AC	Auditory cortex
BIC	Nucleus of the brachium of the IC
CN	Cochlear nucleus
d	Deep layer of DCN
DCN	Dorsal cochlear nucleus
DLPO	Dorsolateral periolivary nucleus
DNLL	Dorsal nucleus of the lateral lemniscus
DPO	Dorsal periolivary nucleus
f	Fusiform cell layer of DCN
grca	Granule cell area of the CN
H	Horizontal cell group of NLL
IC	Inferior colliculus
ICc	Central nucleus of the IC
ICd	Dorsal cortex of the IC
ICx	External cortex of the IC
INLL	Intermediate nucleus of the lateral lemniscus
LNTB	Lateral nucleus of the trapezoid body
LSO	Lateral superior olivary nucleus
m	Molecular layer of DCN
MG	Medial geniculate nucleus
MGd	Dorsal division of MG
MGm	Medial division of MG

B.R. Schofield (✉)
Department of Anatomy and Neurobiology, Northeastern Ohio Universities
Colleges of Medicine and Pharmacy, Rootstown, OH 44272, USA
e-mail: bschofie@neoucom.edu

D.K. Ryugo et al. (eds.), *Auditory and Vestibular Efferents*,
Springer Handbook of Auditory Research 38, DOI 10.1007/978-1-4419-7070-1_9,

MGv	Ventral division of MG
MSO	Medial superior olivary nucleus
MTNB	Medial nucleus of the trapezoid body
NLL	Nuclei of the lateral lemniscus (including DNLL, INLL, VNLL)
PLZ	Paralemniscal zone
Sag	Sagulum
scc	Small cell cap of the VCN
SOC	Superior olivary complex
SPF	Subparafascicular nucleus of the thalamus
SPN	Superior paraolivary nucleus
TR	Thalamic reticular nucleus
V	Layer V of auditory cortex
VCN	Ventral cochlear nucleus
VI	Layer VI of auditory cortex
VNLL	Ventral nucleus of the lateral lemniscus
VNTB	Ventral nucleus of the trapezoid body

9.1 Introduction

The presence of descending pathways in the auditory system has been known for a long time. These pathways may improve the system's ability to extract information from acoustic stimuli and perhaps control what auditory information reaches consciousness. The past two decades have seen a tremendous growth in interest in the central descending system for several reasons. First, there has been considerable progress in understanding the olivocochlear system and the central effects on this system (this volume; Chapter 2, Brown; Chapter 3, Guinan; Chapter 4, Sewell; Chapter 5, Katz et al.; Chapter 8, Glowatzki; Chapter 9, Kõppl). Second, physiological studies have revealed a host of effects of corticofugal projections on processing in many lower auditory centers (Suga et al., Chap. 11). Third, studies in animals and humans have bolstered and extended long-held views that higher "cognitive" functions can affect early auditory processing (e.g., Perrot et al. 2006; Delano et al. 2007; Rinne et al. 2008). Finally, the development of increasingly sensitive anatomical tracers has led to the discovery of numerous central pathways (e.g., Feliciano et al. 1995).

The present chapter provides a brief description of the ascending auditory pathway as a framework for discussing the descending system. A broad view of the descending system encompasses interactions between hearing and functions such as emotion, language, attention, etc. and involves descending projections from areas ranging from primary auditory cortex (AC) to frontal cortex to components of the limbic system. Because of space limitations, the current discussion focuses on descending pathways that originate and terminate within the regions most closely associated with the ascending auditory pathways.

9.2 Overview of Central Auditory Structures and the Ascending Pathways

Figure 9.1 shows a simplified schematic of the ascending auditory pathways. Information from the cochlea enters the brain and synapses in the cochlear nucleus (CN). The CN has dorsal and ventral divisions that give rise to multiple ascending pathways. The ascending pathways are distributed primarily to nuclei of the superior olivary complex (SOC), nuclei of the lateral lemniscus (NLL) and the inferior colliculus (IC). The SOC is a prominent site for analysis of binaural information. The SOC is a source of bilateral ascending pathways to the NLL and IC. The NLL include three main nuclei: ventral, intermediate and dorsal nuclei (VNLL, INLL, DNLL) that are surrounded by fibers of the lateral lemniscus, the large ascending fiber tract carrying almost all the ascending auditory fibers to the IC (Fig. 9.1, oval labeled "1").

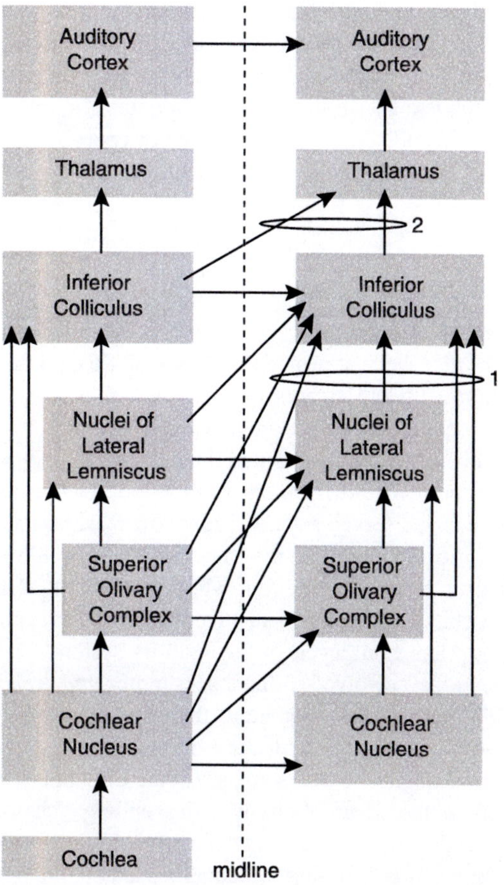

Fig. 9.1 Schematic of ascending pathways. The major ascending auditory pathways are drawn from the perspective of the left cochlea. Commissural pathways, which connect similar nuclei on the two sides, are also shown as it is believed that these connections influence the ascending information. Connections that cross the midline are shown originating from the nuclei on the left; similar pathways originate from the nuclei on the right, but are not shown for the sake of clarity. See Abbreviations. Numbers identify selected fiber tracts: *1* lateral lemniscus; *2* brachium of the IC

The VNLL and INLL are involved primarily with monaural processing and project primarily to the ipsilateral IC. The DNLL is associated with binaural processing and projects bilaterally to the IC. The IC is a midbrain center notable for the many ascending fibers it receives from a large number of lower nuclei (including nonauditory nuclei as well as the auditory nuclei already mentioned). The IC is the main source of ascending projections to the thalamus, with a large ipsilateral projection and a much smaller contralateral projection. The ascending projections form a fiber tract called the brachium of the IC (Fig. 9.1, oval labeled "2"). The main auditory nucleus of the thalamus is the medial geniculate body (MG). The MG can be divided into several parts that have different connections. In addition, there are several nearby thalamic nuclei that have extensive connections with the auditory system; these are described later. The auditory thalamus projects to the AC. For the present discussion, the AC is viewed as the broad expanse of temporal cortex that receives input from auditory thalamic nuclei and contains cells that respond to sounds. Commissural connections, which connect similar nuclei on the two sides of the brain, occur at most levels of the auditory pathways; they are important in ascending and perhaps also descending auditory pathways. Additional details, including additional subcortical nuclei or subdivisions that have special significance for the descending pathways, are described in the text that follows as appropriate. The reader is referred to several reviews for more details on the ascending pathways (e.g., previous volumes in this series; Smith and Spirou 2002; Winer and Schreiner 2005).

9.3 Brief Historical View of the Descending System

The discovery of projections from the SOC to the ear can be taken as a starting point for the modern view of the descending auditory system. Before the discovery of the olivocochlear bundle, the SOC was believed by some to connect with various motor nuclei of the brainstem and thus serve as a center for auditory reflexes (discussed in Rasmussen 1946). Others thought the SOC projected primarily into the lateral lemniscus and served a "relay" function. Rasmussen (1946) described the olivocochlear bundle and identified the SOC as its source. He associated the olivocochlear bundle with a feedback function, while maintaining that the SOC may also serve a relay function. The concept of "feedback" is now routinely associated with descending pathways at all levels of the auditory system. Current understanding is that the SOC is, in fact, involved in feedback, motor *and* "relay" functions (to higher centers). In addition, studies over more than 50 years have suggested that "higher functions" usually associated with cerebral cortex (e.g., attention), can affect the auditory system at its first central synapse (in the CN) and even in the cochlea. Such observations led to the assumption of a descending "chain," or multisynaptic pathway, that extends from AC to the cochlea (Huffman and Henson 1990). Interestingly, many studies of descending systems focused on one or the other of two scenarios: (1) feedback systems, in the form of circuit loops or (2) chains, in the form of multisynaptic descending pathways. Spangler and Warr (1991) provided a review of the descending pathways that directly addressed the issue: "The descending auditory system: regional feedback loops or a descending chain?" (Spangler and Warr 1991, p. 27). The authors concluded that there was good evidence for loops. A major point was that the loops are regional. One set of loops is located at the upper end of the system, including the AC, thalamus, and IC (Fig. 9.2a). A second group of loops can be drawn among the brain stem nuclei (Fig. 9.2b). At each level, there are multiple parallel loops; for example, separate loops can be drawn between specific cortical areas and different nuclei within the thalamus and/or different subdivisions of the IC. Comparison of the "upper" and "lower" system loops shows that the IC is the only structure in common to the two groups. The extent to which the upper and lower loops interact, even within the IC, is unknown.

The status of descending chains was much less clear at the time of Spangler and Warr's (1991) review. In 1990, Huffman and Henson reviewed the role of the IC in the descending pathways. They illustrated a possible trisynaptic pathway from AC to the cochlea (Fig. 9.2c). As described previously, the existence of a chain was supported by earlier physiological studies. The chain proposed by Huffman and Henson was the most reasonable hypothesis given the anatomical data at the time. However, none of the proposed links in the chain had actually been demonstrated at the cellular level; for example, evidence that cortical axons contact IC cells that

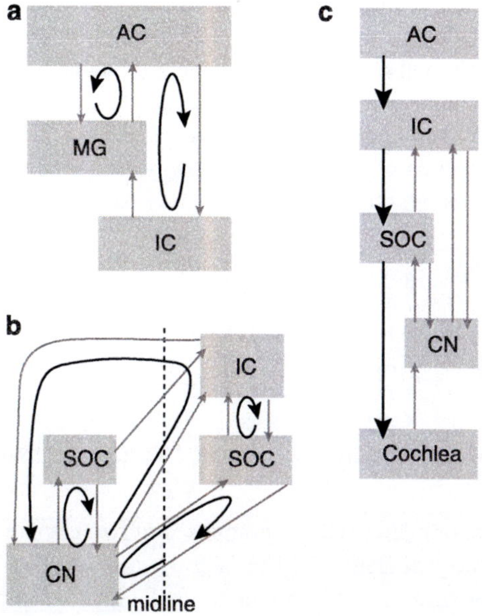

Fig. 9.2 Schematic diagrams summarizing some of the main circuits reviewed in the early 1990s. (**a**, **b**) Loops among some of the auditory nuclei at the upper end (**a**) and lower end (**b**) of the auditory pathways, as described by Spangler and Warr (1991). *Gray arrows* indicate direct connections; *black arrows* indicate several of the loops that are formed. See text for discussion. (**c**) A descending chain (*black arrows*) from auditory cortex (AC) to the cochlea, as described by uffman and Henson (1990). *Gray arrows* indicate additional brain stem connections

project to the SOC was limited to the observation that cortical axons terminate in the same part of the IC in which there are cells that project to the SOC (the axons and the cells were revealed in separate studies). A similar gap in the data concerned projections from IC cells to olivocochlear cells. In their review a year later, Spangler and Warr (1991) concluded, "If a descending chain of neurons exists, much work remains before its component neurons can be defined from the cortex to the inner ear" (Spangler and Warr 1991, p. 41).

What is new since the review by Spangler and Warr (1991)? Soon after the review was published, two studies provided direct anatomical evidence for descending chains in the form of projections from cells in the IC to olivocochlear cells (Thompson and Thompson 1993; Vetter et al. 1993). The discussion that follows highlights data for numerous additional descending chains. Another discovery was that of substantial projections from the AC to subcollicular targets (Feliciano et al. 1995). Even the perspective on cortical projections to the IC has undergone modification. The termination of this pathway has long been seen as focused on the areas of the IC outside its central nucleus. This view still stands, but recent studies have led to a greater appreciation for direct cortical projections to the central nucleus (e.g., Saldaña et al. 1996; Bajo and Moore 2005). In addition, the cellular origins of the corticocollicular projections, traditionally seen as originating from cortical layer V cells, are now seen to include a significant contribution from cortical layer VI cells (Schofield 2009). Finally, there are new studies on the cellular origins of ascending and descending projections. These studies show that, in general, ascending and descending projections arise from different populations of cells, and thus could contribute differentially to loops or chains.

In Sect. 9.4, the descending projections from each major area are reviewed. This perspective highlights the divergence of descending projections. Section 9.5 then describes the convergence of descending pathways within each major region and the identity of the cells targeted by the descending pathways.

9.4 Divergent Descending Projections from Specific Auditory Regions

9.4.1 Projections from the Superior Olivary Complex

The superior olivary complex (SOC) is composed of two large principal nuclei – the medial superior olivary nucleus and the lateral superior olivary nucleus – and a population of cells surrounding the principal nuclei that are known collectively as the periolivary nuclei[1] (Thompson and Schofield 2000). The principal nuclei and most of the periolivary nuclei are sources of ascending projections to the IC and to

[1]The medial nucleus of the trapezoid body is sometimes considered a "principal" nucleus of the SOC; it is considered one of the periolivary nuclei here because of similarities with other periolivary nuclei, especially as related to descending auditory pathways.

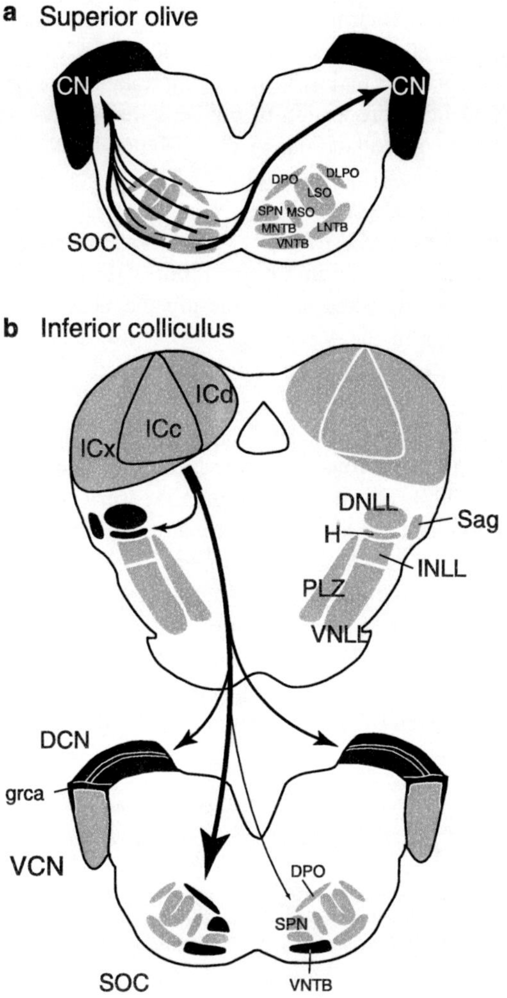

Fig. 9.3 Descending projections from the superior olivary complex (SOC) and the inferior colliculus (IC). (**a**) Schematic cross section indicating the projections from nuclei of the left SOC to the left and right cochlear nuclei (CN). (**b**) Projections from the IC. In both panels, nuclei are indicated in *gray*; those that are targets of the illustrated pathway are *shaded black*. The *arrows* indicate projections from individual nuclei of the SOC; *arrow* thickness indicates relative size of each projection as reflected by the number of projecting cells in each nucleus. In this and subsequent figures, the sections and nuclei were traced from a guinea pig brain. Similar projections arise from left and right nuclei; for clarity, only those from the left nuclei are illustrated. See Abbreviations

the NLL. In addition, the SOC nuclei are sources of descending pathways to the cochlea and to the CN. The projections to the cochlea – the olivocochlear system – are covered in other chapters of this volume. The remainder of the discussion focuses on the projections to the CN.

The nuclei of the SOC project bilaterally to the CN (Fig. 9.3a). The details vary across species, but in general all the periolivary nuclei project to the ipsilateral CN.

In addition, several nuclei (ventral and lateral nuclei of the trapezoid body [VNTB, LNTB], superior paraolivary nucleus [SPN], and dorsal periolivary nucleus) project to the contralateral CN. The majority of the contralateral projections arise from cells in the VNTB. A large proportion of the cells that project to the contralateral CN also project, via axon collaterals, to the ipsilateral CN (Schofield and Cant 1999); most of these cells are located in the VNTB.

The olivary cells that project to the CN use a variety of neurotransmitters. The SOC is the main source of GABAergic, glycinergic, and cholinergic inputs to the CN, and a minor source of glutamatergic inputs (Helfert and Aschoff 1997). The glycinergic cells are scattered across the olivary nuclei, whereas GABAergic cells are most numerous in the VNTB. Cells that use acetylcholine include the medial olivocochlear cells, which have collateral projections to the cochlea and the CN, as well as a group of small cells in the VNTB that project to the CN but not to the cochlea (Sherriff and Henderson 1994).

As a whole, olivary projections terminate throughout the CN (Schwartz 1992; Helfert and Aschoff 1997). This does not imply that every periolivary nucleus projects evenly throughout the CN; in fact, Warr and Beck (1996) have shown that projections from the VNTB, which extend bilaterally to the CN, terminate to different degrees in different parts of the CN and that the patterns are different in the ipsilateral and contralateral CN. The projection from the VNTB to the contralateral dorsal cochlear nucleus (DCN) is organized tonotopically; such organization was not apparent in VNTB projections to the ipsilateral DCN or to the VCN on either side. The cell types targeted by the olivary inputs are discussed later.

9.4.2 Projections from the Nuclei of the Lateral Lemniscus

The nuclei of the lateral lemniscus (NLL) include dorsal, intermediate and ventral nuclei (DNLL, INLL, VNLL; Oertel and Wickesberg 2002). In addition, there are areas surrounding the nuclei that have been characterized to varying degrees. The sagulum is located lateral to the DNLL. Other areas will be described as appropriate.

The NLL are known primarily as sources of ascending projections to the IC (Oertel and Wickesberg 2002). There are reports of descending projections from the NLL to the cochlea and to the CN (e.g., Robertson et al. 1987). The cells giving rise to these projections are relatively few and are scattered in the VNLL and, less so, in the INLL. The cells that project to the cochlea are considered part of the medial olivocochlear system and thus would be expected to project to outer hair cells. The cellular targets of NLL projections to the CN are unknown.

9.4.3 Projections from the IC

The IC can be divided into several subdivisions, some of which may be species specific (Oliver 2005). It is possible to distinguish a central nucleus (ICc), an external cortex (ICx) and a dorsal cortex (ICd) that appear to be comparable across species. The IC is a source of descending projections to the NLL, the SOC, and the CN (Fig. 9.3b) (Thompson 2005). The majority of descending projections from the IC are believed to be glutamatergic (Saint Marie 1996).

All three subdivisions of the IC project ipsilaterally to the NLL or the surrounding regions. The ICd projects to the sagulum. The ICx projects to the sagulum and to the horizontal cell group ("H," a group of cells lying between the DNLL and INLL; Caicedo and Herbert 1993). The ICc projects to the nuclei just described as well as to the DNLL and to the region just medial to the NLL (the paralemniscal zone, PLZ). Sparse projections to the VNLL and INLL have been described in some species. The projections from the ICc to the DNLL are organized tonotopically (i.e., mapped according to the frequency of an acoustic stimulus that would activate the cells); such organization is not apparent in the other collicular projections to the area of the lateral lemniscus.

The IC is a source of dense projections to the SOC (Thompson and Schofield 2000; Thompson 2005). The projections originate from all three IC subdivisions, although the ICc and ICx are more prominent sources in most species studied. In rats, a distinct cluster of olivary-projecting cells has been described in the dorsomedial region of the IC. This cluster is associated with the commissure of the IC and extends rostrally into the superior colliculus (Faye-Lund 1986; Saldaña et al. 2007). The main target of the IC projection is the ipsilateral VNTB, with additional projections to other periolivary nuclei on the ipsilateral side and, less so, to some of the periolivary nuclei (especially the VNTB) on the contralateral side (Fig. 9.3b). The projection from the ICc to the ipsilateral VNTB is organized tonotopically; such organization is not apparent in the other projections.

The IC also projects to the CN (Saldaña 1993; Thompson 2005). Unlike IC projections to the lateral lemniscal area (exclusively ipsilateral) or SOC (predominantly ipsilateral), the collicular projections to the CN are more balanced to the two sides. The projections to the CN arise from the ICc and ICx and less so from the ICd. The cells of origin include multipolar cells in each IC subdivision and "flat" cells in the ICc (Schofield 2001). Despite the similarities of the cells of origin, the ipsilateral and contralateral projections arise from two different populations of IC cells (Schofield 2001; Okoyama et al. 2006). The projections terminate in the granule cell area (grca) and the DCN on both sides. In the DCN, the projections terminate densely in the fusiform cell and deep layers, and sparsely in the molecular layer. The projections to the DCN from the ICc, but not other IC subdivisions, are organized tonotopically.

9.4.4 Projections from the Thalamus and Nearby Areas

The medial geniculate body contains a ventral nucleus (MGv), a dorsal nucleus (MGd), and a medial nucleus (MGm). A lateral part of the posterior thalamic nucleus has been included in the auditory thalamus of some species (Rouiller 1997), but it has not been implicated in descending pathways, so is not considered further here. The intralaminar nuclei of the thalamus project to cerebral cortex and terminate in a diffuse pattern that, along with differences in cortical layers targeted, distinguishes the intralaminar nuclei from the MG. For the current discussion, the relevant intralaminar nuclei are the subparafascicular nucleus (SPF), the posterior intralaminar nucleus, and the posterior limitans nucleus.

Descending projections from the thalamus terminate bilaterally in the IC, SOC, and CN (Fig. 9.4a). Projections from the MG to the IC have been described in cats, rats, gerbils, mice, and monkeys (Winer et al. 2002). The MGm is a source of these projections in all species studied; the MGd is a source in gerbils, rats, and mice (Frisina et al. 1998; Kuwabara and Zook 2000; Senatorov and Hu 2002); and the MGv is a source in mice (Frisina et al. 1998). The projections terminate in the ICx and, in rats, in the ICc (Senatorov and Hu 2002).

The SPF projects to the AC and receives inputs from both the IC and the AC (Yasui et al. 1990). The SPF projects bilaterally to the IC, with an ipsilateral dominance. The terminations are heaviest in the ICx and ICd; few fibers terminate in the ICc. The SPF also projects bilaterally to the SOC. The major target is the ipsilateral SPN, with less dense terminations in the ipsilateral VNTB and the contralateral

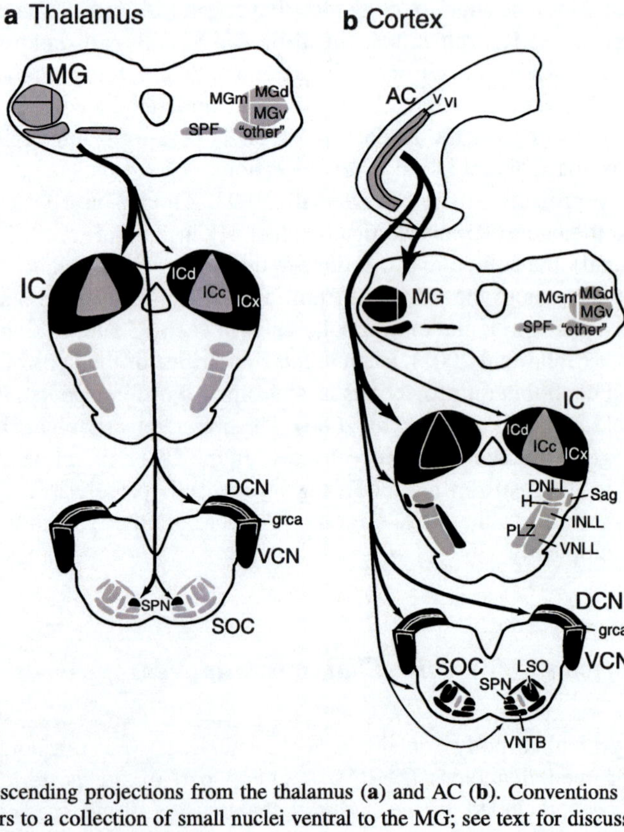

Fig. 9.4 Descending projections from the thalamus (**a**) and AC (**b**). Conventions as in Fig. 9.3. "Other" refers to a collection of small nuclei ventral to the MG; see text for discussion

VNTB and SPN. There is also a small, bilateral projection from the SPF to dorsal and ventral subdivisions of the CN. In contrast to the SPF projections to the IC and SOC, those to the CN are somewhat denser on the contralateral side. Two additional intralaminar nuclei of the thalamus, the posterior intralaminar and posterior limitans nuclei, have descending projections to the ICx and ICd similar to those of the SPF.

The nucleus of the brachium of the IC (BIC) is a group of cells embedded within the brachium of the IC. The BIC is not properly a part of the thalamus; rather it is in the midbrain and extends rostrally from the lateral part of the ICx (see discussion of the BIC by Oliver 2005). The BIC receives ascending input from the IC and projects to the MG and the AC (Le Doux et al. 1985). In addition, the BIC is a major source of input to the superior colliculus, and its function is most often discussed in the context of orienting movements controlled by the superior colliculus. The BIC is a source of descending projections to the ICx and ICc; the projections are bilateral with an ipsilateral dominance (Senatorov and Hu 2002).

9.4.5 Projections from the Auditory Cortex

The AC projects directly to all lower levels of the auditory system except the cochlea (Fig. 9.4b). The projections to the thalamus and IC are the largest of the descending pathways. For many years, these projections were considered the full extent of AC projections to lower auditory centers, even though there were isolated reports of AC projections to, for example, the CN (Zimmerman et al. 1964). A landmark study by Mugnaini and colleagues (Feliciano et al. 1995) took advantage of newly developed sensitive tracers to describe in detail prominent projections from the AC to numerous subcollicular auditory nuclei in rats. The targets include the CN and several nuclei in and around the SOC and the NLL. Similar pathways have been described in other species (discussed later).

9.4.5.1 Auditory Cortical Projections to the Thalamus

All areas of the AC project to the thalamus and, as a group, terminate in all subdivisions of the MG as well as the intralaminar nuclei discussed earlier (Fig. 9.4b; Rouiller and Welker 2000; Smith and Spirou 2002; Winer 2005). Many of the corticothalamic projections are topographic. In some cases, the projections arise from tonotopically organized areas, and the topography of the connections provides for frequency-specific interactions. In other cases, the connections arise from, or terminate in, regions that are not organized tonotopically. The significance of topography in these connections has yet to be identified.

The corticothalamic projections originate from pyramidal cells in cortical layers V and VI. The layer VI cells are by far the most numerous. Layer VI axons terminate in the thalamus as thin axons with small boutons that contain round vesicles and form asymmetric synapses. The synapses excite their target cells, which include thalamocortical cells, GABAergic interneurons (within the target nucleus), and GABAergic neurons in the thalamic reticular nucleus.[2] A second set of corticothalamic projections arises from cortical layer V cells. These cells give rise to axons and boutons in the thalamus that are larger than those of the layer VI cells. They form excitatory synapses with thalamic cells. Additional aspects of the corticothalamic circuitry are discussed in the text that follows, in the context of the cellular targets of the cortical axons (Sect. 9.5.5).

9.4.5.2 Auditory Cortical Projections to the IC

Multiple areas of the AC project bilaterally to the IC (Huffman and Henson 1990; Winer et al. 1998, 2002). The contributions of different cortical areas vary by species; in general, there are heavy projections from the core, primary-like cortical areas, whereas projections from the nonprimary areas are more variable.

The corticocollicular pathway terminates ipsilaterally in all IC subdivisions and contralaterally in the ICd and, in some species, in the ICx (Fig. 9.4b; Huffman and Henson 1990; Winer et al. 2002). On the ipsilateral side, the terminations are dense in the ICd and ICx. A projection to the ICc has been controversial. Such a projection has been described anatomically in rats (Saldaña et al. 1996), monkeys (Fitzpatrick and Imig 1978), cats (Winer et al. 1998), gerbils (Budinger et al. 2000; Bajo and Moore 2005), and ferrets (Bajo et al. 2007), and both anatomically and physiologically in guinea pigs (Lim and Anderson 2007; Peterson and Schofield 2007). In most cases (excepting only rats), the projection to the ICc is described as much less dense than that to ICd or ICx.

Projections from tonotopically organized cortical areas to the IC are topographic. Both anatomical (Bajo et al. 2007) and physiological (Lim and Anderson 2007) studies have demonstrated that the projections to the ICc are tonotopic, suggesting a possibility for frequency-specific effects. Cortical stimulation leads to frequency-specific effects in ICc cells (Suga et al., Chap. 11), though it is not clear that these are due to direct cortical projections to the ICc (an alternative possibility is that AC projections to the ICx contact cells that project to the ICc, establishing a disynaptic pathway from AC to ICc; Jen et al. 2001).

Less is known about the functional organization of projections from AC to ICd and ICx. The cortical axons can terminate in diffuse clouds or in small patches; the significance of the "patchiness" is unknown. Divergence is seen in that an individual cortical area can project to multiple regions of the IC (multiple subdivisions or multiple patches within a subdivision). Convergence is also seen as multiple cortical areas can project to a single part of IC. The significance of this divergence and convergence has not been identified (Winer 2005).

[2] The thalamic reticular nucleus is a group of GABAergic cells that act as inhibitory "interneurons" for thalamic nuclei. The auditory sector of the thalamic reticular nucleus receives inputs from the IC, the MG, and the AC. Cells of the thalamic reticular nucleus project to the thalamocortical cells in the MG.

The ICd is layered, and this layering provides another organizational scheme for the cortical projections (Winer 2005). In general, tonotopically-organized cortical areas terminate primarily in the deeper layers of ICd, whereas the nontonotopically organized cortical areas usually project most heavily to the superficial layer of the ICd.

Most corticocollicular cells are located in cortical layer V. This includes a large population of cells that project ipsilaterally to the IC and a smaller population that project contralaterally (or bilaterally) to the IC (Coomes et al. 2005). The layer V cells include several morphological varieties of pyramidal cell (Winer and Prieto 2001; Bajo and Moore 2005). Bajo and Moore (2005) have distinguished layer V cells according to the presence or absence of a dendritic tuft in cortical layer I. Those with a tuft were observed in numerous cortical areas and projected to the ICd and ICx. Cortical layer V cells without a tuft were identified in core areas of AC (the tonotopically organized A1 and anterior auditory field in gerbils); these cells project to the ICc in a tonotopic pattern. Physiologically, two classes of cortical pyramidal cells have been distinguished by their response to intracellular current injections (Hefti and Smith 2000). One class responds with a burst of action potentials whereas the other class produces a regular series of spikes. The corticocollicular projections appear to include members of both bursting and regular-spiking classes (Lu et al. 2007). The physiologically defined classes have morphological differences as well (e.g., Hefti and Smith 2000). The physiological classes have not yet been characterized in gerbils, but it seems likely that the nontufted cells described by Bajo and Moore (2005) are members of the physiologic regular-spiking class, and at least some of the tufted cells may be members of the bursting class.

Additional corticocollicular cells are located in layer VI. The projections terminate ipsilaterally in the ICd and ICx (Schofield 2009). Such projections have been noted in several species (rats: Games and Winer 1988; Doucet et al. 2003; hedgehog tenrec: Künzle 1995; gerbil: Bajo and Moore 2005; guinea pig: Schofield 2009), but have generally been given little further attention. The layer VI cells constitute about 10% of the cortical cells that project to the ipsilateral IC (Schofield 2009); this number is similar to the number of layer V cells that project to the contralateral IC, and suggests that the layer VI projection could play a significant role in corticocollicular projections. The layer VI cells include pyramidal cells and a variety of nonpyramidal cells.

The boutons of cortical axons are small compared to boutons on ascending axons to the IC (Jones and Rockel 1973). The cortical terminals contain round vesicles and form asymmetric synapses with dendritic spines and, occasionally, dendritic shafts. They most likely release glutamate (Feliciano and Potashner 1995) and lead to excitation of the collicular cells (Mitani et al. 1983).

9.4.5.3 Auditory Cortical Projections to Nuclei Below the IC

Cortical projections to the area of the lateral lemniscus terminate primarily in the areas surrounding the NLL (Fig. 9.4b; Feliciano et al. 1995; Budinger et al. 2000). The projections are exclusively ipsilateral. The targets include the sagulum, the "horizontal cell group," and the paralemniscal zone along the medial margin of all three NLL. These projections have been described in rats and gerbils; an additional projection to the DNLL appears to be present in gerbils.

The projections to the lateral lemniscal area arise from several auditory cortical areas in gerbils (anterior auditory field and some nonprimary fields; Budinger et al. 2000). In rats, the projections arise from Te1 (temporal area 1, an area defined cytoarchitectonically (Zilles 1985) that appears to include both primary AC and one or more additional areas; Doron et al. 2002; Rutkowski et al. 2003). The laminar origins and cell types that give rise to the projections have not been described.

Cortical projections to the SOC have been described in rats, gerbils, and guinea pigs (Feliciano et al. 1995; Budinger et al. 2000; Doucet et al. 2002; Coomes and Schofield 2004b). Details of termination patterns have been described in rats (Feliciano et al. 1995) and guinea pigs (Coomes and Schofield 2004b). The projections terminate much more densely in the ipsilateral SOC than in the contralateral SOC. In rats, the axons terminate heavily in three areas: the LSO, the VNTB, and a "dorsal ribbon" at the dorsal border of the SOC. In guinea pigs, the cortical axons terminate ipsilaterally in all olivary nuclei except the MSO. In both species, there are projections to the contralateral SOC that are less dense than the ipsilateral projections but terminate in a similar pattern. In both species (and gerbil as well; Budinger et al. 2000), the VNTB is the primary target of AC projections. The projections to the SOC arise from primary and nonprimary AC areas (Budinger et al. 2000; Doucet et al. 2002; Coomes and Schofield 2004b). The projections arise from pyramidal cells in layer V.

Bilateral projections from AC to the CN have been described in rats, mice, and guinea pigs (Weedman and Ryugo 1996a; Jacomme et al. 2003; Schofield and Coomes 2005a; Meltzer and Ryugo 2006). The patterns of termination are similar on the two sides. Boutons of the cortical axons contain round vesicles and form asymmetric junctions, suggesting an excitatory effect on their targets (Weedman and Ryugo 1996b). In each of the species described, the majority of cortical terminals are located in the granule cell area of the CN. In mice, additional terminals are scattered in DCN. In guinea pigs, a substantial number of terminals is found in the small cell cap (a region of the VCN; see discussion in Schofield and Coomes 2005a) and the fusiform cell layer of DCN, with additional terminals in other DCN layers and throughout much of the VCN. The cortical projections to the CN originate from primary AC in rats, mice, and guinea pigs and from additional cortical areas in guinea pigs (Weedman and Ryugo 1996a; Jacomme et al. 2003; Schofield and Coomes 2005a; Meltzer and Ryugo 2006; Schofield et al. 2006). The cells of origin are pyramidal cells in cortical layer V.

9.5 Convergence of Descending Pathways and Targets in Individual Nuclei

This section describes the convergence of descending projections in each major subdivision of the auditory pathways. One objective is to review the data from the perspective of loops and chains. To this end, it is necessary to identify the projections of the cells targeted by the descending axons. This type of data will be supplemented with other characteristics that identify the target cells.

9.5.1 Projections to the CN

Projections to the CN originate from the SOC, the NLL, the IC, the thalamus, and the AC (Fig. 9.5a). Virtually all parts of the CN receive descending inputs, but the combinations of inputs vary by target area. A variety of neurotransmitters have been implicated (e.g., glutamate, acetylcholine, GABA, glycine), suggesting diverse effects.

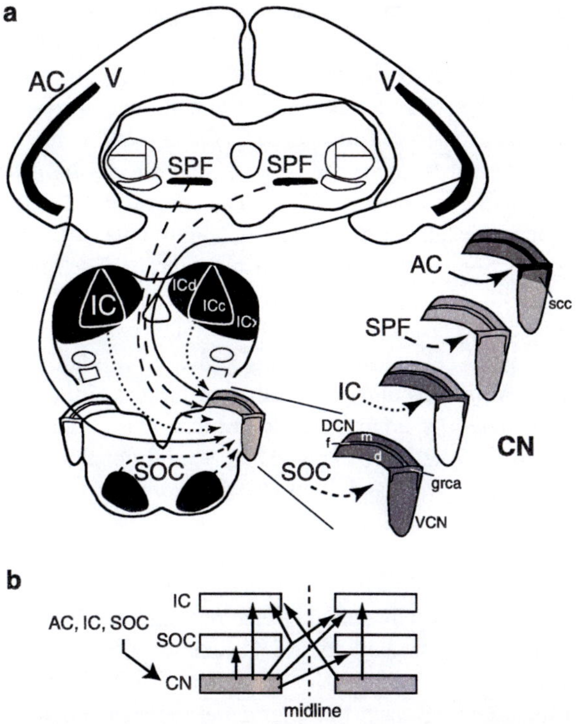

Fig. 9.5 Descending pathways to the cochlear nucleus (CN). (**a**) Summary of descending pathways to the right CN. Origins of the pathways are indicated in *black*, and *distinct arrow* types are used for projections from each level. The regions of the CN innervated by each projection are summarized in the *insets* to the *right*, showing the right CN enlarged. In the *insets*, the shading indicates the relative strength of terminations from the indicated source to the different parts of the CN; *white* means little or no projection, *black* indicates the densest termination, and two levels of *gray* indicate intermediate levels. (**b**) Schematic diagram showing the ascending pathways from the left and right CN (*shaded*) for which the cells of origin appear to be contacted directly by descending projections from the left AC, IC, or SOC. The *branched arrow* indicates cells in the CN that have collateral projections to the left and right IC

The CN has no known descending projections to the cochlea, so inputs to the CN do not form chains to lower structures. However, inputs from higher centers contact intrinsic CN circuits and also form loops by contacting numerous ascending pathways. Much of the current understanding is based on the considerable progress that has been made in identifying CN cell types and their projections (Cant and Benson 2003). The following discussion reviews specific sources of input to CN cells with known projections.

SOC projections contact several cell types in the VCN. One target, globular bushy cells, have stereotypical projections (particularly to the MNTB and LNTB) that relate them to binaural analysis and sound localization (review: Cant and Benson 2003). Globular bushy cells receive direct input from principal cells in the MNTB (in the ipsilateral SOC), establishing a circuit that potentially could enhance binaural processing (Schofield 1994). Stellate (multipolar) cells receive inputs from olivocochlear cells (Brown, Chap. 2), but the projections of these stellate cells are not known (stellate cells as a class project to numerous targets). Cochlear root neurons (cf. Cant and Benson 2003 for review) receive input from cholinergic cells in the ipsilateral and contralateral VNTB (Gómez-Nieto et al. 2008). Cochlear root neurons project to nuclei associated with the startle reflex (and not directly with the ascending auditory pathways) and are not considered further here. Finally, VCN multipolar cells that project to the contralateral IC appear to be contacted directly by descending axons from the AC (Schofield and Coomes 2005b). The cortical inputs can arise from the same side or the opposite side relative to the VCN multipolar cell.

Descending projections to the DCN are also numerous. The major output of the DCN is to the IC, with a large projection to the contralateral IC and a smaller one to the ipsilateral IC. The projections arise from many fusiform cells and a much smaller number of giant cells. Multilabeling studies have identified likely inputs from the AC to fusiform cells and giant cells that project to the IC; these include cells that project ipsilaterally, contralaterally or bilaterally to the IC (Schofield and Coomes 2005b). Fusiform cells are likely to receive additional inputs from the SOC (particularly the VNTB; Warr and Beck 1996) and from the IC (Caicedo and Herbert 1993; Alibardi 1999); it is likely that both of these inputs form loops with cells that project to the IC.

Finally, the granule cells of the CN are noteworthy for several reasons. They are probably the most numerous CN cell, and give rise exclusively to intrinsic projections. These circuits modify the activity of DCN fusiform cells and thus the ascending pathway from the DCN to the IC (Oertel and Young 2004). Many areas (auditory and nonauditory) project to the granule cell area and are likely to influence DCN outputs (Oertel and Young 2004). Ryugo and colleagues (Weedman and Ryugo 1996a, b; Meltzer and Ryugo 2006) have identified direct synaptic contacts between the axons of AC cells and the granule cells in the CN. Other inputs to the granule cell area include cells in the SOC and IC (Shore and Moore 1998); whether these inputs contact granule cells has not been determined (e.g., Benson and Brown 1990 have shown that olivocochlear axon collaterals that end in the granule cell area actually contact the dendrites of VCN stellate [multipolar] cells). Figure 9.5b summarizes the ascending pathways from the CN that appear to be contacted by descending projections to the CN.

9.5.2 *Projections to the SOC*

Descending projections to the SOC originate from the IC, thalamus, and AC. As described above, all the olivary nuclei except the MSO receive projections from the AC, and numerous nuclei receive additional inputs from the IC and/or the thalamus and nearby areas (Fig. 9.6a). The VNTB is notable because it receives the heaviest projections from both the IC and the AC, and significant projections from the SPF. The SPN also receives projections from all three sources of descending inputs. The SOC contains many intrinsic circuits, and also projects to many targets, including the cochlea, CN, NLL, and IC (Thompson and Schofield 2000). For the most part, the ascending projections to the IC arise from a population of cells distinct from those that project to the CN or the cochlea (Aschoff and

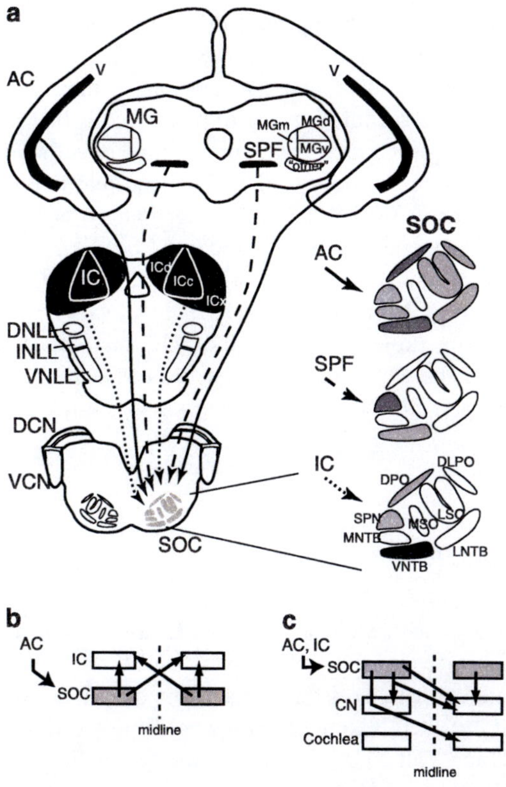

Fig. 9.6 Descending pathways to the superior olivary complex (SOC). (**a**) Origins of projections from different levels to the right SOC. *Inset*: Patterns of termination from each level, shown in enlargements of the right SOC. Conventions as in Fig. 9.5b. (**b**) Schematic diagram showing the ascending pathways from the left and right SOC that appear to be contacted by descending projections from the left AC. (**c**) Descending projections from the left and right SOC (*shaded*) that appear to be contacted by projections from the left AC and/or left IC. The *branched arrow* indicates cells in the SOC that have collateral projections to the left and right CN

Ostwald 1988; Schofield 2002). As such, descending inputs to the SOC could contact intrinsic circuits as well as descending projections (forming chains) or ascending projections (forming loops).

The ascending projection from the SOC to the IC originates from almost all olivary nuclei. Projections from the AC appear to contact olivo-collicular cells ipsilateral and contralateral to the cortical cells of origin (Fig. 9.6b; Peterson and Schofield 2007). The targeted cells project to the IC on either side, suggesting that the cortico-olivary projection could influence olivary projections to the IC on both sides. Most contacts were observed in the periolivary nuclei (especially the VNTB), with additional contacts in the LSO. Interestingly, the descending projections from the IC to the SOC terminate in some of the same areas as the cortico-olivary axons; however, there is no evidence that collicular axons contact olivo-collicular cells (Faye-Lund 1986).

Descending projections from the SOC extend to the cochlea and the CN. The medial olivocochlear cells appear to receive descending inputs from both the IC and the AC (Thompson and Thompson 1993; Vetter et al. 1993; Mulders and Robertson 2000, 2002). Both the AC and the IC project to the ipsilateral SOC, where they contact medial olivocochlear cells that project to the contralateral cochlea. There are also descending inputs to olivary cells that project to the CN. The IC projects to the ipsilateral SOC to contact olivary cells that project ipsilaterally, contralaterally, or bilaterally to the CN (Schofield and Cant 1999). Most of these contacts occur in the VNTB. A similar pattern is seen for projections from the AC to the ipsilateral SOC (Schofield and Coomes 2006). In addition, the small AC projection to the contralateral SOC also contacts CN-projecting cells; these cells project to the CN on the same side as the olivary cell. Figure 9.6c summarizes the descending projections from the SOC that appear be contacted by descending inputs to the SOC. In summary, the descending projections to the SOC form both loops (with ascending projections to the IC) and chains (to both the cochlea and the CN).

9.5.3 Projections to the NLL

Descending projections to the areas of the NLL arise from the IC and the AC (Fig. 9.7). The IC projects to the main nuclei (DNLL and, less so, to INLL and VNLL). The targets of these projections have not been identified, but the major projection from all the NLL is to the IC (bilateral from the DNLL, mostly ipsilateral from the INLL and VNLL). It is likely that the descending projections from the IC form loops by contacting NLL cells that project to one or the other IC. Both collicular and cortical descending projections terminate in the sagulum and regions surrounding the NLL. Again, the target cells have not been identified. The projections of the cells in these areas (especially the paralemniscal zone medial to the NLL) are quite varied, and not limited to auditory nuclei. Further data is needed for identification of loops or chains.

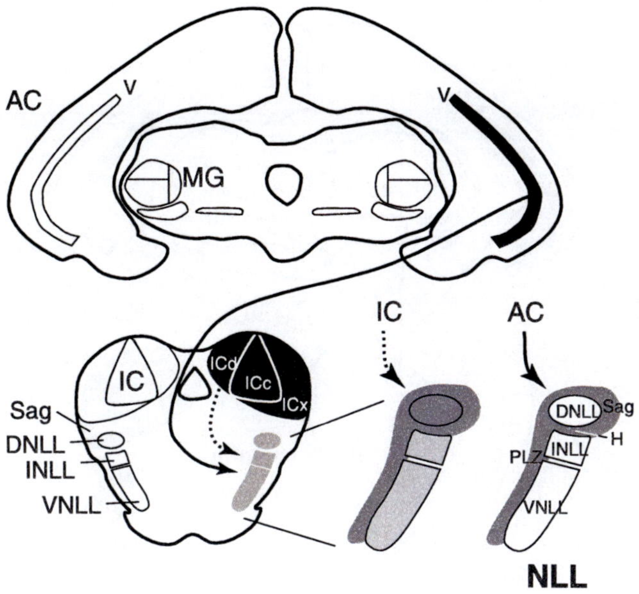

Fig. 9.7 Descending pathways to the nuclei of the lateral lemniscus (NLL) and surrounding areas. Origins of projections to the *right* side are indicated in *black*. *Insets* show the distribution of terminations from the inferior colliculus (IC) and AC. Conventions as in Fig. 9.5

9.5.4 Projections to the IC

The IC receives descending projections from the AC and from the MG and nearby nuclei (Fig. 9.8a). The heaviest projections go to the ICc and ICx; the extent to which different descending pathways converge on single IC cells is unknown. In any case, the IC is the origin of ascending auditory pathways to the ipsilateral and contralateral thalamus and descending pathways to the NLL ipsilaterally and the SOC and CN bilaterally. It has been shown that the descending pathway to the CN originates from a different population of cells than the ascending pathway (Coomes and Schofield 2004c; Okoyama et al. 2006); preliminary data also suggest that the descending pathway to the SOC originates from a different population than the ascending pathway (Schofield 2000). Thus, it appears that most of the descending projections from the IC originate from different populations of cells than provide the origins of the ascending projections.

Cortical projections to the IC form loops via cortical axons that contact cells in the ipsilateral IC that project ipsilaterally or contralaterally to the thalamus (Fig. 9.8b; Mitani et al. 1983; Peterson and Schofield 2007). Cortical projections to the IC also form descending chains (Fig. 9.8c). Cortical projections to the ipsilateral IC contact cells in the IC that project ipsilaterally to the CN or the SOC or contralaterally to the CN (Schofield and Coomes 2006). Cortical projections to the contralateral IC contact cells that project to either CN (Schofield and Coomes 2006).

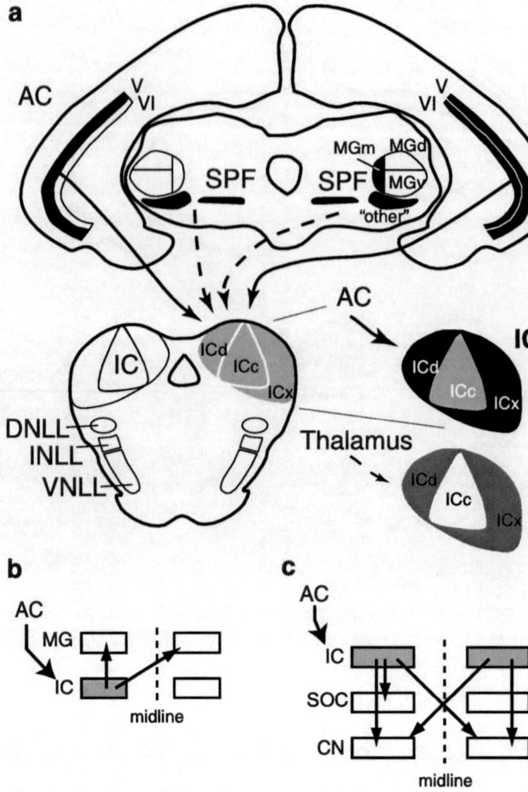

Fig. 9.8 (**a**) Descending pathways to the inferior colliculus (IC). Origins are shown in *black*. *Insets*: Distribution of terminations from the AC and the thalamus (SPF and "other" nuclei; see text for discussion). Conventions as in Fig. 9.5. (**b**) Schematic diagram showing the ascending pathways from the left IC (*shaded*) that appear to be contacted by descending projections from the left AC. (**c**) Descending pathways from the left and right IC (*shaded*) that appear to be contacted by descending projections from the left AC

9.5.5 Projections to the Thalamus from the AC

All parts of the AC project to auditory nuclei in the thalamus. The number of thalamic cells that have descending projections is dwarfed by the number that project to the cerebral cortex. In general, corticothalamic cells project directly to thalamocortical cells (as well as other targets), setting up loops. There appear to be ample opportunities for corticothalamic axons to contact thalamic cells that project to lower centers, thus forming chains; however, data for such connections has not been reported. The following section discusses several perspectives on cortico-thalamo-cortical connections.

One perspective considers the corticothalamic projections in the context of three parallel systems reflecting the organization of the ascending pathways (Fig. 9.9a;

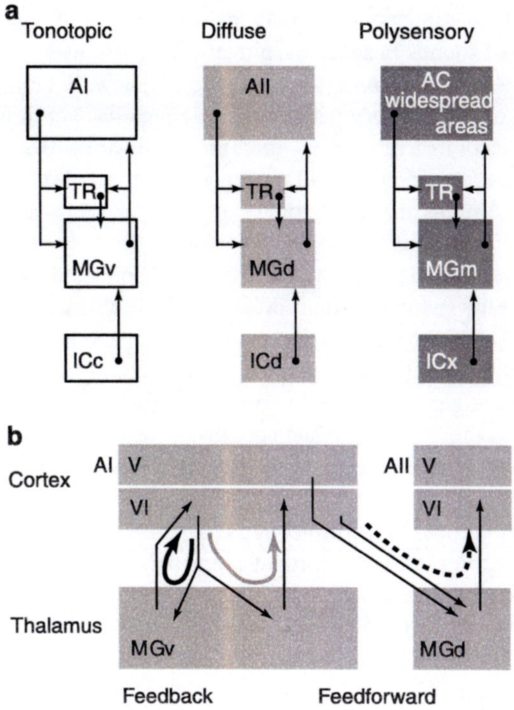

Fig. 9.9 Descending projections to the thalamus. (**a**) Schematic diagrams illustrating three parallel systems for describing corticothalamic as well as ascending auditory connections. See text for discussion. (**b**) Feedback loop (*thick black arrow*) and feedforward loops (*thick gray* and *thick dotted arrow*) in the thalamo-cortico-thalamic pathways. *Gray boxes* represent thalamic nuclei and cortical layers V and VI. A1, AII: auditory cortical areas. See text for discussion

Rouiller 1997). A "tonotopic" (or "lemniscal") system originates in the ICc and projects through the MGv into primary AC (A1). As its name implies, cells and connections all along this pathway preserve frequency tuning. The "diffuse" (or nontonotopic) system originates in the ICd and relays through the MGd. Whereas the ICd does receive ascending auditory input from lower centers, the primary input to the ICd has been considered the AC (Huffman and Henson 1990). A behavioral study in which the ICd was selectively lesioned led the authors to suggest that the ICd is particularly involved in auditory attention (Jane et al. 1965). The diffuse system has been associated with cortical area AII, based on work in cats (where much of the anatomical work on thalamo-cortico-thalamic circuits has been done). It is presumed that other mammals have a cortical area analogous to cat AII; whether or not an equivalent to AII can be agreed upon, the concept of a diffuse system could still apply broadly. The third system is the "polysensory" system. This system originates in the ICx and projects to the MGm, which projects very broadly across the AC. This system is so named because both the ICx and the MGm receive somatosensory inputs in addition to auditory inputs. The functional significance of this sensory convergence is not well understood. It may provide, in part, somatosensory inputs to allow localization of sounds in space in spite of changes in body and/or ear position. In fact, there are also somatosensory inputs to the CN; how the somatosensory inputs to so many levels of the auditory pathways are related has yet to be determined. In any event, the projections from the MGm to the cortex are widespread, and equally wide areas of cortex provide descending projections to the MGm. While the tonotopic and diffuse systems maintain some parallel characteristics, the polysensory system clearly overlaps extensively with the other two.

A second perspective on cortico-thalamo-cortical connections focuses on details of the cellular origins (which cortical layers) and their form of termination (e.g., axon morphology) in the thalamus (Sherman and Guillery 1996; Smith and Spirou 2002). A prominent difference between the layer V and layer VI projections is that the layer V axons are generally able to "drive" the thalamic cells (i.e., elicit action potentials), whereas the layer VI cells do not drive the target cells but rather "modulate" the responses of thalamic cells to ascending sensory input (Sherman and Guillery 1996; Smith and Spirou 2002). The distinctions between drivers and modulators and differences in the projections of their thalamic targets have implications for whether cortico-thalamo-cortical loops perform feedback or feedforward functions. Reciprocal connections, associated with a feedback function, can occur at the level of a single cortical column and the corresponding group of thalamic cells that project to that column (Rouiller and Welker 2000). This type of connection is commonly seen with cortical layer VI projections to MG (Fig. 9.9b, thick black arrow). Layer VI cortical cells also project to thalamic cells that project to a more distant site within the cortical area occupied by the corticothalamic cell (Fig. 9.9, thick gray arrow). This type of circuit could provide a feedforward signal, allowing one column of cortex to alter the ascending information to another column. A third pattern, generally associated with layer VI or layer V corticothalamic cells, involves projections from one area of cortex to thalamic cells that project to another area of cortex (e.g., Fig. 9.9, thick dotted arrow). The example illustrated could provide for interactions between the tonotopic and diffuse systems; other examples involving many cortical areas are possible. This latter example has also been called "feedforward." The various concepts that have been associated with corticothalamic and thalamocortical connections can be generalized to some extent across species and sensory (and even motor) systems (e.g., Sherman and Guillery 1996; Rouiller and Welker 2000; Smith and Spirou 2002; Sherman 2005; see also Smith et al. 2007 for a discussion of how these concepts may or may not apply to the "other" thalamic nuclei).

9.6 Loops, Chains, and Branches

How do the discoveries since Spangler and Warr's (1991) review affect the concepts of loops and chains? First, the newly discovered pathways, particularly those from the cortex to lower brain stem nuclei, provide opportunities for many more loops.

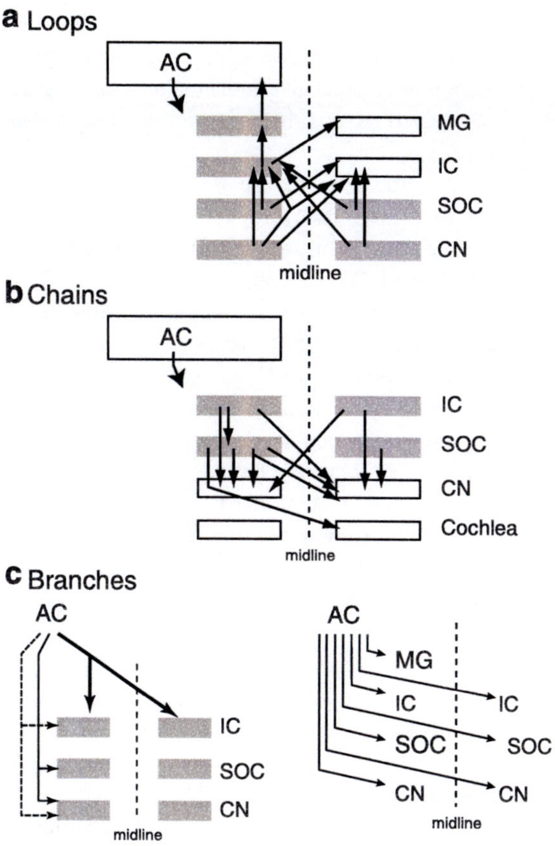

Fig. 9.10 Loops, chains, and branches in the descending auditory pathways. (**a**) Summary of loops formed by cortical axons contacting cells in lower centers that give rise to ascending pathways. Each *upward arrow* indicates a single ascending pathway, originating in the nucleus where the arrow starts (*gray boxes*) and ending in the nucleus at the *arrow tip*. The *branched arrow* indicates a projection by cells that have axon collateral projections to the indicated targets. The projections are drawn from the perspective of the left AC; a similar pattern exists for descending projections from the right AC. (**b**) Summary of descending chains formed by cortical axons contacting cells in lower centers that give rise to descending pathways (conventions as in (**a**)). (**c**) Branching patterns in the descending corticofugal system. *Left*: Examples of branched cortical projections, in which single axons send collaterals to two different brain stem targets. *Right*: The predominant corticofugal projection pattern, with each cell innervating a single subcortical target

Figure 9.10a summarizes the subcortical ascending pathways that appear to be contacted directly by descending projections from AC. The cells of origin of these ascending pathways occur at every level of the central auditory system. Further, the effects are bilateral, suggesting that the projections from AC on one side could affect not only its own ascending input but also the ascending input to the contralateral cortex.

How are loops significant? A loop is, by definition, bidirectional. The feedback perspective leads to the view that, for example, AC projections to ascending pathways allow AC to modulate its own input. As a loop, it can also be appreciated that the ascending pathways are in a position to modulate the descending input they receive (this perspective has been most appreciated in the thalamo-cortical system, where the loops are sometimes described as reverberating circuits, but could apply to other cortical projections as well).

Descending chains, hypothesized before the review by Spangler and Warr (1991), have now been described at several levels of the auditory pathways. As described previously, projections from the IC form chains by contacting medial olivocochlear cells as well as SOC cells that project to the CN. The majority of the descending chains, however, originate with projections from the AC. Figure 9.10b summarizes the descending pathways in the brain stem that appear to be contacted directly by descending projections from AC. In this case, the targets include medial olivocochlear cells, providing a two-neuron pathway from cortex to the cochlea. The data on descending axons that contact descending pathways provide evidence for disynaptic chains; it is beyond current technical abilities to identify the trisynaptic chain from cortex to cochlea that Huffman and Henson drew (shown in Fig. 9.2c). Despite the technical limitations, there is evidence for multiple parallel descending chains. The multiplicity and parallel nature suggest a multitude of functions while also complicating experimental approaches to dissect the functional roles of any particular pathway.

In addition to loops and chains, there is evidence for a third pattern of connections – branches – within the descending system. This pattern refers to projections in which axons branch to innervate two or more lower targets. Evidence for collateral branching occurs in several components of the descending systems. The best known example is the collateral projection of medial olivocochlear cells to the CN (Brown, Chap. 2; Robertson and Mulders, Chap. 10). Branching is also present in the descending projections from the AC. Two points are important in regard to the cortical projections. First, branching in cortical projections is a minority pattern; the majority of corticofugal cells appear to innervate a single subcortical target (Doucet et al. 2003). However, a number of corticofugal cells innervate two or more targets. Quantitative data are available for the corticocollicular projection (Coomes et al. 2005). In this pathway, about 10% of the corticocollicular cells project to the contralateral IC. The majority of these cells appear to project *also* to the ipsilateral IC; double-label experiments show that up to 81% of the contralaterally projecting corticocollicular cells were also labeled from the ipsilateral IC (for methodological reasons, 81% is likely to be an underestimate; Schofield et al. 2007). The implication is twofold: (1) the majority of corticocollicular cells project to the ipsilateral IC (a single target), and (2) a minority of cells projects bilaterally, sending axonal branches to both colliculi. Certainly the function of the contralateral projection must be considered in the context of being part of a bilateral projection to both colliculi. A branching pattern may also apply to other corticofugal projections. Feliciano et al. (1995) described individual cortical axons with collaterals that could be traced to the CN and the VNTB. This observation was based on tracing of individual axons, so the percentage of cells with collateral projections was not clear. In other experiments, many cortical cells that project to one CN were found to project to one IC; a number of cortical cells appear to project to three targets (both ICs and one CN; Coomes and Schofield 2004a). Branching projections may characterize only a small percentage of corticofugal cells, but given the number of such cells, even a small percentage can constitute a substantial pathway.

Branching projections (axon collaterals) are seen in many areas of the brain. They are often suggested as a means for coordinating activity in the targets (perhaps to synchronize activity) or for having a broad effect on, say, the sensitivity of the targets to sensory inputs. In contrast, the single-target projection of many of the corticofugal cells is well-suited to exerting effects that are specific to that target (Fig. 9.10c, right panel).

9.7 Summary and Questions for Future Research

Recent studies have greatly expanded knowledge of the central descending auditory pathways. Most prominently, the projections of the AC are much more extensive than previously believed. The cortical projections extend to all levels of the central auditory pathways. They also converge with other descending pathways; for example, cortical projections terminate in the CN along with descending projections from the thalamus, IC, and SOC. Some of the latter projections are themselves contacted by cortical axons, establishing parallel descending chains that terminate in the CN. There is a similar convergence of descending projections in other regions, including the SOC and IC. The plethora of descending pathways suggest that they could affect a wide array of brain stem pathways, and thus a wide range of auditory functions. The convergence of descending pathways with each other and with ascending pathways and modulatory (e.g., serotoninergic, dopaminergic, noradrenergic) systems could provide the basis for a wide variety of effects.

The new data have extended our views of both loops and chains in the descending system. Feedback loops have been a consistent concept in discussions of the descending auditory system. Loops form a dominant theme in conceptualizing the connections between cortex and thalamus. In the brain stem, loops have been applied primarily to the olivocochlear system, but will clearly be incorporated into broader concepts of brain stem auditory processing. The concept of descending chains was inherent in many early studies, and has slowly gathered supportive data. Once again, the projections of the cortex have provided the greatest source of new data. Though earlier concepts considered a single descending chain from cortex to cochlea, it is clear now that there are numerous descending chains that are likely to affect the cochlea as well as other brain stem auditory nuclei. Finally, collateral branching of descending axons has been observed in a variety of descending systems. Branching axons may provide a mechanism for coordinating activity at multiple sites or for exerting broad effects, such as adjusting sensitivity, at multiple points in a circuit.

There are many questions about the descending pathways that are yet to be addressed. Understanding the functions of these pathways will require elucidation of their synaptic organization and the neurotransmitters and receptors associated with them. How are the descending pathways related to other pathways? A huge gap in our understanding concerns interactions between descending pathways and the intrinsic circuitry in many regions. There is increasing appreciation for the dependence of cortical responses on behavioral state or task performance (e.g., Hromádka and Zador 2007; Nuñez and Malmierca 2007; Scheich et al. 2007); certainly these factors would be reflected in the descending projections from cortex. Finally, there is increasing interest in the roles of descending projections in permitting, or perhaps directing, plasticity of responses in the lower auditory centers (e.g., Suga et al., Chap. 11; de Boer and Thornton 2008).

Acknowledgments Special thanks to Susan Motts and Kyle Nakamoto for comments on an earlier draft of this chapter. The work described here that was completed in the author's laboratory was supported by the National Institutes of Health DC04391 and DC05277. It is regrettable that space limitations prevent citation of many original studies. The author acknowledges those researchers here, with gratitude for their understanding.

References

Alibardi L (1999) Fine structure, synaptology and immunocytochemistry of large neurons in the rat dorsal cochlear nucleus connected to the inferior colliculus. J Hirnforsch 39:429–439

Aschoff A, Ostwald J (1988) Distribution of cochlear efferents and olivo-cochlear neurons in the brainstem of rat and guinea pig. Exp Brain Res 71:241–251

Bajo VM, Moore DR (2005) Descending projections from the auditory cortex to the inferior colliculus in the gerbil, *Meriones unguiculatus*. J Comp Neurol 486:101–116

Bajo VM, Nodal FR, Bizley JK, Moore DR, King AJ (2007) The ferret auditory cortex: descending projections to the inferior colliculus. Cereb Cortex 17:475–491

Benson TE, Brown MC (1990) Synapses formed by olivocochlear axon branches in the mouse cochlear nucleus. J Comp Neurol 295:52–70

Budinger E, Heil P, Scheich H (2000) Functional organization of auditory cortex in the Mongolian gerbil (*Meriones unguiculatus*). IV. Connections with anatomically characterized subcortical structures. Eur J Neurosci 12:2452–2474

Caicedo A, Herbert H (1993) Topography of descending projections from the inferior colliculus to auditory brainstem nuclei in the rat. J Comp Neurol 328:377–392

Cant NB, Benson CG (2003) Parallel auditory pathways: projection patterns of the different neuronal populations in the dorsal and ventral cochlear nuclei. Brain Res Bull 60:457–474

Coomes DL, Schofield BR (2004a) Projections from individual cortical cells to multiple targets in the auditory brainstem in guinea pigs. Assoc Res Otolaryngol Abstr 27:120

Coomes DL, Schofield BR (2004b) Projections from the auditory cortex to the superior olivary complex in guinea pigs. Eur J Neurosci 19:2188–2200

Coomes DL, Schofield BR (2004c) Separate projections from the inferior colliculus to the cochlear nucleus and thalamus in guinea pigs. Hear Res 191:67–78

Coomes DL, Schofield RM, Schofield BR (2005) Unilateral and bilateral projections from cortical cells to the inferior colliculus in guinea pigs. Brain Res 1042:62–72

de Boer J, Thornton AR (2008) Neural correlates of perceptual learning in the auditory brainstem: efferent activity predicts and reflects improvement at a speech-in-noise discrimination task. J Neurosci 28:4929–4937

Delano PH, Elgueda D, Hamame CM, Robles L (2007) Selective attention to visual stimuli reduces cochlear sensitivity in chinchillas. J Neurosci 27:4146–4153

Doron NN, Ledoux JE, Semple MN (2002) Redefining the tonotopic core of rat auditory cortex: physiological evidence for a posterior field. J Comp Neurol 453:345–360

Doucet JR, Rose L, Ryugo DK (2002) The cellular origin of corticofugal projections to the superior olivary complex in the rat. Brain Res 925:28–41

Doucet JR, Molavi DL, Ryugo DK (2003) The source of corticocollicular and corticobulbar projections in area Te1 of the rat. Exp Brain Res 153:461–466

Faye-Lund H (1986) Projection from the inferior colliculus to the superior olivary complex in the albino rat. Anat Embryol 175:35–52

Feliciano M, Potashner SJ (1995) Evidence for a glutamatergic pathway from the guinea pig auditory cortex to the inferior colliculus. J Neurochem 65:1348–1357

Feliciano M, Saldaña E, Mugnaini E (1995) Direct projections from the rat primary auditory neocortex to nucleus sagulum, paralemniscal regions, superior olivary complex and cochlear nuclei. Aud Neurosci 1:287–308

Fitzpatrick KA, Imig TJ (1978) Projections of auditory cortex upon the thalamus and midbrain in the owl monkey. J Comp Neurol 177:537–555

Frisina RD, Walton JP, Lynch-Armour MA, Byrd JD (1998) Inputs to a physiologically characterized region of the inferior colliculus of the young adult CBA mouse. Hear Res 115:61–81

Games KD, Winer JA (1988) Layer V in rat auditory cortex: projections to the inferior colliculus and contralateral cortex. Hear Res 34:1–26

Gómez-Nieto R, Rubio ME, Lopez DE (2008) Cholinergic input from the ventral nucleus of the trapezoid body to cochlear root neurons in rats. J Comp Neurol 506:452–468

Hefti BJ, Smith PH (2000) Anatomy, physiology, and synaptic responses of rat layer V auditory cortical cells and effects of intracellular GABA(A) blockade. J Neurophysiol 83:2626–2638

Helfert RH, Aschoff A (1997) Superior olivary complex and nuclei of the lateral lemniscus. In: Ehret G, Romand R (eds) The central auditory system. Oxford University Press, New York, pp 193–258

Hromádka T, Zador AM (2007) Toward the mechanisms of auditory attention. Hear Res 229:180–185

Huffman RF, Henson OW Jr (1990) The descending auditory pathway and acousticomotor systems: connections with the inferior colliculus. Brain Res Rev 15:295–323

Jacomme AV, Nodal FR, Bajo VM, Manunta Y, Edeline JM, Babalian A, Rouiller EM (2003) The projection from auditory cortex to cochlear nucleus in guinea pigs: an in vivo anatomical and in vitro electrophysiological study. Exp Brain Res 153:467–476

Jane JA, Masterton RB, Diamond IT (1965) The function of the tectum for attention to auditory stimuli in the cat. J Comp Neurol 125:165–191

Jen PH, Sun X, Chen QC (2001) An electrophysiological study of neural pathways for corticofugally inhibited neurons in the central nucleus of the inferior colliculus of the big brown bat, *Eptesicus fuscus*. Exp Brain Res 137:292–302

Jones EG, Rockel AJ (1973) Observations on complex vesicles, neurofilamentous hyperplasia and increased electron density during terminal degeneration in the inferior colliculus. J Comp Neurol 147:93–118

Künzle H (1995) Regional and laminar distribution of cortical neurons projecting to either superior or inferior colliculus in the hedgehog tenrec. Cereb Cortex 5:338–352

Kuwabara N, Zook JM (2000) Geniculo-collicular descending projections in the gerbil. Brain Res 878:79–87

Le Doux JE, Ruggiero DA, Reis DJ (1985) Projections to the subcortical forebrain from anatomically defined regions of the medial geniculate body in the rat. J Comp Neurol 242:182–213

Lim HH, Anderson DJ (2007) Antidromic activation reveals tonotopically organized projections from primary auditory cortex to the central nucleus of the inferior colliculus in guinea pig. J Neurophysiol 97:1413–1427

Lu Y, Gao H, Waterman JD, Schofield BR (2007) Physiology, morphology and connections of guinea pig auditory cortical layer V pyramidal neurons. Soc for Neurosci, Program Number 278. 14, 2007 Neuroscience Meeting Planner. San Diego, CA: Soc. for Neurosci, Online

Meltzer NE, Ryugo DK (2006) Projections from auditory cortex to cochlear nucleus: a comparative analysis of rat and mouse. Anat Rec A Discov Mol Cell Evol Biol 288:397–408

Mitani A, Shimokouchi M, Nomura S (1983) Effects of stimulation of the primary auditory cortex upon colliculogeniculate neurons in the inferior colliculus of the cat. Neurosci Lett 42:185–189

Mulders WH, Robertson D (2000) Morphological relationships of peptidergic and noradrenergic nerve terminals to olivocochlear neurones in the rat. Hear Res 144:53–64

Mulders WH, Robertson D (2002) Inputs from the cochlea and the inferior colliculus converge on olivocochlear neurones. Hear Res 167:206–213

Nuñez A, Malmierca E (2007) Corticofugal modulation of sensory information. Adv Anat Embryol Cell Biol 187:1–75

Oertel D, Wickesberg RE (2002) Ascending pathways through ventral nuclei of the lateral lemniscus and their possible role in pattern recognition in natural sounds. In: Oertel D, Fay RR, Popper AN (eds) Integrative functions in the mammalian auditory pathway. Springer, New York, pp 207–237

Oertel D, Young ED (2004) What's a cerebellar circuit doing in the auditory system? Trends Neurosci 27:104–110

Okoyama S, Ohbayashi M, Ito M, Harada S (2006) Neuronal organization of the rat inferior colliculus participating in four major auditory pathways. Hear Res 218:72–80

Oliver DL (2005) Neuronal organization in the inferior colliculus. In: Winer JA, Schreiner CE (eds) The inferior colliculus. Springer, New York, pp 69–114

Perrot X, Ryvlin P, Isnard J, Guenot M, Catenoix H, Fischer C, Mauguiere F, Collet L (2006) Evidence for corticofugal modulation of peripheral auditory activity in humans. Cereb Cortex 16:941–948

Peterson DC, Schofield BR (2007) Projections from auditory cortex contact ascending pathways that originate in the superior olive and inferior colliculus. Hear Res 232:67–77

Rasmussen GL (1946) The olivary peduncle and other fiber projections of the superior olivary complex. J Comp Neurol 14:141–219

Rinne T, Balk MH, Koistinen S, Autti T, Alho K, Sams M (2008) Auditory selective attention modulates activation of human inferior colliculus. J Neurophysiol 100:3323–3327

Robertson D, Cole KS, Harvey AR (1987) Brainstem organization of efferent projections to the guinea pig cochlea studied using the fluorescent tracers fast blue and diamidino yellow. Exp Brain Res 66:449–457

Rouiller EM (1997) Functional organization of the auditory pathways. In: Ehret G, Romand R (eds) The central auditory system. Oxford University Press, New York, pp 3–96

Rouiller EM, Welker E (2000) A comparative analysis of the morphology of corticothalamic projections in mammals. Brain Res Bull 53:727–741

Rutkowski RG, Miasnikov AA, Weinberger NM (2003) Characterisation of multiple physiological fields within the anatomical core of rat auditory cortex. Hear Res 181:116–130

Saint Marie RL (1996) Glutamatergic connections of the auditory midbrain: selective uptake and axonal transport of D-[^{3}H]aspartate. J Comp Neurol 373:255–270

Saldaña E (1993) Descending projections from the inferior colliculus to the cochlear nuclei in mammals. In: Merchan MA (ed) The mammalian cochlear nuclei: organization and function. Plenum Press, New York, pp 153–165

Saldaña E, Feliciano M, Mugnaini E (1996) Distribution of descending projections from primary auditory neocortex to inferior colliculus mimics the topography of intracollicular projections. J Comp Neurol 371:15–40

Saldaña E, Viñuela A, Marshall AF, Fitzpatrick DC, Aparicio MA (2007) The TLC: a novel auditory nucleus of the mammalian brain. J Neurosci 27:13108–13116

Scheich H, Brechmann A, Brosch M, Budinger E, Ohl FW (2007) The cognitive auditory cortex: task-specificity of stimulus representations. Hear Res 229:213–224

Schofield BR (1994) Projections to the cochlear nuclei from principal cells in the medial nucleus of the trapezoid body in guinea pigs. J Comp Neurol 344:83–100

Schofield BR (2000) Distinct origins of ascending and descending projections from the inferior colliculus in guinea pigs. Assoc Res Otolaryngol Abstr 23:35

Schofield BR (2001) Origins of projections from the inferior colliculus to the cochlear nucleus in guinea pigs. J Comp Neurol 429:206–220

Schofield BR (2002) Ascending and descending projections from the superior olivary complex in guinea pigs: different cells project to the cochlear nucleus and the inferior colliculus. J Comp Neurol 453:217–225

Schofield BR (2009) Projections to the inferior colliculus from layer VI cells of auditory cortex. Neuroscience 159:246–258

Schofield BR, Cant NB (1999) Descending auditory pathways: projections from the inferior colliculus contact superior olivary cells that project bilaterally to the cochlear nuclei. J Comp Neurol 409:210–223

Schofield BR, Coomes DL (2005a) Auditory cortical projections to the cochlear nucleus in guinea pigs. Hear Res 199:89–102

Schofield BR, Coomes DL (2005b) Projections from auditory cortex contact cells in the cochlear nucleus that project to the inferior colliculus. Hear Res 206:3–11

Schofield BR, Coomes DL (2006) Pathways from auditory cortex to the cochlear nucleus in guinea pigs. Hear Res 216–217:81–89

Schofield BR, Coomes DL, Schofield RM (2006) Cells in auditory cortex that project to the cochlear nucleus in guinea pigs. J Assoc Res Otolaryngol 7:95–109

Schofield BR, Schofield RM, Sorensen KA, Motts SD (2007) On the use of retrograde tracers for identification of axon collaterals with multiple fluorescent retrograde tracers. Neuroscience 146:773–783

Schwartz IR (1992) The superior olivary complex and lateral lemniscal nuclei. In: Webster DB, Popper AN, Fay RR (eds) The mammalian auditory pathway: neuroanatomy. Springer, New York, pp 117–167

Senatorov VV, Hu B (2002) Extracortical descending projections to the rat inferior colliculus. Neuroscience 115:243–250

Sherman SM (2005) Thalamic relays and cortical functioning. Prog Brain Res 149:107–126

Sherman SM, Guillery RW (1996) Functional organization of thalamocortical relays. J Neurophysiol 76:1367–1395

Sherriff FE, Henderson Z (1994) Cholinergic neurons in the ventral trapezoid nucleus project to the cochlear nuclei in the rat. Neuroscience 58:627–633

Shore SE, Moore JK (1998) Sources of input to the cochlear granule cell region in the guinea pig. Hear Res 116:33–42

Smith PH, Spirou GA (2002) From the cochlea to the cortex and back. In: Oertel D, Fay RR, Popper AN (eds) Integrative functions in the mammalian auditory pathway. Springer, New York, pp 6–71

Smith PH, Bartlett EL, Kowalkowski A (2007) Cortical and collicular inputs to cells in the rat paralaminar thalamic nuclei adjacent to the medial geniculate body. J Neurophysiol 98:681–695

Spangler KM, Warr WB (1991) The descending auditory system. In: Altschuler RA, Bobbin RP, Clopton BM, Hoffman DW (eds) Neurobiology of hearing: the central auditory system. Raven Press, New York, pp 27–45

Thompson AM (2005) Descending connections of the auditory midbrain. In: Winer JA, Schreiner CE (eds) The inferior colliculus. Springer, New York, pp 182–199

Thompson AM, Schofield BR (2000) Afferents of the superior olivary complex. Microsc Res Tech 51:330–354

Thompson AM, Thompson GC (1993) Relationship of descending inferior colliculus projections to olivocochlear neurons. J Comp Neurol 335:402–412

Vetter DE, Saldaña E, Mugnaini E (1993) Input from the inferior colliculus to medial olivocochlear neurons in the rat: a double label study with PHA-L and cholera toxin. Hear Res 70:173–186

Warr WB, Beck JE (1996) Multiple projections from the ventral nucleus of the trapezoid body in the rat. Hear Res 93:83–101

Weedman DL, Ryugo DK (1996a) Pyramidal cells in primary auditory cortex project to cochlear nucleus in rat. Brain Res 706:97–102

Weedman DL, Ryugo DK (1996b) Projections from auditory cortex to the cochlear nucleus in rats: synapses on granule cell dendrites. J Comp Neurol 371:311–324

Winer JA (2005) Three systems of descending projections to the inferior colliculus. In: Winer JA, Schreiner CE (eds) The inferior colliculus. Springer, New York, pp 231–247

Winer JA, Prieto JJ (2001) Layer V in cat primary auditory cortex (AI): cellular architecture and identification of projection neurons. J Comp Neurol 434:379412

Winer JA, Schreiner CE (2005) The inferior colliculus. Springer, New York

Winer JA, Larue DT, Diehl JJ, Hefti BJ (1998) Auditory cortical projections to the cat inferior colliculus. J Comp Neurol 400:147–174

Winer JA, Chernock ML, Larue DT, Cheung SW (2002) Descending projections to the inferior colliculus from the posterior thalamus and the auditory cortex in rat, cat, and monkey. Hear Res 168:181–195

Yasui Y, Kayahara T, Nakano K, Mizuno N (1990) The subparafascicular thalamic nucleus of the rat receives projection fibers from the inferior colliculus and auditory cortex. Brain Res 537:323–327

Zilles K (1985) The cortex of the rat. Springer, Berlin

Zimmerman EA, Chambers WW, Liu CN (1964) An experimental study of the anatomical organization of the cortico-bulbar system in the albino rat. J Comp Neurol 123:301–324

Chapter 10
Central Effects of Efferent Activation

Donald Robertson and Wilhelmina H.A.M. Mulders

10.1 Introduction

The action of the medial olivocochlear system (MOCS) in the auditory periphery is well established and is described in detail elsewhere in this volume (Guinan, Chap. 3; Sewell, Chap. 4; Katz et al., Chap. 5). The major peripheral effect of activation of the MOCS is a reduction in gain of the outer hair cell (OHC) cochlear amplifier and a consequent reduction in sensitivity of the primary afferent neurons to tones at their most sensitive, or characteristic frequency (CF). This action on the OHCs has different functional consequences for the responses of primary afferent neurons to tone stimuli, depending on whether these are observed in quiet or in the presence of background noise.

In quiet, the MOCS action generally results in a simple rightward shift of the single afferent input–output curves to CF tones (Fig. 10.1 left). There is little effect on the maximum discharge rate. This is because the OHC mechanical amplifier has a limited operating range, and therefore its contribution to basilar membrane vibration (and hence neural firing) is saturated more than about 60 dB above neural threshold (Patuzzi et al. 1984; Patuzzi and Rajan 1992; Rajan and Patuzzi 1992). In a small proportion of the most sensitive neurons, medial efferent activation results in a minor reduction in spontaneous firing rate of primary afferents, probably because of the small (maximum of 3 mV) drop in endocochlear potential that results from the MOCS-induced increase in the basolateral wall conductance of the OHCs (Fex 1967; Desmedt and Robertson 1975). The reduction in the endocochlear potential reduces the standing current through the inner hair cells, hyperpolarizing them and lowering the basal rate of neurotransmitter release (Sewell 1984).

D. Robertson (✉)
The Auditory Laboratory, Discipline of Physiology, M311 School of Biomedical Biomolecular and Chemical Sciences, The University of Western Australia, Crawley, WA 6009, Australia
e-mail: drobed@cyllene.uwa.edu.au

D.K. Ryugo et al. (eds.), *Auditory and Vestibular Efferents*,
Springer Handbook of Auditory Research 38, DOI 10.1007/978-1-4419-7070-1_10,

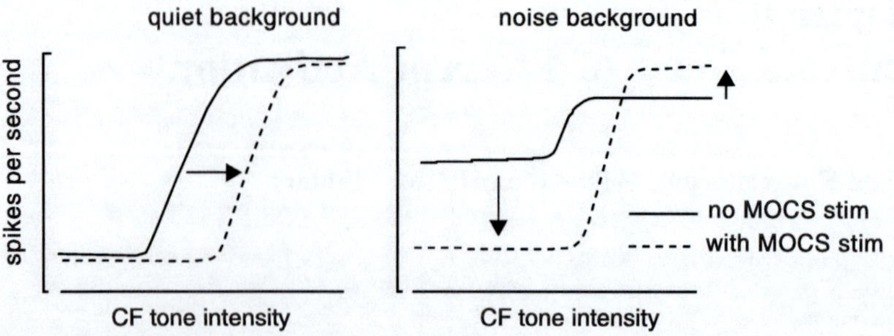

Fig. 10.1 Idealized representation of effects of MOCS activation on input–output responses of single primary afferent neuron to a range of CF tone intensities. *Left*, responses in quiet background; *right*, responses in presence of continuous background noise. Note elevated basal firing rate and reduced maximum rate in noise without MOCS stimulation

In the presence of background noise, the input–output functions of primary afferents to CF tones are compressed for several reasons (Fig. 10.1right). The continuous response to the background noise raises the primary afferent neurons' basal firing rate. In addition, the resulting adaptation reduces the maximum discharge rate. These two effects of the background noise lead to a reduced range of output firing rates. In addition, there is a "jamming" effect of the noise on the cochlear amplifier function of the OHCs and this shifts the input–output function to the right. It has been elegantly shown (Winslow and Sachs 1987; Kawase et al. 1993) that these effects cause a reduced ability of the primary afferents to encode intensity differences in the tonal stimuli. When the MOCS is stimulated, there is a preferential, level-dependent suppression of the response to the background noise and less effect on responses to higher level tonal stimuli. The reduced rate of background firing results in a release from adaptation, so the maximum discharge rate returns towards normal, and the net result is a partial restoration of the dynamic range of the response to CF tones. This "antimasking" effect of the MOCS has received much attention because of its potentially important role in enhancing auditory signal detection (Winslow and Sachs 1987; Kawase et al. 1993; Mulders et al. 2008; Seluakumaran et al. 2008b).

From a simplistic perspective, both of these peripheral effects of MOCS activation might be expected to result in comparable "upstream" changes in the activity of central nervous neurons in quiet and in background noise. If, for example, the sensitivity of primary afferents to CF tones in quiet were reduced by 20 dB, one might expect that a central neuron with the same CF would show a comparable reduction in sensitivity. This question was addressed many years ago by Desmedt (1962), who measured click-evoked field potentials in anesthetized cats at several stages of the central auditory pathways and compared the effects of MOCS activation on these central responses with those produced on the cochlear compound action potential to the same stimuli. Desmedt quantified the MOCS effect by titrating the amount of suppression as an equivalent reduction in acoustic stimulus intensity, using the amplitude/intensity function for the evoked potentials. Using this method, the amount of suppression was found to be very similar at all stages of the pathway and also very similar to the equivalent suppression of the cochlear nerve action potential.

However, there are several reasons why this issue of the central consequences of efferent activation has been revisited by others and why it still remains a matter deserving further investigation. The broadband clicks employed in the early study by Desmedt (1962) tell us little about the nature of responses to tones of particular frequency, for example, at the neurons' CF. In addition, field potential recordings may obscure subtle effects on neuronal subpopulations. The ascending pathways are highly complex, with a bewildering range of neuronal subtypes and circuitry involving multiple interneuronal connections, both excitatory and inhibitory, that shape each neuron's dynamic range and response area. A simple reduction in afferent drive from the cochlea caused by MOCS stimulation might not necessarily be translated faithfully as a linear reduction in the operation of all elements of this complex central circuitry. Finally, it needs to be borne in mind that the olivocochlear neurons terminate not only within the cochlea, but as shown in Brown (Chap. 2), they also send collaterals to other auditory brainstem regions, notably the cochlear nucleus (CN) and superior olivary complex (SOC). As discussed further in this chapter, the precise action of these collateral pathways is still contentious, but their very existence opens up the possibility that MOCS activation may produce a variety of peripheral and central actions that may interact in ways that produce surprising outcomes for central neuronal firing behavior.

For several reasons, it of interest to know more about the central effects of olivocochlear activation, and in particular, to know if they differ from those predicted simply from the effects on the primary afferent neurons. First, the postulated role of olivocochlear neurons in auditory signal processing (antimasking) is based to a large extent on their effects documented in primary afferents. To be meaningful in terms of behavioral output, it is a necessary condition that these effects are also apparent in the behavior of at least some of the higher order neurons of the pathway. Second, a detailed study of variations in the effects of olivocochlear activation on the diverse neuronal response types in the central pathways could generate insights into the various roles that these different neurons may play in signal processing.

This chapter reviews the evidence for effects of olivocochlear activation on neurons in central auditory structures of the more common animal models. The broad questions addressed are:

1. Are effects seen in central neurons that are not readily predictable from peripheral afferent changes? (Sect. 10.2)
2. What mechanisms might be responsible for such "nonclassic" effects? (Sects. 10.3– 10.5)
3. What are the implications of central effects for understanding auditory signal processing and the role of the olivocochlear efferents? (Sect. 10.6)

In addressing these questions, the chapter considers only the effects of activation of the MOCS on single-neuron responses in the cochlear nucleus (CN) and inferior colliculus (IC). Some of the technical obstacles to obtaining clear answers as to the origin of these effects are highlighted. The effects of lateral olivocochlear system activation (LOCS) will not be discussed in view of the paucity of data and the difficulties of achieving selective and reliable activation of this component of the olivocochlear system. A number of other aspects of central efferent effects are not considered, such as the role of other efferent systems innervating the brain stem (see, e.g., Pickles 1976a; Klepper and Herbert 1991; Thompson et al. 1995; Ebert 1996; Mulders and Robertson 2001; Shore et al. 2003; Mulders and Robertson 2005). Aspects of corticofugal pathways and detailed overarching anatomy of the efferent pathways from cortex to cochlea are considered elsewhere in this volume (Brown, Chap. 2; Schofield, Chap. 9; Suga et al., Chap. 11).

A number of methods have been employed to study the effects of MOCS activation on central neurons. These can be broadly classified as follows:

1. Single-neuron recordings in vivo while activating MOCS. Both intra- and extracellular recording methods have been used.
2. Single-neuron recordings in vitro (brain slices) while applying putative MOCS agonists and antagonists.
3. Behavioral experiments of auditory processing while manipulating MOCS function.

Each of these experimental approaches yields different levels of understanding of the problem, and each suffers from particular advantages and disadvantages that are discussed in turn in the text that follows.

10.2 Single-Neuron Recordings In Vivo

10.2.1 Technical Issues

The most commonly used experimental approach to activate the MOCS, illustrated schematically in Fig. 10.2, involves electrical stimulation of the olivocochlear efferents by placing stimulating electrodes at the point of decussation of the medial axons at the floor of the IVth ventricle. This activation of the efferents is combined with single-neuron recording from different central nuclei, using either extracellular recording of action potential firing, or intracellular recordings that enable direct observation of inhibitory and excitatory synaptic events. Electrical stimulation has the virtue of conceptual simplicity, and it guarantees activation of the olivocochlear efferents that can be verified by monitoring the signature peripheral changes: suppression of the cochlear nerve action potential and an increase in the externally recorded cochlear microphonic potential (Mulders and Robertson 2000). Collateral pathways of MOCS axons to CN and elsewhere will also be activated at the same time as the axons traveling to the cochlea.

However, this method has some serious limitations and potential sources of error. First, electrical stimulation at the floor of the IVth ventricle generally results in activation of the entire olivocochlear bundle, which courses as compact fascicles at

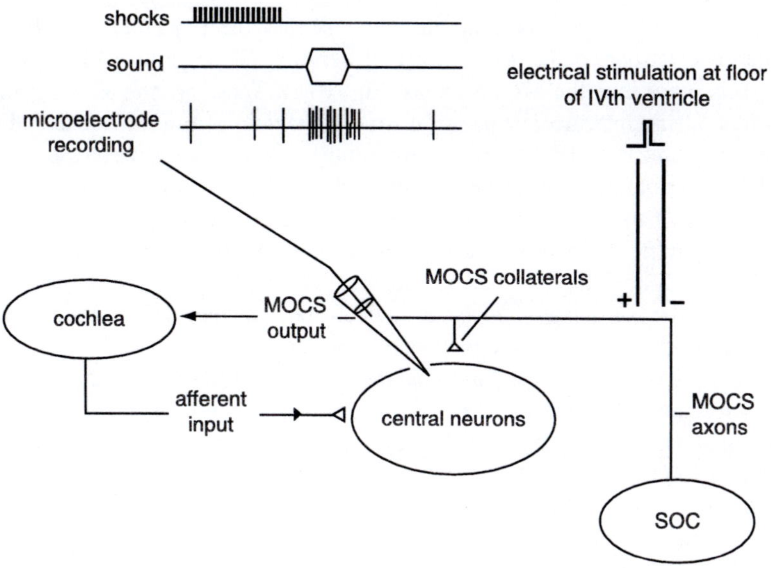

Fig. 10.2 Schematic representation of the basic stimulation and recording arrangement used to study central effects of MOCS activation in vivo. Typical relationship between acoustic stimulus and shock train to MOCS shown in upper part of figure

the level of the facial genua. Thus the effects on cochlear neural output span a frequency range limited only by the extent of efferent innervation of the organ of Corti. This may mimic to a reasonable degree, the efferent activation resulting from physiological stimuli such as broadband noise, but it is not routinely possible with this method to measure the effects on central neural responses of more limited activation of smaller numbers of efferents innervating limited cochlear regions. It is conceivable that very focal stimulation might be able to be used to activate small fascicles of the MOCS axons, but so far this has not been achieved (see Suga et al., Chap. 11, with regard to the application of this technique to selective activation of corticofugal pathways).

Second, the electrical stimulation regimen used makes a detailed investigation of the time course of efferent effects difficult. In the classic paradigm for studying MOCS effects, test acoustic stimuli are usually delivered 5–10 ms after the end of a train of shocks delivered to the MOCS axons (Fig. 10.2). This is because the peripheral effect of MOCS activation caused by single shocks is immeasurably small, and because the size of electrical shock artifacts picked up by the recording microelectrode often makes accurate recording of physiological responses difficult during the shock train itself. As a consequence, the precise onset of effects during the shock train often cannot be accurately estimated, and rapid, short-lasting effects may be missed altogether. In addition, probing for effects after stimulation has ceased means that one is measuring effects that are generally in the process of dissipating. This poses problems for quantifying effects when test stimuli of appreciable duration are used, as is usually the case in studies of central neurons.

The final and most important limitation is that the placement of electrical stimulating electrodes at the floor of the IVth ventricle is dangerously close to major fiber tracts other than the MOCS axons (Fig. 10.3). These are the ascending fiber tracts (dorsal and intermediate acoustic striae) emanating from the dorsal and posteroventral cochlear nuclei, as well as the commissural pathway between the cochlear nuclei. The ascending tracts project to the nuclei of the lateral lemniscus and to the IC. The commissural pathway is believed to comprise principally the axons of large glycinergic multipolar neurons in the cochlear nucleus that correspond to neurons described physiologically as "onset choppers." These glycinergic neurons mediate inhibition in the contralateral cochlear nucleus and elsewhere. Unless great care is taken to limit the spread of electrical stimulation to the olivocochlear bundle, activation of these other tracts can confound the effects seen. As indicated in Fig. 10.3, these confounding effects can arise from orthodromic propagation of impulses providing synaptic direct input to the IC, which in turn could activate descending projections to the SOC or CN. Antidromic and/or orthodromic activation of recurrent collateral circuitry within the CN regions constitutes a further possible complication. In either case, the effects on single neurons from which recordings are being obtained might be wrongly concluded to arise from MOCS activation.

Because of this problem of inappropriate activation of non-olivocochlear tracts, considerable care has to be exercised with the choice of stimulating electrodes and their placement. Bipolar stimulating electrodes with close separation between the

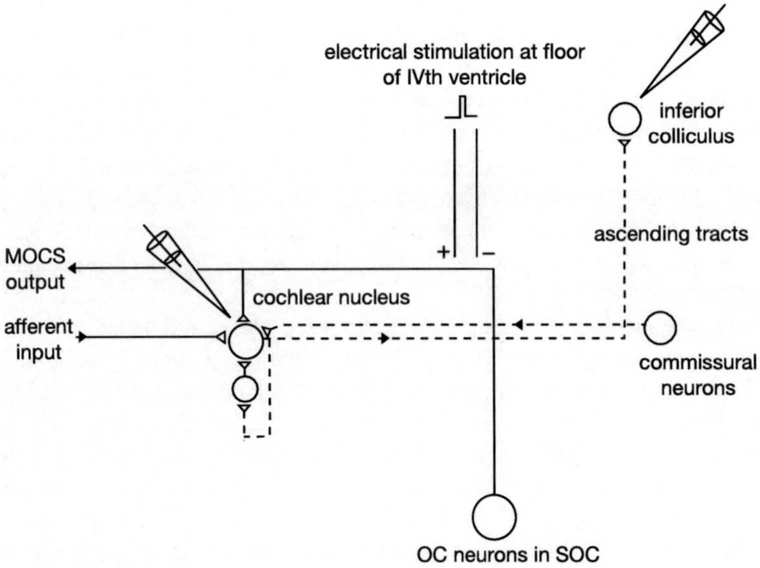

Fig. 10.3 Schematic representation showing ascending and commissural fiber tracts that run close to MOCS axons and are at risk of uncontrolled stimulation by shocks to MOCS. *Dotted lines*, tracts whose activation could give rise to spurious effects in cochlear nucleus and/or inferior colliculus, either by orthodromic or antidromic conduction of action potentials. *Arrows* indicate direction of action potential conduction

bipolar pair, with good insulation and small exposed tips, are required. Some sense of the spatial spread of current can be obtained by eliciting medial olivocochlear (MOCS) effects in the cochlea and observing the impact of raising or lowering the electrodes relative to the floor of the IVth ventricle. To ensure correct electrode placement, careful mapping of the sensitivity of facial nerve signs (whisker and eyebrow twitch) to shocks delivered through the stimulating electrodes is used (Seluakumaran et al. 2008a). Precise placement at the point of lowest facial nerve threshold at least ensures that the electrodes are placed between the facial genuae. Provided current strengths are kept below a critical amount, and the electrodes are not advanced deeper into the brain stem, selective olivocochlear activation can be achieved. In guinea pig, it appears that placement of the electrodes slightly rostral to the lowest facial nerve threshold point maximizes the possibility of selective activation of MOC axons (Seluakumaran et al. 2008a).

Ideally, this approach needs an independent method of verification, and this can be obtained by monitoring field potentials in the IC. If shocks to the floor of the ventricle are kept well below the strength that elicits measurable field potential responses in the IC, this provides additional assurance that major ascending tracts have not been activated as well as the MOCS axons (Seluakumaran et al. 2008a). Despite these precautions however, it cannot be ruled out that some unintended activation of non-olivocochlear tracts occurs. The large "onset chopper" neurons whose axons form a commissural pathway between the two cochlear nuclei may be especially prone to this problem. In a detailed study of the responses of onset chopper neurons to midline electrical stimulation in guinea pig (Mulders et al. 2007), it has been found that even at shock strengths that do not activate the axons of other major output neurons of the cochlear nucleus, some onset choppers show clear evidence of antidromic activation, indicating that their large-diameter myelinated axons in the intermediate acoustic stria may have lower thresholds than other axons to shocks intended to activate only MOCS axons.

A final technical consideration is that electrical stimulation in the brain stem is likely to activate a variety of motor pathways, including those innervating the middle ear muscles. Activation of these motor pathways, by generating masking noise and by affecting middle ear sound conduction, can confound the results and hence paralysis of skeletal muscle, elimination of middle ear muscle contraction, and artificial ventilation are obligatory requirements in all such studies.

10.2.2 Early Work

Reference has already been made to the pioneering study of Desmedt (1962), who recorded field potentials in response to acoustic clicks from brain stem, midbrain, and cortex and measured the effects of electrical stimulation of the olivocochlear axons at the floor of the IVth ventricle, in cats. Comis and Whitfield (1968) were the first to study the effects of electrical stimulation of the SOC on activity of single cochlear nucleus neurons. Unlike all subsequent studies, they used a ventral approach to the SOC, instead of selectively activating the olivocochlear fibers at the floor of the IVth ventricle. The aim of these authors was in fact not to stimulate the MOCS (the division of the olivocochlear efferents into medial and lateral components was not well known until the mid-1970s), but instead, to activate a separate non-olivocochlear projection from the SOC to the CN. However, the placement of their stimulating electrodes deep within the SOC makes it likely that at least part of the olivocochlear projection was activated in their experiments, although no monitoring of effects in the cochlea was performed. These authors found numerous examples of neurons in both ventral and dorsal subdivisions of the CN that were excited by SOC stimulation. They used local iontophoretic injection of atropine to demonstrate that this excitation was probably cholinergic. Despite the substantial limitations of this early study, it does provide the first single-neuron evidence suggesting that olivocochlear activation might not result in exclusively inhibitory effects in central nuclei.

In a remarkably complete study in decerebrate cats, Starr and Wernick (1968) used electrical stimulation of the olivocochlear fibers at the floor of the IVth ventricle (Fig. 10.2). Importantly, these authors verified the correct placement of the stimulating electrodes by monitoring effects on the cochlear compound action potential and cochlear microphonic, although these effects were not routinely quantified. They reported a variety of effects of olivocochlear stimulation on both spontaneous and sound-evoked activity in all major subdivisions of CN. Approximately 50% of neurons showed no change, and about equal numbers showed either an increase or a decrease in spontaneous firing. Many neurons also showed complex and long-lasting temporal patterns of spontaneous firing change as a result of olivocochlear stimulation that appeared to be rather different from the small, short-lasting drops in spontaneous rate seen in primary afferent fibers (Wiederhold 1970; Wiederhold and Kiang 1970). Complex excitatory and inhibitory effects were also observed when tone-evoked responses were examined. Some neurons could show different effects depending on tone level, something that the authors noted was not observed in primary afferents. Importantly, these authors also showed that changes in spontaneous activity were still observed in about 40% of CN neurons after destruction of the ipsilateral cochlea, eliminating all peripheral effects of olivocochlear stimulation. Although it was not stated explicitly by the authors, this latter observation presumably applied to neurons in the dorsal subdivision of CN, because all spontaneous firing in the ventral CN is probably dependent on input from the cochlea (Koerber et al. 1966). Regardless, this result clearly implicated olivocochlear collateral pathways in the effects observed in CN neurons. Starr and Wernick speculated that in the intact system, direct olivocochlear actions in the CN and the effects of peripheral afferent suppression interacted within complex central circuitry to generate the diversity of central effects seen.

By demonstrating the presence of excitation as well as inhibition, and by showing effects in the absence of intact cochlear efferent innervation, these two initial studies in CN provided the first evidence that effects of olivocochlear activation in central nuclei are not always readily predictable from the established effects in the cochlea. However, in neither of the studies were stringent attempts made to rule out activation of non-olivocochlear pathways, nor was particular attention paid to the relationship between the effects seen and neuronal response type classification.

10.2.3 Recent Studies in CN and IC

In a series of recent studies in guinea pig and rat, the effect of olivocochlear activation on single CN and IC neurons has been investigated in some detail, both in quiet and in the presence of background noise (Mulders et al. 2002, 2003, 2008; Seluakumaran et al. 2008a, b). The studies in guinea pig employed the necessary precautions to reduce the likelihood of confounding factors described in Sect. 10.2.1. In addition, quantitative comparison was made of the peripheral effects of MOCS stimulation (assessed by cochlear action potential suppression) and the changes in central neuron responses.

The overall results of these studies can be summarized by stating that in both of these higher centers, about 50% of neurons show MOCS effects on responses to CF tones in quiet and in noise backgrounds, which are similar to those described in primary afferents, while the remainder show a variety of effects that are not normally observed in primary afferents.

10.2.3.1 Effects in Quiet

In quiet, MOCS stimulation causes a rightward shift in the input–output curve to CF tones with little or no change in maximum firing rate in a variety of different response types in both CN and IC (schematically illustrated in Fig. 10.4). These effects are qualitatively predictable from the effects on primary afferents. Attempts to quantify the degree of rightward shift in comparison to the peripheral neural sensitivity changes in the same animals are complicated by the fact that the peripheral effect is based on the change in compound action potential generated at the onset of the test tone, whereas the single-neuron input–output curves are generated from spike counts collected over the duration of the entire 50-ms tone bursts. In many such cases, however, the central and peripheral threshold shifts appear to be roughly equivalent.

In ventral and posteroventral CN, in which single neuron classification is considerably more standardized than in the IC, it can be stated with some confidence that such classic MOCS effects are seen in primary-like neurons, and all classes of chopper neurons, including onset choppers. In the IC, neurons classified on the basis of their input–output curves as "monotonic" and "nonmonotonic" also showed these classic effects.

However, not all neurons show such simple, predictable effects. In some CN neurons, strong suppression of firing is seen across the full range of CF tone intensities employed (Fig. 10.5c, d). Although some small suppression of maximum discharge rate can be seen in a proportion of primary afferents, it is not as strong as

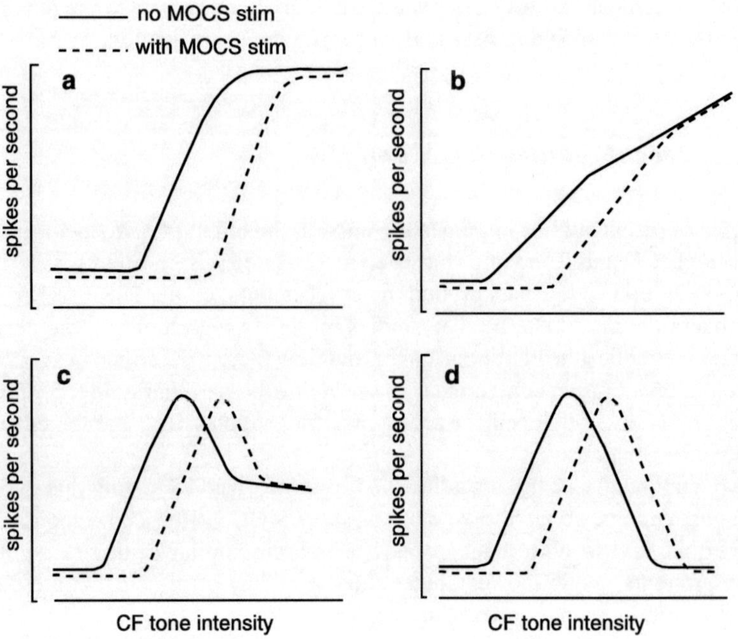

Fig. 10.4 Schematic examples of CF input–output curves of CN and IC neurons in quiet background illustrating effects of MOCS activation that resemble effects seen in primary afferents (i.e. rightward shift with little or no change in maximum discharge rate; compare with Fig. 10.1a). (**a**, **b**) saturating and nonsaturating types of monotonic input–output curves commonly found in CN and IC. (**c**, **d**) nonmonotonic types of input–output curves found in IC. Note that shapes of input–output curves are schematic only and do not accurately depict diversity of types seen in each central structure

that seen in these central nervous system neurons, implying that an additional inhibitory mechanism is present. In agreement with this notion, in animals in which removal of peripheral MOCS action is achieved either by intracochlear perfusion of the efferent blocker strychnine or destruction of OHCs, some CN neurons still showed inhibition and rightward shifts of their input–output curves caused by MOCS activation (Mulders et al. 2002).

A small number of transient and sustained choppers in CN show increases in maximum discharge rate, despite a rightward shift in their input–output curves (Fig. 10.5a). This appears to be similar to the level-dependent changes reported by Starr and Wernick (1968) and contrasts with the effects on primary afferents. In a small number of transient and sustained choppers, CF input–output curves are not altered at all by MOCS stimulation (Fig. 10.5e), even though measurement of cochlear responses shows the presence of substantial suppression of the peripheral afferent neural response to the same tone frequencies. These latter results imply the existence of centrally mediated excitation that offsets the effects of peripheral suppression.

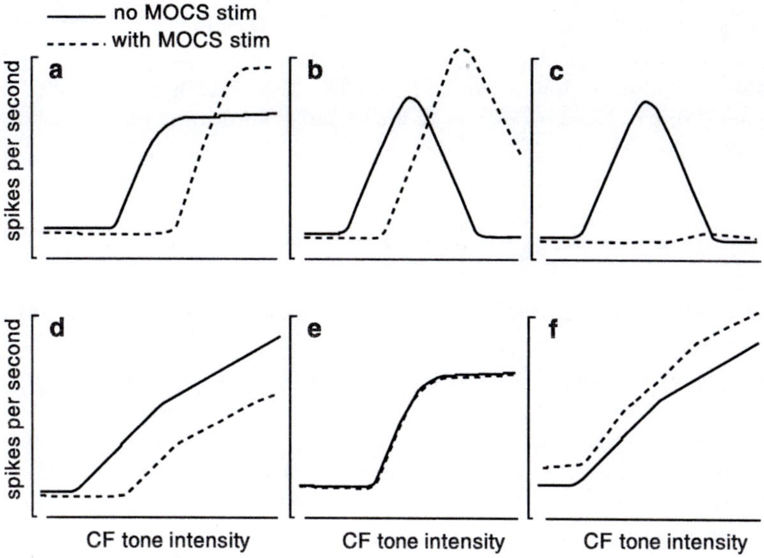

Fig. 10.5 Schematic examples of input–output curves of CN and IC neurons in quiet background illustrating effects of MOCS activation that differ from those seen in primary afferents. (**a**, **b**) rightward shift with increase in maximum rate; (**c**) inhibition that is much stronger than seen in primary afferents, (**d**) rightward shift accompanied by; substantial drop in maximum rate (**e**), no change despite presence of peripheral threshold change; (**f**) increase in firing rate across extent of input–output curve. Note that shapes of input–output curves are schematic only and do not accurately depict diversity of types seen in each central structure

Direct evidence of excitation in input–output curves of CN neurons by MOCS stimulation was less common in these more recent guinea pig studies compared to the earlier reports of Starr and Wernick (1968) and Comis and Whitfield (1968) in the cat. Mulders et al. (2002) reported that onset-like neurons and other neurons whose responses were not classifiable showed a leftward shift (increased responsiveness) in their CF input–output curves, and in some cases, elevations of spontaneous firing rates were also observed (Fig. 10.5f). However, a more recent study (Mulders et al. 2008) with more stringent cell classification failed to reveal obvious leftward shifts in input–output curves, or increases in spontaneous firing rates in any response types, including well classified onset chopper neurons.

In the IC, effects of MOCS stimulation on single neuron responses have been reported only in the central nucleus contralateral to the cochlea to which test sounds are presented (Seluakumaran et al. 2008a). Typical afferent-like effects in quiet are commonly observed for various response types differing in the shape of their input–output curves (Fig. 10.4a, c, d). However, as in CN, a range of other effects are seen. Some neurons show marked reduction in maximum firing rate as well as rightward shifts in their input–output curves (Fig. 10.5c, d). In some cases, the degree of rightward shift of the input–output curve was far greater than the threshold change of the cochlear nerve action potential, again implying the existence of additional inhibitory mechanisms operating on central neurons. As in the CN, other neurons exhibited large increases in maximum rate, at the same time as rightward shifts in their input–output curves (Fig. 10.5a, b). No instances were seen in this nucleus of direct excitatory effects near threshold (leftward shifts in input–output curves).

10.2.3.2 Effects in Background Noise

The picture with regard to effects of MOCS stimulation in the presence of background masking noise is equally complex, both in CN and IC. However, a key point is that in many neurons, antimasking effects are seen that are qualitatively similar to those observed in primary neurons (Fig. 10.6). As in primary afferent neurons, the net effect is a significant restoration of the neurons' output dynamic range, with an accompanying improvement in measures of signal discrimination (Mulders et al. 2008; Seluakumaran et al. 2008b). In many instances, this improvement in output dynamic range is achieved, as it is in primary afferents, by a reduced background firing to the noise and increased maximum discharge rate. However, an interesting variant is seen in IC (Fig. 10.7a), in which some neurons under masked conditions shift their input–output curve dramatically to the right but do not show the usual increase in background firing (Fig. 10.7a). These neurons have zero or very low spontaneous firing rates under both quiet and masked conditions. Nonetheless, in these neurons, MOCS stimulation still dramatically improves the maximum discharge rate.

A notable exception to the antimasking effect of MOCS stimulation in CN is the onset chopper neuron category, in which MOCS stimulation appears to have either no effect on masked input–output curves or in fact causes a further rightward shift

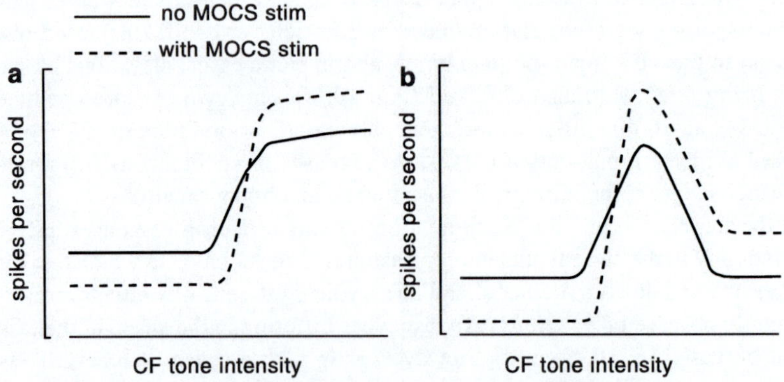

Fig. 10.6 Schematic examples of monotonic and nonmonotonic input–output curves of CN and IC neurons in presence of continuous background noise illustrating effects of MOCS activation that resemble effects seen in primary afferents (i.e., restoration of output dynamic range by drop in basal firing and increase in maximum rate; compare to Fig. 10.1b)

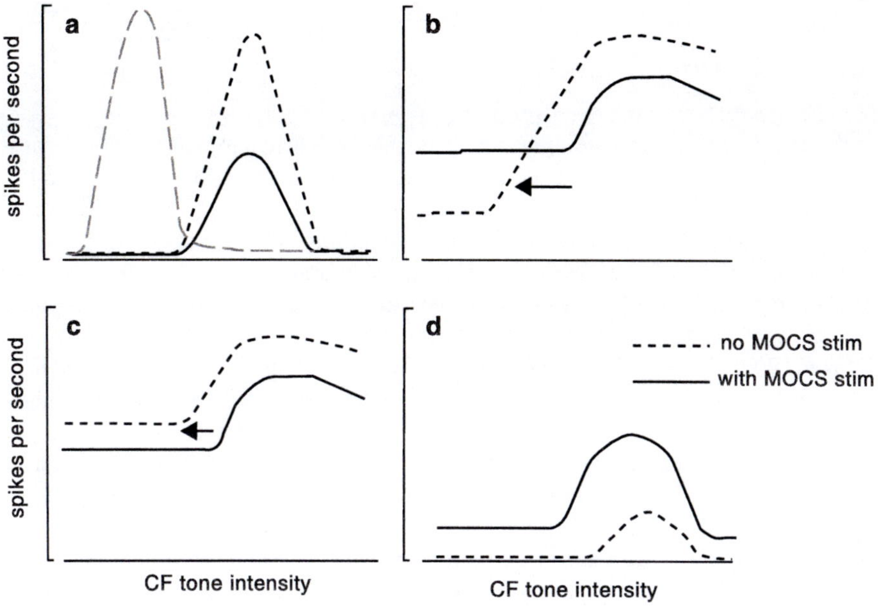

Fig. 10.7 Schematic examples of input–output curves of CN and IC neurons in presence of background noise illustrating effects of MOCS activation that differ from those seen in primary afferents. *Coarse dashed line* in (**a**) shows input–output curve in unmasked condition (quiet background) for comparison with masked condition (*fine dashed line*). (**a**) Nonmonotonic neuron in IC. Note release from masking despite no basal firing to masker; (**b**) improvement in threshold caused by leftward shift of curve and drop in basal firing rate; (**c**) improvement in threshold caused by leftward shift accompanied by overall elevation in firing rate; (**d**) increased suppression, that is, lack of antimasking effect

and reduction in firing rate. Examples of this phenomenon are also seen in IC (notably in neurons with nonmonotonic input–output curves) in which MOCS stimulation causes a further deterioration in masked responses, rather than an antimasking effect (Fig. 10.7d). Other variants are also seen in both central nuclei, such as increases in both background and maximum rate, and reductions in background rate without changes in maximum rate.

A finding of considerable interest is the fact that in both CN and IC, examples have been found (Fig. 10.7b, c) of neurons whose input–output curves to CF tones in the presence of background noise are shifted to the left by MOCS stimulation, that is, their thresholds to tones were actually improved. This novel effect was seen in transient and sustained chopper neurons of CN and in IC neurons with CF input–output curves showing varying degrees of nonmonotonicity. It is distinct from the classic antimasking effect which, although it expands a neuron's output dynamic range by lowering background firing rate and restoring maximum rate, is not accompanied by any substantial leftward shift of the CF input–output curve (Fig. 10.1b).

10.2.4 In Vivo Evidence for MOCS Collateral Involvement in Novel Central Effects

A critical issue in understanding how effects of MOCS activation on central neurons differ from those observed in the primary afferent neurons concerns the cellular targets and action of the MOCS collaterals. Collateral branches of MOCS axons have been described within the SOC itself, but nothing is known about their role in modifying activity of interneurons in this important brain stem region (Brown et al. 1988; Brown 1993). Most detailed studies have focused on the collaterals that terminate in the CN (Brown et al. 1988, 1991; Winter et al. 1989; Benson and Brown 1990; Benson et al. 1996) and see Brown, Chap. 2). In the CN, MOCS collaterals are thought to terminate primarily on large multipolar neurons (although these are not the only targets in the cochlear nucleus). The MOCS collateral terminals on cochlear nucleus neurons have been shown to exhibit ultrastructural features typical of excitatory synapses. As a consequence, it has been suggested that these collaterals may be excitatory in their action in the cochlear nucleus, in contrast to the overall suppressive effects of the peripheral MOCS terminals on cochlear neural output. As such, these collaterals are possible candidates for generating reported excitatory effects of MOCS stimulation on cochlear nucleus neurons, and they may also explain the presence of central neurons that show no suppression despite the presence of peripheral MOCS-mediated suppression.

As mentioned previously, Starr and Wernick (1968) and Mulders et al. (2002) combined extracellular single neuron recording in CN with various methods of eliminating cochlear effects of the MOCS. Acute perfusion of the cochlea with strychnine to block the OHC receptors for the efferent neurotransmitter (Mulders et al. 2002) is an especially powerful technique because it permits normal sound-evoked responses to be studied and enables the effects on the same CN neuron to be observed with and without peripheral MOCS-mediated suppression. Although many cochlear nucleus neurons showed no residual effects of MOCS stimulation after elimination of peripheral suppression, numerous examples were found of clear persistent excitation or inhibition in a range of chopper and onset neuron types as well as a number of unclassifiable neurons. Such effects must presumably be mediated by MOCS collaterals, although it is not possible with such methods to conclude whether the effects are direct or multisynaptic in origin.

Another approach to the problem of isolating the effects of MOCS collaterals has been to make intracellular recordings from cochlear nucleus neurons and to search for evidence of direct synaptic events (excitatory postsynaptic potentials [EPSPs] and inhibitory postsynaptic potentials [IPSPs]) as a result of MOCS stimulation. Presumably if such events are observed, they must originate from an action of the MOCS collaterals on CN neurons, rather than being indirectly mediated through MOCS action on the cochlea. Using this method in rat CN, Mulders et al. (2003) observed EPSPs in onset neurons. Examples were also found of inhibitory events (IPSPs) in chopper neurons. This study was limited by the fact that peripheral measures of the effectiveness of MOCS stimulation were not made, and because of uncertainty about the precise classification of neuronal response type.

The latency of both excitatory and inhibitory events was surprisingly long (in excess of 3 ms in most cases, and occasionally two sequential EPSPs were seen in response to single midline electrical stimuli. These data suggest that the effects seen may have been indirect, mediated by MOCS collateral action on neurons other than those being recorded from.

In a more recent study Mulders et al. (2007) used extracellular recordings with stringent classification of neuronal response type and tight control over MOCS stimulation site and current strength. An action potential collision technique was used in an attempt to differentiate between antidromically generated action potentials and ones generated by excitatory synaptic input (MOCS collateral input). In about 50% of onset chopper neurons the results were compatible with direct synaptic excitation by orthodromic activation of MOCS collaterals. However, the fact that in the remaining onset chopper neurons the results of collision assays were compatible with antidromic activation of onset chopper axons in the intermediate acoustic stria highlights the difficulty of precise and selective activation of MOCS axons. The results of all such studies should probably be regarded with some caution, as it is probably impossible to completely rule out antidromic activation of recurrent pathways in the central nuclei.

Whether or not the action of MOCS collaterals in CN and SOC has significant upstream consequences for activity in structures such as the IC has not been intensively investigated. In one study (Seluakumaran et al. 2008a), systemic injection of gentamicin was used to eliminate MOCS effects in the cochlea and the responses of IC neurons to MOCS stimulation was recorded. No residual effects were seen, but the small sample size and the possibility that gentamicin might have central as well as peripheral effects means that this result is far from conclusive.

10.3 In Vitro Studies

The issue of MOCS collateral action has also been addressed using brain slice recording methods. Fujino and Oertel (2001) studied the effects of bath application of cholinergic agonists and antagonists on patch-clamped CN neurons in slices of mouse brain stem. They found that so-called D-stellate cells (named for their morphology and dorsally directed axons) were not affected by drugs presumed to activate the postsynaptic receptors associated with cholinergic MOCS collaterals. D-stellate neurons in brain slices are believed to correspond to onset chopper neurons classified by their in vivo properties, and this result suggests that the multipolar cell targets of MOCS collaterals are not onset choppers. In addition, Fujino and Oertel (2001) found that so-called T-stellate cells, presumed to correspond to sustained and transient chopper in vivo response categories, were strongly excited by the same pharmacological agents. These data are seemingly at variance with the available in vivo evidence for MOCS collateral action on onset type neurons, but there are many possible explanations for such discrepancies. For example, bath application of cholinomimetics may not be physiologically equivalent to synaptic release of acetylcholine by MOCS collaterals at specific synapses. It might also be that D-stellate cells in the mouse are not the analog of onset chopper neurons defined by physiological criteria in other species. With regard to the excitatory effects on T-stellate cells, it is possible that these neurons receive input from other cholinergic pathways such as the direct SOC to CN pathway (Sherriff and Henderson 1994) that was the presumed subject of investigation by Comis and Whitfield (1968). This pathway would not be activated by MOCS stimulation in the intact animal at the floor of the IVth ventricle, but its postsynaptic receptors would be accessible to bath application of cholinomimetics.

10.4 Mechanisms of Nonclassic MOCS Effects in Central Neurons

The diverse effects of MOCS stimulation seen in central nuclei may have several explanations. The presence of MOCS collaterals in the cochlear nucleus that may be excitatory or inhibitory on subpopulations of neurons obviously creates a substrate for modifying the effects of reduced afferent input. Examples of neurons in which frank excitation is seen, or in which no rightward shift in I/O curves occurs despite a reduction in peripheral sensitivity, may be the result of a direct excitatory action of these collaterals, counteracting the effects of peripheral suppression. Similarly, suppression that exceeds that usually seen in primary afferents might also be explained by an inhibitory effect of collaterals. These effects need not be mediated by direct collateral input to the neurons from which recordings are being made, because the complexity of interneuronal circuitry in central nuclei also means there is a large repertoire of possible indirect effects as well.

In addition to possible collateral action, however, other factors need to be considered when seeking explanations for the complex central effects of MOCS activation. The response properties of many neurons in CN and IC are derived from an integration of multiple inputs from widespread cochlear regions. These inputs can be either excitatory or inhibitory, presumably with differing synaptic weights and different dependence on acoustic stimulus intensity and frequency. In addition, the method of MOCS stimulation employed in the in vivo studies excites many MOCS axons that collectively innervate a large length of the organ of Corti. Central neurons receiving very restricted place-specific afferent input from the peripheral receptor might be expected to exhibit similar MOCS-mediated effects to primary afferents. However, neurons that integrate widespread input from across the cochlea might be expected to exhibit more complex effects depending on how MOCS stimulation alters the balance of excitatory and inhibitory inputs under a given set of stimulus conditions. Further, MOCS-mediated effects in the periphery are themselves dependent on the acoustic stimulus frequency and intensity. Suppression of primary neural output at CF will be greatest for low-level stimulation because of the saturation of the OHC active process at higher stimulus levels. An additional factor is that MOCS effects on primary afferent neural sensitivity are maximum near CF for off-CF frequencies that are more and more distant from the tuning curve tip,

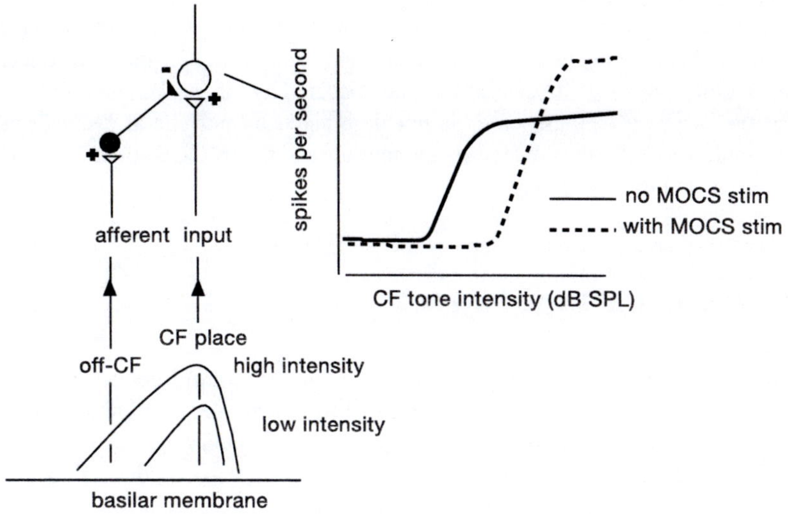

Fig. 10.8 Schematic representation of a possible simple circuit to explain effect of MOCS activation on input–output curve of a single central neuron in quiet background such as that illustrated in Fig. 10.5a. Inhibitory interneuron shown in black. Positive and negative signs indicate excitatory and inhibitory synapses

MOCS effects are reduced because these responses are relatively uninfluenced by the OHC active process.

Figure 10.8 attempts to provide one illustration of how some of these factors might interact in central circuitry to produce nonclassic effects on CF input output curves. In this example, a central neuron receives excitatory CF input and inhibitory input from adjacent off-CF cochlear regions. The neuron's firing rate at low stimulus levels is dominated by the excitatory CF input, but at high levels of a CF tone firing rate will be determined by a combination of inhibition and excitation as the excitation pattern on the basilar membrane spreads to recruit the off-CF region. MOCS stimulation will produce inhibition of the response at low levels of the CF tone (rightward shift of input–output curve) but at high levels, this effect will become less significant because of the saturation of the OHC active process at CF. At the same time, however, MOCS stimulation could release the neuron from off-CF inhibition, providing the off-CF regions that provide the inhibition are still sufficiently close to the tuning curve tip to be subject to MOCS effects. The result would be an increase in firing rate at high tone levels.

Similar logic might be used to model other types of central neural responses to MOCS stimulation. Detailed modeling would be especially challenging in the case of responses in the presence of masking noise, but it could reveal useful information not only about the mechanism of the diverse central MOCS effects, but also about central afferent circuitry in general. One striking example is provided by neurons that show a marked restoration of the maximum discharge rate (see, e.g., Fig. 10.7a) as a result of MOCS stimulation even though the masking noise itself does not cause an increase in background firing rate (albeit there is a marked rightward shift in threshold induced by the masker). The release from masking must be, in these cases, due to some other factor than a MOCS-induced reduction in background firing rate, because this is not elevated by the masker. One possibility is that the masking in these cases is not dominated by peripheral events, but by central inhibitory effects activated by the masker and that MOCS stimulation selectively turns off this inhibition.

10.5 Behavioral Experiments

Several studies have attempted to elucidate the role of central connections of the MOCS using a combination of behavioral measurements and manipulations of the MOCS system, either pharmacologically or by genetic modification. The focus in these studies has been on signal detection in noise. In early studies (Pickles and Comis 1973; Pickles 1976b), indwelling cannulae were used to infuse a blocker of acetylcholine receptors into the cochlear nucleus of awake cats. Tone thresholds in masking noise were markedly more elevated than absolute thresholds, suggesting that a cholinergic pathway in the cochlear nucleus is involved in antimasking. Such experiments, however, provide no proof that it is the olivocochlear collaterals that are involved, because, as already mentioned, there are other cholinergic pathways from SOC to CN.

More recent studies (May et al. 2002) have used transgenic knockout mice lacking a crucial subunit of the peripheral intracochlear acetylcholine receptor (α9 nicotinic acetylcholine receptor). Such animals could still perform as well as normal animals in detection and discrimination of tones in noise. MOCS collateral pathways to CN have recently been shown to remain intact in such animals (Brown and Vetter 2009), and one interpretation of the result is therefore that the collaterals contribute substantially to antimasking, independent of any peripheral action of the MOCS. May et al. suggest that this might be mediated through excitatory collateral effects on chopper-type neurons corresponding to the T-stellate cell category described in mouse CN slices. Some caution needs to be exercised in the interpretation of these and similar studies because of a number of concerns that are well discussed by May et al. (2002). In particular, it has recently been reported that the design of behavioral testing regimens that incorporate substantial training components may permit animals to recruit other mechanisms for enhancing signal detection in noise, possibly obscuring initial effects of loss of components of MOCS action (May et al. 2004).

10.6 Functional Significance

The first and arguably most important result that emerges from the studies discussed in this chapter is that despite the complexities of central circuitry and the presence of MOCS collaterals, many central neurons, both in CN and IC, exhibit effects of MOCS stimulation that are very similar to those reported in primary afferents.

This is especially significant with regard to the antimasking effects in the presence of background noise. This phenomenon, which forms the main basis for theories of peripheral MOCS involvement in signal discrimination in background noise, had previously been demonstrated only in primary afferents. A new element is the finding that in addition to the classical release from masking by which MOCS stimulation extends the output dynamic range, a number of central neurons also show real improvements in their masked tone thresholds. These results together strengthen the case for MOCS involvement in enhancement of signal detection and discrimination in noisy environments.

The functional significance of the many other "nonclassic" effects that differ from those seen in primary afferents is unclear and must remain speculative at this stage. The finding that in some neurons, MOCS stimulation in quiet background causes no change in their CF input–output curves even though peripheral sensitivity is reduced at the same frequency, is intriguing. This seems to represent a form of "efference copy" in which central sensitivity is adjusted to compensate for peripheral sensitivity changes caused by MOCS activation. By comparing activity between such neurons and others with rightward shifts in their input–output curves, the brain could deduce that firing rate changes were not the result of a change in acoustic stimulus intensity. Further study will be required to determine which neuronal categories show such sensitivity adjustments.

Excitatory effects in which sensitivity to tones in quiet is increased are also intriguing. It is conceivable that such effects could serve to enhance responses of some neural populations to important signals in selective attention tasks, while inhibitory effects might operate to reduce responses of other neural classes to the same stimuli. The coarse mode of MOCS stimulation employed in these studies is perhaps not well suited to unraveling such mechanisms.

The particular case of onset choppers is, at one and the same time, interesting and confusing. In CN, these are the only neurons in which the traditional form of MOCS stimulation appears to reduce further the sensitivity to CF tones in the presence of background noise. Onset choppers are believed to generate broadband inhibition in the cochlear nuclei, and this MOCS effect may serve to reduce the inhibitory action of these cells on other neural populations. The purpose of such reduced inhibition is, however, unclear. However, other in vivo evidence suggests that single shocks to MOCS, in contrast to the long trains of shocks traditionally used, can cause excitatory effects in onset choppers (Mulders et al. 2003, 2007; Seluakumaran et al. 2008b). Such short-acting excitation of onset choppers could enhance their postulated role in some forms of release from masking (Pressnitzer et al. 2001; Verhey et al. 2003).

10.7 Summary

Forty years of research has provided solid evidence that the effects of MOCS activation on central neurons are complex and diverse. In many cases, sensitivity reduction in quiet and antimasking in noise are observed, in a manner qualitatively predictable from the effects of MOCS on primary afferent responses. However, in many other neurons, effects are seen that are not readily predictable from the peripheral action of the MOCS. These effects may be variously explained by complex interactions within the ascending neuronal network, by MOCS collateral action separate from the peripheral effects, or by a combination of both.

Despite this work, important questions remain. Although more recent work has used various neuronal classification schemes, there is still no clear picture of how tightly the diverse MOCS effects are segregated according to neuronal response type. The effects of more naturalistic activation of the MOCS need to be investigated, in contrast to activation of the entire efferent bundle with shock trains. The precise action of the central collaterals of the MOCS is still unclear, as is the physiological identity of their cellular targets. For this matter to be clarified, the apparent conflict between results obtaincd with different methodologies needs to be resolved.

Acknowledgments This work was supported by grants from the NHMRC (Australia), the Medical Heath and Research Infrastructure Fund, The Royal National Institute for Deaf People, and The University of Western Australia.

References

Benson TE, Brown MC (1990) Synapses formed by olivocochlear axon branches in the mouse cochlear nucleus. J Comp Neurol 295:52–70

Benson TE, Berglund AM, Brown MC (1996) Synaptic input to cochlear nucleus dendrites that receive medial olivocochlear synapses. J Comp Neurol 365:27–41

Brown MC (1993) Fiber pathways and branching patterns of biocytin-labeled olivocochlear neurons in the mouse brainstem. J Comp Neurol 337:600–613

Brown MC, Vetter DE (2009) Olivocochlear neuron central anatomy is normal in alpha9 knockout mice. J Assoc Res Otolaryngol 10:64–75

Brown MC, Liberman MC, Benson TE, Ryugo DK (1988) Brainstem branches from olivocochlear axons in cats and rodents. J Comp Neurol 278:591–603

Brown MC, Pierce S, Berglund AM (1991) Cochlear-nucleus branches of thick (medial) olivocochlear fibers in the mouse: a cochleotopic projection. J Comp Neurol 303:300–315

Comis SD, Whitfield IC (1968) Influence of centrifugal pathways on unit activity in the cochlear nucleus. J Neurophysiol 31:62–68

Desmedt JE (1962) Auditory-evoked potentials from cochlea to cortex as influenced by activation of the olivocochlear bundle. J Acoust Soc Am 34:1478–1496

Desmedt JE, Robertson D (1975) Ionic mechanism of the efferent olivo-cochlear inhibition studied by cochlear perfusion in the cat. J Physiol 247:407–428

Ebert U (1996) Noradrenalin enhances the activity of cochlear nucleus neurons in the rat. Eur J Neurosci 8:1306–1314

Fex J (1967) Efferent inhibition in the cochlea related to hair-cell dc activity: study of postsynaptic activity of the crossed olivocochlear fibres in the cat. J Acoust Soc Am 41:666–675

Fujino K, and Oertel D (2001) Cholinergic modulation of stellate cells in the mammalian ventral cochlear nucleus. J Neurosci 21:7372–7383

Kawase T, Delgutte B, Liberman MC (1993) Antimasking effects of the olivocochlear reflex. II. Enhancement of auditory-nerve response to masked tones. J Neurophysiol 70:2533–2549

Klepper A, Herbert H (1991) Distribution and origin of noradrenergic and serotonergic fibers in the cochlear nucleus and inferior colliculus of the rat. Brain Res 557:190–201

Koerber KC, Pfeiffer RR, Warr WB, Kiang NY (1966) Spontaneous spike discharges from single units in the cochlear nucleus after destruction of the cochlea. Exp Neurol 16:119–130
May BJ, Prosen CA, Weiss D, Vetter D (2002) Behavioral investigation of some possible effects of the central olivocochlear pathways in transgenic mice. Hear Res 171:142–157
May BJ, Budelis J, Niparko JK (2004) Behavioral studies of the olivocochlear efferent system: learning to listen in noise. Arch Otolaryngol Head Neck Surg 130:660–664
Mulders WH, Robertson D (2000) Effects on cochlear responses of activation of descending pathways from the inferior colliculus. Hear Res 149:11–23
Mulders WH, Robertson D (2001) Origin of the noradrenergic innervation of the superior olivary complex in the rat. J Chem Neuroanat 21:313–322
Mulders WH, Robertson D (2005) Catecholaminergic innervation of guinea pig superior olivary complex. J Chem Neuroanat 30:230–242
Mulders WH, Winter IM, Robertson D (2002) Dual action of olivocochlear collaterals in the guinea pig cochlear nucleus. Hear Res 174:264–280
Mulders WH, Paolini AG, Needham K, Robertson D (2003) Olivocochlear collaterals evoke excitatory effects in onset neurones of the rat cochlear nucleus. Hear Res 176:113–121
Mulders WH, Harvey AR, Robertson D (2007) Electrically evoked responses in onset chopper neurons in guinea pig cochlear nucleus. J Neurophysiol 97:3288–3297
Mulders WH, Seluakumaran K, Robertson D (2008) Effects of centrifugal pathways on responses of cochlear nucleus neurons to signals in noise. Eur J Neurosci 27:702–714
Patuzzi R, Rajan R (1992) Additivity of threshold elevations produced by disruption of outer hair cell function. Hear Res 60:165–177
Patuzzi R, Johnstone BM, Sellick PM (1984) The alteration of the vibration of the basilar membrane produced by loud sound. Hear Res 13:99–100
Pickles JO (1976a) The noradrenaline-containing innervation of the cochlear nucleus and the detection of signals in noise. Brain Res 105:591–596
Pickles JO (1976b) Role of centrifugal pathways to cochlear nucleus in determination of critical bandwidth. J Neurophysiol 39:394–400
Pickles JO, Comis SD (1973) Role of centrifugal pathways to cochlear nucleus in detection of signals in noise. J Neurophysiol 36:1131–1137
Pressnitzer D, Meddis R, Delahaye R, Winter IM (2001) Physiological correlates of comodulation masking release in the mammalian ventral cochlear nucleus. J Neurosci 21:6377–6386
Rajan R, Patuzzi RB (1992) Additivity of threshold losses produced by acute acoustic trauma. Hear Res 60:216–230
Seluakumaran K, Mulders WH, Robertson D (2008a) Effects of medial olivocochlear efferent stimulation on the activity of neurons in the auditory midbrain. Exp Brain Res 186:161–174
Seluakumaran K, Mulders WH, Robertson D (2008b) Unmasking effects of olivocochlear efferent activation on responses of inferior colliculus neurons. Hear Res 243:35–46
Sewell WF (1984) The relation between the endocochlear potential and spontaneous activity in auditory nerve fibres of the cat. J Physiol 347:685–696
Sherriff FE, Henderson Z (1994) Cholinergic neurons in the ventral trapezoid nucleus project to the cochlear nuclei in the rat. Neuroscience 58:627–633
Shore SE, El Kashlan H, Lu J (2003) Effects of trigeminal ganglion stimulation on unit activity of ventral cochlear nucleus neurons. Neuroscience 119:1085–1101
Starr A, Wernick JS (1968) Olivocohlear bundle stimulation: effect on spontaneous and tone-evoked activities of single units in cat cochlear nucleus. J Neurophysiol 31:549–564
Thompson AM, Moore KR, Thompson GC (1995) Distribution and origin of serotoninergic afferents to guinea pig cochlear nucleus. J Comp Neurol 351:104–116
Verhey JL, Pressnitzer D, Winter IM (2003) The psychophysics and physiology of comodulation masking release. Exp Brain Res 153:405–417
Wiederhold ML (1970) Variations in the effects of electric stimulation of the crossed olivocochlear bundle on cat single auditory-nerve-fiber responses to tone bursts. J Acoust Soc Am 48:966–977
Wiederhold ML, Kiang NY (1970) Effects of electric stimulation of the crossed olivocochlear bundle on single auditory-nerve fibers in the cat. J Acoust Soc Am 48:950–965

Winslow RL, Sachs MB (1987) Effect of electrical stimulation of the crossed olivocochlear bundle on auditory nerve response to tones in noise. J Neurophysiol 57:1002–1021
Winter IM, Robertson D, Cole KS (1989) Descending projections from auditory brainstem nuclei to the cochlea and cochlear nucleus of the guinea pig. J Comp Neurol 280:143–157

Chapter 11
Corticofugal Modulation and Beyond for Auditory Signal Processing and Plasticity

Nobuo Suga, Weiqing Ji, Xiaofeng Ma, Jie Tang, Zhongju Xiao, and Jun Yan

11.1 Introduction

The auditory system consists of the ascending and descending (corticofugal) systems. The corticofugal system forms multiple feedback loops with the ascending system, so that the neural mechanisms for auditory signal processing cannot be fully understood without the exploration of the function of the corticofugal system. Corticofugal modulation (changes in response properties evoked by the corticofugal system) of subcortical neurons for auditory signal processing is one of the most important functions of the auditory cortex in the cerebrum. The function of the corticofugal auditory system has been studied by examining the changes in the subcortical auditory neurons evoked by electric stimulation of, drug applications to, and/or cooling of the primary auditory cortex (AI). Focal cortical inactivation experiments tell us what the corticofugal system is doing in the control condition, and focal cortical activation experiments tell us how auditory signal processing is corticofugally modulated as AI receives a specific auditory signal. These mutually complementary experiments are absolutely necessary to explore corticofugal function. Focal activation or inactivation of AI evokes changes not only in the subcortical auditory nuclei, but also in the activated or inactivated AI and other cortical auditory areas on the ipsilateral and contralateral cerebral hemispheres. These cortical areas in turn evoke changes in the subcortical auditory nuclei. The effect of the cortical activation or inactivation does not end in these nuclei, because the corticofugal auditory system forms multiple feedback loops together with the ascending auditory system, and the corticofugally produced subcortical changes are in turn carried up to the cortex through the feedback loops. The shortest feedback loop is the thalamo–cortico–thalamic loop, and the longest one is the cochleo–cortico–cochlear (through multiple auditory nuclei) loop. The colliculo–thalamo–cortico–collicular loop is the

N. Suga (✉)
Department of Biology, Washington University, St. Louis, MO 63130, USA
e-mail: suga@biology.wustl.edu

D.K. Ryugo et al. (eds.), *Auditory and Vestibular Efferents*,
Springer Handbook of Auditory Research 38, DOI 10.1007/978-1-4419-7070-1_11,

intermediate length and has been the most extensively studied. The changes occurring in the auditory cortex are looped back to the cortex through these multiple feedback loops. If the gain of these feedback loops is larger than 1.0, a small change occurring in the cortex would become very large within a brief time. This problem appears to be prevented just before the auditory signal reaches the cortex by the thalamic reticular nucleus, which perhaps acts as a selective adaptive filter.

Corticofugal modulation occurs through complex multiple pathways, so it is difficult to fully explore its function. We thus far understand it in a simplistic way. First, we would like to make it clear that cortical changes occur and are transferred to the subcortical auditory nuclei and then additional changes occur there. However, we do not know exactly what the purely cortical changes are and what the purely subcortical changes are that occur in individual nuclei, because of the multiple feedback loops.

11.2 Necessity of Multiparametric Corticofugal Modulation

Research on auditory neurophysiology had mostly focused on auditory processing in the frequency domain, studying response patterns, frequency tuning, and a tonotopic (i.e., frequency) map in the auditory system, because the cochlea through the auditory cortex is tonotopically organized, that is, neurons tuned to specific single frequencies are systematically arranged in a nucleus or cortex. Sounds used by animals are commonly characterized by multiple parameters such as frequency, frequency modulation, amplitude, amplitude modulation, amplitude spectrum, duration, etc. Animals frequently emit identical or different sounds in specific combinations or sequences, so there are additional parameters characterizing the sounds such as time delay, time interval, repetition rate, etc. There are three important acoustic parameters for sound localization: interaural time and level differences and spectral notches. Therefore, there are many parameters of sounds that should be analyzed by the auditory system. Neuroethological research has focused on the responses of neurons to behaviorally relevant acoustic stimuli that mimic species-specific sounds or sounds produced by predators and prey. Such research indicates that the auditory system has different types of cortical and subcortical neurons tuned to the specific values of the parameters characterizing the behaviorally relevant sounds (information-bearing parameters), that they are tuned to multiple parameters, and that their tuning in one parameter is influenced by another parameter or parameters. These subcortical neurons are subject to corticofugal modulation in the frequency, amplitude, time and/or spatial domains. Accordingly, corticofugal modulation should be studied in terms of the neural responses to multiple parameters. For the multiparametric corticofugal modulation, a large number of corticofugal fibers are necessary. As a matter of fact, the number of corticofugal descending fibers is much larger than that of thalamocortical ascending fibers.

11.3 Research Performed Before 1995

Since the late 1950s, many studies on corticofugal modulation of thalamic and collicular neurons have been undertaken in anesthetized animals. In these studies, strong activation or inactivation of AI evoked excitation and/or inhibition of these subcortical neurons. These physiological data were contradictory: some authors found only, or predominantly, inhibitory corticofugal modulation (Massopust and Ordy 1962; Watanabe et al. 1966; Amato et al. 1969; Sun et al. 1996); others found only, or predominantly, excitatory corticofugal modulation (Andersen et al. 1972; Orman and Humphrey 1981; Villa et al. 1991); and some others found roughly equal levels of excitatory and inhibitory modulation (Ryugo and Weinberger 1976; Syka and Poplear 1984; Jen et al. 1998). The apparent contradiction between these studies might be resolved if the frequency dependence of excitation and inhibition and the relationship in tuning between stimulated and recorded neurons are considered. Regardless of the excitatory and/or inhibitory modulation, these data indicate that one of the corticofugal functions can be nonspecific gain control. However, the corticofugal system must have much more elegant functions than simple gain control, because there is a much larger number of corticofugal fibers than thalamocortical fibers.

11.4 Experimental Philosophy and Methodology

11.4.1 Experimental Philosophy

To study the function of the corticofugal system, one should not ignore that cortical and subcortical neurons both are tuned to specific values of an acoustic parameter. Therefore, electrical stimulation or a drug application for activation or inactivation should be highly focal except for the initial phase of corticofugal research, and corticofugal effects on subcortical neurons should be evaluated with regard to the relationship in tuning between stimulated or inactivated cortical neurons and recorded subcortical neurons.

The experiments designed on this philosophy led us to several findings in the mustached bat, *Pteronotus parnellii parnellii*; the big brown bat, *Eptesicus fuscus*; the Mongolian gerbil, *Meriones unguiculatus*; and the house mouse, *Mus domesticus*. In most of these experiments, unanesthetized animals were used, and the tuning curves of both the cortical neurons to be electrically stimulated or affected by a drug and the subcortical neurons to be corticofugally modulated were first measured. With this initial characterization in hand, electrical pulses or drugs were applied to the characterized cortical neurons, and their effects on the subcortical neurons were evaluated, focusing on the relationship in tuning between them and on the frequency (or other parameter) dependence of facilitation and inhibition. Using this experimental procedure, it was found that corticofugal modulation occurs in a specific and systematic way for the improvement and adjustment of auditory signal processing in the frequency, amplitude, time, and spatial domains.

11.4.2 Electric Stimulation of the Primary Auditory Cortex

For focal electric stimulation of AI, electric pulses should be very weak. When the electric pulse was 50 nA and 0.2 ms, subcortical changes (corticofugal modulation) were hardly observed, so Suga and his collaborators have been using a 100-nA, 0.2-ms monophasic constant electric current pulse as their standard stimulus. Such an electric pulse has been delivered at a rate of 6/s for 11 min in the research on the *Pteronotus parnellii parnellii* (Yan and Suga 1996; Zhang et al. 1997; Xiao and Suga 2002b; Tang et al. 2007). However, in their research on *Eptesicus fuscus*, the standard electric stimulus has been a 6.2-ms-long train of four electric pulses (100 nA, 0.2-ms long, 2.0 ms interval) delivered at a rate of 10/s for 30 min so as to evoke subcortical changes lasting up to 3.5 h (Yan and Suga 1998; Chowdhury and Suga 2000; Ma and Suga 2001a, b), although the above stimulus lasting for 2 min can evoke small subcortical changes (Ma and Suga 2001a). In their research on *Meriones unguiculatus*, it has also been a 6.2-ms-long train of four electric pulses (100 nA, 0.2-ms long, 2.0-ms interval) delivered at a rate of 10/s for 30 min (Sakai and Suga 2001, 2002). In the research on *Mus domesticus* by Yan and his collaborators, the electric stimulation of AI has been a 500-nA, 1.0-ms electric pulse delivered at a rate of 4/s for 7 min (Yan and Ehret 2001, 2002; Yan et al. 2005).

11.4.3 Drug Applications to the Primary Auditory Cortex

Activation or inactivation of AI by a drug should be very focal. In experiments by Suga and his collaborators, lidocaine (Zhang and Suga 2000, 2005), muscimol (Gao and Suga 1998, 2000; Xiao and Suga 2002a), bicuculline methiodide (Xiao and Suga 2002b, 2004, 2005; Ma and Suga 2004), acetylcholine, or atropine (Ji et al. 2001, 2005; Ji and Suga 2003; Ma and Suga 2005) was applied to the surface of AI or the primary somatosensory cortex (SI) via the dura mater which had a few holes (20–50 μm in diameter) made by electrode penetrations. The drug diffused into AI or SI through these holes. Therefore, the area affected by the drug was apparently quite limited. For example, muscimol (0.4 μg) applied to SI selectively blocked the development of the conditioning-dependent cortical best frequency (BF) shift without affecting the auditory responses and frequency tuning of cortical neurons (Gao and Suga 2000). Muscimol (0.1–0.2 μg) applied to one part of the auditory cortex disrupts frequency, but not time-delay discrimination, whereas muscimol applied to the other part of the auditory cortex disrupts time delay, but not frequency discrimination (Riquimaroux et al. 1991). These physiological and behavioral data indicate that muscimol applied to SI did not diffuse to AI to affect AI neurons, and that muscimol applied to one part of the auditory cortex did not diffuse to the other part of the auditory cortex. The diffusion of muscimol in the above condition is quite different from that of muscimol injected into the brain (Edeline et al. 2002).

An inactivation experiment with a drug should be accompanied with an activation experiment with electric stimulation. For example, bilateral inactivation of the primary somatosensory cortex blocks the conditioning-dependent cortical and collicular best frequency (BF) shifts (Gao and Suga 2000), but its activation by electric stimulation augments the BF shifts (Ma and Suga 2001a, 2003). Such a pair of activation and inactivation experiments mutually strengthens the conclusion derived from the activation or inactivation experiment.

As described later, electric stimulation of AI neurons facilitates the auditory responses of "BF-matched" neurons in other areas or nuclei (i.e., the BF of a recorded subcortical neuron is the same as that of the stimulated cortical neurons), but suppresses the responses and shifts the BF of "BF-unmatched" neurons. The cortical activation by electric stimulation increases the positive feedback, and the cortical inactivation by a drug reduces this positive feedback. Therefore, any drug that affects cortical auditory responses would reduce or increase the subcortical auditory responses, depending on whether it reduces or increases the activity of matched or unmatched cortical neurons, without diffusing to the subcortical auditory nuclei.

11.5 Research After 1995 (General)

Since 1995, many experiments have been performed to explore the function of the corticofugal auditory system, adapting the experimental philosophy and methodology described in the preceding text. These experiments have indicated that the corticofugal system sharpens and shifts the tuning of subcortical neurons in the frequency, amplitude, time and spatial domains and plays a key role in the reorganization of the auditory system of an adult normal animal according to auditory experience. In other words, the corticofugal auditory system improves and adjusts (reorganizes) the cortical input for the processing of auditory signals according to the excitation and inhibition in the auditory cortex elicited by the auditory signals, that is, according to auditory experience.

Tuning shifts of subcortical neurons occur either toward or away from the tuning of activated cortical auditory neurons. These shifts are respectively defined as "centripetal" or "centrifugal" tuning shifts. The centripetal shifts cause an increased (i.e., expanded) neural representation of the values of acoustic parameters to which the activated cortical neurons are tuned. On the other hand, the centrifugal shifts cause a decreased (i.e., compressed) representation of those values to which the activated cortical neurons are tuned (Fig. 11.1c, d). Since these two types of tuning shifts occur systematically according to the relationship in tuning between the recorded and stimulated neurons, the auditory system is reorganized

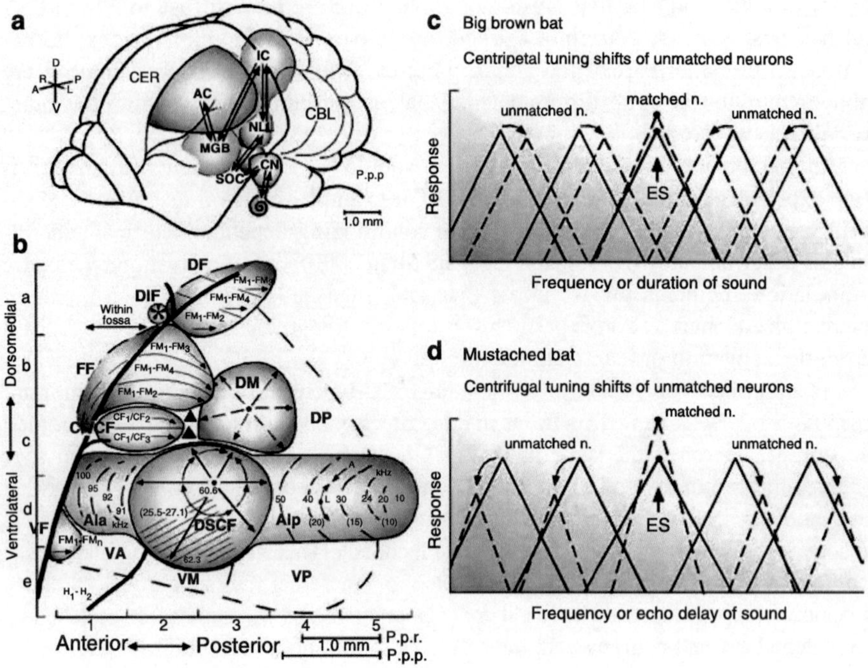

Fig. 11.1 The auditory pathway, auditory cortex and tuning shifts. (**a**) The dorsolateral view of the brain of *Pteronotus parnellii parnellii*. The *arrows* indicate the ascending and descending (corticofugal) systems. (**b**) A neurophysiological map of the auditory cortex (AC) of the *Pteronotus parnellii parnellii*. The *numbers* and *lines* in the anterior (AIa) and posterior (AIp) divisions and the Doppler-shifted constant frequency (DSCF) area in the primary AC (AI) indicate iso-best-frequency lines. The area of the AC sensitive to combinations of constant-frequency signals (CF/CF) consists of two subdivisions that contain a Doppler-shift (velocity) axis. Neurons in the frequency modulation-frequency modulation (FF) (the FF area had been called the FM–FM area because it consists of FM–FM combination-sensitive neurons. Both the DF and VF areas, subsequently found, also consist of FM–FM neurons. Hence the FM–FM area is now called the FF area), dorsal fringe (DF) and ventral fringe (VF) areas are sensitive to combinations of frequency-modulated signals (FM–FM). Each area consists of three subdivisions. These areas contain an echo-delay (range) axis. *CBL* cerebellum; *CER* cerebrum; *CN* cochlear nucleus; *DIF* dorsal intrafossa area; *DM* dorsomedial area; *DP* dorsoposterior area; H_1–H_2 first and second harmonic combination-sensitive area; *IC* inferior colliculus; *MGB* medial geniculate body; *NLL* nucleus of the lateral lemniscus; *SC* superior colliculus; *SOC* superior olivary complex; *VA* ventroanterior area; *VM* ventromedial area; *VP* ventroposterior area. Focal electrical stimulation of the tonotopically organized auditory cortex evokes facilitation, inhibition and best frequency (BF) shifts in the AC and subcortical auditory nuclei. There are two types of BF shifts of BF-unmatched neurons: centripetal (**c**) and centrifugal (**d**). In (**c**, **d**), the *unbroken* and *broken triangular curves* represent the frequency-response curves in the control and shifted conditions, respectively. BF-matched neurons do not show BF shifts, but facilitation of their auditory responses. [(**c**) Is based upon the data obtained by Ma and Suga (2001a) and Chowdhury and Suga (2000); (**d**) is based upon the data obtained by Zhang et al. (1997)]. Centripetal and centrifugal tuning shifts have also been found in the time domain: centripetal duration tuning shifts (Ma and Suga 2001b) and centrifugal delay tuning shifts (Yan and Suga 1996; Xiao and Suga 2004) (from Suga and Ma 2003)

in two different ways: expanded and compressed reorganizations. The "expanded" reorganization has been found in different species of animals and different sensory systems, whereas the "compressed" reorganization thus far has been found only in the auditory subsystems of the *Pteronotus parnellii parnellii*, which are highly specialized for echolocation.

Because cortical electric stimulation used to explore the corticofugal function is unnatural, plastic changes in cortical and subcortical auditory neurons evoked by cortical electric stimulation were compared with those elicited by a long-lasting repetitive tone burst stimulation or by auditory fear conditioning. Then, it was found that, for the conditioning, the corticofugal system, together with AI, SI, and the cholinergic basal forebrain (nucleus basalis: NB), play a key role in eliciting tone-specific plasticity, represented by BF shifts, and reorganization of the central auditory system.

This chapter reviews the research on corticofugal modulation and plasticity for auditory signal processing performed since 1995 and the neurophysiological data obtained mostly from AI in the cerebral cortex, the central nucleus of the inferior colliculus (ICc) in the midbrain, the ventral division of the medial geniculate body (MGBv) in the thalamus, and the cochlea after focal electric stimulation of or focal drug application to AI and/or auditory fear conditioning. Since 1995, many experiments have also been performed without adapting the experimental philosophy and methodology described in the preceding text (e.g., Syka and Poplear 1984; Sun et al. 1996; Jen et al. 1998). These experiments are not reviewed in our current chapter, but in a review article written by Jen et al. (2002).

11.6 Corticofugal Modulation in the Frequency Domain

11.6.1 Frequency-Dependent Facilitation and Inhibition and Best Frequency Shifts

Electric stimulation of cortical auditory neurons evokes both facilitation and inhibition of the auditory responses of collicular and thalamic neurons as well as cortical neurons. The amount of facilitation and inhibition varies with the frequency of a tone burst to which the neurons respond, and also with the relationship in frequency tuning between the stimulated and recorded neurons. The response threshold of a neuron is usually lowest at a certain frequency. This frequency is defined as the neuron's BF (Fig. 11.2a, b). When a recorded subcortical neuron is matched in BF to the stimulated cortical neurons, the response of the matched neuron is augmented at its BF and is inhibited at frequencies lower and/or higher than the BF. As a result, its frequency tuning is sharpened. When unmatched, the neuron is inhibited in the response at its BF and is facilitated in the responses at non-BFs. As a result, its

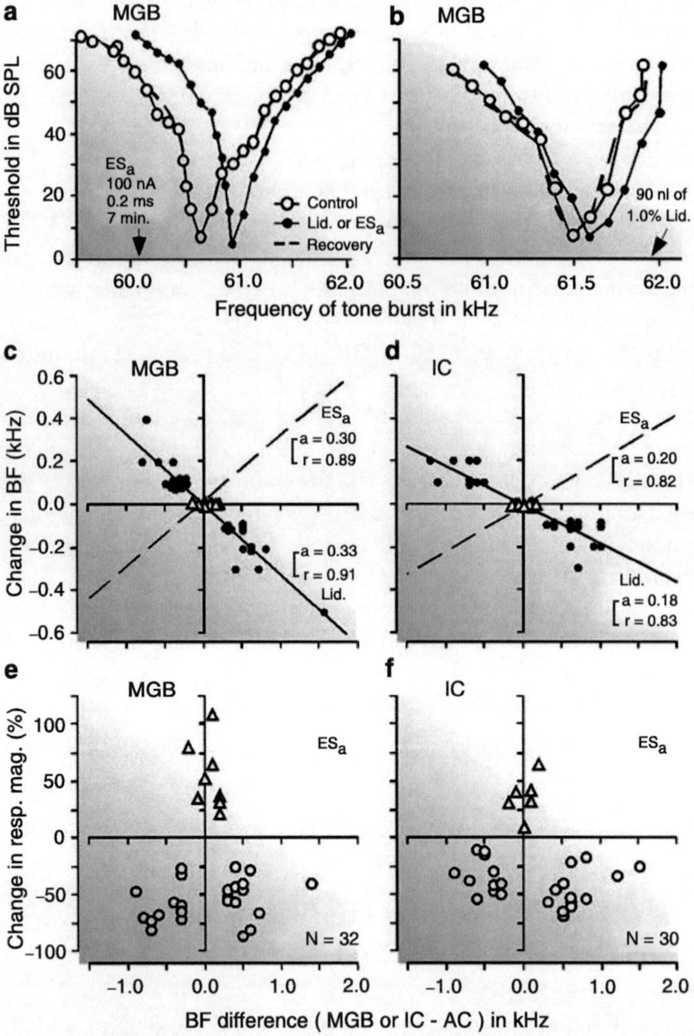

Fig. 11.2 Changes in tuning curve, best frequency, and response magnitude of thalamic and collicular DSCF neurons evoked by focal activation or inactivation of cortical DSCF neurons in *Pteronotus parnellii parnellii*. (**a**, **b**) Shifts in the frequency-tuning curves of two thalamic (MGBv) neurons evoked by an activation (**a**) or inactivation (**b**) of cortical neurons: activation by electric stimulation of a 0.2-ms, 100-nA electric pulse delivered at a rate of 5/s for 7 min (ES_a) and inactivation by 90 nl of 1.0% lidocaine (lid.). The best frequencies (BFs) of the activated or inactivated cortical neurons are indicated by the *arrows*. The curves were measured before (control; *open circles*), during (*closed circles*), and after (recovery; *dashed lines*) the cortical inactivation or activation. Most of the data points for the recovery are not shown because almost all of them overlapped with those for the control. (**c**, **d**) Changes in the BFs of thalamic (**c**) and collicular (**d**) neurons, i.e., reorganization of the tonotopic axis, evoked by a focal activation (*dashed lines*) or inactivation (*solid lines and filled circles*) of cortical neurons. The abscissae represent the differences in BF between the stimulated cortical (AI) and recorded thalamic (MGBv) or collicular (ICc) neurons in the control

frequency tuning is shifted. The shift of a frequency-tuning curve is always accompanied with a shift in BF, so that such a shift has been simply called a BF shift (Fig. 11.2a, b). Because the BF shift of an unmatched neuron is the most prominent and easily recordable change elicited by focal electric stimulation of AI, this term is used to represent the corticofugally elicited changes. One should not forget that the BF-matched neurons are corticofugally facilitated in response and are sharpened in tuning. These frequency-dependent effects of cortical neurons on the matched and unmatched subcortical neurons improve the input to the stimulated cortical neurons and the subcortical and cortical representations of the stimulus parameters to which the stimulated cortical neurons are tuned (Zhang et al. 1997; Yan and Suga 1998; Ma and Suga 2001a; Sakai and Suga 2002; Suga et al. 2002, for review). This corticofugal function has been named egocentric selection (Yan and Suga 1996). The changes resulting from egocentric selection are larger in thalamic neurons than in collicular ones (Fig. 11.2c–f; Zhang et al. 1997; Zhang and Suga 1997), and the changes in the collicular neurons are larger than those in the cochlear nuclear ones (Luo et al. 2008). Focal inactivation of AI by lidocaine evokes the BF shifts, which are opposite to those evoked by focal electric stimulation of AI (Fig. 11.2b; Zhang et al. 1997; Zhang and Suga 2000). The inactivation-dependent changes in the auditory responses and BF shifts are larger for the thalamic neurons than for the collicular ones, and indicate that in the control condition, the corticocollicular and the corticothalamic feedbacks enhance the collicular and thalamic auditory responses, respectively. On the other hand, nonfocal cortical inactivation including the matched and unmatched neurons does not evoke BF shifts, although it evokes a large reduction in the auditory responses of subcortical auditory neurons (Zhang and Suga 1997; Yan and Suga 1999). To elicit a BF shift, therefore, it is necessary to create a focally uneven distribution of neural activity across AI.

For focal electric stimulation of AI with a 6.1-ms-long train of four electric pulses (100 nA, 0.2-ms long, 2.0-ms interval) delivered at a rate of 10/s for 30 min, the BF shifts develop to the peak at 0.5 h after the onset of the stimulation and disappear within 3.0 h after the peak (the time courses of the BF shifts are further described later. See Fig. 11.10B).

Fig. 11.2 (continued) condition. The cortical BF was ~61.2 kHz on the average. The *triangles* and *circles* represent the data obtained from matched and unmatched subcortical neurons, respectively. The *regression lines*, their slopes (*a*), and correlation coefficients (*r*) are shown in the graphs. The BF shift is centrifugal for the cortical activation, but centripetal for the cortical inactivation. (**e, f**) The abscissae are the same as those in (**c, d**). The ordinates represent percent change in the response magnitude (number of pulses per tone burst) of thalamic (**e**) and collicular (**f**) neurons evoked by the cortical activation. The *triangles* and *circles*, respectively, represent percent changes in the response magnitude of matched and unmatched subcortical neurons at the BFs of individual neurons in the control condition. To measure response magnitudes, tone bursts were set at the best amplitude of each neuron in the control condition. Changes in BF (**c, d**) and response magnitude (**e, f**) both are larger in the MGBv than in the ICc (Zhang et al. 1997; Zhang and Suga 2000)

11.6.2 Expanded and Compressed Reorganizations of Tonotopic Maps

Focal electric stimulation of AI evokes two types of BF shifts of unmatched neurons: centripetal and centrifugal. Centripetal BF shifts are the shifts toward the BF of electrically stimulated cortical neurons or the frequency of a stimulus tone (Fig. 11.1c), whereas centrifugal BF shifts are the shifts away from the stimulated cortical BF or the tone frequency (Fig. 11.1d). These two types of BF shifts for reorganization of a tonotopic map occur in a specific spatial pattern in the ICc, MGBv, and AI. The spatial pattern of BF shifts is basically the same in the ICc, MGBv, and the AI (Figs. 11.2c, d and 11.3a, c; Zhang et al. 1997; Gao and Suga 1998; Chowdhury and Suga 2000; Ma and Suga 2001a; Xiao and Suga 2002b; Yan and Ehret 2002; Jen and Zhou 2003). Unlike centripetal BF shifts, centrifugal BF shifts are prominent only in the highly specialized auditory subsystem of the *Pteronotus parnellii parnellii* (Fig. 11.3c). (Please note that neurons do not shift in location, but their frequency-tuning curves do, presumably, because of changes in the morphology of axons and dendrites.)

Figure 11.3 shows "the BF shift-difference curves," each of which represents the relationship between BF shifts and differences in BF between the recorded and stimulated neurons studied in either *Eptesicus fuscus* (a), *Pteronotus parnellii parnellii* (b, c), *Meriones unguiculatus* (d), or *Mus domesticus* (e). The BF shift-difference curve indicates how the tonotopic axis changes around the BF of the stimulated neurons. In *Eptesicus fuscus*, the centripetal BF shift at around 37 kHz, which is the BF of the stimulated cortical neurons, is much more prominent for frequencies higher than the 37 kHz (a). In *Meriones unguiculatus*, this is also the case at around 1.1 kHz, which is the BF of the stimulated cortical neurons (d). In the posterior division of the primary auditory cortex (AIp) of the *Pteronotus parnellii parnellii*, however, the centripetal BF shift at around 39 kHz, being the BF of the stimulated cortical neurons, is much more prominent for frequencies lower than the 39 kHz (b). The Doppler-shifted constant frequency (DSCF) area of the *Pteronotus parnellii parnellii* over-represents sounds at ~61.0 kHz with extraordinarily sharply tuned neurons. Electric stimulation of 61.0-kHz-tuned cortical neurons evokes the centrifugal BF shifts of neurons, which are equal in amount on both sides of the BF of the stimulated neurons (c). In *Mus domesticus*, the centripetal BF shift equally occurs on both sides of 16 kHz, which is the BF of electrically stimulated cortical neurons (e). It is expected that the shape of the BF shift-difference curve changes according to the BFs of stimulated cortical neurons and that the amount of the BF shift is small for the frequencies that are over-represented in the tonotopic map. The arrows in Fig. 11.3a, b, d, e indicate that the small centrifugal BF shifts occur at both ends of the distribution of centripetal BF shifts. The BF shift-difference curves in Fig. 11.3a, b, d, e indicate that centripetal BF shifts occur in a large area surrounding the matched neurons and small centrifugal BF shifts occur in the narrow zone surrounding this large centripetal area. That is, they indicate that focal cortical electric stimulation evokes "center-surround" reorganization in the ICc and AI (Sakai and Suga 2002; Ma and Suga 2004).

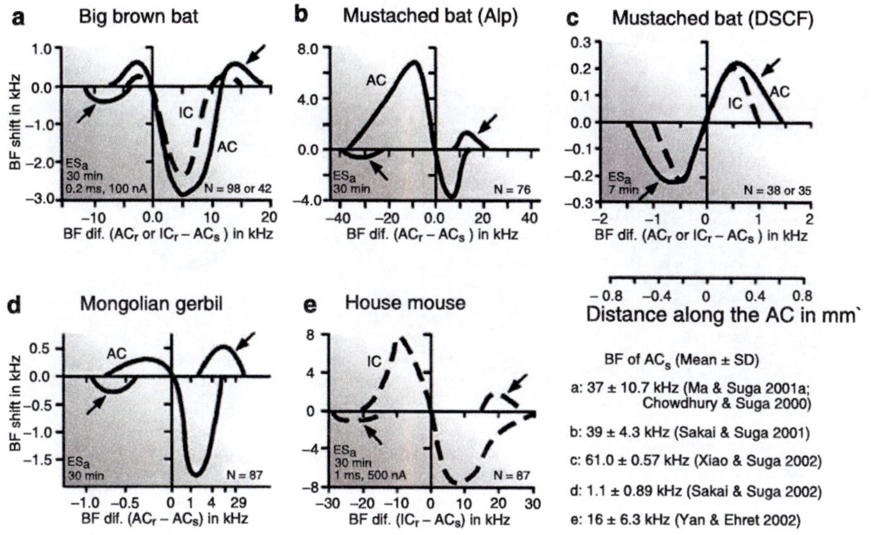

Fig. 11.3 "BF-shift-difference" curves obtained from the primary auditory cortex (AC) or the central nucleus of the inferior colliculus (IC) of four species of mammals. The BF shift changes as a function of the difference in BF between the recorded collicular (IC_r, *dashed curves*) or cortical (AC_r, *solid curves*) neurons and the electrically stimulated cortical neurons (AC_s). Each BF-shift-difference curve encompasses a scatter plot of BF shifts of many neurons studied (N). Note the differences in the curves between species and between different areas of the same species. (**a**, **b**, **d**, **e**) Centripetal BF shifts, except where indicated by *arrows*. A prominent centripetal BF shift occurs at ~5 kHz higher than the stimulated cortical BF in *Eptesicus fuscus* (**a**) and at ~1 kHz higher than that in *Meriones unguiculatus* (**d**). By contrast, the prominent centripetal BF shift occurs at ~10 kHz lower than the stimulated cortical BF in the posterior division of AI (AIp) of *Pteronotus parnellii parnellii* (**b**). In *Mus domesticus*, prominent centripetal BF shifts occur at ~9 kHz higher and lower than the stimulated cortical BF (**e**). (**c**) Prominent centrifugal BF shifts occur at ~0.5 kHz higher and lower than the stimulated cortical BF in the Doppler-shifted constant frequency (DSCF) area of *Pteronotus parnellii parnellii*. The shape of these BF-shift-difference curves might change with the mean BF of stimulated cortical neurons (AC_s). The mean and standard deviation of the BFs of stimulated cortical neurons and references are shown at the *bottom right* for each of (**a–e**). The characteristics of the electrical stimulation (ES_a) were 0.2 ms, 100 nA pulses for (**a–d**) and 1 ms, 500 nA pulses for (**e**). More intense electrical stimulation presumably increases both BF shifts and the frequency range at which BF shifts occur (Suga and Ma 2003)

The distribution of BF shifts in a tonotopic map was studied in detail in AIs, rather than the ICcs, of *Meriones unguiculatus* (Sakai and Suga 2002) and *Eptesicus fuscus* (Ma and Suga 2004), because AIs are flat, show columnar organization, and can be more easily mapped than the ICcs. In these AIs, the centripetal BF shift occurs in a large area surrounding the stimulated cortical neurons (Fig. 11.4a) and the small centrifugal BF shift in a narrow zone surrounding this large area. Different from the centripetal area, many neurons in the surrounding narrow zone do not show BF shifts (Fig. 11.4b). The major reorganization in these AIs is thus due to the centripetal BF shift. The BF shift is largest for neurons located along the tonotopic

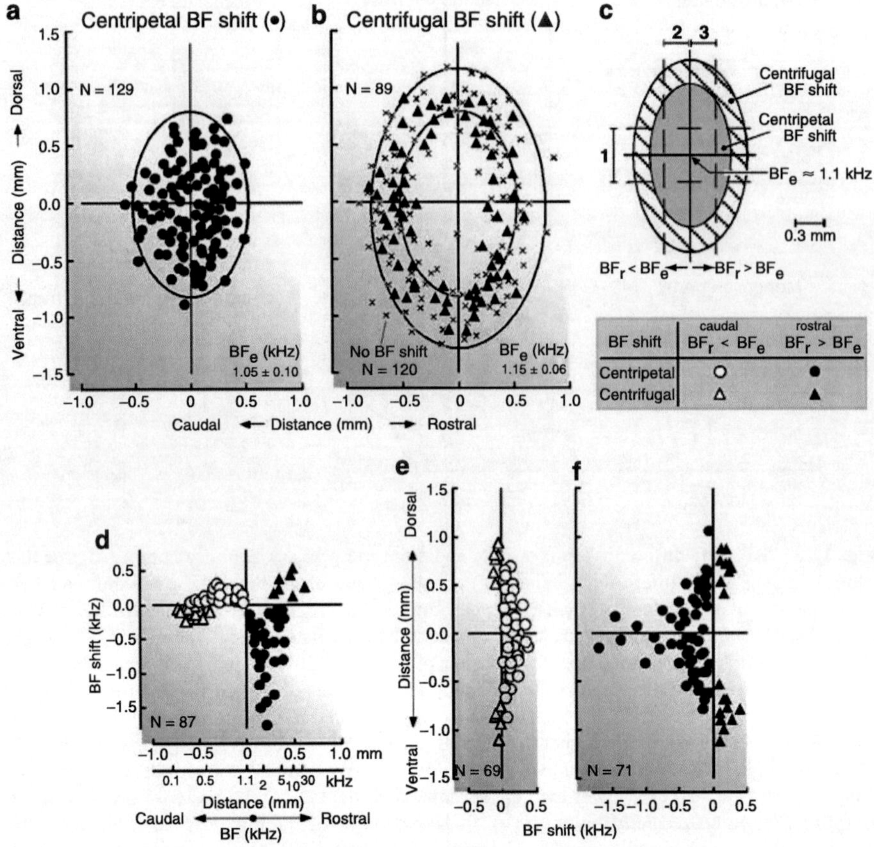

Fig. 11.4 Distribution of centripetal and centrifugal BF shifts in the primary auditory cortex (AI) evoked by focal electric stimulation of AI in *Meriones unguiculatus*. Electric stimulation of 1.1-kHz-tuned neurons in AI evokes centripetal (**a**, *circles*) and centrifugal (**b**, *triangles*) BF shifts of other AI neurons. Locations of recorded neurons along the cortical surface are plotted relative to that of the stimulated cortical neurons at the origin of the coordinates. *x*- and *y*-axes: directions parallel and orthogonal to the tonotopic axis of AI, respectively. *X*s in (**b**) indicate neurons that showed no BF shifts. Data are pooled from 16 hemispheres of 11 animals. *Confidence ellipses* are shown for neurons that showed centripetal (**a**) or centrifugal (**b**) BF shifts. The amounts of BF shifts were measured in a zone parallel (*1*) or orthogonal (*2* and *3*) to the tonotopic axis of AI (**c**). The directions and amounts of BF shifts of neurons in the rostro-caudal (*1* in **c**) and dorso-ventral (*2* and *3* in **c**) zones are respectively plotted in (**d–f**) as a function of distance along the cortical surface. BF_e, BF of electrically stimulated AI neurons; BF_r, BF of recorded AI neurons. See the *inset* at the middle *right* for symbols (Sakai and Suga 2002)

axis crossing the stimulated cortical neurons (Fig. 11.4e, f; Sakai and Suga 2002; Ma and Suga 2004). The BF shift-difference curve shown in Fig. 11.3d is the envelope of the distribution of the BF shifts shown in Fig. 11.4d.

The centripetal BF shift results in an increase in the number of neurons responding to the frequency equal to the stimulated cortical BF or the frequency of a stimulus tone (Yan and Suga 1998; Gao and Suga 1998, 2000; Ma and Suga 2001a). Therefore, such reorganization is called "expanded" reorganization. (It should be noted that the major portion of the center-surround reorganization is the expanded reorganization occurring at the center and that the reorganization at the surround is minor.) On the other hand, the centrifugal BF shift results in a reduced representation that is associated with the augmentation of responses and the sharpening of the tuning curves of the matched neurons (Zhang et al. 1997). Therefore, such reorganization is named "compressed" reorganization (Suga et al. 2002). These two types of reorganizations evoke expanded and compressed representations of acoustic signals, respectively.

The preceding description of the two types of reorganizations is based on the major change in BF. It should be noted that the expanded reorganization results in the over-representation of one frequency occurring at the cost of the under-representation of other frequencies. Therefore, for center-surround reorganization, there is a large over-representation (due to centripetal BF shifts) at the center, a small under-representation (due to centrifugal BF shifts) at the surround, and a small over-representation (resulting from centrifugal BF shifts) just outside of the surround. Likewise, the compressed reorganization results in a large under-representation (due to centrifugal BF shifts) near the center and a small over-representation just outside of the area for the centrifugal BF shifts. If the over-representation increases sensitivity and/or discrimination, the major increase due to expanded reorganization would occur at the stimulated BF and the minor increase, at two frequencies higher and lower than the stimulated BF. The behavioral experiments related to expanded and compressed representations remain to be performed, but we may speculate that compressed representation increases the contrast in the neural representation of an auditory signal, so that it is more suited for the improvement of the discrimination of acoustic signals than expanded representation.

11.6.3 Role of Excitation and Inhibition in Producing Two Types of Reorganizations

When excitation is stronger and more widespread to neighboring unmatched neurons than inhibition, it evokes centripetal BF shifts in the cortical and subcortical neurons. On the contrary, when inhibition is stronger and more widespread to neighboring unmatched neurons than excitation, it evokes centrifugal BF shifts (Suga et al. 2000). An application of bicuculline methiodide, BMI, (an antagonist of GABA-A receptors) to the cortical stimulation site changes centrifugal BF shifts into centripetal BF shifts (Fig. 11.5), and augments centripetal BF shifts. In other words, compressed reorganization changes into expanded reorganization when cortical inhibition is removed (Xiao and Suga 2004, 2005; Ma and Suga 2004). The reorganization (therefore, organization) of a tonotopic map is apparently based upon the balance between excitation and inhibition.

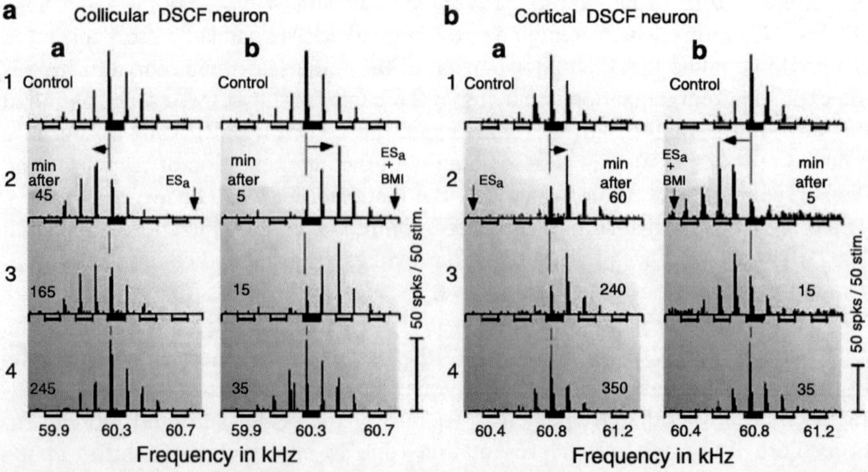

Fig. 11.5 Changes in the direction of BF shifts of a collicular and a cortical DSCF neuron of *Pteronotus parnellii parnellii* evoked by an antagonist of GABA-A receptors, bicuculline methiodide (BMI).The arrays of PST histograms display frequency-response curves of a collicular (**a**) and a cortical (**b**) DSCF neuron, showing changes in the curves evoked by electric stimulation (ES_a) or ES_a+BMI applied to cortical DSCF neurons. ES_a evokes a centrifugal BF shift (*a*), whereas ES_a+BMI evokes a centripetal BF shift (*b*). The vertical and horizontal arrows respectively indicate the BFs of cortical DSCF neurons receiving ES_a or ES_a+BMI and centrifugal or centripetal BF shifts of the recorded neurons. The amplitude of tone bursts was set at 10 dB above the minimum threshold of a given neuron. ES_a, 0.2 ms, 100-nA electric pulses delivered at a rate of 5/s for 7.0 min; BMI, 1.0 nl of 5 mM BMI (Xiao and Suga 2002b)

In the highly specialized cortical auditory areas, inhibition is stronger and more widespread than excitation and evokes compressed reorganization, whereas in the less specialized cortical auditory areas, excitation is stronger and more widespread than inhibition and evokes expanded reorganization.

11.6.4 Corticofugal Modulation of Cochlear Hair Cells

In *Pteronotus parnellii parnellii*, the receptor potential called the cochlear microphonic response (CM) is sharply tuned to ~61 kHz. Electric stimulation of cortical DSCF neurons at a high rate (33/s) evokes a short-term centrifugal BF shift of the contralateral CM, although the stimulation at a low rate (5/s), which is effective in evoking collicular and cortical BF shifts, does not evoke the BF shift of the CM. The BF of the CM systematically shifts as much as 0.25 kHz around 61.0 kHz, according to the BF and the location of the stimulated cortical DSCF neurons (Fig. 11.6; Xiao and Suga 2002a). This means that the frequency tuning of cochlear hair cells shifts. The corticofugal modulation of the frequency tuning of cochlear hair cells systematically occurs according to the location of the focal stimulation of the tonotopic map of the cortical DSCF area.

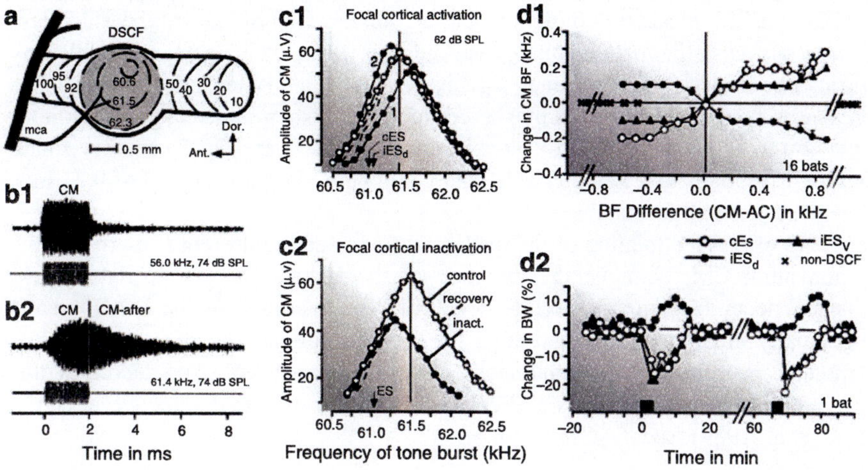

Fig. 11.6 Doppler-shifted constant frequency (DSCF) area over-representating ~61 kHz, cochlear microphonic responses (CM) sharply tuned to ~61 kHz, and corticofugal modulation of the CM by activation or inactivation of DSCF neurons. (**a**) The primary auditory cortex. The *dashed lines* and *numbers* indicate iso-BF lines. The DSCF area at the center over-represents a sound between 60.6 and 62.3 kHz. *mca* middle cerebral artery. (**b**) CMs evoked by 2.0-ms tone burst stimuli. The frequencies and amplitudes of the tone bursts are indicated in kHz and dB SPL, respectively. The envelope of the CM is usually identical to that of a stimulus tone burst (*b1*). However, it is quite different from that of a stimulus tone at ~61 kHz, because the cochlea has an extremely sharply tuned resonator to ~61 kHz (*b2*). The long-lasting "CM-after" is due to the damped oscillation of this sharply-tuned resonator (Suga and Jen 1977). The resonance frequency of the resonator changes when DSCF neurons are stimulated or inactivated (**c–d**). (**c**) Changes in CM evoked by electric stimulation, ES (*c1*) or muscimol (*c2*) applied to cortical DSCF neurons tuned to 61.0 kHz. *Open circles*: the frequency-response curves of the CM in the control condition measured with tone bursts at 62 dB SPL. *Filled circles*: "shifted" condition evoked by the activation by ES (*c1*) or inactivation by muscimol (*c2*). *Dashed curves*: recovered condition. Data points for the recovery curves are not plotted because of overlap with those for the control curves. The BFs of activated or inactivated neurons are indicated by the *arrows*. Contralateral ES (cES), regardless of the dorsal or ventral portion of the DSCF area, evokes a centrifugal BF shift (*c1*, curve *1*), whereas ipsilateral ES (iES) delivered to the dorsal portion of the DSCF area, evokes a centripetal BF shift (*c1*, curve *2*) (for simplicity, the resonance frequency is called the BF). Muscimol applied to the contralateral or ipsilateral DSCF area reduces the CM at its BF and evokes the centripetal BF shift toward 61.0 kHz (*c2*, *filled circles*). (**d**) Changes in the BF of the CM evoked by electric stimulation of contralateral (cES), ipsilateral dorsal (iES_d) and ipsilateral ventral (iES_v) DSCF neurons. In *d1*, the crosses indicate that the CM showed no change for electric stimulation of non-DSCF neurons. Each *vertical bar* attached to a symbol indicates a standard error; some are hidden by symbols. The abscissa indicates the difference in BF between the CM and stimulated cortical neurons. The data were obtained from 16 bats. The mean BF of the CM was 61.36±0.133 kHz ($n=16$). *d2* shows the time courses of changes in the width of the tuning curve of the CM evoked by electric stimulation of contralateral (*open circles*), ipsilateral dorsal (*filled circles*) and ipsilateral ventral (*filled triangles*) DSCF neurons. The second electric stimulation (*rectangle* at the *bottom*) was delivered 65 min after the first. The BF of the CM was 61.40 kHz and the stimulated cortical BF was 61.00 kHz (Xiao and Suga 2002a, b)

Without any electric stimulation, the BF of the CM of the awake stationary *Pteronotus parnellii parnellii* changes as much as 0.15 kHz in an unpredictable way during the emission of biosonar signals (Goldberg and Henson 1998). Such a change is presumably evoked by the corticofugal system and related to auditory attention to echoes. (The 0.25-kHz shift around 61 kHz appears to be very small, but it corresponds to a Doppler shift that could be evoked by a speed of 0.71 m/s, the speed of small insects.)

Corticofugal modulation of the ipsilateral CM is quite different from that of the contralateral CM. Namely, the direction of the BF shift of the ipsilateral CM changes depending on whether the dorsal or ventral portion of the DSCF area is electrically stimulated (Xiao and Suga 2002a). Corticofugal modulation of cochlear hair cells apparently occurs in humans because otoacoustic emission decreases during cortical electric stimulation (Perrot et al. 2006) or visual selective attention (Puel et al. 1989) (see the Sect. 11.6.5.).

11.6.5 Ipsilateral vs. Contralateral Corticofugal Modulation

In *Eptesicus fuscus*, the contralateral collicular and cortical BF shifts tend to be somewhat smaller than, but otherwise are basically the same as, the ipsilateral BF shifts (Ma and Suga 2001a). We had speculated that regardless of the species of mammal, the contralateral ascending auditory system is corticofugally modulated in the same way as the ipsilateral one for bilaterally balanced corticofugal modulation. However, totally unexpected data were obtained from the DSCF area in AI of *Pteronotus parnellii parnellii* which is highly specialized for fine frequency analysis for processing Doppler-shifted echoes (Fig. 11.1b).

The DSCF area is large (Suga and Jen 1976; Suga et al. 1987) and consists of two subdivisions: dorsal (DSCFd) and ventral (DSCFv). The DSCFd contains ipsilaterally-inhibited and contralaterally excited (IE) neurons tuned to intense sounds, whereas the DSCFv contains bilaterally excited (EE) neurons tuned to weak sounds (Manabe et al. 1978). The DSCFv is bilaterally connected by commissural fibers through the corpus collosum, but the DSCFd is not (Liu and Suga 1997).

Electric stimulation of the cortical DSCFd or DSCFv always evokes the centrifugal BF shifts of ipsilateral cortical and collicular DSCF neurons and of contralateral hair cells. On the other hand, electric stimulation of cortical DSCFd and DSCFv neurons respectively evokes the centripetal and centrifugal BF shifts of contralateral cortical DSCFd neurons, and the centrifugal and centripetal BF shifts of contralateral cortical DSCFv neurons. That is, the direction of a BF shift flip-flops depending on the stimulation sites (Xiao and Suga 2005). The collicular DSCF neurons are clustered in the dorsoposterior division (DPD) of the ICc (Zook et al. 1985). The cortical DSCFd and DSCFv stimulation, respectively, evokes the centrifugal and centripetal BF shifts of dorsally located contralateral DPD neurons, and the centripetal and centrifugal BF shifts of ventrally located contralateral DPD neurons (Xiao and Suga 2002a, 2005).

When bicuculline methiodide applied to the DSCF area blocks synaptic inhibition in this area, all the BF shifts in the DSCF area, DPD, and cochlea on both sides become centripetal. Therefore, centrifugal BF shifts originate from inhibition occurring in the DSCF area (Xiao and Suga 2002a, 2005). In the DPD, two types of binaural neurons, as well as monaural neurons, are clustered separately (Wenstrup et al. 1986). Because the DSCFd and DSCFv respectively contain IE and EE neurons, they probably modulate IE and EE neurons in the DPD differently in both the frequency and spatial domains. It remains to be explored why such complex modulation of BF shifts occurs.

The FF area (previously called the FM–FM area; "F" stands for frequency; see Fig. 11.1) of the auditory cortex of *Pteronotus parnellii parnellii* is highly specialized for processing echo delays, and focal electric stimulation of this area evokes centrifugal best delay (BDe) shifts of ipsilateral cortical (Xiao and Suga 2004) and collicular (Yan and Suga 1996) delay-tuned neurons and also contralateral cortical delay-tuned neurons (Tang et al. 2007). There is no difference in the centrifugal BDe shifts between the ipsilateral and contralateral delay-tuned neurons.

11.7 Multiparametric Corticofugal Modulation

As described in the introduction, animal sounds, including human speech sounds, are characterized by multiple parameters such as frequency, amplitude (intensity), duration, time interval between sounds, etc. The central auditory system produces various types of neurons tuned to behaviorally relevant acoustic parameters, that is, "information-bearing parameters" (Suga 1982, 1989) other than a single frequency (Suga 1972, 1973, 1984, 1994; Covey and Casseday 1999; and Rauschecker and Tian 2000, for reviews). Corticofugal modulation occurs for different types of subcortical neurons and is multiparametric (Suga and Ma 2003, for review).

11.7.1 Modulation of Duration Tuning in Eptesicus fuscus

Duration-tuned neurons show the maximal response to a specific duration of a sound ("best duration: BDu"), as well as to its frequency (Pinheiro et al. 1991; Casseday et al. 1994; Ehrlich et al. 1997; Galazyuk and Feng 1997; Ma and Suga 2001b). In the ICc of *Eptesicus fuscus*, the BDu changes along the tonotopic axis; the lower the BF, the shorter the BDu (Jen and Wu 2006). When cortical duration-tuned neurons are electrically stimulated, a collicular duration-tuned neuron matched (i.e., the same) in BDu and BF with the stimulated cortical neurons is augmented and its duration-tuning is sharpened (Fig. 11.7a), whereas an unmatched collicular duration-tuned neuron is either shifted (Fig. 11.7b) or broadened in duration-tuning. The BDu shift and broadening occur toward the BDu of the stimulated cortical neurons: within a certain range, the larger the BDu difference, the

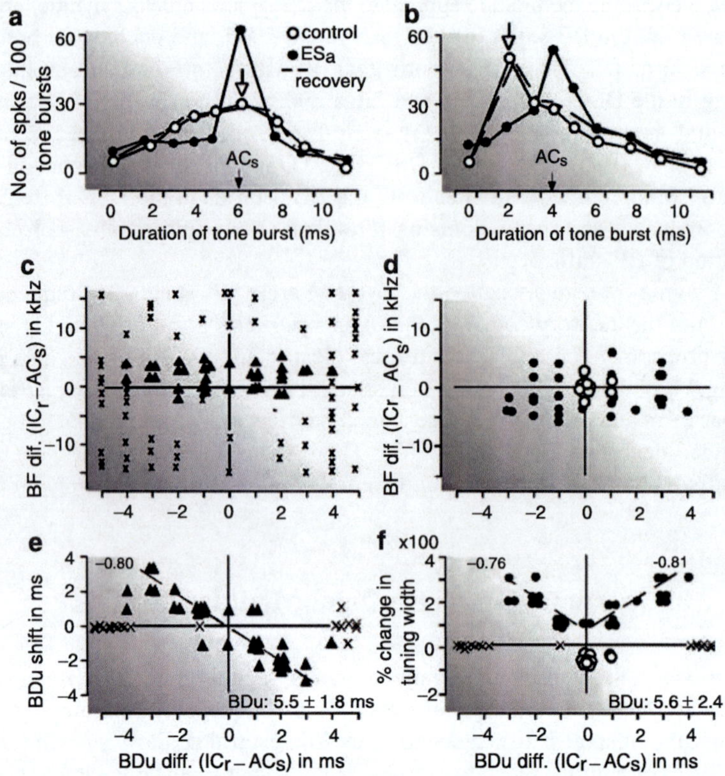

Fig. 11.7 Corticofugal modulation of duration-tuned collicular neurons evoked by electrical stimulation of duration-tuned cortical neurons. **(a, b)** The stimulated cortical (AC_S) and recorded collicular (IC_R) neurons were matched **(a)** or unmatched **(b)** in both best frequency (BF) and best duration (BDu). The *arrows* indicate the BDu of IC_R and AC_S neurons. Cortical stimulation sharpened **(a)** or shifted **(b)** duration-response curves. **(c, d)** Distributions of three types of changes in the duration-response curves: BDu shifts (*triangles*), sharpening (*filled circles*) and broadening (*open circles*). The abscissa and ordinate, respectively, represent BDu and BF differences between IC_R and AC_S neurons. Each *triangle* in **(c)** represents a BDu and a BF difference between paired AC_S and BDu-shifted neurons. Each *filled* or *open circle* in **(d)**, respectively, represents a BDu and a BF difference of paired AC_S and IC_R neurons that showed sharpening (*filled circles*) or broadening (*open circles*) of the duration-response curve. Crosses in **(c–f)** mark neurons that showed no changes in BDu and width of the duration-tuning curve. Note that changes in duration-tuning occur only when the BF and BDu differences are less than 6 kHz and 4 ms, respectively. **(e, f)** Distributions of the BDu shifts **(e)** and width changes **(f)** in duration-response curves. The extent of change depends on the BDu differences between IC_R and AC_S neurons. Note that the larger the BDu difference, the larger the change. The correlation coefficient is shown for each regression line. The mean BDu of AC_S neurons is also shown in **(e, f)** (Ma and Suga 2001b)

larger the BDu shift (Fig. 11.7e) and the broadening (Fig. 11.7f). In addition to these centripetal changes in duration tuning, the BFs of the duration-tuned neurons show a centripetal shift if their BFs are unmatched with those of the stimulated neurons. All these changes occur only when BDu and BF differences between the recorded collicular and stimulated cortical neurons are respectively less than 4 ms and less than 6 kHz (Fig. 11.7c, d; Ma and Suga 2001b, 2007).

11.7.2 *Modulation of Delay Tuning in Pteronotus parnellii parnellii*

In echolocation, the delay of an echo from the sound (biosonar pulse or, simply, pulse) emitted by a bat carries target-distance information. In *Pteronotus parnellii parnellii,* "delay-tuned" neurons are located in the cortical area called the FF area (Suga 1994, for review). Electric stimulation of cortical delay-tuned neurons augments the response at the "best delay (BDe)" of a cortical and a subcortical delay-tuned neuron matched in BDe to the stimulated cortical neurons and sharpens its delay tuning without shifting its BDe. It simultaneously suppresses the responses at the BDes of unmatched cortical and subcortical delay-tuned neurons and shifts their BDes away from the BDe of the stimulated cortical neurons. That is, cortical electric stimulation evokes centrifugal BDe shifts. The amounts of BDe shifts are proportional to the differences in BDe, within a certain range, between the stimulated and recorded delay-tuned neurons (Fig. 11.8a; Yan and Suga 1996; Xiao and Suga 2004).

In the auditory cortex of *Pteronotus parnellii parnellii*, delay-tuned neurons are clustered not only in the FF area, but also in the dorsal fringe (DF) and ventral fringe (VF) areas (Fig. 11.1b). The DF and VF areas are at higher hierarchical levels than the FF area, and these three areas are strongly interconnected (Fitzpatrick et al. 1998). Focal electric stimulation of the FF area evokes centrifugal BDe shifts not only in the ipsilateral FF area and ICc (Yan and Suga 1996), but also in the contralateral FF, ipsilateral DF, and ipsilateral VF areas (Tang et al. 2007; Tang and Suga 2008, 2009). The amount of the centrifugal BDe shift is the same for all of the ipsilateral and contralateral FF areas and the ipsilateral DF and VF areas. Focal electric stimulation of the DF or VF area, however, evokes centripetal BDe shifts in the ipsilateral FF and DF areas. The amount of the centripetal BDe shifts is 2.5 times larger than that of the centrifugal BDe shifts elicited by the FF stimulation. These cortico-cortical interactions indicate that the FF-to-DF and FF-to-VF feed-forward and FF-to-FF lateral projections shape the highly selective neural representation of the echo delay (target distance) exciting FF neurons, and that the DF-to-FF and VF-to-FF feedback and DF-to-DF lateral projections enhance the representation of the echo delay exciting the FF neuron to focus information processing on the echo delay information carried by the excited FF neurons (Tang and Suga 2009).

11.7.3 *Modulation of Response Latencies in Eptesicus fuscus*

The response latency of auditory neurons typically shortens with an increase in stimulus intensity. However, certain cortical and collicular neurons of the little

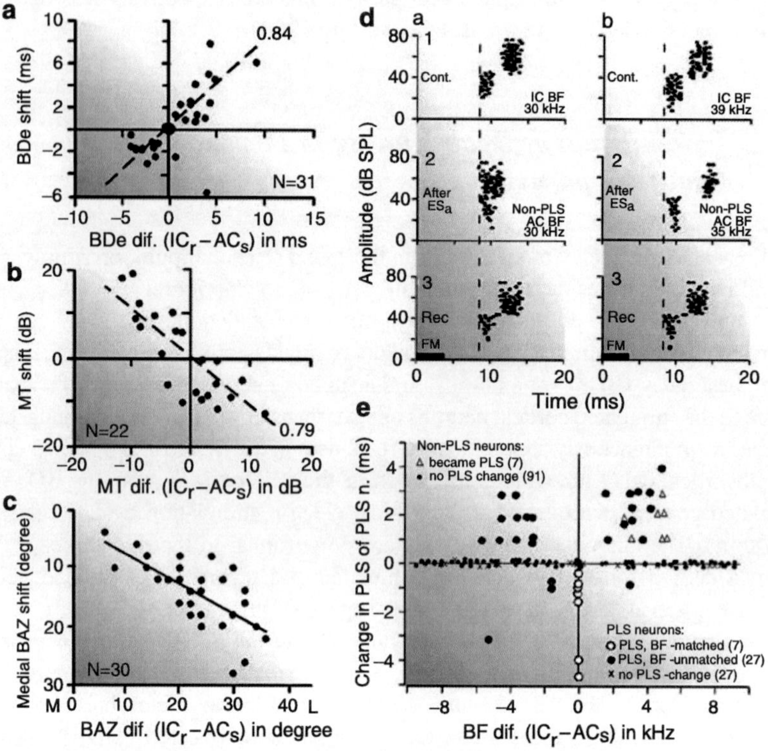

Fig. 11.8 Corticofugal modulation of collicular neurons in nonfrequency domains. (**a**) Centrifugal best-delay (BDe) shifts of delay-tuned collicular neurons in *Pteronotus parnellii parnellii* (Yan and Suga 1996; Xiao and Suga 2002a, b). (**b**) Centripetal minimum-threshold (MT) shifts of BF-matched collicular neurons in *Mus domesticus* (Yan and Ehret 2001, 2002). (**c**) Centripetal best azimuth (BAZ) shifts of collicular neurons sensitive to the contralateral auditory fields in *Eptesicus fuscus*. Their BAZs shift toward the midline, i.e., toward the BAZs of the stimulated cortical neurons. *L* lateral; *M* medial (Zhou and Jen 2005). The extent of BDe, MT and BAZ shifts are linearly related to the difference in BDe, MT and BAZ between the recorded collicular (IC_R) and electrically stimulated cortical (AC_S) neurons, respectively. The correlation coefficient is shown for each regression line. *N*, number of neurons studied. (**d**) Corticofugal modulation of two collicular paradoxical latency-shift (PLS) neurons whose BFs matched (*a*) or unmatched (*b*) the BFs of the stimulated cortical neurons. The responses of the PLS neurons to 5-ms FM sounds are shown as a function of stimulus amplitudes. Each action potential is shown by a *dot*. In *a*, the PLS disappears after the focal cortical electric stimulation (ES_a). In *d*, the PLS became larger after the cortical electric stimulation. *1–3* show the spike discharges obtained before (control), 30 min after and 60 min after the onset of the ES_a. *ICBF*, the BF of the recorded collicular neuron; *ACBF*, the BF of the stimulated cortical neurons. (**e**) Changes in the PLS of collicular PLS neurons evoked by the ES_a are different depending on the BF differences between the recorded collicular and stimulated cortical neurons: BF-matched PLS neurons reduce a PLS (*open circles*), whereas BF-unmatched PLS neurons mostly increase it (*filled circles*). Some non-PLS neurons showed PLSs after the ES_a (*open triangles*). Small *x*, no PLS change in PLS neurons. *Small dot*, no latency change in non-PLS neurons. The *numbers in parentheses* are the number of neurons studied (Ma and Suga 2008)

brown (Sullivan 1982a,b; Galazyuk et al. 2005; Wang et al. 2007), big brown (Ma and Suga 2008), and Mexican free-tailed (Klug et al. 2000) bats show a "paradoxical latency-shift (PLS)": long latencies to intense sounds but short latencies to weak sounds (Fig. 11.8d). These neurons presumably are involved in the processing of target distance information, because they show facilitation of a response when an intense sound corresponding to a biosonar pulse is followed by a weak sound corresponding to an echo (Sullivan 1982a, b). Electric stimulation of cortical auditory neurons evokes two types of changes in the collicular PLS neurons, depending on the relationship in BF between the stimulated cortical and recorded collicular neurons. When the BF is matched between them, the cortical stimulation does not shift the BFs of the collicular neurons and shortens their response latencies at intense sounds, so that the PLS becomes smaller (Fig. 11.8d). When the BF is unmatched, however, the cortical stimulation shifts the BFs of the collicular neurons and lengthens their response latencies at intense sounds, so that the PLS becomes larger (Fig. 11.8d). Cortical electric stimulation also modulates the response latencies of non-PLS neurons. These corticofugal modulations of the collicular responses occur only when the BF difference between the recorded and stimulated neurons is less than 6 kHz (Fig. 11.8e). The cortical electric stimulation produces an inhibitory frequency tuning curve or curves of a collicular neuron (Ma and Suga 2008). BMI applied to a collicular PLS neuron eliminates its PLS (Galazyuk et al. 2005), so that corticofugal feedback is certainly involved in shaping the temporal patterns of the responses of subcortical auditory neurons, presumably through inhibition.

11.7.4 Modulation of the Minimum Threshold in *Mus domesticus* and *Eptesicus fuscus*

A "frequency-threshold (tuning)" curve is based on many thresholds measured as a function of frequency. The lowest threshold to a tone burst stimulus, that is, the threshold at the BF of a given neuron is called the "minimum" threshold (MT). In the central auditory system, a MT and a BF both differ from neuron to neuron. In *Mus domesticus*, focal cortical electric stimulation evokes shifts in both the MT and BF of a collicular neuron. A collicular neuron matched to stimulated cortical neurons in BF but not in MT shows no BF shift, but a MT shift toward the MT of the stimulated neurons. The larger the MT difference, the larger the centripetal MT shift is. On the other hand, a collicular neuron unmatched in both BF and MT increases its MT regardless of MT and BF differences (Fig. 11.8b; Yan and Ehret 2002). In corticofugally inhibited neurons of *Eptesicus fuscus*, both BF-matched and -unmatched neurons show centripetal MT shifts: the larger the MT difference, the larger the MT shift (Jen and Zhou 2003). Corticofugal modulation of MT is somewhat different between *Mus domesticus* and *Eptesicus fuscus*.

11.7.5 Modulation of Spatial Tuning in *Eptesicus fuscus*

The spatial tuning (i.e., directional sensitivity) of the ear varies with the frequency of a stimulus tone. In the central auditory system, binaural interactions produce neurons whose spatial tunings are different from those determined by the ear. In *Eptesicus fuscus*, cortical electric stimulation sharpens the spatial tuning curves of corticofugally inhibited neurons and broadens those of corticofugally facilitated neurons (Jen et al. 1998). Collicular neurons show centripetal "best azimuth (BAZ)" shifts for cortical electric stimulation only when the difference in BF between the stimulated and recorded neurons is less than 6 kHz. The larger the difference in BAZ between the stimulated and recorded neurons, the larger the centripetal BAZ shift (Fig. 11.8c; Zhou and Jen 2005).

11.7.6 Important Principles of Corticofugal Modulation Emerged in *Eptesicus fuscus*

What is particularly important among the findings through the research on *Eptesicus fuscus* is (1) matched neurons are corticofugally augmented in response and sharpened in tuning in the frequency, amplitude (intensity), time (duration), and spatial domains; (2) tuning curves of unmatched neurons in the frequency, amplitude, time, and spatial domains all show systematic centripetal shifts for expanded representation of the values of the parameters to which electrically stimulated cortical neurons are tuned, that is, for expanded representation of the information bearing parameters (Suga 1982, 1989) of an auditory signal frequently stimulating the animal; (3) these changes evoked by cortical electric stimulation last up to 3.5 h after the onset of stimulation; and (4) the shifts in duration- and spatial-tuning curves occur only when the BF difference between the stimulated and recorded neurons is less than 6 kHz.

11.8 Tone-Specific Plasticity (BF shift) Elicited by Auditory Fear Conditioning

Does acoustic stimulation evoke cortical and subcortical plasticity as cortical electric stimulation does? Thus far only in *Eptesicus fuscus*, the BF shifts of cortical and collicular neurons have been studied with three types of stimuli: a tone burst stimulus repetitively delivered to the animal, focal electric stimulation of AI (Yan and Suga 1998; Gao and Suga 1998; Chowdhury and Suga 2000; Ma and Suga 2001a), and auditory fear conditioning (Fig. 11.9a, a; Gao and Suga 1998, 2000; Ji et al. 2001). Hereafter, auditory fear conditioning is simply called conditioning. Repetitive tone burst stimulation lasting 30 min evokes the small short-lasting cortical and collicular BF shifts. The focal AI stimulation lasting 30 min evokes the

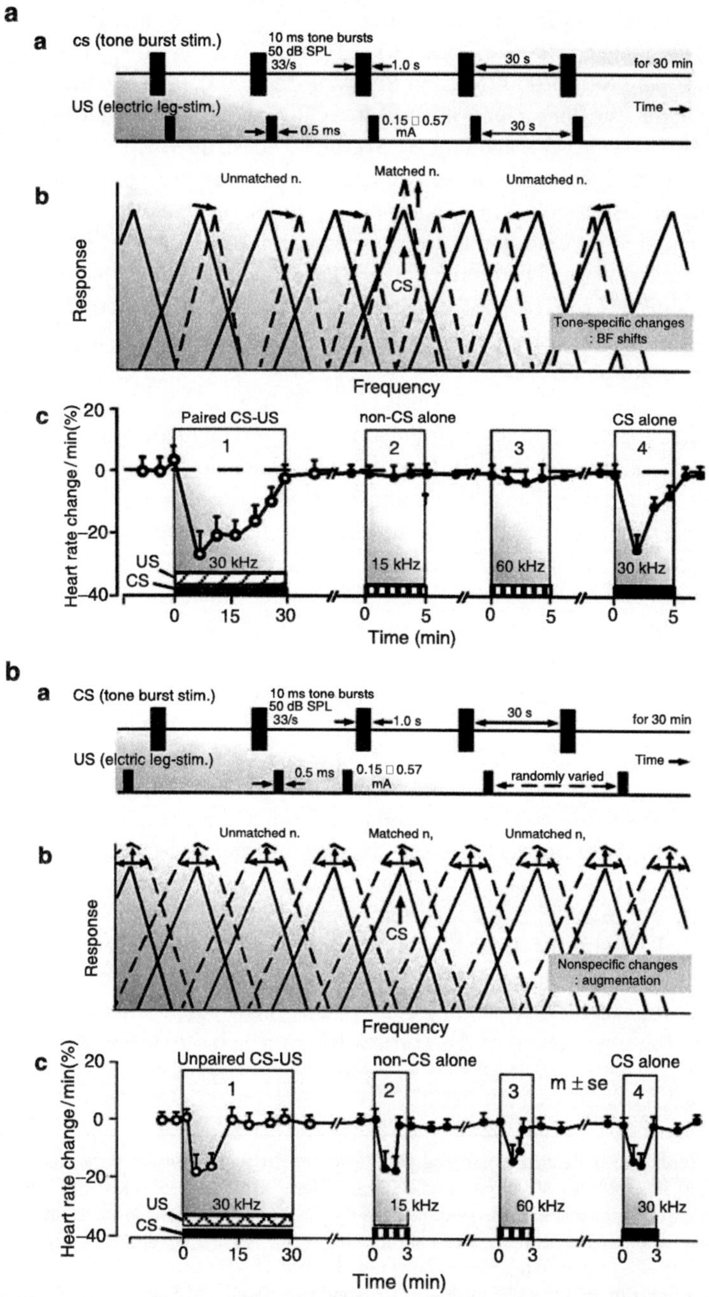

Fig. 11.9 Conditioning, pseudo-conditioning, frequency-tuning changes and heart rate change. (**a**, *a*) Paired conditioning (CS) and unconditioned (US) stimuli and the parameter values of the CS (tone bursts) and US (electric leg-shock) used for the experiments on *Eptesicus fuscus*. (**a**, *b*) Egocentric selection elicited by the conditioning in the central auditory system: facilitation of the

large short-lasting cortical and collicular BF shifts. The cortical BF shift tends to last slightly longer than the collicular BF shift (Fig. 11.10B, curves a and b; Chowdhury and Suga 2000; Ma and Suga 2001a, 2003). The conditioning lasting 30 min elicits the large long-lasting cortical and large short-lasting collicular (Fig. 11.10A and curves c and d in B) BF shifts. These three types of stimuli evoke basically the same BF shifts in the ICc and AI, although the tone burst stimulation alone is much less effective in evoking the BF shifts than the others. The frequency of a tone burst or a conditioning tone corresponds to the BF of electrically stimulated cortical neurons. Therefore, a BF-matched neuron means that the BF of the recorded neuron is equal to the frequency of the tone burst or the conditioning tone, and the term the BF-shift-difference curve (see Fig. 11.3) is also applied to a tone burst or a conditioning tone.

The cortical and collicular BF shifts observed in *Eptesicus fuscus* are evoked by neither short repetitive acoustic stimulation (conditioning stimulus: CS) alone, nor electric leg-stimulation (unconditioned stimulus: US) alone, but by paired stimulation (CS-US), that is, by the conditioning. Backward conditioning (US-CS) does not evoke these BF shifts. The BF shifts are specific to the frequency of the CS (Fig. 11.9b in a). In other words, the BF shifts are "tone-specific" plasticity. The collicular BF shift develops to the peak ~30 min after the onset of the conditioning, then gradually decreases and disappears in 210 min (Fig. 11.10B, curve c), as does the BF shift evoked by the cortical electric stimulation (Fig. 11.10B, curve a). On the other hand, the cortical BF shift gradually increases and reaches a plateau after the termination of the conditioning. This plateau is sustained for a long period of time, more than 26 h (Fig. 11.10B, curve d; Gao and Suga 2000; Suga et al. 2000; Ji et al. 2001). The long-term cortical BF shift does not represent a saturation of the BF shift, because the second conditioning session 3.5 h after the first elicits another short-term collicular BF shift and a further increase in the long-term cortical BF shift (Fig. 11.10B, curves e and f; Gao and Suga 2000). The short-term collicular BF shift contributes to evoking the large long-term cortical BF shift (Ji et al. 2001). The long-term cortical BF shift is elicited by an increase in the cortical acetylcholine (ACh) level resulting from the activation of the cholinergic basal forebrain: nucleus basalis (NB; Ji et al. 2001; Suga et al. 2003; Ji and Suga 2003; Ma and Suga 2003). Therefore, the time course of the cortical BF shift is quite different depending on

Fig. 11.9 (continued) response and sharpening of the frequency-tuning curve of BF-matched neurons and BF shifts of BF-unmatched neurons. Here "BF-matched" means that the BF of a recorded neuron is the same as the frequency of the CS. (**a**, *c*) *Eptesicus fuscus* shows a decrease in heart rate to the paired CS-US, i.e., conditioning. (*1*) After the conditioning, the conditioned autonomic response (heart rate change) does not occur to 15-kHz (*2*) and 60-kHz (*3*) tone bursts, but to the 30-kHz tone bursts (CS) used for the conditioning (Ji and Suga 2007). When the US is unpaired with the CS by randomizing it (**b**, *a*), nonspecific augmentation (sensitization) of neurons in the central auditory system is elicited (**b**, *b*). *Eptesicus fuscus* shows a heart-rate decrease to the unpaired CS-US, i.e., pseudo-conditioning. After the pseudo-conditioning, it shows a heart rate decrease not only to the 30-kHz tone bursts used for the pseudo-conditioning, but also to 15 and 60-kHz tone bursts (**b**, *c*; Ji and Suga 2008)

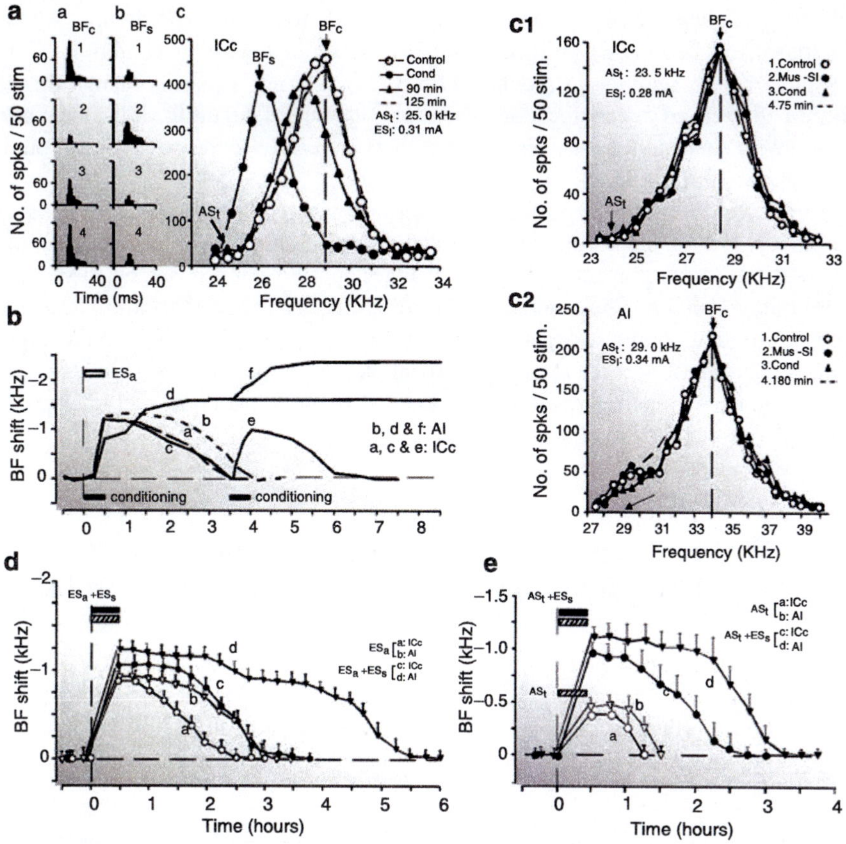

Fig. 11.10 The BF shifts of collicular and cortical neurons in *Eptesicus fuscus* elicited by auditory fear conditioning or by electric stimulation of the auditory and/or somatosensory cortices or by long repetitive acoustic stimulation. (**a**) Changes in the responses (*a* and *b*) and frequency-response curve (*c*) of a collicular neuron caused by a 30-min-long conditioning session consisting of 60 pairs of a train of acoustic stimuli (AS_t = CS) and an electric leg-stimulation (ES_l = US). All of the data were obtained with tone bursts fixed at 10 dB above the minimum threshold of the neuron. The CS was 25.5 kHz, and the BF of the collicular neuron was 29.5 kHz. The data were obtained before (*1*, control condition), immediately after (2), 90 min after (*3*), and 125 min after the conditioning (*4*). The peri-stimulus-time histograms in the left columns (*a* and *b*) show the changes in the responses at the best frequencies in the control (BF_c) and shifted (BF_s) conditions, respectively, which are indicated by *vertical arrows* in (*c*). The BF shift disappeared (i.e., BF_s shifted back to BF_c) 125 min after the conditioning (Gao and Suga 1998). (**b**) Time courses of the BF shifts of collicular (curves *a* and *c*) and cortical (*b* and *d*) neurons evoked by cortical electrical stimulation, ES_a (*a* and *b*, *dashed curves*) or the conditioning (c and d, *undashed curves*). A second conditioning session 3.5 h after the first also evoked collicular (*e*) and cortical (*f*) BF shifts. Note that the time course of the collicular BF shift evoked by electrical stimulation (*a*) is identical to that caused by the conditioning (*c*), whereas the time course of the cortical BF shift evoked by electrical stimulation (*b*) is different from that caused by the conditioning (*d*). The *horizontal bars* indicate the electrical stimulation or conditioning of 30 min duration. Each curve is the mean of 10–15 curves obtained from different neurons (Ma and Suga 2001a, b; Chowdhury and Suga 2000; Gao and Suga 2000). (**c**) Bilateral inactivation of the somatosensory

whether it is elicited by focal cortical electric stimulation or the conditioning. All the findings of the cortical BF shift in *Eptesicus fuscus* elicited by the conditioning are the same as the findings in rodents. The BF shift is not a plastic change that is specifically related to echolocation. Therefore, the neural circuit for it is apparently shared with bats and rodents. Like rodents, bats are easily conditioned to the conditioning tone (Fig. 11.9a, c).

Here, we operationally define short-term plasticity as changes that recover within 3.5 h after the onset of the cortical electrical stimulation or conditioning, whereas long-term plasticity refers to changes that do not recover by more than 10% after 3.0–3.5 h. The amount of the BF shifts depends on the strength of the electric stimulation: the stronger the stimulus, the larger the BF shift, and the larger the BF shift, the slower the recovery. A focal electric stimulation of AI lasting only 2 min can evoke small cortical and collicular BF shifts that completely disappear 75 min after the stimulus (Ma and Suga 2001a).

11.9 Findings Important for the Understanding of the Neural Circuit Eliciting Tone-Specific Plasticity (the BF Shifts)

Several activation and inactivation experiments (Suga et al. 2000, 2002, for reviews) indicate that the conditioning elicits the collicular BF shift via the corticofugal system, as does cortical electric stimulation. Unlike the cortical electric stimulation, the conditioning activates not only the auditory system, but also the somatosensory

Fig. 11.10 (continued) cortex (SI) with 0.4 μg of muscimol applied to its surface abolishes development of the conditioning-dependent BF shifts of a collicular (*C1*) and a cortical (*C2*) neuron, but does not change their responses and frequency-response curves. (Because of the SI inactivation, the data shown in C1 and C2 are quite different from those shown in (**a**).) BF_c, BF in the control condition. Frequency-response curves were obtained before the conditioning (*1*, control); during bilateral SI inactivation (*2*); immediately after the conditioning under SI inactivation (*3*); and 75 or 180 min after the conditioning (*4*). The frequencies of the CS and the current of the US are listed in *C2*. The responses and frequency-response curves of the neurons are affected by neither muscimol applied to the somatosensory cortex nor the conditioning (Gao and Suga 2000). (**d**) Collicular and cortical BF shifts evoked by a short train of electric stimuli of AI (ES_a) are augmented by electrical stimulation of the somatosensory cortex (ES_s). ES_s was delivered 1.0 s after ES_a, mimicking the conditioning. Curves *a* and *b* represent the time courses of collicular and cortical BF shifts, respectively, evoked by ES_a alone. Curves *c* and *d* represent the time courses of collicular and cortical BF shifts, respectively, evoked by ES_a followed by ES_s. These stimuli were delivered over 30 min (*horizontal bars*). (**e**) ES_s following a train of acoustic stimuli (AS_t) augments the collicular and cortical BF shifts evoked by AS_t. Curves *a* and *b*, respectively, represent the time courses of the collicular and cortical BF shifts evoked by AS_t. Curves *c* and *d*, respectively, represent the time courses of the collicular and cortical BF shifts evoked by $AS_t + ES_s$. In (**d, e**), means and standard errors are based on the data obtained from the number of neurons ranging between 12 and 20, as indicated by N. (**d, e**) show that ES_s has a larger augmenting effect on the cortical BF shift than on the collicular BF shift (Ma and Suga 2003) system, amygdala, cholinergic neuromodulator, etc. These systems are mutually connected to produce the conditioning-dependent cortical and collicular BF shifts. The corticofugal auditory system is thus part of the large neural circuit for plastic changes related to learning, memory and attention. Therefore, its function should be understood in relation to this large neural circuit. In the following, we describe how different portions of the auditory and other systems are involved in the BF shifts elicited by the conditioning and by focal electric stimulation of the auditory system.

11.9.1 The Corticofugal Auditory System

For the conditioning, first, the time course of the cortical BF shift is different from that of the collicular BF shift, as described in the preceding text. Second, atropine (an antagonist of muscarinic ACh receptors) applied to the AI surface abolishes the cortical BF shift and reduces the collicular BF shift by 25% without affecting the collicular auditory responses and frequency tuning. Third, atropine applied to the IC abolishes the collicular BF shift, reduces the cortical BF shift by 38%, and changes the cortical BF shift from long-term to short-term without affecting the cortical auditory responses and frequency tuning. Fourth, inactivation of AI by muscimol abolishes the collicular BF shift without affecting the collicular responses and frequency tuning (Gao and Suga 1998, 2000). These findings made in *Eptesicus fuscus* indicate, for the conditioning, that ~25% of the collicular BF shift is due to the cortical BF shift carried down by the corticofugal system, that the remaining collicular BF shift (~75%) is produced in the IC by utilizing ACh and corticofugal signals, and that the collicular BF shift contributes to producing the large long-term cortical BF shift, although the IC cannot produce the long-term BF shift (Ji et al. 2001). These conclusions obtained from the conditioning experiments are supported by electric stimulation experiments: electric stimulation of the ICc (Zhang and Suga 2005) or the MGBv (Wu and Yan 2007) evokes the collicular BF shift via AI and corticofugal feedback. Electric stimulation of the MGBv evokes the cortical BF shift (Jafari et al. 2007; Ma and Suga 2009). It is thus clear that the corticofugal system plays a role in evoking the subcortical BF shift and that the subcortical BF shift contributes to eliciting the large cortical BF shift.

11.9.2 The Primary Auditory Cortex, AI

The cortical and collicular BF shifts evoked by the cortical electric stimulation are basically the same as those elicited by the conditioning except for the time course of the cortical BF shift (Fig. 11.10B; Gao and Suga 2000). The direction of the cortical and collicular BF shifts evoked by the focal electric stimulation of AI changes when an antagonist, bicuculline methiodide (Xiao and Suga 2002b; Ma and Suga 2004), or an agonist, muscimol (Ma and Suga 2004), of GABA-A receptors is applied to the site of the electric stimulation. A local anesthetic, lidocaine, focally applied to the AI surface also changes the directions of the thalamic and collicular BF shifts (Zhang et al. 1997). These findings clearly indicate that the neural circuit within AI plays an essential role in evoking the BF shifts, that AI evokes the thalamic and collicular BF shifts (see "MGBv vs. MGBm"), and that CS-US association in the medial division of the medial geniculate body (MGBm) and the posterior intralaminar nucleus (PIN) is not required for the cortical BF shift. (See "MGBv vs. MGBm" later.)

11.9.3 The Thalamic Auditory Nuclei: MGBv vs. MGBm

MGBv neurons project specifically to AI, are sharply tuned in frequency, and do not habituate. On the contrary, MGBm neurons project widely over cortical auditory areas including AI, have a broad or multipeaked frequency-tuning curve, and habituate after several stimulus presentations (Aitkin and Webster 1972; Calford 1983; Bordi and LeDoux 1994a, b). Therefore, the MGBm is not suited for evoking the tone-specific cortical plasticity (BF shift), but presumably is suited for evoking the nonspecific cortical plasticity which is described later.

Unlike the MGBv, the MGBm is the nonlemniscal nucleus. It receives the projection from the external nucleus of the IC (ICx) that carries auditory, somatosensory, and visual signals. The MGBm projects to the amygdala as well as to the auditory cortex. The MGBm-to-amygdala projection is essential in eliciting conditioned behavioral responses (LeDoux 1993, for review; Lanuza et al. 2004).

Focal electric stimulation of the MGBv evokes the cortical (Jafari et al. 2007) and collicular (Wu and Yan 2007) BF shifts. Muscimol applied to AI abolishes this collicular BF shift (Zhang et al. 2005). The conditioning elicits short-term BF shifts in both the MGBv and MGBm (Edeline and Weinberger 1991, 1992). Therefore, it is clear that the MGBv is involved in evoking the cortical BF shift. Weinberger (1998, 2007) hypothesized that the CS-US association occurs in the MGBm/PIN and then it evokes the cortical BF shift. However, neither activation nor inactivation experiments have been performed to prove that the MGBm/PIN indeed evokes the conditioning-dependent cortical BF shift. The conditioning-elicited MGBm change is blocked by a lesion of the amygdala (Maren et al. 2001; Poremba and Gabriel 2001). Therefore, an interesting question is, Where is the origin of the change in the MGBm? What kinds of cortical plastic changes are evoked by it? A most recent experiment indicates that focal electric stimulation of the MGBm evokes nonspecific cortical augmentation, but not the cortical BF shift (Ma and Suga 2009). Therefore, one may conclude that the MGBv and MGBm are respectively involved in evoking the cortical BF shift and nonspecific augmentation.

Electric stimulation of AI facilitates MGBv neurons, but inhibits MGBm neurons (Yu et al. 2004). The cholinergic nuclei in the midbrain tegmentum project to the MGB (Motts et al. 2008). ACh depolarizes MGBv neurons and hyperpolarizes MGBm neurons (Mooney et al. 2004). Suga (2008) therefore hypothesized that there is a "corticofugal differential gating" for auditory information processing and cortical plasticity. This gating selects either the lemniscal or the nonlemniscal input for cortical auditory signal processing and works together with different neuromodulatory systems for cortical plasticity. The hypothesis states that when the cortical BF shift is evoked by the MGBv, cortical nonspecific augmentation evoked by the MGBm is suppressed through the thalamic reticular nucleus, and that unlike the BF shift, the nonspecific augmentation is augmented by noncholinergic neuromodulators, because tone-specific plasticity is incompatible with nonspecific plasticity (Suga 2008).

11.9.4 ICc: The Central Nucleus of the Inferior Colliculus

The collicular BF shift elicited by focal electric stimulation of either AI, MGBv, or ICc is due to the BF shift being transmitted down from AI to the ICc by the corticofugal system, as previously described. The conditioning-elicited collicular BF shift, however, consists of the corticofugally elicited BF shift and the intrinsically evoked collicular BF shift (Ji et al. 2001). ACh released into the ICc by the cholinergic midbrain tegmentum nuclei (Hallanger et al. 1987), together with corticofugal signals, presumably plays a role in evoking the intrinsically evoked collicular BF shift. As previously described, the short-term collicular BF shift contributes to producing the large long-lasting cortical BF shift.

11.9.5 The Primary Somatosensory Cortex

Bilateral inactivation of the primary somatosensory cortex (SI) by an agonist of GABA-A receptors (muscimol) does not affect the auditory responses and frequency tuning curves of cortical and collicular neurons and the development of nonspecific augmentation of these neurons due to pseudo-conditioning, but selectively abolishes the development of the cortical and collicular BF shifts elicited by the conditioning (Fig. 11.10C; Gao and Suga 1998, 2000) or pseudo-conditioning (Ji and Suga 2008, 2009). Electric stimulation of the SI after, but not before, tone burst stimulation or focal electric stimulation of AI augments the cortical and collicular BF shifts. This augmentation particularly lengthens the duration of the cortical BF shift (Fig. 11.10D, E). It does not occur when the nucleus basalis is lesioned (Ma and Suga 2003). This pair of inactivation and activation experiments clearly indicates that the SI, through the cholinergic basal forebrain, plays an essential role in the development of the conditioning-elicited BF shifts.

11.9.6 The Cholinergic Neuromodulator

Focal electric stimulation of AI evokes the short-term cortical and collicular BF shifts. When it is accompanied with ACh applied to AI, the large long-term cortical

BF shift and the large short-term collicular BF shift are evoked, as are those elicited by the conditioning (Ma and Suga 2005). These findings indicate that AI, corticofugal feedback and ACh together play a key role in evoking the long-term cortical BF shift, and that the long-term cortical BF shift can be evoked without CS-US association in the MGBm and PIN.

Electric stimulation of the nucleus basalis augments the development of the cortical and collicular BF shifts evoked by the focal electric stimulation of AI or by tone burst stimulation (Ma and Suga 2003). It also evokes the large long-lasting cortical BF shift when it is delivered together with tone burst stimulation (Bakin and Weinberger 1996; Bjordahl et al. 1998; Kilgard and Merzenich 1998; Ma and Suga 2003; Yan and Zhang 2005; Zhang et al. 2005). These findings indicate that ACh released in AI by the nucleus basalis augments the cortical and collicular BF shifts evoked by both the neural net in AI and corticofugal feedback and makes the cortical BF shift long-term. These findings confirm that the BF shifts of cortical neurons can be evoked without CS-US association in the MGBm and PIN.

11.9.7 The Amygdala: Inputs from the Sensory Thalamus and Cortex

The amygdala receives the projections from the primary sensory and association cortices, multisensory thalamic nuclei, etc. (LeDoux 1993, for review; Romanski and LeDoux 1993). Both the thalamic and cortical projections to the amygdala are important to fear conditioning (Lanuza et al. 2004). The amygdala projects to the nucleus basalis, which consists of cholinergic and GABAergic neurons. It shows the neural responses related to the conditioning and plays an essential role in eliciting conditioned behavioral responses (Phelps and LeDoux 2005, for review). However, activation and inactivation experiments have not yet been performed to demonstrate whether it also plays an essential role in eliciting the cortical and collicular BF shifts. It has been shown that electrical stimulation of the PIN, not the MGBm, paired with a tone burst could evoke a heart-rate change, as does the conditioning (Cruikshank et al. 1992), and that a lesion of the MGBm impairs conditioning and blocks associative plasticity in the amygdala (Poremba and Gabriel 1997).

11.9.8 The Prefrontal Cortex

The primary sensory cortices project to the prefrontal cortex which, in turn, projects to the nucleus basalis (Rasmusson et al. 2007). The nucleus basalis augments the BF shift, as already described (Bakin and Weinberger 1996; Bjordahl et al. 1998; Kilgard and Merzenich 1998; Ma and Suga 2003; Yan and Zhang 2005; Zhang et al. 2005; Sarter et al. 2006; Parikh et al. 2007). The prefrontal cortex, which plays an important role in attention (Dalley et al. 2004), presumably plays a role in focusing on a specific sound by causing its over-representation through BF shifts.

11.9.9 The Ascending Reticular Activating System

The ascending reticular activating system (ARAS), beginning at the reticular formation, consists of the multisensory thalamic nuclei, histaminergic hypothalamus, and nucleus basalis, all of which diffusely project to the cerebral cortex (Siegel 2002). The ARAS evokes cortical general augmentation for arousal (Magoun 1963) and may augment the cortical BF shift.

11.10 The Neural Circuit for Tone-Specific Plasticity: Working Mode

As described in the preceding text, the corticofugal system, AI, MGBv, ICc, somatosensory cortex, amygdala, and nucleus basalis all are involved in eliciting the cortical and subcortical BF shifts. So, Gao and Suga (1998) proposed the neural circuit model to explain the cortical and subcortical BF shifts elicited by the conditioning. This model is not intended to explain either the conditioned behavioral responses or discharge rate changes, but the BF shifts elicited by the conditioning. The Gao–Suga model, which was elaborated upon by Suga et al. (2000, 2002) and then by Suga and Ma (2003), states that the subthreshold cortical and collicular BF shifts specific to short tone burst stimulation (CS) are evoked by the neural net intrinsic to AI and the corticofugal feedback system activated by CS alone. [For the measurement of a BF shift, the frequency of a tone burst is scanned by, e.g., 0.5-kHz steps. Then, the BF shift less than 0.5 kHz may be defined as a subthreshold BF shift for convenience. The presence of the subthreshold BF shift is evident, because it is easily changed into a large BF shift by ACh or NMDA applied to AI (Ji et al. 2001, 2005) or by electric stimulation of the nucleus basalis (Bakin and Weinberger 1996; Bjordahl et al. 1998; Kilgard and Merzenich 1998; Ma and Suga 2003; Yan and Zhang 2005; Zhang et al. 2005). A short train of tone bursts such as CS evokes the subthreshold cortical and collicular BF shifts, which can also be changed into the small BF shifts by lengthening the train duration (Yan and Suga 1998; Gao and Suga 1998, 2000; Chowdhury and Suga 2000; Ma and Suga 2001a, b, 2003)].

When CS is paired with electric leg-stimulation (US), auditory and somatosensory signals (i.e., CS and US) are sent to the amygdala through the auditory and somatosensory cortices and association cortex. CS-US association takes place in the amygdala. In addition, it may also take place in the association cortex. The amygdala sends the "associated" signal to the nucleus basalis in the forebrain, which in turn releases ACh in AI. Then, the subthreshold cortical BF shift evoked by CS is augmented and changed into a large long-term BF shift. The subthreshold collicular BF shift is increased by ACh presumably released by the cholinergic midbrain tegmental nuclei and by the augmented cortical BF shift through the corticofugal feedback system. Such a collicular BF shift contributes to the development of the large long-term cortical BF shift. The gain of this positive feedback is presumably controlled by the thalamic reticular nucleus. The Gao–Suga model

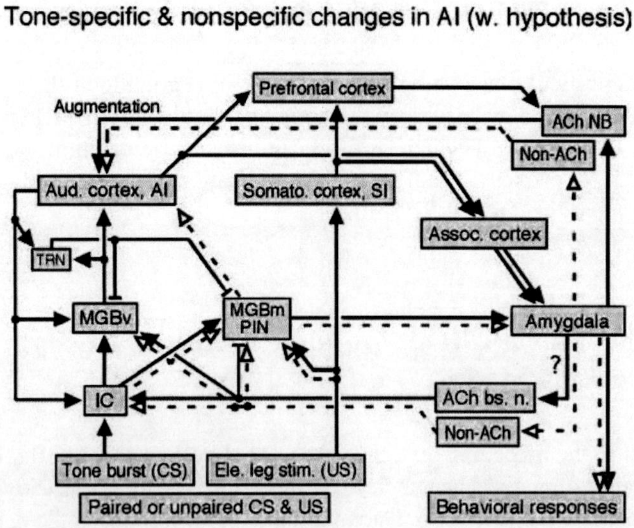

Fig. 11.11 Simplified neural circuits for tone-specific plasticity elicited by paired CS-US and for nonspecific plasticity elicited by unpaired CS-US (working hypothesis). The *solid arrows* indicate the neural circuit for tone-specific plasticity (BF shifts) elicited by paired CS-US (auditory fear conditioning). The *dashed arrows* indicate the neural circuit for nonspecific plasticity (nonspecific augmentation or sensitization) elicited by unpaired CS-US (pseudo-conditioning). *ACh* acetylcholine; *CS* conditioning stimulus (tone bursts); *IC* inferior colliculus; *MGBm* medial division of the medial geniculate body; *MGBv* ventral division of the medial geniculate body; *NB* nucleus basalis (basal forebrain); *PIN* posterior intralaminar nucleus in the thalamus; *TRN* thalamic reticular nucleus; *US* unconditioned stimulus (electric leg-shock). Projections with a short bar at the end mean that they particularly evoke inhibition. When the conditioning elicits the cortical BF shift through the cortical neural net, corticofugal feedback and cholinergic nucleus basalis, the primary auditory cortex, through the TRN, probably reduces the signals from the MGBm and PIN that evoke the cortical nonspecific augmentation. The cholinergic system augments the BF shift, whereas noncholinergic modulatory systems may augment the nonspecific augmentation (Suga 2008)

described above is illustrated in the block diagram in Fig. 11.11 (solid arrows). For simplicity, the collicular BF shift evoked by corticocollicular feedback is described above. Needless to say, the thalamic BF shift evoked by corticothalamic feedback would occur in addition to the collicular BF shift and would also contribute to the development of the cortical BF shift.

The Gao–Suga model is quite different from the Weinberger model (1998). In the Weinberger model, CS-US association occurs only in the multisensory thalamic nuclei: the MGBm and PIN. Then, the MGBm/PIN sends the "associated" signal to AI and evokes a small short-term cortical BF shift. At the same time, the MGBm/PIN sends the associated signal to the amygdala. The amygdala then sends the associated signal to the nucleus basalis, which releases ACh in AI. Then, the small short-term cortical BF shift is augmented and changed into long-term. As already described, neurons in the MGBv and MGBm show BF shifts (Edeline and Weinberger 1991, 1992). MGBv neurons play an essential role in evoking the cortical BF shift. In contrast, MGBm neurons play an essential role in evoking conditioned responses. They evoke nonspecific augmentation, not the BF shift, in AI (Ma and Suga 2009). The origin of the BF shift in the MGBm has been unclear (Maren and Quirk 2004). It can be the IC, because it has been demonstrated that the IC shows the short-term BF shift (Gao and Suga 1998, 2000; Ji et al. 2001). Furthermore, there have been many neurophysiological findings indicating that the cortical BF shift can be evoked without the CS-US association in the MGBm/PIN (Bakin and Weinberger 1996; Kilgard and Merzenich 1998; Gao and Suga 1998; Ma and Suga 2001a, 2003; Zhang and Suga 2005; Yan and Zhang 2005; Puckett et al. 2007; Jafari et al. 2007; Wu and Yan 2007). Therefore, the Weinberger model has little neurophysiological support (Suga et al. 2000 and Suga and Ma 2003, for reviews).

11.11 Corticofugal Differential Gating for Cortical Plasticity: Nonspecific Plasticity Elicited by Pseudo-Conditioning

Different from the conditioning, unpaired CS and US, that is, pseudo-conditioning, evokes nonspecific plasticity (augmentation or sensitization) of cortical auditory neurons (Bakin et al. 1992). (Hereafter, unpaired CS and US are called CSu and USu.) Nonspecific plasticity is a plastic change which is incompatible with tone-specific plasticity (BF shifts). Therefore, Suga (2008) hypothesized the presence of a corticofugal differential gating mechanism. The Suga model speculates the following neurophysiological events: (1) the MGBm is involved in the nonspecific plasticity (nonspecific augmentation or sensitization) of cortical neurons elicited by pseudoconditioning; (2) when the USu or strong CSu is randomly delivered to the animal, the response of the MGBm neurons to it shows little habituation; (3) when the cortical BF shift is elicited by the conditioning, the MGBm is somewhat suppressed by descending signals from AI through the thalamic reticular nucleus, so that nonspecific plasticity is not augmented; (4) when nonspecific plasticity is elicited by the pseudo-conditioning, the BF shift is not augmented; and (5) nonspecific augmentation of cortical neurons depends on noncholinergic neuromodulators. The black diagram in Fig. 11.11 (dashed arrows) is a simplified neural circuit model for nonspecific augmentation. It should be noted that nonspecific augmentation also occurs in the subcortical auditory nuclei.

MGBm neurons show responses to CS and US and conditioning-dependent BF shifts (Edeline and Weinberger 1992). However, they are not involved in evoking the cortical BF shift, but in evoking nonspecific augmentation (Ma and Suga 2009). The most recent works on pseudo-conditioning indicate that, unlike the cortical BF shift, the cortical nonspecific augmentation depends on neither the somatosensory cortex nor ACh (Ji and Suga 2008), and that unlike the collicular BF shift, the collicular nonspecific augmentation depends on neither the corticofugal system nor ACh (Ji and Suga 2009).

11.12 Reorganization of the Tonotopic Map Caused by a Cochlear Lesion

A partial cochlear lesion results in a "permanent" partial inactivation of the central auditory system and the frequency-map reorganization (BF shifts) of the central auditory system. Such reorganization has been examined "during the inactivation," usually several or many days after the lesion (e.g., Irvine and Rajan 1996; Harrison et al. 1996). On the other hand, the frequency-map reorganization elicited by the conditioning or focal activation or inactivation of AI has been examined "after the termination" of the conditioning, activation or inactivation which lasted usually less than 60 min. Unlike the cochlear lesion, therefore, they elicit reversible reorganization. The neural mechanism for the reorganization induced by the cochlear lesion is related more to that elicited by focal cortical inactivation rather than that elicited by the conditioning, because the cochlear lesion is not related to associative learning but to the changes in the activities of AI neurons and the corticofugal feedback. The BF shift evoked by focal electric stimulation of AI is short term. However, the BF shift lasts as long as the stimulus lasts (Ma and Suga 2001a, b), so that AI and the corticofugal system could work continuously for the reorganization after the lesion. It remains to be studied whether the lesion-induced reorganization is due to the interaction between neurons in the ascending auditory system or the corticofugal feedback or both.

11.13 Concluding Remarks: Corticofugal Modulation Shared by Different Animal Species and Different Sensory Systems

The corticofugal auditory system sharpens and shifts the tuning of subcortical neurons in the frequency, amplitude, time, and spatial domains and plays a key role in the reorganization of the auditory system of an adult normal animal according to auditory experience. In other words, it improves and adjusts (reorganizes) the cortical input for auditory signal processing in multiparametric domains. The corticofugal system is tonotopically organized, so that the reorganization of the central auditory system in the frequency domain is systematically related to activated or inactivated cortical auditory neurons. Such systematical reorganization also occurs in the time, amplitude and spatial domains, regardless of whether the central auditory system has neural representational maps for nonfrequency domains (Suga and Ma 2003, for review).

Cortical and subcortical reorganization due to corticofugal modulation occurs according to an uneven distribution of neural activity over the auditory cortex that can be evoked by acoustic stimulation, focal electric stimulation, focal drug application, or focal cochlear lesion. The auditory cortex, corticofugal feedback loops, somatosensory cortex, amygdala and cholinergic nucleus basalis together play a key role in producing tone-specific plasticity of the auditory system elicited by auditory fear conditioning. The corticofugal system is incorporated into a large neural net for learning, memory, and attention.

The auditory system can be reorganized in two different ways: expanded and compressed reorganizations. The former has been found in different species of animals and different sensory systems, whereas the latter thus far has been found only in the auditory subsystems of *Pteronotus parnellii parnellii*, which are highly specialized for echolocation (Suga and Ma 2003, for review). In terms of neurophysiology, what is particularly important is that for a behaviorally insignificant sound repetitively delivered to an animal, cortical auditory neurons change their response properties via the cortical neural net and corticofugal feedback loops, and that the cortical changes become large and long term when the sound becomes behaviorally significant to the animal through associative learning (Gao and Suga 1998, 2000). Therefore, a neuroethological approach makes sense for the exploration of the cortical auditory mechanisms necessary for processing species-specific sounds and sounds produced by prey and predators.

One may doubt whether the auditory mechanisms found in bats are relevant to those in other species, and may write "Here, the data obtained from bats are excluded or not considered because their auditory system is specialized for echolocation." As shown in Figs. 11.3 and 11.4 and reviewed by Suga and Ma (2003), the corticofugal function for egocentric selection and plasticity, which is first explored in bats, is shared not only with the auditory systems of non-bat species but also with the visual and somatosensory systems. What is important is to identify the common and specialized functions and to explore mechanisms for specialization deriving from common ones. Shifts in frequency and delay tuning in the highly specialized subsystems of *Pteronotus parnellii parnellii* are all centrifugal. However, as shown in Fig. 11.5, this uniqueness in the specialized subsystems has derived from the common neural mechanism in others such as the auditory systems of *Eptesicus fuscus*, *Meriones unguiculatus*, and *Mus domesticus* by a quantitative, not qualitative, change in the balance between facilitation and inhibition. In other words, the uniqueness in the corticofugal modulation in the subsystems of the *Pteronotus parnellii parnellii* derives from the common corticofugal neural mechanisms that strengthen inhibition in the auditory cortex. Therefore, one may conclude that even these *Pteronotus parnellii parnellii* subsystems share the corticofugal function and mechanisms with the auditory systems of non-bat species of mammals.

Acknowledgments Our work has been supported by a research grant from the National Institute on Deafness and Other Communication Disorders (DC000175). Suga's collaborators active in the field of auditory physiology are listed as coauthors of this article. In addition to them, Syed M. Chowdhury, Enquan Gao, Masashi Sakai, Wei Yan, Yongkui Zhang, and Yungfeng Zhang contributed to our research on corticofugal modulation. We thank Ms. Sally E. Miller for editing this article.

References

Aitkin LM, Webster WR (1972) Medial geniculate body of the cat: organization and responses to tonal stimuli of neurons in ventral division. J Neurophysiol 35:365–380

Amato G, La Grutta V, Enia F (1969) The control exerted by the auditory cortex on the activity of the medial geniculate body and inferior colliculus. Arch Sci Biol (Bologna) 53:291–313

Andersen P, Junge K, Sveen O (1972) Cortico-fugal facilitation of thalamic transmission. Brain Behav Evol 6:170–184

Bakin JS, Weinberger NM (1996) Induction of physiological memory in the cerebral cortex by stimulation of the nucleus basalis. Proc Natl Acad Sci USA 93:11219–11224

Bakin JS, Lepan B, Weinberger NM (1992) Sensitization induced receptive field plasticity in the auditory cortex is independent of CS-modality. Brain Res 577:226–235

Bjordahl TS, Dimyan MA, Weinberger NM (1998) Induction of long-term receptive field plasticity in the auditory cortex of the waking guinea pig by stimulation of the nucleus basalis. Behav Neurosci 112:467–479

Bordi F, LeDoux JE (1994a) Response properties of single units in areas of rat auditory thalamus that project to the amygdala. I. Acoustic discharge patterns and frequency receptive fields. Exp Brain Res 98:261–274

Bordi F, LeDoux JE (1994b) Response properties of single units in areas of rat auditory thalamus that project to the amygdala. II. Cells receiving convergent auditory and somatosensory inputs and cells antidromically activated by amygdala stimulation. Exp Brain Res 98:275–286

Calford MB (1983) The parcellation of the medial geniculate body of the cat defined by the auditory response properties of single units. J Neurosci 3:2350–2364

Casseday JH, Ehrlich D, Covey E (1994) Neural tuning for sound duration: role of inhibitory mechanisms in the inferior colliculus. Science 264:847–850

Chowdhury SA, Suga N (2000) Reorganization of the tonotopic map of the auditory cortex evoked by cortical electrical stimulation in the big brown bat. J Neurophysiol 83:1856–1863

Covey E, Casseday JH (1999) Timing in the auditory system of the bat. Annu Rev Physiol 61:457–476

Cruikshank SJ, Edeline JM, Weinberger NM (1992) Stimulation at a site of auditory-somatosensory convergence in the medial geniculate nucleus is an effective unconditioned stimulus for fear conditioning. Behav Neurosci 106:471–483

Dalley JW, Cardinal RN, Robbins TW (2004) Prefrontal executive and cognitive functions in rodents: neural and neurochemical substrates. Neurosci Biobehav Rev 28:771–784

Edeline JM, Weinberger NM (1991) Subcortical adaptive filtering in the auditory system: associative receptive field plasticity in the dorsal medial geniculate body. Behav Neurosci 105:154–175

Edeline JM, Weinberger NM (1992) Associative retuning in the thalamic source of input to the amygdala and auditory cortex: receptive field plasticity in the medial division of the medial geniculate body. Behav Neurosci 106:81–105

Edeline JM, Hars B, Hennevin E, Cotillon N (2002) Muscimol diffusion after intracerebral microinjections: a reevaluation based on electrophysiological and autoradiographic quantifications. Neurobiol Learn Mem 78:100–124

Ehrlich D, Casseday JH, Covey E (1997) Neural tuning to sound duration in the inferior colliculus of the big brown bat, *Eptesicus fuscus*. J Neurophysiol 77:2360–2372

Fitzpatrick DC, Olsen JF, Suga N (1998) Connections among functional areas in the mustached bat auditory cortex. J Comp Neurol 391:366–396

Galazyuk AV, Feng AS (1997) Encoding of sound duration by neurons in the auditory cortex of the little brown bat, *Myotis lucifugus*. J Comp Physiol A 180:301–311

Galazyuk AV, Lin W, Llano D, Feng AS (2005) Leading inhibition to neural oscillation is important for time-domain processing in the auditory midbrain. J Neurophysiol 94:314–326

Gao E, Suga N (1998) Experience-dependent corticofugal adjustment of midbrain tonotopic map in bat auditory system. Proc Natl Acad Sci USA 95:12663–12670

Gao E, Suga N (2000) Experience-dependent plasticity in the auditory cortex and the inferior colliculus of bats: role of the corticofugal system. Proc Natl Acad Sci USA 97:8081–8086

Goldberg RL, Henson OW Jr (1998) Changes in cochlear mechanics during vocalization: evidence for a phasic medial efferent effect. Hear Res 122:71–81

Hallanger AE, Levey AI, Lee HJ, Rye DB, Wainer BH (1987) The origins of cholinergic and other subcortical afferents to the thalamus in the rat. J Comp Neurol 262:105–124

Harrison RV, Ibrahim D, Stanton SG, Mount RJ (1996) Reorganization of tonotopic maps in chinchilla auditory midbrain after long-term basal cochlear lesions induced at birth. In: Salvi RJ, Hendersen D, Fiorino F, Colletti V (eds) Auditory system plasticity and regeneration. Thieme, New York, pp 238–255

Irvine DRF, Rajan R (1996) Injury- and use-related plasticity in the primary sensory cortex of adult mammals: possible relationship to perceptual learning. Clin Exp Pharmacol Physiol 23:939–947

Jafari MR, Zhang Y, Yan J (2007) Multiparametric changes in the receptive field of cortical auditory neurons induced by thalamic activation in the mouse. Cereb Cortex 17:71–80

Jen PH, Wu CH (2006) Duration selectivity organization in the inferior colliculus of the big brown bat, *Eptesicus fuscus*. Brain Res 1108:76–87

Jen PH, Zhou X (2003) Corticofugal modulation of amplitude domain processing in the midbrain of the big brown bat, *Eptesicus fuscus*. Hear Res 184:91–106

Jen PH, Chen QC, Sun XD (1998) Corticofugal regulation of auditory sensitivity in the bat inferior colliculus. J Comp Physiol A 183:683–697

Jen PH, Zhou X, Zhang J, Chen QC, Sun X (2002) Brief and short-term corticofugal modulation of acoustic signal processing in the bat midbrain. Hear Res 168:196–207

Ji W, Suga N (2003) Development of reorganization of the auditory cortex caused by fear conditioning: effect of atropine. J Neurophysiol 90:1904–1909

Ji W, Suga N (2007) Serotonergic modulation of plasticity of the auditory cortex elicited by fear conditioning. J Neurosci 27:4910–4918

Ji W, Suga N (2008) Tone-specific and nonspecific plasticity of the auditory cortex elicited by pseudoconditioning: role of acetylcholine receptors and the somatosensory cortex. J Neurophysiol 100:1384–1396

Ji W, Suga N (2009) Tone-specific and nonspecific plasticity of inferior colliculus elicited by pseudoconditioning: role of acetylcholine and auditory and somatosensory cortex. J Neurophysiol 102:941–952

Ji W, Gao E, Suga N (2001) Effects of acetylcholine and atropine on plasticity of central auditory neurons caused by conditioning in bats. J Neurophysiol 86:211–225

Ji W, Suga N, Gao E (2005) Effects of agonists and antagonists of NMDA and ACh receptors on plasticity of bat auditory system elicited by fear conditioning. J Neurophysiol 94:1199–1211

Kilgard MP, Merzenich MM (1998) Cortical map reorganization enabled by nucleus basalis activity. Science 279:1714–1718

Klug A, Khan A, Burger RM, Bauer EE, Hurley LM, Yang L, Grothe B, Halvorsen MB, Park TJ (2000) Latency as a function of intensity in auditory neurons: influences of central processing. Hear Res 148:107–123

Lanuza E, Nader K, Ledoux JE (2004) Unconditioned stimulus pathways to the amygdala: effects of posterior thalamic and cortical lesions on fear conditioning. Neuroscience 125:305–315

LeDoux JE (1993) Emotional memory systems in the brain. Behav Brain Res 58:69–79

Liu W, Suga N (1997) Binaural and commissural organization of the primary auditory cortex of the mustached bat. J Comp Physiol A 181:599–605

Luo F, Wang Q, Kashani A, Yan J (2008) Corticofugal modulation of initial sound processing in the brain. J Neurosci 28:11615–11621

Ma X, Suga N (2001a) Plasticity of bat's central auditory system evoked by focal electric stimulation of auditory and/or somatosensory cortices. J Neurophysiol 85:1078–1087

Ma X, Suga N (2001b) Corticofugal modulation of duration-tuned neurons in the midbrain auditory nucleus in bats. Proc Natl Acad Sci USA 98:14060–14065

Ma X, Suga N (2003) Augmentation of plasticity of the central auditory system by the basal forebrain and/or somatosensory cortex. J Neurophysiol 89:90–103

Ma X, Suga N (2004) Lateral inhibition for center-surround reorganization of the tonotopic map of bat auditory cortex. J Neurophysiol 92:3192–3199

Ma X, Suga N (2005) Long-term cortical plasticity evoked by electric stimulation and acetylcholine applied to the auditory cortex. Proc Natl Acad Sci USA 102:9335–9340

Ma X, Suga N (2007) Multiparametric corticofugal modulation of collicular duration-tuned neurons: modulation in the amplitude domain. J Neurophysiol 97:3722–3730
Ma X, Suga N (2008) Corticofugal modulation of the paradoxical latency shifts of inferior collicular neurons. J Neurophysiol 100:1127–1134
Ma X, Suga N (2009) Specific and nonspecific plasticity of the primary auditory cortex elicited by thalamic auditory neurons. J Neurosci 29:4888–4896
Magoun HW (1963) The waking brain. Charles C Thomas, Springfield
Manabe T, Suga N, Ostwald J (1978) Aural representation in the Doppler-shifted-CF processing area of the auditory cortex of the mustache bat. Science 200:339–342
Maren S, Quirk GJ (2004) Neuronal signalling of fear memory. Nat Rev Neurosci 5:844–852
Maren S, Yap SA, Goosens KA (2001) The amygdala is essential for the development of neuronal plasticity in the medial geniculate nucleus during auditory fear conditioning in rats. J Neurosci 21:RC135
Massopust LC Jr, Ordy JM (1962) Auditory organization of the inferior colliculi in the cat. Exp Neurol 6:465–477
Mooney DM, Zhang L, Basile C, Senatorov VV, Ngsee J, Omar A, Hu B (2004) Distinct forms of cholinergic modulation in parallel thalamic sensory pathways. Proc Natl Acad Sci USA 101:320–324
Motts SD, Slusarczyk AS, Sowick CS, Schofield BR (2008) Distribution of cholinergic cells in guinea pig brainstem. Neuroscience 154:186–195
Orman SS, Humphrey GL (1981) Effects of changes in cortical arousal and of auditory cortex cooling on neuronal activity in the medial geniculate body. Exp Brain Res 42:475–482
Parikh H, Marzullo TC, Yazdan-Shahmorad A, Gage GJ, Kipke D (2007) Laminar characterization of spiking activity in the rat motor cortex. Conf Proc IEEE Eng Med Biol Soc 2007:4735–4738
Perrot X, Ryvlin P, Isnard J, Guénot M, Catenoix H, Fischer C, Mauguière F, Collet L (2006) Evidence for corticofugal modulation of peripheral auditory activity in humans. Cereb Cortex 16:941–948
Phelps EA, LeDoux JE (2005) Contributions of the amygdala to emotion processing: from animal models to human behavior. Neuron 48:175–187
Pinheiro AD, Wu M, Jen PH (1991) Encoding repetition rate and duration in the inferior colliculus of the big brown bat, *Eptesicus fuscus*. J Comp Physiol A 169:69–85
Poremba A, Gabriel M (1997) Medial geniculate lesions block amygdalar and cingulothalamic learning-related neuronal activity. J Neurosci 17:8645–8655
Poremba A, Gabriel M (2001) Amygdalar efferents initiate auditory thalamic discriminative training-induced neuronal activity. J Neurosci 21:270–278
Puckett AC, Pandya PK, Moucha R, Dai W, Kilgard MP (2007) Plasticity in the rat posterior auditory field following nucleus basalis stimulation. J Neurophysiol 98:253–265
Puel JL, Rebillard G, Bonfils P, Pujol R (1989) Effect of visual selective attention on otoacoustic emissions. In: Wilson JP, Kemp DT (eds) Cochlear mechanisms. Plenum, New York, pp 315–321
Rasmusson DD, Smith SA, Semba K (2007) Inactivation of prefrontal cortex abolishes cortical acetylcholine release evoked by sensory or sensory pathway stimulation in the rat. Neuroscience 149:232–241
Rauschecker JP, Tian B (2000) Mechanisms and streams for processing of "what" and "where" in auditory cortex. Proc Natl Acad Sci USA 97:11800–11806
Riquimaroux H, Gaioni SJ, Suga N (1991) Cortical computational maps control auditory perception. Science 251:565–568
Romanski LM, LeDoux JE (1993) Information cascade from primary auditory cortex to the amygdala: corticocortical and corticoamygdaloid projections of temporal cortex in the rat. Cereb Cortex 3:515–532
Ryugo DK, Weinberger NM (1976) Corticofugal modulation of the medial geniculate body. Exp Neurol 51:377–391
Sakai M, Suga N (2001) Plasticity of the cochleotopic (frequency) map in specialized and nonspecialized auditory cortices. Proc Natl Acad Sci USA 98:3507–3512
Sakai M, Suga N (2002) Centripetal and centrifugal reorganizations of tonotopic map of the auditory cortex in gerbils. Proc Natl Acad Sci USA 99:7108–7112

Sarter M, Gehring WJ, Kozak R (2006) More attention must be paid: the neurobiology of attentional effort. Brain Res Rev 51:145–160
Siegel J (2002) The neural control of sleep and waking. Springer, New York
Suga N (1972) Analysis of information-bearing elements in complex sounds by auditory neurons of bats. Audiology 11:58–72
Suga N (1973) Feature extraction in the auditory system of bats. In: Moller AR (ed) Basic mechanisms in hearing. Academic, New York, pp 675–742
Suga N (1982) Functional organization of the auditory cortex. In: Woolsey C (ed) Cortical sensory organization. Humana, Clifton, pp 157–218
Suga N (1984) The extent to which biosonar information is represented in the bat auditory cortex. In: Edelman GM, Gall WE, Cowan WM (eds) Dynamic aspects of neocortical function. Wiley, New York, pp 315–373
Suga N (1989) Principles of auditory information-processing derived from neuroethology. J Exp Biol 146:277–286
Suga N (1994) The processing of auditory information carried by species-specific complex sounds. In: Gazzaniga MS (ed) The cognitive neurosciences. MIT, Cambridge, pp 295–313
Suga N (2008) Role of corticofugal feedback in hearing. J Comp Physiol A 194:169–183
Suga N, Jen PH (1976) Disproportionate tonotopic representation for processing CF-FM sonar signals in the mustache bat auditory cortex. Science 194:542–544
Suga N, Jen PH (1977) Further studies on the peripheral auditory system of "CF-FM" bats specialized for fine frequency analysis of Doppler-shifted echoes. J Exp Biol 69:207–232
Suga N, Ma X (2003) Multiparametric corticofugal modulation and plasticity in the auditory system. Nat Rev Neurosci 4:783–794
Suga N, Niwa H, Taniguchi I, Margoliash D (1987) The personalized auditory cortex of the mustached bat: adaptation for echolocation. J Neurophysiol 58:643–654
Suga N, Gao E, Zhang Y, Ma X, Olsen JF (2000) The corticofugal system for hearing: recent progress. Proc Natl Acad Sci USA 97:11807–11814
Suga N, Xiao Z, Ma X, Ji W (2002) Plasticity and corticofugal modulation for hearing in adult animals. Neuron 36:9–18
Suga N, Ma X, Gao E, Sakai M, Chowdhury SA (2003) Descending system and plasticity for auditory signal processing: neuroethological data for speech scientists. Speech Commun 41:189–200
Sullivan WE III (1982a) Neural representation of target distance in auditory cortex of the echolocating bat *Myotis lucifugus*. J Neurophysiol 48:1011–1032
Sullivan WE III (1982b) Possible neural mechanisms of target distance coding in auditory system of the echolocating bat *Myotis lucifugus*. J Neurophysiol 48:1033–1047
Sun X, Chen QC, Jen PH (1996) Corticofugal control of central auditory sensitivity in the big brown bat, *Eptesicus fuscus*. Neurosci Lett 212:131–134
Syka J, Poplear J (1984) Inferior colliculus in the rat: neuronal responses to stimulation of the auditory cortex. Neurosci Lett 51:235–240
Tang J, Suga N (2008) Modulation of auditory processing by cortico-cortical feed-forward and feedback projections. Proc Natl Acad Sci USA 105:7600–7605
Tang J, Suga N (2009) Corticocortical interactions between and within three cortical auditory areas specialized for time-domain signal processing. J Neurosci 29:7230–7237
Tang J, Xiao Z, Suga N (2007) Bilateral cortical interaction: modulation of delay-tuned neurons in the contralateral auditory cortex. J Neurosci 27:8405–8413
Villa AE, Rouiller EM, Simm GM, Zurita P, de Ribaupierre Y, de Ribaupierre E (1991) Corticofugal modulation of the information processing in the auditory thalamus of the cat. Exp Brain Res 86:506–517
Wang X, Galazyuk AV, Feng AS (2007) FM signals produce robust paradoxical latency shifts in the bat's inferior colliculus. J Comp Physiol A 193:13–20
Watanabe T, Yanagisawa K, Kanzaki J, Katsuki Y (1966) Cortical efferent flow influencing unit responses of medial geniculate body to sound stimulation. Exp Brain Res 2:302–317
Weinberger NM (1998) Physiological memory in primary auditory cortex: characteristics and mechanisms. Neurobiol Learn Mem 70:226–251

Weinberger NM (2007) Associative representational plasticity in the auditory cortex: a synthesis of two disciplines. Learn Mem 14:1–16

Wenstrup JJ, Ross LS, Pollak GD (1986) Binaural response organization within a frequency-band representation of the inferior colliculus: implications for sound localization. J Neurosci 6:962–973

Wu Y, Yan J (2007) Modulation of the receptive fields of midbrain neurons elicited by thalamic electrical stimulation through corticofugal feedback. J Neurosci 27:10651–10658

Xiao Z, Suga N (2002a) Modulation of cochlear hair cells by the auditory cortex in the mustached bat. Nat Neurosci 5:57–63

Xiao Z, Suga N (2002b) Reorganization of the cochleotopic map in the bat's auditory system by inhibition. Proc Natl Acad Sci USA 99:15743–15748

Xiao Z, Suga N (2004) Reorganization of the auditory cortex specialized for echo-delay processing in the mustached bat. Proc Natl Acad Sci USA 101:1769–1774

Xiao X, Suga N (2005) Asymmetry in corticofugal modulation of frequency tuning in mustached bat auditory system. Proc Natl Acad Sci USA 102:19162–19167

Yan J, Ehret G (2001) Corticofugal reorganization of the midbrain tonotopic map in mice. Neuroreport 12:3313–3316

Yan J, Ehret G (2002) Corticofugal modulation of midbrain sound processing in the house mouse. Eur J Neurosci 16:119–128

Yan J, Suga N (1996) Corticofugal modulation of time-domain processing of biosonar information in bats. Science 273:1100–1103

Yan W, Suga N (1998) Corticofugal modulation of the midbrain tonotopic map in the bat auditory system. Nat Neurosci 1:54–58

Yan J, Suga N (1999) Corticofugal amplification of facilitative auditory responses of subcortical combination-sensitive neurons in the mustached bat. J Neurophysiol 81:817–824

Yan J, Zhang Y (2005) Sound-guided shaping of the receptive field in the mouse auditory cortex by basal forebrain activation. Eur J Neurosci 21:563–576

Yan J, Zhang Y, Ehret G (2005) Corticofugal shaping of freqeucny tuning curves in the central nucleus of the inferior colliculus of mice. J Neurophysiol 93:71–83

Yu YQ, Xiong Y, Chan YS, He J (2004) Corticofugal gating of auditory information in the thalamus: an in vivo intracellular recording study. J Neurosci 24:3060–3069

Zhang Y, Suga N (1997) Corticofugal amplification of subcortical responses to single tone stimuli in the mustached bat. J Neurophysiol 78:3489–3492

Zhang Y, Suga N (2000) Modulation of responses and frequency tuning of thalamic and collicular neurons by cortical activation in mustached bat. J Neurophysiol 84:325–333

Zhang Y, Suga N (2005) Corticofugal feedback for collicular plasticity evoked by electric stimulation of the inferior colliculus. J Neurophysiol 94:2676–2682

Zhang Y, Suga N, Yan J (1997) Corticofugal modulation of frequency processing in bat auditory system. Nature 387:900–903

Zhang Y, Hakes JJ, Bonfield SP, Yan J (2005) Corticofugal feedback for auditory midbrain plasticity elicited by tones and electrical stimulation of basal forebrain in mice. Eur J Neurosci 22:871–879

Zhou X, Jen PH (2005) Corticofugal modulation of directional sensitivity in the midbrain of the big brown bat, *Eptesicus fuscus*. Hear Res 203:201–215

Zook JM, Winer JA, Pollak GD, Bodenhamer RD (1985) Topology of the central nucleus of the mustache bat's inferior colliculus: correlation of single unit properties and neuronal architecture. J Comp Neurol 231:530–546

Index

D.K. Ryugo et al. (eds.), *Auditory and Vestibular Efferents*,
Springer Handbook of Auditory Research 38, DOI 10.1007/978-1-4419-7070-1,

N

O